Lecture Notes in Educational Technology

Series Editors

Ronghuai Huang, Smart Learning Institute, Beijing Normal University, Beijing, China

Kinshuk, College of Information, University of North Texas, Denton, TX, USA

Mohamed Jemni, University of Tunis, Tunis, Tunisia

Nian-Shing Chen, National Yunlin University of Science and Technology, Douliu, Taiwan

J. Michael Spector, University of North Texas, Denton, TX, USA

D1800039

The series Lecture Notes in Educational Technology (LNET), has established itself as a medium for the publication of new developments in the research and practice of educational policy, pedagogy, learning science, learning environment, learning resources etc. in information and knowledge age, – quickly, informally, and at a high level.

Abstracted/Indexed in:

Scopus, Web of Science Book Citation Index

More information about this series at https://link.springer.com/bookseries/11777

Ronghuai Huang · Bing Xin · Ahmed Tlili ·
Feng Yang · Xiangling Zhang · Lixin Zhu ·
Mohamed Jemni
Editors

Science Education in Countries Along the Belt & Road

Future Insights and New Requirements

 Springer

Editors
Ronghuai Huang
Smart Learning Institute
Beijing Normal University
Beijing, China

Ahmed Tlili
Smart Learning Institute
Beijing Normal University
Beijing, China

Xiangling Zhang
Beijing Institute of Education
Beijing, China

Mohamed Jemni
Cultural and Scientific
Organization—ALECSO
Arab League Educational
Tunis, Tunisia

Bing Xin
Children and Youth Science Center
of CAST
Beijing, China

Feng Yang
Children and Youth Science Center
of CAST
Beijing, China

Lixin Zhu
Smart Learning Institute
Beijing Normal University
Beijing, China

ISSN 2196-4963   ISSN 2196-4971   (electronic)
Lecture Notes in Educational Technology
ISBN 978-981-16-6957-6   ISBN 978-981-16-6955-2   (eBook)
https://doi.org/10.1007/978-981-16-6955-2

Preface

The field of science education has gained significant attention during the past few decades, especially on the curriculum design and the usage of digital resources. United Nations Educational, Scientific, and Cultural Organization (UNESCO) claims that the world is increasingly influenced by technology, as well as science. Therefore, it is crucial to not only spread science education, but also cultivate learners' interests in science. Most of the books published in this domain focused on a specific region, without providing a global and comprehensive overview. To cover this gap, our book *Science Education in countries along the Belt and Road: Future Insights and New Requirements* were assembled to not only provide science education details in many regions from the Belt and Road, but also to discuss the needs of the development of science education, and to put forth suggestions to advance science education for Belt and Road countries.

This book offers a detailed and up-to-date overview about science education from an international perspective. Specifically, our book covered 29 countries distributed from Arabia and Africa, Asia and Europe. Each country's information and science education progress are demonstrated in a separate chapter, with the following five significant areas:

- Overview of the country
- Overview of the education development (i.e., education system and policy, statistics on the national education and educational research and international collaboration)
- Current situation of science education (i.e., policies and standards, curriculums, digital resources and teacher training, student assessment and achievement, science and technology venues and centers, and utilizing emerging technologies)
- Requirements for future development of science education
- Discussion and conclusion.

The last chapter was written to summarize the previous contents and provide insightful suggestions, such as promoting the construction of educational informatization and open educational resources. To summarize, this book goes beyond simply clarifying the status of science education in the Belt and Road. It provides a road map to facilitate science education adoption along the Belt and Road.

To have a book with precise information from each country, every single chapter was written by reputable scholar(s) from that country. Altogether, 63 authors participated in the writing process. 30 reviewers were then responsible for the peer-review process, before the editors compiled all the chapters in this important book.

Beijing, China   Ronghuai Huang
Beijing, China   Bing Xin
Beijing, China   Ahmed Tlili
Beijing, China   Feng Yang
Beijing, China   Xiangling Zhang
Beijing, China   Lixin Zhu
Tunis, Tunisia   Mohamed Jemni

Acknowledgment

We would like to first thank all of the authors for their valuable contributions to this book by sharing the status of science education in their countries. These studies and their reported findings definitely give readers stepping stones for further research and development about science education adoption.

We would also like to thank all of the reviewers who accept to review the submitted chapters and give their constructive comments and suggestions for the authors to further enhance the quality of their book chapters, hence enhancing the overall quality of this book. We really appreciate them for giving their reviews in a timely manner that helps our book to meet the production timeline.

Special thanks also go to our colleagues in the Smart Learning Institute of Beijing Normal University, China, and the Children and Youth Science Center of China Association for Science and Technology for their support to finish this book project.

Beijing, China	Ronghuai Huang
Beijing, China	Bing Xin
Beijing, China	Ahmed Tlili
Beijing, China	Feng Yang
Beijing, China	Xiangling Zhang
Beijing, China	Lixin Zhu
Tunis, Tunisia	Mohamed Jemni

Contents

Part I
Introduction

Chapter 1
Introduction to Science Education and the Belt and Road

Lixin Zhu, Yao Song, Zhenyu Cai, Xiangling Zhang, Ahmed Tlili, and Ronghuai Huang

Abstract This book aims to highlight science education in countries along the Belt and Road. This chapter introduces the background and context of the book. Specifically, the book consists of 31 chapters divided into three sections, namely Arab and African, European and Asian countries. Each chapter provides an in-depth analysis of the problems faced by the country's current science education research and practice.

Keywords Science education · Belt and road · Roadmap

1.1 Introduction

The 2030 Agenda for Sustainable Development covers the 17 Sustainable Development Goals (SDG), which are an urgent call for action by all countries (https://sdgs.un.org/goals). Science plays an important role in the process of achieving SDGs. Hence, science education is of vital importance to train scientific and technical personnel, as well as improve the quality of science education of a given

L. Zhu · Y. Song · Z. Cai · A. Tlili · R. Huang
Beijing Normal University, No. 19 Xinjiekouwai Street, Haidian District, Beijing 100875, China
e-mail: 11112018113@bnu.edu.cn

Y. Song
e-mail: songyao@mail.bnu.edu.cn

Z. Cai
e-mail: zhenyu_cai@163.com

A. Tlili
e-mail: ahmedtlili@ieee.org

R. Huang
e-mail: huangrh@bnu.edu.cn

X. Zhang (✉)
Beijing Institute of Education, ShiFang Street No. 2, Huangsi Street, Xicheng District, Beijing 100120, China
e-mail: zhangxiangling@bjie.ac.cn

country (https://en.wikipedia.org/wiki/Science_education). Since the 1990s, traditional science education in schools has been difficult to cope with challenges, which has put forward requirements for the development of a global perspective of science education and the internationalization of science education research. Existing research shows that the occurrence of problems has played an important role in various key science education research fields, but has not yet become the main focus of science education research (Chiu & Duit, 2011). The "Belt and Road" initiative, on the other hand, is the practice of building a community with a shared future for mankind, and promotes the interconnection of the Eurasian and African continents with the Pacific, Indian and Atlantic, thereby promoting mutual benefit and win-win results. This initiative is conducive to the development of cooperation in education and personnel training, and is conducive to the establishment of a cooperative mechanism.

1.2 Background

1.2.1 Science Education

Science education cultivates scientific and technological talents and improves the scientific quality of the whole nation, which is the basic content of school education and the foundation of modern social civilization and progress. It is relative to the famous cultural philosopher Bernard who believes that science education with modern significance only gradually entered the classroom after the European Renaissance. It was initially mathematics and astronomy. Geography, and later physics, chemistry, biology and other disciplines. With the continuous development of transformational science and technology and the promotion of the industrial revolution in the 18–19 period, the disciplines have further distinct differences, and finally formed another complete science education system as it is now.

Science education has always been an important part of the work of moral education. Science education is the teaching and learning of science to non-scientists, such as school children, college students, or adults within the public. The field of science education includes scientific content, scientific process, scientific method, social science and pedagogical work. Science education standards provide expectations for students to develop understanding throughout K-12 education and beyond. The traditional topics covered by the science education standards are physics, life, earth, space, and humanities. Science education can stimulate students' curiosity and proactive inquiry ability, cultivate students' various skills and cognitive abilities (Kostas Dimopoulos & Vasilis Koulaidis, 2003), allow students to cultivate innovative consciousness in the process of hands-on practice, cultivate students' cooperation and problem-solving skills, and attach importance to the cultivation of humanistic soft power, as well as develop critical thinking skills (Doganay Ahmet & Ozturk Ayse, 2011). In today's era, science education has a more fundamental role in enhancing the scientific quality of all people and building the foundation of an innovative country. Nowadays, science and technology are changing with each passing day, which is profoundly changing people's production and lifestyle. Because of this, every person living in the era of rapid development of science and technology must have scientific quality. Elementary school science education is of great significance for stimulating and protecting children's curiosity and thirst for knowledge from an early age, and cultivating students' scientific spirit and practical innovation ability. From this point of view, the cultivation of scientific literacy must start from an early age and do a good job at an early age.

Science education is not limited to theoretical education. The goal it pursues is not only to master scientific knowledge, but also to integrate the learning of scientific knowledge in the process of hands-on practice under the magnificent natural and social landscape, in the context of the relationship between science and society. The new curriculum standard emphasizes that primary school science is a practical and comprehensive course. The curriculum takes inquiry activities as an important way for students to learn science, comprehensively using knowledge and methods in different fields to understand natural phenomena and solve practical problems. These requirements are obvious for the improvement of students' practical ability, scientific thinking, and scientific spirit. As the main force, schools and teachers need to make changes and innovative attempts in teaching concepts and teaching practices to open the door to the guidance of elementary school science education. Various high-quality science education institutions, such as universities, research institutes, and popular science venues, should also take active actions to create conditions to ensure the implementation of science courses, so that the skylight of elementary science education can pass through the sky of science.

1.2.2 Belt and Road

The construction of the "Belt and Road" is a major measure for China to expand its opening to the outside world. It is the world's largest international cooperation platform and the most popular international public product. It is a great practice for building a community with a shared future for mankind. Over the past seven years, the "Belt and Road" initiative has continued to increase its international influence from vision to action, from concept to consensus, from laying foundations, pillars and beams to comprehensive and in-depth development. This sudden new crown pneumonia epidemic has posed a serious threat to the lives and health of people in all countries, and caused a serious impact on the world economy. In this context, insisting on the high-quality development of the "Belt and Road" to promote the construction of a community with a shared future for mankind has profound time value and practical significance.

The joint construction of the "Belt and Road" aims to promote the orderly and free flow of economic factors, efficient allocation of resources and deep market integration, and jointly create a new framework for open, inclusive, balanced and inclusive regional economic cooperation. The basic connotation of "Belt and Road" is to closely integrate the new situation of the in-depth development of economic globalization and regional economic integration, better coordinate the domestic and international situations, better coordinate domestic development and opening up, and make full use of the international and domestic markets. Two kinds of resources, adhere to open development, cooperative development, and win-win development, adhere to bilateral, multilateral, and regional and sub-regional open cooperation, with policy communication, facility connectivity, unimpeded trade, financial connectivity and people-to-people connectivity as the main content and strong grasp. We will expand the strategic points of convergence and convergence of interests with countries along the route, orderly promote the construction of a large logistics channel for land-sea coordination and east-west commodity resources, accelerate the interconnection of infrastructure with neighboring countries and regions, and focus on promoting bilateral and multilateral economic and trade investment. Increase the level and level of cooperation, actively promote the negotiation of investment agreements and free trade agreements with countries and regions along the route, promote regional trade liberalization and investment facilitation, and form a "Belt and Road" as the two wings, with neighboring countries as the basis, and along the route. The country is a key, global-oriented high-standard free trade zone network that provides favorable conditions for realizing regional economic integration and laying a solid foundation for the Asia-Pacific free trade zone.

1.2.3 Science Education Initiative for the Belt and Road Countries

The continuing decline of enrollment in science among young people is worrisome, and it is in this endeavor that UNESCO's work in science education seeks to make a difference, ensuring it's not a just privilege. UNESCO identifies Women and Africa as priority areas for development, so that science education aims not only to produce more science-oriented youth, but also pays particular attention to the education of girls (UNESCO 2017). It also hopes to have a positive impact on economic and social development by influencing teachers and curriculum planners.

Serving national development is an important mission of education. It is the common goal of all countries to focus on improving the quality of education and promoting education equity. Whether it is responding to the epidemic or restoring the economy, the destinies of all countries are closely linked. Mankind is a community of shared destiny in the same boat. We need to help each other and work together to promote international cooperation in building the "Belt and Road" to play an important role in effectively responding to global crises and achieving sustainable development.

1.3 Structure of the Book

This book includes three main sections, namely Arab and African, European and Asian countries. The first section covers science and education research in 13 countries. The second section covers 9 countries, while the third section covers 5 sections. Each chapter includes several sections related to each country, such as the overview of the country, overview of education development, current situation of science education. The wide range of topics covered in this book provides a comprehensive account of the state of science education in countries along the Belt and Road.

In the aspect of overview of the country, firstly, the geographical location, population and current political system of the country are presented. The current situation of economy, technologies and cultural development, which provides a basis for the analysis of education status, were also discussed.

Education development emphasizes education systems and policies, as well as educational research and international cooperation. The education system can be divided into institutionalized education and non-institutionalized education. Specifically, it can be divided into educational purposes, educational content, educational methods, educational activities, educational media, educational facilities, educational environment, students, teachers, teaching management personnel and other elements. These elements are mutually independent, interrelated and interact to form an organic whole. Educational policy is an action standard on education formulated by a political party and a country in order to realize the educational development goals and tasks in a certain historical period, according to the basic tasks and principles of the party and the country in a certain historical period. Education policy includes five basic characteristics: political and principle, purpose and feasibility, stability and discontinuity, legitimacy and authority, systematisms and versatility.

Scientific literacy generally refers to the knowledge of key scientific concepts and the understanding of scientific processes, which includes the application of science to cultural, political, social and economic issues. At present, the issue of scientific literacy is becoming more and more important in education. In current school education, students have gradually tended to learn through exploratory learning rather than memorizing facts. This also means that understanding the process of science and the application of scientific concepts is one of the central goals of current education. As the concept of core literacy has been put forward and deepened, countries around the world are paying more and more attention to scientific literacy. How to evaluate the scientific literacy of students scientifically and rationally is one of the important challenges facing science education, and the current internationally authoritative scientific literacy evaluation the project accordingly provides us with a reliable basis for evaluating the current state of science education in various countries and the scientific literacy of students. Current situation of science education discusses how to implement the requirements of science education better, as well as future strategies for the challenges and problems.

1.4 Objectives

Through literature research, understanding the current development status and challenges of international science education, especially the status and problems of science education in the "Belt and Road" countries, so as to better provide suggestions for the coordinated development of science education in these countries.

Deeply understand and compare the differences in science education among the countries of the "Belt and Road" and promote the coordinated development of science education in the "Belt and Road" countries through multi-round and multi-level research iterations, the depth and height of comparative inquiry will be enhanced, and the discovery of more valuable research insights will be promoted.

A comparative analysis of the status quo and needs of science education in countries along the "Belt and Road" from different dimensions to reveal the general laws and special manifestations of education, so as to discover the commonalities and differences in the development status of science education in the "Belt and Road" countries.

References

https://sdgs.un.org/goals

https://en.wikipedia.org/wiki/Science_education

http://www.unesco.org/new/en/natural-sciences/special-themes/science-education/

Chiu, M. H., & Duit, R. (2011). Globalization: Science education from an international perspective. *Journal of Research in Science Teaching, 48*(6), 553–566.

Dimopoulos, K., & Koulaidis, V. (2003). Science and technology education for citizenship: The potential role of the press. *Science Education, 87*(2), 241–256.

Doganay, A., & Ozturk, A. (2011). An investigation of experienced and inexperienced primary school teachers' teaching process in science and technology classes in terms of metacognitive strategies. *Educational Sciences: Theory & Practice, 11*(3), 1320–325

Lixin Zhu is working as a senior engineer, Beijing Normal University. He is also the Co-director, Learning Environment Designing and Assessment & Evaluation lab, the National Engineering Lab for CyberLearning and Intelligent Technology. His research focuses on educational technology, STEM education.

Yao Song is currently studying for a bachelor's degree at Beijing Normal University, majoring is Statistics and minoring in Computer Science.

Zhenyu Cai is currently a graduate student at Faculty of Education, Beijing Normal University. His research interest includes the field of technology application in education as well as STEM education.

Xiangling Zhang is working as a lecturer of teacher professional development at Beijing Institute of Education. She finished her post-doctoral studies at Beijing Normal University in 2020. Her research focuses on educational technology, STEM education.

Ahmed Tlili is the Co-Director of the OER Lab, Smart Learning Institute of Beijing Normal University. He is also the Associate Researcher of Beijing Normal University. His research areas include game-based learning, smart learning environments, technology enhanced learning, learner modeling, adaptive learning systems, learning analytics, and educational psychology.

Ronghuai Huang is a Professor in Faculty of Education and Dean of Smart Learning Institute in Beijing Normal University. He received "Chang Jiang Scholar" award in 2017, which is the highest academic award issued to an individual in higher education by the Ministry of Education in China. He now serves as the Director of the National Engineering Lab for Intelligent Cyber-Learning Technology, and Director of Beijing Key Laboratory for Educational Technology. He is also the president of International Association of Smart Learning Environments, Editor-in-Chief of Springer's Smart Learning Environments and Journal of Computers in Education, as well as Editor-in-Chief of Springer's series Lecture Notes in Educational Technology and Smart Computing and Intelligence.

Part II
Arab and African Countries

Chapter 2
Science Education in Algeria

Hafed Zarzour

Abstract Throughout history, science education has played a vital role in developing and modernizing the countries. The education in Algeria has been developing for the last years as a result of several reforms undertaken for enhancing the quality of learning and teaching in the whole education system, ranging from the primary school to higher education. Hence, this book chapter attempts to present the science education in Algeria. It starts by providing some information about the geographical location, population, and political system, as well as outlining the economic, technologies, and cultural development in the country. It then presents an overview of the education development and the current situation of science education in Algeria. The present chapter further explores the requirements for future development of science education. Finally, challenges and strategies, reflections and issues, and future pathways are discussed in the hope of improving the leaning and teaching for tomorrow's world.

Keywords Science education · Education in Algeria · Teaching and learning in Algeria · Educational technology · Education system

2.1 Overview of the Country

2.1.1 Geographical Location, Population and Political System

Algeria is the 10th largest country in the world, and the largest country in Africa, in the Arab world, and in the Mediterranean Basin with 2,381,741 km^2 of area. It has an important geostrategic position as it is located in the center of North Africa facing Europe, bordered by Morocco in the west, Western Sahara, Mauritania, and Mali in the southwest, Niger in the southeast, Libya in the east, Tunisia in the northeast, and the Mediterranean sea in the north. The climate in Algeria is transitional between

H. Zarzour (✉)
Laboratory of Computer Science and Mathematics, Faculty of Science and Technology, University of Souk Ahras, 41000 Souk Ahras, Algeria

maritime in the north and semi-arid to arid in the middle and south, respectively, with an average annual temperature of 12 °C (Stambouli, 2011). The country is divided into 48 provinces (Wilayas) and its capital is Algiers, which is the biggest and most populous city.

According to the National Office of Statistics (National Office of Statistics, 2019a), the population in Algeria was estimated to be about 43 million of inhabitants in 2019. The natural increase recorded during the same year reached 837,000 people with a natural growth rate of 1.93%. The distribution of births by sex gave 104 boys for 100 girls and the total fertility rate was 3.0 children per woman. One third of the Algerian population is youth. The median age was 27.7 years, which is considerably lower compared to that of some other population countries such as the EU population (43.1 years). Algeria is a People's Democratic Republic, where the Arabic and Tamazight are the two national and official languages of the state.

2.1.2   Current Situation of Economic, Technologies and Cultural Development

Algeria is classified by the World Bank as an upper-middle income country and ranked as the third important economy in the region of the Middle East and North Africa. With about 4.5 trillion cubic meters of the proven natural gas reserves, Algeria is considered as the 10th largest natural gas reserves in the world, the second largest in Africa, and the 6th largest gas exporter (Abada & Bouharkat, 2018; World Bank Group, 2019).

In 2018, the Gross Domestic Product (GDP) growth reached 1.5% and sustained at the same percentage in Q1-2019. Recently, the growth in non-hydrocarbon sectors, including construction and public works, commercial services, industrial, and agriculture sectors reached a slight increase. However, the economy of Algeria is mainly depending on the exports of petroleum products, particularly natural gas, oil and other hydrocarbons. The renewable energy is potential that can contribute to the economy as the Sahara represents 86% of the total land.

The country has known many advances in the last years with the respect to the use of new technologies in various contexts. For example, the National Center for Space Technology (CNTS) has launched a microsatellite called Alsat-1, which is considered as the first step in developing the national space infrastructure (Bentoutou, 2012). The mission objective of Alsat-1 was to manage the natural disasters as well as other remote sensing utilizations. Next, many other satellites named Alsat-1B, Alsat-2A, Alsat-2B, lsat-1 N, and Alcomsat-1 were launched for different purposes, ranging from monitoring the agricultural and disaster to providing the broadcast television and high-speed Internet access.

Moreover, the Internet usage is increasing rapidly in Algeria. Currently, three telecommunication companies are providing mobile and wireless communication services including 4G Internet access with mobile phones. These companies are:

Mobilis, Ooredoo, Djezzy. Cultural development in Algeria contributes to the promotion of culture in all its dimensions and in all its forms with the objective of enhancing any national and local cultural systems.

2.2 Overview of the Education Development

2.2.1 *Education System and Policy*

Education is one of the most important priorities for the Algerian government. It is free for all at all stages in government schools and compulsory from the ages of 6–15, in which each Algerian is required to receive the basic education. Officially, the compulsory school entrance age is 6 and the academic year begins in September and ends in June.

The education system of the country is structured into several levels: preparatory, primary school lasts for 5 years (Ages 6–10), lower secondary school lasts for 4 years (Ages 11–14), upper secondary school lasts for 3 years (Ages 15–17), vocational training, and higher education. State policy in the primary and secondary education, higher education, and professional training and continuing education are implemented by the Ministers of National Education, Higher Education and Scientific Research, and Professional Education and Training, respectively.

After bringing the compulsory education level, students who leave school can get the vocational training they need in order to develop the necessary skills and build a career. On the other hand, to pursue higher education, students must obtain the national exam named Baccalauréat, or an equivalent foreign qualification. In 2004, Algeria adopted the LMD system, which is composed of three grades: 3-years License (Bachelor), 2-year Master, and Doctorate lasting 3 years. However, some studies like medicine, dentistry, and pharmacy are still taught following to the classical educational system. The LMD system is distinguished by the contents of innovative educational programs, evaluation and accreditation of education programs, reorganization of the teachings, and new education architecture (Benouar, 2013).

The reform aiming at using the LMD system in higher education is one of the major forms of the governmental policies that were initiated to improve education in Algeria since gaining the independence in 1962. In addition, investing a significant part of the general budget in this sector and making education free of charge at all levels in public institutions are other critical components of the government's policy to enhance education in the country.

Table 2.1 Main indicators for the national education sector in Algeria for the 2019–2020 academic year (National Office of Statistics, 2019b)

Level	Pupils	Teachers	Educational establishments
Preparatory	495,481	17,791	19,037
primary	4,513,749	199,850	
Lower secondary	2,979,737	159,065	5,512
Upper secondary	1,222,673	102,279	2,433
Total	9,211,640	478,985	26,982

2.2.2 Statistics on the National Education

The main indicators for the Algerian national education sector in the 2019–2020 school year are shown in Table 2.1. It can be observed that the pupils enrolled in primary school, including preparatory school, represent 54.40% of the total number of pupils, while those enrolled in lower secondary and upper secondary represent 32.34% and 13.26%, respectively. Of the total number of teachers found (478,985), 21,7641 are in preparatory and primary schools, 159,065 are in middle schools, and the rest in high schools. Regarding to school level indicator for the 2019–2020 school year, the sector of national education has a total of 26,982 schools, including 19,037 primary schools, 5,512 middle schools, and 2,433 high schools.

To obtain the data for the number of pupils per teacher, the total number of pupils in each education level is divided by the total number of teachers in the corresponding education level. The obtained results for the three levels are as follows: 23 pupils/teachers in preparatory and primary schools, 19 pupils/teachers in lower secondary schools, and 12 pupils/teachers in upper secondary schools. Pupil Teacher Ratio (PTR) consisting of the average nationally of the total number of pupils divided by the total number of teachers is 19 for all pupils in the all levels of education. This indicates that on average there is one teacher for every 19 pupils.

In the report published by the Global Out-of-School Children Initiative (Global Out-of-School Children Initiative, 2014), it was highlighted that the out-of-school numbers calculated based on the administrative data are 101,304 pupils for children aged 6–10 years in primary school and 151,879 pupils for children aged 11–14 years in lower secondary. The risk of children dropping out of primary school involves around of 94,271 pupils, while the children at risk of dropping out of school in lower secondary school affects 408,172 pupils.

2.2.3 Educational Research and International Collaboration

The network of higher education institutions covers the whole territory of the country under direct Ministry supervision. This network consists of 50 universities, 20 national higher schools, 13 university centers, 11 higher teacher training colleges,

10 higher schools, and 2 annexes (MESRS, 2020). The number of enrolled students in higher education institutions has increased significantly from 2,881 in 1962–1963 to 1,730,000 in 2017–2018 (Souleh, 2017; Université de Rouen, 2019).

In order to implement the national strategy of research, the General Directorate for Scientific Research and Technological Development was created in 2009 under the authority of the Minister of Higher Education and Scientific Research. Some of its main tasks (DGRSDT, 2020) are to put into action the most appropriate elements related to programming, scientific cooperation, university research, human resource development, as well as research results valorization. The research structures are spread over Algeria and composed of research agencies (5 Thematic Research Agencies and 2 Research Agencies), research centers (12 Research Centers -EPST type- MESRS, 11 Research Centers -EPST type- outside MESRS, 4 Research Centres under the COMENA, 2 Research Centers under the ASAL), research and development centers with 12 centers, and research units (12 Research Units Attached to universities and schools and 14 Research Units Attached to EPST) (DGRSDT, 2020).

Developing international collaboration among educational institutions is the most crucial aspect in regard to acquire access to the latest technologies and information. In this context, many programs of international cooperation and training between institutions of higher education in Algeria and other institutions in other countries ware developed. For example, the program of residential training abroad is a one of international exchange programs that offers scholarships to excellent students for preparing their Master's degree and PhD.

2.3   Current Situation of Science Education

2.3.1   Policies and Standards

The policies governing the education system in Algeria are mainly defined by the Algerian constitution. It is reported, notably in article 65 edition 2016, that the right to education is guaranteed and the education is compulsory and free for all children under the age of 16 years old. The State, moreover, organizes the national education and protects equal access to schooling and vocational training. The management of staff and educational institutions is decentralized. Law 08/04 of 23 January 2008 on the national education guidance enshrines, through articles 10, 11, 12, 13 and 14, the guarantee of the right to education. For example, in article 10, it is indicated that the State ensures the right to education to all Algerians without discrimination based on sex, social or geographical origins. In article 14, it is reported that education is free at all levels in public institutions of national education. In addition, the State provides support for the education of underprivileged students by enabling them to benefit from multiple aids, particularly in terms of scholarships, textbooks and school supplies, food, accommodation, transport, and school health (Ministry of National Education, 2020a).

2.3.2   Curriculums, Digital Resources and Teacher Training

Because the education in Algeria falls under the responsibility of ministries, the curriculums that are standardized for each field of study are approved by the corresponding Ministry. All primary, lower secondary, and upper secondary school deliver the same textbooks and curriculum as designed by the Ministry of National Education. Since 2003, Algeria has engaged gradually in a series of reforms at all levels of education. These reforms were initiated with the aim of improving the effectiveness of the education system and responding to increases in socio-economic advances. The educational curriculum is, therefore, reformed in terms of skills and textbooks, to integrate learners-centered approaches rather than the traditional approaches, which are based on teachers-centred pedagogical pattern during the learning process.

The integration of digital resources into education has a direct impact on the quality of learning and teaching, and therefore on the national education system. The use of digital resources in education started decades of years ago. Some of major initiatives launched by the Algerian government include (Guemide & Benachaiba, 2012): the project of the Ministry of Education to equip all schools with computers by 2005; the distance education project; the virtual university project; and the research network to be put in place by the Ministry of Higher Education and Scientific Research.

There are many teachers training schools and colleges in Algeria that enable to enhance learning by training teachers to acquire the skills and competencies needed to further teaching in schools. Primary school, lower secondary, and upper secondary teachers are usually trained in one of the higher normal schools spread over the country. For example, the higher normal school of Elbachir Elibrahim (https://www.ens-kouba.dz/) aims at training secondary school teachers for the sector of national education in different fields such as: computer science, natural sciences, mathematics, physics, and chemistry.

2.3.3   Student Assessment and Achievement

For any education system, the evaluation of students in schools and classrooms is a relevant topic because the assessment process can describe not only what students have learned, but also how well they have acquired the knowledge. In Algerian education system, the students' assessment and achievement as well as their advancement from one level to another are based on some pedagogical principles. Indeed, there are different forms for the evaluation such as the ongoing assessment and the annual assessment (Ministry of National Education, 2020b).

In the ongoing assessment form, teachers can get continuous feedback on the performance of students' learning by gathering information using formal and informal classroom observations about their learning activity and behavior. In a such form of assessment, a variety of tools can be used, ranging from simple to more complex tests covering one or more concepts, or one or more units of the program.

On the other hand, the annual assessment is a kind of summative assessment in which the teachers can determine the learning outcomes and the profile for the students at the end of the year. This allows them to have evidence about what their students have learned during that year, as well as to measure the students' abilities for determining whether they can succeed in achieving the higher level of education. The students' move from one year to another is, therefore, made on the basis of the results obtained in the final exam at the school and the teachers' council decision. Additionally, the exams for transitions from primary to lower secondary school, from lower secondary to upper secondary school, and from upper secondary to higher education are common and organized at the national level.

Algeria participated to Trends in International Mathematics and Science Study (TIMSS), that is considered as the enabler to provide reliable trend data on the mathematics and science achievement of fourth and eighth-grade students over the world. In 2015, Algeria joined the Programme for International Student Assessment (PISA) (Kartianom & Ndayizeye, 2017), which aims to measure the ability of 15-year-olds in using their reading, mathematics and science knowledge and skills for meeting the challenges of the real-life. For example, the average score in mathematics learning achievement was 360 in 2015.

2.3.4 Science and Technology Venues and Centers

Almost all universities have a faculty of science and technology in which the subject of science and technology is taught to students due to its relevance to their lives. More specifically, there are two universities specializing in the field of science and technology, that are the university of Science and Technology Houari Boumediene (USTHB) (https://www.usthb.dz/) and the University of Sciences and Technology—Mohamed Boudiaf (USTO-MB) (https://www.univ-usto.dz/).

The first one was founded in Algiers in 1974. It is one of the most prestigious universities in Algeria. It has over 20,000 students and more than 1,500 teachers and workers. With 8 faculties, USTHB offers education in various fields, including biological science, Physics, chemistry, Mathematics, civil engineering, electronics and computer science, mechanical engineering and engineering processes, and earth science and country planning. The second university USTO-MB established in 1975 in Oran, provides training for students in different areas of learning, ranging from natural science and life to mathematics and computer science. It has about 26,000 enrolled students taught by 1000 teachers, preforming both academic and research tasks.

Science, Technology, Engineering, Art, and Mathematics (STEAM) is the first STEAM center in Algeria launched in 2016. The center has trained more than 900 students and 25 teachers from across the country after its first 16 months (World Leaning, 2020).

2.3.5   Utilizing Emerging Technologies

The future of any education system is highly linked with the utilization of emerging technologies that can influence the way in which the education institutions teach and students learn. Over the past years, many initiatives have been undertaken in numerous studies to deal with the challenges of using emerging technologies in learning and teaching. For example, Mostefaoui et al. (2017) developed a remote electronic laboratory having the features of a low-cost alternative solution and flexible lab. It uses only the open source hardware and software products and requires a little maintenance. The graphical user interface of the remote lab can be used remotely with a low bandwidth Internet connectivity. The experimental results showed that the students who learned with this solution realized their task in a short time and had slightly better scores than those learned with the traditional method in a hands-on laboratory.

Soltani et al. (2018) presented an innovative framework based on one of artificial intelligence applications, that is the facial emotion detection used in Massive Open Online Courses (MOOCs) with the aim to help teachers to stay aware of the emotions of learners and their evolution during the learning activity. The framework was able to provide adaptive learning contents based on students' emotional states and profiles. It was developed based on three principles: modeling the learner using the MOOC; using of pedagogical agents during the learning activities; and capturing and interpreting the facial emotion of the students.

Furthermore, students from Algeria have participated in the project related to an education program at the Surrey Space Center, and instruments on board the satellite equipped with a Space Mag-PV Boom, magnetometer, RadFET radiation monitors, C3D2 camera, and Thin Film Solar Cel (Siebrits, 2019).

2.4   Requirements for Future Development of Science Education

The development of science education plays an important role in developing and modernizing the countries. The future development of science education in Algeria enabling the transition from the traditional education to the modern education is a major task that requires the involvement of the whole society. New technologies such as Artificial Intelligence, Internet of things (IoT), Virtual Reality (VR), 5G are expected to drive country growth in the coming years and change how we think, learn and live. This then demands a rethinking of all existing learning and teaching policies and standards to build a new curriculum that could help in preparing students for further complex issues.

In addition, the science education should be developed in the perspective of supporting the growth and the diversity of national economy with an appropriate adaptation to dynamically changing conditions.

Educators and teachers from whole country should learn from others in other countries. Godek (2004) suggested that the science education should be practical, relevant and appropriate, and the technical knowledge should be taken from developed countries but it must be suitable to the society and their needs. Because the future is uncertain and complex, it is closely important to adopt the strategy of future-focused to make considerable changes while keeping the current education system working.

2.5   Discussion and Conclusion

The education in Algeria has been developing for the last years as a result of several reforms undertaken for improving the learning and teaching in the whole education system, ranging from the primary school to higher education to vocational training.

While the Algerian government has placed great importance on the education development, there are still some challenges to overcome. The curriculum taught today at school does not meet tomorrow's employability skills such as the use of emerging technologies like Artificial Intelligence. Thus, there is a need of national strategy based on innovative programs preparing the students for successful transitions to tomorrow's world and employment by teaching Artificial Intelligence the at all level of education; AI needs to be considered as a pillar for any further reform of the education system. Moreover, the higher teacher training colleges should predict the evolution in the society to develop learning programs to allow students not only to reproduce the knowledge, but also to create it. As the country's main source of revenues is relied to gas and oil sector, there is an urgent need to develop new strategies to diversify the economy by investing more in innovation and education.

During the pandemic of Corona Virus Disease 2019 (COVID-19), the students and teachers have faced problems in remote education due to the slow internet connections and a lack of Information and Communications Technology (ICT) infrastructure (Bozkurt et al., 2020). Data transfer speeds and Internet connectivity should be improved to better provide students with opportunities to learn at any given time and space in the digital age.

References

Abada, Z., & Bouharkat, M. (2018). Study of management strategy of energy resources in Algeria. *Energy Reports, 4*, 1–7.

Benouar, D. (2013). Algerian experience in education, research and practice. *Procedia-Social and Behavioral Sciences, 102*, 361–367.

Bentoutou, Y. (2012). A real time EDAC system for applications onboard earth observation small satellites. *IEEE Transactions on Aerospace and Electronic Systems, 48*(1), 648–657.

Bozkurt, A., Jung, I., Xiao, J., Vladimirschi, V., Schuwer, R., Egorov, G., Lambert, S., Al-Freih, M., Pete, J., Olcott, Jr. D., & Rodes, V. (2020). A global outlook to the interruption of education due to COVID-19 pandemic: Navigating in a time of uncertainty and crisis. *Asian Journal of Distance Education, 15*(1), 1–126.

DGRSDT. (2020). *Research structures.* Retrieved from http://www.dgrsdt.dz/v1/index.php?fc=St_RSDT.

Global Out-of-School Children Initiative. (2014). *Summary Algeria country report on out-of-school children.* Retrieved from https://www.unicef.org/mena/media/6526/file/Algeria%20Country%20Report%20on%20OOSC%20Summary_EN.pdf%20.pdf.

Godek, Y. (2004). The development of science education in developing countries. *GU KirsehirEgitini FakultesiDergisi, 5,* 1–9.

Guemide, B., & Benachaiba, C. (2012). Exploiting ICT and E-learning in teacher's professional development in Algeria: The case of English secondary school teachers. *Turkish Online Journal of Distance Education, 13*(3), 33–49.

Kartianom, K., & Ndayizeye, O. (2017). What's wrong with the Asian and African students' mathematics learning achievement? The multilevel PISA 2015 data analysis for Indonesia, Japan, and Algeria. *Jurnal Riset Pendidikan Matematika, 4*(2), 200–210.

MESRS. (2020). Universities. Retrived from https://www.mesrs.dz/universites.

Ministry of National Education. (2020a). *Principes, objectifs généraux de l'éducation et organisation du cursus.* Retrieved from https://www.education.gov.dz/fr/systeme-educatif-algerien/principes-et-objectifs-generaux-de-leducation.

Ministry of National Education. (2020b). Retrieved from https://www.education.gov.dz/.

Mostefaoui, H., Benachenhou, A., & Benattia, A. A. (2017). Design of a low cost remote electronic laboratory suitable for low bandwidth connection. *Computer Applications in Engineering Education, 25*(3), 480–488.

National Office of Statistics. (2019a). Retrieved from https://www.ons.dz/IMG/pdf/demographie2019_bis.pdf.

National Office of Statistics. (2019b). *Les principaux indicateurs du secteur de l'éducation nationale année scolaire 2018–2019.* Retrieved from https://www.ons.dz/IMG/pdf/education_nat2018-2019.pdf.

Siebrits, A. (2019). Algeria. In *Integrated space for African society* (pp. 113–142). Springer, Cham.

Soltani, M., Zarzour, H., & Babahenini, M. C. (2018). Facial emotion detection in massive open online courses. In *World Conference on Information Systems and Technologies*, March (pp. 277–286). Springer, Cham.

Souleh, S. (2017). High education and scientific research sector in Algeria. *European Scientific Journal.*

Stambouli, A. B. (2011). Algerian renewable energy assessment: The challenge of sustainability. *Energy Policy, 39*(8), 4507–4519.

Université de Rouen. (2019). *ESAGOV L'Enseignement Supérieur Algérien à l'heure de la Gouvernance Universitaire.* Projet Erasmus + Rapport Final 2019. Retrieved from https://www.uni-med.net/wp-content/uploads/2020/04/ESAGOV-rapport-WP11.pdf

World Bank Group. (2019). *Algeria's economic update — October 2019.* Retrieved from http://pubdocs.worldbank.org/en/226791570663165545/EN-MPO-OCT19-Algeria.pdf

World Learning. (2020). Algiers STEAM Center. Retrieved from https://www.worldlearning.org/program/algiers-steam-center

Hafed Zarzour received his Ph.D degree in Computer Science from Annaba University, Algeria. He is currently an associate professor of Computer Science at Souk Ahras University, Algeria. He has published several research papers in International Journals and Conferences of high repute including Elsevier, Springer, Wiley, IEEE, ACM, Taylor and Francis, IGI Global, Inderscience, etc. His research focuses on Educational Technology, Technology Enhanced Learning, Artificial Intelligence, and Deep Learning.

Chapter 3
Science Education in Egypt—Intelligent Technology in Education Development

Ola Hosny, Ghada Barsoum, Ashraf Darwish, and Aboul Ella Hassanien

Abstract Enduring personal achievement throughout one's life while serving the community's needs into the future require sustainable changes for the education system based on the Artificial Intelligence and emerging technology. This entails forcing science education as a whole-of-society approach and building an enabling intelligent environment that allow all societal actors to critically act towards cohesive educational and research goals. Once this is done, and research is centralized at the decision-making process, citizens shouldn't miss any opportunities to explore knowledge of and about science, its culture and values, and put forward future plans that sets scientific thinking as a solid foundation for future success. This chapter describes the current status of education in Egypt and provide the future trends in this field.

Keywords Science education · STEM · Evidence-based research · Place-based learning · Fourth generation of Universities · Artificial intelligence · Digital transformation

O. Hosny (✉)
Professional Educator Diploma, School of Humanities and Social Sciences, The American University in Cairo, New Cairo, Egypt
e-mail: olahosny@aucegypt.edu

G. Barsoum
Public Policy and Administration Department (PPAD), School of Global Affairs and Public Policy, The American University in Cairo, New Cairo, Egypt
e-mail: gbarsoum@aucegypt.edu

A. Darwish
Faculty of Science, Helwan University, Cairo, Egypt
e-mail: ashraf.darwish.eg@ieee.org

A. E. Hassanien
Faculty of Computers and Artificial Intelligence, Cairo University, Cairo, Egypt

© The Author(s), under exclusive license to Springer Nature Singapore Pte Ltd. 2022   23
R. Huang et al. (eds.), *Science Education in Countries Along the Belt & Road*,
Lecture Notes in Educational Technology,
https://doi.org/10.1007/978-981-16-6955-2_3

3.1 Overview of the Country

3.1.1 Geographical Location, Population and Political System

The Arab Republic of Egypt is positioned northerly at the African Continent with slight part in the Asian Continent i.e. Sinai desert. It is of 1,002,450 km^2 and acknowledged as the 31st largest country in the world. Being strategically located, Egypt has seashores on the White and Red Seas with the Nile River running through it. It is divided into two sections: Upper Egypt in the south and Lower Egypt in the north (Baker et al., 2020). The Arab Republic of Egypt has borders with Israel, Sudan, the Gaza Strip, and Libya, being identified with unique resourceful geography, culture, and history. The Arab Republic of Egypt has three-layered administrative system i.e. 27 governorates, divided into marakiz or aqsam, comprising districts and villages (Osman, 2016). Cairo, the Arab Republic of Egypt's capital, is the largest city in both Africa and the Middle East and an important political and cultural focal point in the region. (Sims, n.a.).

In 2020, the Arab Republic of Egypt reached the largest population in the Arab Region, with its population surpassing 100 million inhabitants documenting 101,122,996 (CAPMAS, 2020). Of its total population, 51.6% are males and 48.4% are females, with a median age of 24.6 years old. Around 57.8% of the population live in rural areas, while 42.2% live in urban areas (CAPMAS, 2020).

Since 1952, Egypt's system has been "democratic republic based on citizenship and the rule of law. The political system is based on political and partisan multiplicity, the peaceful transfer of power, the separation and balance of powers, authority going with responsibility, and respect for human rights and freedoms, as set out in the Constitution" (International IDEA, 2020, p. 12).

3.1.2 Current Situation of Economic, Technologies and Cultural Development

The Arab Republic of Egypt's "economic system aims at achieving prosperity in the country through sustainable development and social justice to guarantee an increase in the real growth rate of the national economy, raising the standard of living, increasing job opportunities, reducing unemployment rates and eliminating poverty" (International IDEAL, 2020, p. 16). The economic system is driven by precise criteria, this included; promoting transparency and governance, effecting competitiveness, encouraging investment, preventing monopolistic practices, guaranteeing ownership; and protecting workers and consumers' rights (International IDEAL, 2020). The economic system is socially committed to closing any income gaps and developing balanced wage and pension systems to ensure a decent life (International IDEA, 2020).

Egypt's science and innovation system is shaped by the Ministry of Scientific Research (MOSR), represented in its Academy of Scientific Research and Technology (ASRT) work. The ASRT is largely responsible for; assessing Egypt's critical issues, their impact and drawing up Science, Technology and Information (STI) strategies to tackle the identified issues. In addition, the ASRT acts as a think tank, coordinating Egypt's research programs, through which scientists, experts, research institutes, NGOs and the private sector are brought together in a collaborative work to debate Egypt's grand challenges and design needed research studies that inform the policymaking process. Moreover, in cooperation with the Ministry of Higher Education and Scientific Research, MOSR also works on developing the national research policy and the research strategy with close alliance with the public universities and research institutes, ensuring harmonized activities that again support identified critical issues (Bond, n.a.).

Egypt's cultural development is uniquely identified by emblematic signs. It embraces 80 pyramids, among which is the Great Pyramid of Giza. According to the UNESCO, Egypt comprises seven of the world's heritage sites; the Wadi al-Hitan, the St. Catherine Monastery, Abu Simbel temples, many temples among Philae, The Pyramids of Giza Complex, Old Cairo, Luxor, and Abu Mena. More importantly (Business Monitor International, 2018). In addition, Egypt is the home country to Mo Salah, playing for Liverpool Team, and the fourth best-paid soccer player in the world in 2020 (Hopwood, 2020).

3.2 Overview of the Education Development

3.2.1 Education System and Policy

The Egyptian education system is divided into two phases. First, the pre-university education, comprising the basic and secondary levels supported by multiple educational tracks. Second, the higher education, comprising universities and higher institutes with different areas of studies. Both systems comprise public and private institutions across all governorates (The Egyptian Center for Economic Studies, 2020). By constitution, in Egypt every citizen has the right to education, in which a free education is made available in the two levels by law, with a commitment of spending no less than 4% of the GDP for education. Egypt's educational system is fostered to maintain global quality criteria, cultivating; logical thinking, cultural morals, and innovation and creativity (International IDEA, 2020).

In 2018/2019, of the total enrolled students in the education system, 88% are in the pre-university stage, and the remaining (12%) are in higher education. In rural, private schools represents one fourth of the public schools, while it is only half in the urban areas. In rural areas, classrooms fall short compared to the number of enrolled students, accumulating 162 students per class. On the other side, in urban areas students' density reach only 119 (The Egyptian Center for Economic Studies,

2020). The students-teacher ratio in rural areas is around 25.5, compared to 21.1 in urban areas (The Egyptian Center for Economic Studies, 2020).

In 2020, a new education system has been launched, cancelling all exams in the early stages of education and allowing 1.2 million students in grades 10 & 11 to take their exams electronically. In addition, distance learning was adopted for secondary level students supported by the knowledge bank and e-library (The Egyptian Center for Economic Studies, 2020). New experiences' evaluations have not been completed yet, but enhancement plans are expected to be developed once the process is completed.

Currently, educational systems in Egypt is changed to depend on the online and virtual class systems and smart educational environment (Economist Intelligence Unit, 2019). This environment will depend on the recent technologies such as artificial intelligence revolution in education and emerging technologies to design the fourth generation of Universities. This kind of Universities can increase the Egyptian Universities ranking and will satisfy the quality assurance and accreditation criteria.

3.2.2   Statistics on the National Education

The national pre-university education for both public and private schools serves over 22 million students. The public schools serve 89.6% of the students, while the private schools serve only 10.4%. In 2019/2020, public schools increased twice private schools, while students' numbers increased five times in public schools. Statistics for schools, classes & students of pre-university education *by sector and educational stage* for the academic year 2018/2019 as shown in Table 3.1.

Interestingly, there is no significant difference between males and females; enrollment rates. Statistics of schools, classes & students *by gender* of pre-university education, *by educational stage* for the academic year 2018/2019 as shown in Table 3.2.

The Greater Cairo (comprising the Cairo, Giza and Qaliubiya governorates) documents the largest enrollment rates (i.e. almost 20%). This high enrollment rate is carried over in governorates that are near Cairo, in both sides; Upper Egypt (e.g. Menia and Sohag), and Lower Egypt (e.g. Sharkia and Behera). Alexandria, the second-largest city in Egypt, also serves a high percentage of students (5.5%) compared to its population (5 million habitant). Statistics of schools, classes & students of pre-university education *by governorate* for the academic year 2018/2019 as shown in Table 3.3.

On the other side, the Egyptian system has higher teachers' rate for the primary stage in response to the high students' enrollment rates compared to other stages. Statistics of teachers *by educational stage and **sex*** for the academic year 2018/2019 as shown in Table 3.4.

Correspondingly, the teacher-students ratio in the primary stage is the highest, followed by the pre-primary stage. Class density is also the highest in the primary stage, followed by the preparatory and the secondary stages, aligning with the high

Table 3.1 Statistics by sector and educational stage for the academic year 2018/2019

Educational stage	Private sector			Governmental sector		
	Students	Class room	Schools	Students	Class room	Schools
Pre-primary	351393	11612	2516	1038549	26848	9549
primary	1197889	35198	2248	11002210	213926	16514
Preparatory	375401	12159	1888	4636903	98880	10387
General secondary	257373	8198	1298	1496539	35142	2360
Industrial secondary	2562	100	8	899769	25023	1219
Agricultural secondary	–	–	–	225481	5029	256
Commercial secondary[a]	147016	2922	197	649430	15373	708
Societal education[b]	–	–	–	133007	5048	5048
Handicapped education	418	95	16	39441	4615	1002
Total	2332052	70284	8171	20121329	429884	47043

[a] Commercial Secondary includes Hotel Secondary
[b] Include single class (Mixed) + improving crafts for Girls
Source CAPMAS (2020), p. 146

Table 3.2 Statistics by gender of pre-university education, by educational stage for the academic year 2018/2019

Educational stage	Total	Females	Males	Classes	Schools
Pre-primary	1389942	673291	716651	38460	12065
primary	12200099	5928755	6271718	249124	18762
Preparatory	5012304	2440586	2571718	111039	12275
General secondary	1753912	957246	796666	43340	3658
Industrial secondary	902331	331333	570998	25123	1227
Agricultural secondary	225481	34218	191263	5029	256
Commercial secondary[a]	796446	457505	338941	18295	905
Societal education[b]	133007	91605	41402	5048	5048
Handicapped education	39859	14615	25244	4710	1018
Total	22453381	10929154	11524227	500168	55214

[a] Excluding AL-Azhar Education
[b] Include single class (Mixed) + improving crafts for Girls
Source CAPMAS (2020), p. 141

enrollment rates at these stages. Statistics of teachers-students ratio, class density *by educational stage* as shown in Table 3.5.

Looking into the dropout rates, the Egyptian educational system suffers from high dropout rates in the preparatory stage compared to other stages. Statistics of number & percentages dropout for primary stage by sex and governorate for academic year 2018–2019 as shown in Table 3.6.

Table 3.3 Statistics by governorate for the academic year 2018/2019, Excludes (single Class, Handicapped Schools—Al-Azhar & Experimental Education)

Governorate	2019/2018			2018/2017		
	Students	Rooms	Schools	Students	Rooms	Schools
Cairo	2265 194	53556	4970	2182785	53906	4865
Alexandria	1 230 856	25711	2 409	1169 307	25325	2268
Port-Said	165 379	4 391	449	157 581	4259	441
Suez	174940	4 363	406	168817	4266	393
Damietta	378547	8815	907	371 306	8552	853
Dakahlia	1415119	30570	2 945	1361 891	30 138	3167
Sharkia	1625 501	35595	3 889	1546 479	35 053	3 807
Kalyoubia	1365107	28371	2252	1306 480	27 464	2195
Kafr-El-Sheikh	771919	17 158	1987	729 008	16759	1949
Gharbia	1119 704	23234	2274	1065370	22 626	2202
Menoufia	1035555	22 084	2 245	991065	22 033	2145
Behera	1446356	30716	3 480	1362817	29 902	3378
Ismailia	331190	7999	903	311 325	7965	885
Giza	2163 821	42822	3 384	2 063435	41 407	3267
Beni-Suef	734113	16044	1727	701 480	15654	1673
Fayoum	785704	16400	1436	764 696	16 449	1389
Menia	1308185	27851	2 596	1 242 525	27689	2537
Asyout	1040 552	22 002	2191	989 980	21 546	2127
Suhag	1112570	24558	2 443	1074 205	24 050	2392
Qena	741 598	16961	1771	701536	16 700	1736
Aswan	345 224	9405	1233	329076	9 199	1208
Luxor	274 131	6447	806	259 834	6 433	775
Red Sea	104676	2810	372	98882	2691	357
El-Wadi El-Gidid	65 560	2328	427	61286	2267	416
Matrouh	131257	3414	507	120 691	3299	490
North Sinai	111170	3533	547	108822	3564	533
South Sinai	36587	1488	293	33120	1419	285
Total	22280 515	488 626	48 849	21273799	480 615	4733 47

Source CAPMAS (2020), p. 144

3.2.3 *Educational Research and International Collaboration*

In 2014, the Egyptian Knowledge Bank (EKB), a national online library archive and resource was developed, providing access to learning resources and tools for educators, researchers, students, and the general public. With the support of the EKB, in September 2018, the Ministry of Education lead an educational reform

Table 3.4 Statistics of teachers by educational stage and sex for the academic year 2018/2019

Educational Stage	Teachers			
	% of total female	Total	Female	Male
Pre-primary	99.6	58342	58136	206
primary	64.2	445797	286011	159786
Preparatory	97.2	8280	8046	234
General secondary	53.7	261009	140116	120893
Industrial secondary	42.7	106574	45484	61090
Agricultural secondary	45.0	93599	42073	51526
Commercial secondary[1]	39.6	13276	5252	8024
Societal education[2]	59.5	41999	24969	17030
Handicapped education	60.2	9791	5892	3899
Total	59.3	1038667	615979	422688

Source CAPMAS (2020), p. 149
[1]Commercial Secondary includes Hotel Secondary
[2]Include Single Class (mixed) + Improving Crafts for Girls

Table 3.5 Statistics of teachers-students ratio, class density *by educational stage*

Educational Stage	Students per teacher		Class Density		Girls %	
	19/18	18/17	19/18	18/17	19/18	18/17
Pre-primary	23.8	26.1	36.1	35.7	48.4	48.3
primary	27.4	26.6	49	47.5	48.6	48.5
Preparatory	16.1	19.7	26.3	26.3	68.9	70.5
General Secondary	19.2	18.3	45.1	43.7	48.7	48.9
Industrial Secondary	16.5	16.0	40.5	40.6	54.6	54.1
Agricultural Secondary	9.6	9.5	35.5	35.8	36.7	36.2
Commercial Secondary[1]	17	15.9	44.8	42.4	15.2	15.9
Societal Education[2]	19	17.8	43.5	42.0	57.4	58.0
Handicapped Education	4.1	3.9	8.5	8.3	36.7	36.9

Source CAPMAS (2020), p. 147

under the banner of Education 2.0 (EDU 2.0). This reform aimed to update the education system, prepare youth for industrial revolution, and propel education towards the Sustainable Development Goals of Egypt Vision 2030. Focusing on curriculum changes, teaching and pedagogy advancement, assessment institutionalization, and digital transformation, the new system worked on transforming schooling and learning to global standards. Challenges included; limited resources, equity, quality, etc. (European Union, 2018). Since then, efforts have been and are intended to continue facing the challenges and working around building genuine steps towards sustainability through continuing reforms.

Table 3.6 Statistics of number & percentages dropout for primary stage by sex and governorate for academic year 2018–2019

Governorate	Percentage of dropout[a]			No. of dropout		
	Total	Females	Males	Total	Females	Males
Cairo	0.3	0.5	0.3	3085	1274	1811
Alexandria	0.6	0.3	0.7	3551	1484	2067
Port-Said	0.3	0.2	0.4	228	84	144
Suez	0.1	0.1	0.1	94	38	56
Damietta	0.4	0.2	0.6	770	207	563
Dakahlia	0.4	0.3	0.5	2653	921	1732
Sharkia	0.2	0.2	0.3	1890	803	1087
Kalyoubia	0.5	0.4	0.6	3553	1419	2134
Kafr-El-Sheikh	0.3	0.2	0.3	944	313	631
Gharbia	0.5	0.4	0.6	2705	970	1735
Menoufia	0.6	0.5	0.6	2758	1156	1602
Behera	0.4	0.3	0.5	2745	1109	1636
Ismailia	0.3	0.3	0.3	498	247	251
Giza	0.3	0.3	0.4	3637	1564	2073
Beni-Suef	0.7	0.6	0.9	2620	1007	1613
Fayoum	0.3	0.2	0.4	1205	452	753
Menia	0.4	0.3	0.4	2420	1034	1386
Asyout	0.6	0.5	0.8	3149	1127	2022
Suhag	0.4	0.4	0.5	2384	945	1439
Qena	0.3	0.3	0.3	1037	551	486
Aswan	0.2	0.2	0.2	337	151	186
Luxor	0.2	0.2	0.2	266	122	144
Red Sea	0.1	0.1	0.2	89	32	57
El-Wadi El-Gidid	0.1	0.1	0.1	20	8	12
Matrouh	0.6	0.1	0.3	431	337	94
North Sinai	0.1	0.1	0.1	63	32	31
South Sinai	0.4	0.4	0.3	55	31	24
Total	0.4	0.3	0.5	43187	17418	25769

[a] No. of students who dropped out for two Consecutive years/No. of new students who enrolled with stage two years ago * 100

Source CAPMAS (2020), p. 160

Parallelly, in 2018, the Ministry of Education in partnership with the Social Research Center (SRC) at the American University in Cairo (AUC) have developed the Research and Documentation Project (RDP) to research these fast-track reforms and document historic change in Egyptian education, through grounded research. This partnership provided an opportunity for independent researchers to work together with the ministry to generate evidence in a time of enormous change, challenges, and educational innovation.

The RDP uses four main research approaches:

- *Oral histories*: is a method of gathering, preserving and interpreting the voices and memories of people, communities, and participants in past events
- *Case studies*: is a means to build evidence from the ground with parents, students, teachers and other stakeholders for the purpose of understanding how, and if, the interventions are changing attitudes, practices, and modes of learning.
- *Data analysis*: is the process of cleaning and categorizing data to identify required data
- *Social media analysis:* is the process of collecting the most valuable data from your social media channels and drawing actionable conclusions.

The main aims of the RDP are:

- To generate original, timely, policy-relevant multidisciplinary educational research
- To build capacities in educational research, keeping in mind the need to incorporate new approaches that capture big data, social media, and digital transformation, among other changes in the field
- To support and inform policy discussions on education during Covid-19
- To build an archive of policy change and innovation.

Moreover, the MOE has benefited from the Cairo-based Microsoft Advanced Technology Lab (ATL), which binds technologies such as artificial intelligence (AI) and cloud facilities, to accelerate research and development that support the newly introduced system.

More importantly, the MOE paid special attention to inclusive education through which three products were developed; "the Special Education Curriculum Frameworks, the Guidelines for the Adaptation and Accommodation of Learning Materials for the Children with Sensory Disabilities and the Teachers Guide on Inclusive Education, all developed under the Education 2.0. reform's mandate" (UNICEF, 2020).

3.3 Current Situation of Science Education

Since early eighties and onwards the benefits of science education have been defined multiply, all of which revolve around three main concepts; full engagement, inter-disciplinary aspects, and critical thinking. Amongst the available definitions in the literature, two were much clearer.

First, Hazelkorn (2015) articulated six main benefits for science education, represented in

- Promoting a culture of scientific thinking and evidence-based reasoning;
- Ensuring citizens participation in a progressively multifaceted industrial and high-tech world;
- Developing citizens' competencies for problem-solving, innovation, and analytical and critical thinking;
- Inspiring students to careers in science that can support creative solutions;
- Enabling public private partnerships for innovative approaches;
- Empowering citizen's engagement in community learning discussions, arguments and policymaking.

Second, Lewis and Kelly (1987) complemented the above by confirming science education's purpose to be:

- Promoting agricultural development, industrial production, scientific research and social development
- Providing pupils with a scientific spirit of curiosity and inquiry
- Understanding and change the natural world
- Encouraging people to question and search for data.

With the above defined benefits and purposes and in line with Egypt's needs in investing in the manpower, linking graduates' skills with the labor market needs, and preparing calibers that can face country's economic downturns, the new educational system was introduced (El-Deghaidy, 2012). This system promoted students' critical thinking, social critique and analysis of local contexts, qualifying them to find solutions and suggestions for a better Egyptian future considering the national environmental, social, economic and cultural considerations. In brief, an education for sustainable development (ESD) was targeted, starting effecting needed changes in the community (Hopkinson & James, 2010).

ESD necessitated the integration of disciplines in "a cohesive meaningful experience where science and mathematics (main disciplines in the acronym) are learnt in a personalized context while developing various skills such as problem solving, communication skills, inductive and deductive reasoning and inquiry skills" (El-Deghaidy, 2015, p. 5). To this end, the science education (Wang, 2012) combining four disciplines; science, technology, engineering and management (STEM) was prioritized in the Egyptian education system (Thompson, 2013).

For that, in 2011, the Egyptian STEM educational system was re-designed, renewing the concept of STEM schooling that was already introduced decades ago, to revolve around enhancing research in natural sciences curriculum considering humanities to support Egypt's national development (Enah, 2014). This system was fostered by the project-based learning pedagogy (El-Baz, 2009), in which students are triggered by real-life issues stemmed from Egypt's national grand challenges, supported by an enabling environment, to undertake needed research that can conclude solutions for such challenges considering all STEM's dimensional perspectives (Vasquez et al., 2013).

Currently, there are 15 STEM schools in Egypt, spread all over the governorates as follows: Cairo, Giza, Alexandria, Dakahlia, Kafr El-Sheikh, Assuit, Ismailia, Luxor, Red Sea, Gharbiah, Menoufia, Sharqiah, Obour, Qena, and Beni Suef. More schools are the pipeline, planning to allow more opportunities for outstanding students who can deploy research to push the national status forward.

STEM schools follow a gender & economic sensitive selection criteria. That said females are encouraged to join equally like boys, and any financial burdens are subsided by the government. Conditions stipulated include; attaining a minimum percentage of 98% in the preparatory exam (later lowered to 95%), with an acknowledgment in at least one of the following subjects; math, science, and English. And, scoring at least 290 degree in the in IQ-based entry exam.

3.3.1 Policies and Standards

The science education in Egypt represented in the STEM schools are governed by multiple ministerial degrees, summarized as follows:

First, Ministerial Decision No. (369) for 2011 concerning the establishment of the secondary STEM schools' system. This decision specified the role of the schools to; 1- Supporting the gifted and talented students and promoting their abilities; 2- Developing and teaching advanced curricula in science, mathematics and technology; 3- Developing the use of information technology methods to develop the educational process; 4- Consolidating spiritual and educational values and deepening the values of tolerance and openness to the world; and 5- Opening the door to the creativity potential of students (Ministry of Education, 2011).

Second, Ministerial Decision No. (202) for 2012 concerning awarding Egyptian secondary certificate in science and technology from STEM secondary schools.

Third, Ministerial Decision No. (238) for 2012 concerning the admission system, studying and testing at secondary STEM schools. This included; (A) Limiting the STEM certificate to the third grade of secondary schools for outstanding students in science and technology; (B) Having successful students in the second grade of secondary schools to choose one of the two groups: Sciences or Mathematics; (C) Specifying the areas of studies for "mathematics" to: robotic engineering—electronic engineering, etc., and for "sciences" to: hydraulic—earth and space sciences, etc., and enforcing practical exams at the end of each semester.

Fourth, Ministerial Decision No. (382) for 2012 concerning the admission grade for the STEM schools, to be 95% instead of 98% on the preparatory school final exam.

Fourth, Ministerial Decision No. (308) for 2013, later substituted by Ministerial Decision No. (322) of 2013 in regulation of the STEM schools exams, necessitating the announcement of the results of examinations in mid-July of each year, and specifying the second-round exams to be in the event of absenteeism or failure in one or two subjects, in which students not to be counted more than 60% of the total score of the subject in the event of success in the second round, and in the event of repeated failure of the student, he/she is transferred to the third secondary grade in regular government schools.

Fifth, Ministerial Decision No. (172) for 2014 concerning the establishment of a STEM unit inside the governorate to support the STEM schools on the governorates level.

Sixth, Ministerial Decision No. (313) for 2015 concerning subsidiary committees to support STEM schools on the governorates level. This committee is expected to take over the responsibility of the centralized unit on the governmental level (mentioned above).

3.3.2   Curriculums, Digital Resources and Teacher Training

As argued by Khuyen et al. (2020), teachers' practices are influenced by their perceptions. Thus, STEM teachers' professional development program in Egypt focused on having teachers adopting new pedagogies i.e. students-centered, making them believe that they can effect targeted transformation (El-Deghaidy, 2020, p. 5). This was done through training STEM teachers "to reinforce students' achievement as well as leverage students' pursuance in science-related careers for global goals" (Khuyen et al., 2020, p. 2). Currently the there are two training programs; one for the in-service teachers, and the other for the pre-service teachers. The first, develops teachers' capacity with the various pedagogies applicable to STEM education which emphasizes a student-centered approach and link to meaningful real-life issues. The later, present STEM education integrated modules on pedagogical practices (i.e. place based learning and issue analysis) and integrated curricula design units using the backward design approach and project-based learning (El-Deghaidy, 2020). With that, the teaching quality is expected to be enhanced and mentored with needed support.

More importantly, in partnership with international partners, a four-year STEM undergraduate teaching degree program at five Egyptian public universities were launched so that future teachers receive the coursework and experience to be effective STEM classroom instructors. In addition, a one-year diploma for teachers and school leaders to specialize in STEM education is also made available. Besides, there are more than one international diploma offerings in Egypt training schoolteachers in the

STEM track. The first is offered by the American University in Cairo, and the second is offered by the Ain Shams University in partnership with the European Union.

Moreover, there are multiple initiatives that support the STEM concept of Egypt, among them is the STEM HuB, which aims to equip Egypt's young scientists, innovators, and entrepreneurs with the necessary skills to serve national objectives and lead Egypt into the next era of technical innovation. To that end, the STEM Hub offers programs designed to allow participants to explore, apply, and innovate through hands-on activities and project-based learning approach. This unique learning methodology aims to improve the quality of knowledge acquisition as well as leverage the skills, talents, and leadership potential of Egyptian youth.

3.3.3 Student Assessment and Achievement

To align with the science education challenging setting, the assessment of the STEM schools in Egypt is made based on specific scoring criteria and passing grades, which necessitate student's achievement of a total score of 60% to pass each year, broken down into; 40% on subjects and 60% on the capstone project. This obliges students to: complete bi-monthly journals and the portfolio, representing student's progress over the year, and, to complete posters presented at the end of the year together with the prototype exhibition, representing student's twenty-first century skills (AbdelMaguid, 2017).

In addition, the Program for International Students Assessment (PISA) evaluates students' key knowledge and skills, enabling him/her to join the modern societies' areas of studies (OECD, 2018) has also been considered in the Egyptian context.

3.3.4 Science and Technology Venues and Centers

Admitting the science education is about teaching students to think in disciplinary ways—to think like a scientist, and with the current STEM schooling systems that is being introduced nationally, to protect citizens' rights to have equal opportunity to voice their experiences, and to respect these voices and support them with research-based studies, Egypt has deployed its national research institutes, sub-institutes and private ones to support science education and build an enabling environment of scientists. This was done through Egypt's two main national holistic research institutions.

First, as mentioned earlier, the Academy of Scientific Research & Technology (ASRT) affiliated to MOSR, is the national acknowledged institution for scientific research. ASRT's importance lies in; supporting national bodies to incubate joint system of scientific research, and, increasing the number of qualified researchers who can deploy science in favor of national development.

Second, Egypt's National Research Center (NRC) is a multidisciplinary research center devoted to basic and applied research, covering studies in; industry, health, environment, agriculture, basic sciences and engineering. Besides, other national research centers that are specialized in specific areas, like; Agricultural Research Center Egypt, Egyptian Petroleum Research Institute, Egyptian Atomic Energy Authority, National Authority for Remote Sensing and Space Sciences, National Institute of Oceanography & Fisheries, National Research Institute of Astronomy and Geophysics NRIAG, Electronics Research Institute, National Telecommunication Institute, National Water Research Center Egypt, Egyptian Meteorological Authority, Housing and Building Research Center Egypt, Egyptian Center for Economic Studies, Central Laboratory for Agricultural Expert Systems, Institut Français d'Archéologie Orientale du Caire, Egyptian National Scientific Technical Information Network, Agricultural Extension & Rural Development Research Institute, Agricultural Genetic Engineering Research Institute.

Moreover, Egypt now has Eight technological colleges comprising 45 above average technical institutes that support the practicum component of the STEM schools. Besides, in 2019/2020, new technological universities were established (i.e. University of Technology in New Cairo, the Technological University of Quesna, and the University of Technology in Beni Suef) working towards the same goal. More importantly, five technology universities are in the pipeline (i.e. East Port Said, 6th October, Borg El Arab, New Luxor, and Assiut), directing towards speeding up the targeted integrated development.

These venues and centers are supported by difference strategies and initiatives that give them the needed operating support, like: The National Strategy for Artificial Intelligence; The national strategy for the electric car industry; Egypt's Industrial Initiative; "FUSION" initiative, partnering engineering education and scientific research, and qualifying a new cadre of researchers who can push forward the national plans.

In terms of digital transformation, as mentioned by the State Information Systems (2020), Egypt has many achievements:

- Work is underway to complete the national project to raise the efficiency of the universities 'information infrastructure in line with the state's policy of digital transformation in order to convert it into smart universities.
- Launching of the Geographic Information Portal (GIS) of the Ministry of Higher Education and Scientific Research.
- Establishing 8 technological parks in cooperation with the Ministry of Communications and Information Technology in the following universities: "Menoufia, Mansoura, Sohag, Aswan, Minya (Qena), South Valley, and the Suez Canal", within the framework of the country's strategy towards settling and spreading the culture of creativity and innovation around the Republic, encouraging development and digital economy, and the employment of the latest technology fields in achieving more competition and entrepreneurship in line with the strategy of the Ministry in serving the surrounding community and preparing a generation

capable of creativity and innovation and to integrate information and communications technology in the educational process to achieve the maximum possible benefit from advanced technological systems.

- Activating the electronic complaints system for citizens.
- Activating the system of learning and electronic tests in universities.
- Activating the electronic payment system for different services in universities.

3.3.5 Utilizing Artificial Intelligence and Emerging Technologies in the Era of Digital Transformation

In 2018, Egypt acted as an elected Chair of the Executive Bureau of the Arab Telecommunications and Information Council of Ministers (ATICM), besides, gaining the African Telecommunication Union (ATU) membership. In 2019, Egypt also acquired the membership of the International Telecommunication Union (ITU).

Recently, Egypt established centers and administrations for the digital transformation in most of the universities and established the first faculty for artificial intelligence in the Middle East. Egypt launched a new project to transfer to the fourth generation of Universities which depend on artificial intelligence and emerging technologies such as Internet of Things, big data analytics, cloud and fog computing and blockchain technology.

Currently, digital transformation applies to many areas, including education. Therefore, the educational environment has also begun to change. Indeed, more and more educational programs incorporate digital culture into the curriculum.

With that, and in support of the STEM schooling Egypt technological development allowed it to put forward operating mechanisms that enhance the technological usage across its institutions. Steps completed up till now are:

- Through a public private partnership (Rowad 2030 and Alexandria University), Egypt is currently establishing an incubator in the field of artificial intelligence.
- Egypt has established up till now 19 technological incubators, and 5 are still in the pipeline.
- A total of 68 companies were established resulting from the working incubators, to operate in the market. Besides, 25 startups are being incubated.
- Egypt has currently 13 technological alliances.

In addition, Egypt has given due consideration to capacity building. It focuses on creating "a diversified pool of talents and experts, through launching several new initiatives and programs, including; Masarak initiative, Wazeefa Tech, the Next Technology Leader (NTL), the Arab digital platform for providing free self-training for the Arab young people, the African App Launchpad and others" (Minister of Communications & Information Technology, 2018, p. 11), to enhance youth skills to encourage innovation and entrepreneurship, promoted by the STEM system.

3.4   Requirements for Future Development of Science Education

Speaking of future development of science education requires us to think of the national research plans, and how they can and should be able to enhance teacher practice and curricular resources. That said, research plans should be able to inform that learning environments about; (1) teachers' practices; (2) teaching tools that support multidisciplinary learning; and (3) enabling contexts that support teachers and students' efforts.

This may be exemplified in allowing diversified opportunities for students to learn more about science, and being derived into a discovery learning journey, unfolding any scientific whereabouts. Such multidisciplinary opportunities would allow students to acquire more multifaceted understandings and actions.

Another attempt for future development of science education is in focusing on developing a stronger bond between students and the natural world and supporting students' strong faith of their ability to change and/or influence the natural world. This could help developing a more genuinely bicultural society. To promote this bond, and allow students the needed opportunity, "place-based pedagogies are highly advisable, including; place-based learning, experiential education, community-based education, education for sustainability, environmental education or more rarely, service learning" (McClennen, 2016, p. 4). These pedagogies places students in local heritage, cultures, landscapes, opportunities and experiences, and uses these as a foundation for the study of science education.

Once these raised suggestions are fulfilled, the following benefits will be required; Learning will be grounded nationally; student-centered approach will be incubated; social-emotional learning will be prioritized; lessons will be inclined towards inquiry-based; and, learning will be more relevant and engaging (Smith, 2016). More importantly, "students can be challenged to see the world through ecological, political, economic and social lenses, students can have more agency and autonomy—boosting motivation and persistence, students can meet deeper learning, and finally students can gain better appreciation and understanding of the world around them" (Smith, 2016, p. 8).

3.5   Discussion and Conclusion

With the growing population, advanced digital development, and global pandemics, the utilization of science education became a must, to rely less on country's national resources, and face any expected economic downturns. This can only happen if societies' abilities are enhanced, enjoying smart educate, preparing creative and entrepreneurial individuals to think autonomously and critically, and having everyone engaged in lifelong learning (LLL) journey, where knowledge, social and technological innovation are properly utilized and adapted to advance country's status.

Agreeing with Hazelkorn (2015), science education should be set as a cohesive goal to understand possible opportunities for economic development, and identify possible risks for determination. That said, science education should be an integral approach supporting the education system K-12. By that, students' competencies should be enhanced, in which the four disciplines of STEM education are tuned by Arts. This would necessitate the development of; students' critical thinking, teachers' professional development, and curriculums learning outcomes. Besides, more public-private partnerships should be boosted, ensuring a participatory approach in which every party has a full chance to contribute to the development of the community. Finally, raising the awareness of the public on science intakes and emphasizing needed policy linkages on the national and global levels are expected to draw desirable framework for development. As a result Egypt will be a leader in education in the Middle East and Africa.

References

AbdelMaguid, L. (2017). *The initiative of stem schools in Egypt: Issues of process, teachers' compatibility and governance.* Cairo, Egypt: American University in Cairo.

Baker, R. et al. (2020). *Egypt.* Cairo, Egypt: Encyclopedia Britannica.

Bond, M. et al. (n.a.). *Science and innovation in Egypt.* California, USA: Creative Commons.

Business Monitor International. (2018). *Egypt.* London, UK: Business Monitor International.

CAPMAS. (2020). *Egypt in figures 2020.* Cairo, Egypt: CAPMAS.

Economist Intelligence Unit. (2019). *Country report: Egypt.* London, UK: Economist Intelligence Unit.

El Baz, H. (2009). The effectiveness of a project-based model in the development of learning skills for 1st preparatory. *The Egyptian Society for Science Education, 1*(41), 59–67.

El-Deghaidy, H. (2012). Education for sustainable development: Experiences from action with science teachers. *Discourse and Communication for Sustainable Education, 3*(1), 23–40.

El-Daghaidy, H. (2015). *Science education in Egypt based on integrating ecological needs and steam education.* Cairo, Egypt: The American University in Cairo.

El-Deghaidy, H. (2020). Localising STE^{2}AM Education. Leiden, Netherlands: Brill Publisher.

Enah, K. (2014). *STEM education transfer from the United States: A critical analysis of local STEM practices in customizing STEM education in Egypt.* Unpublished Master Thesis in International Comparative Education, American University in Cairo, Egypt.

European Union. (2018). *A stable Egypt for a stable region: Socio-economic challenges and prospects.* Cairo, Egypt: European Union.

Hazelkorn, E. (2015). *Science education for responsible citizenship.* Brussels, Belgium: European Commission.

Hopkinson, P., & James, P. (2010). Practical pedagogy for embedding ESD in science, technology, engineering and mathematics curricula. *International Journal of Sustainability in Higher Education, 11*(4), 365–379.

Hopwood, D. (2020). Egypt. *Politics and Society, 3*(1), 1945–1990.

International IDEA. (2020). *Egypt's Constitution of 2014.* Cairo, Egypt: International IDEA.

Khuyen, N. et al. (2020). Measuring teachers' perceptions to sustain STEM education development. *Sustainability, 12*(1531), 1–15.

Lewis, L., & Kelly, J. (1987). *Science and technology education and future human needs.* Oxford, USA: Pergamon Press.

McClennen, N. (2016). *Place-based education: Communities as learning environments.* Washington, USA: Getting Smart.

Minister of Communications and Information Technology. (2018). *MCIT year book 2018.* Cairo, Egypt: Minister of Communications and Information Technology.

Ministry of Education. (2011). *Ministerial Decision no. 369 for 2011.* Cairo, Egypt: Ministry of Education

OECD. (2018). *PISA 2015 results in focus.* Washington DC, USA: OECD.

Osman, M. (2016). *Population situation analysis.* Cairo, Egypt: Baseera.

Sims, D. (n.a.). *The case of Cairo, Egypt.* Cairo, Egypt: GTZ.

Smith, J. (2016). *What is place-based education and why does it matter?* Washington, USA: Getting Smart.

State Information Systems. (2020). *Higher education and scientific research sector.* Cairo, Egypt: State Information Systems.

The Economist Intelligence Unit. (2019). *Egypt country report.* London, UK: The Economist Intelligence Unit.

The Egyptian Center for Economic Studies. (2020). *The impact of COVID-19 on Egypt's pre-university education system.* Cairo, Egypt: The Egyptian Center for Economic Studies.

Thompson, R. (2013). *Evaluation of an integrated STEM professional development model into an elementary school.* Munich: Grin Verlag.

UNICEF. (2020). *Ministry of Education and Technical Education and partners celebrate key milestones for inclusive education in Egypt under Education 2.0.* Cairo, Egypt: UNICEF

Vasquez, J., Comer, M., & Sneider, C. (2013). *STEM lesson essentials: Integrating science, technology, engineering, and mathematics.* New Hampshire: Heinemann.

Wang, H. (2012). *A new era of science education: Science teachers' perceptions and classroom practices of science, technology, engineering, and mathematics (STEM) integration.* Doctoral Dissertation, University of Minnesota, Minnesota.

Ola Hosny is an Educational Research and Development Specialist with over 20 years of working experience in conceptualizing, managing and evaluating different development projects working mainly in the areas of youth development, educational development, poverty eradication, gender equality, women empowerment, and entrepreneurship. Besides, Hosny is a part-time instructor at the Professional Educator Diploma of the American University in Cairo. More importantly, Hosny is a Monitoring, Evaluation and Learning (MEL) Specialist working as an External Evaluator for educational development projects. Hosny obtained her MA in International and Comparative Education from the AUC, and is currently obtaining her PhD (last year) in Higher Education: Research, Evaluation and Enhancement from Lancaster University.

Ghada Barsoum is associate professor and currently chair of the Department of Public Policy and Administration at The American University in Cairo (AUC). Barsoum has numerous publications in reputable international peer-reviewed journals including Gender, Work and Organizations; Public Organization Review; International Journal of Social Welfare, and Current Sociology. She is also the author of a book on the employment crisis of female graduates in Egypt, and a number of book chapters, technical reports, published policy papers and encyclopedia entries. She has consulted for the International Labor Organization, UNESCO, UNFPA and UNICEF. While at AUC, she has been the principal investigator of projects with funding from the Ford Foundation, the United Nations Development Programme and Friedreich Ebert Foundation. Prior to joining AUC, Barsoum was research associate at the Population Council, West Asia and North Africa Office, where she spearheaded efforts for a national survey on youth in Egypt. Barsoum obtained her PhD in sociology from the University of Toronto and her master's degree from AUC.

Ashraf Darwish his bachelor and master degree in mathematics from Tanta University in Egypt. He received his Ph.D. in computer science from computer science department at Saint Petersburg State University in 2006 from Russian Federation. He has worked as associate professor and then professor of computer science at the mathematics and computer science department, faculty of science, Helwan University in Cairo. Currently, he is adjunct professor of computer science at ConTech University, CA, USA. Prior to this he was an assistant professor in the same department. From 2017 to 208 he worked as acting department chair. In 2014 he received the prestigious research distinguished award in computer science and information technology. From 2015 till now he is the vice chair of the scientific research group in Egypt (SRGE) in the field of computer science and information technology. He has collaborated actively with researchers in this scientific group in several other disciplines of computer science and its applications in medicine, engineering, agriculture, new and renewable energy, nanotechnology and space science.

Professor Darwish research interests span both computer science and information technology. Much of his work has been focusing on artificial intelligence, mainly through the application of data mining, machine learning deep learning and robotics in different areas of research. Moreover, his research interests include Internet of Things application, sensor networks, cloud and fog computing and cyber physical systems. He is associate editor of toughly twenty five international journals and he was a co-editor of some special issues. He has served as general chair, programme committee, sessions chair and keynote speaker for some of international well known conferences and workshops.

Professor Darwish has a wealth of academic experience and has authored multiple publications which include research papers, editorial papers, book chapters, essays and editing books. Professor Darwish is member of some computing associations such as IEEE and machine intelligence research Labs in USA. In addition, professor Darwish is an expert and reviewer in quality assurance and accreditation systems of education at universities and academic institutes. From 2011 to 2014 Professor Darwish has served in the diplomatic sector as the cultural and educational attaché of Egypt, embassy of Egypt to Kazakhstan.

Aboul Ella Hassanein is the Founder and Head of the Egyptian Scientific Research Group (SRGE) and a Professor of Information Technology at the Faculty of Computer and Information, Cairo University. Professor Hassanien is ex-dean of the faculty of computers and information, Beni Suef University. Professor Hassanien has more than 800 scientific research papers published in prestigious international journals and over 40 books covering such diverse topics as data mining, medical images, intelligent systems, social networks and smart environment. Prof. Hassanien won several awards including the Best Researcher of the Youth Award of Astronomy and Geophysics of the National Research Institute, Academy of Scientific Research (Egypt, 1990). He was also granted a scientific excellence award in humanities from the University of Kuwait for the 2004 Award, and received the superiority of scientific - University Award (Cairo University, 2013). Also He honored in Egypt as the best researcher in Cairo University in 2013. He was also received the Islamic Educational, Scientific and Cultural Organization (ISESCO) prize on Technology (2014) and received the state Award for excellence in engineering sciences 2015. He was awarded the medal of Sciences and Arts of the first class by the President of the Arab Republic of Egypt, 2017.

Chapter 4
Science Education in Jordan

Sereen M. B. Bataineh and Jwan H. M. Ibbini

Abstract Implementing good educational systems is essential for creating future leaders equipped with multiple skills and competencies such as self-learning and development, teamwork, and entrepreneurship qualifications. Therefore, this chapter aims to highlight the current state of science education in Jordan. It considers the country's demography, including population, economy, technology, cultural development, and political system, as these factors highly influence the national science education. Accordingly, the researchers provided a detailed overview of Jordan's education system development and currently employed policies. This chapter gives specific statistics regarding teachers and students number at different educational levels provided by the Ministry of Education and the Department of Statistics. Due to this research's importance, some educational research institutions and programs have been mentioned in this chapter. Education and research are crucial for the development of any country. However, science education is affected by multiple factors such as economy, politics, and global variants like the COVID-19 Pandemic. Therefore, policies and standards should be reviewed and modified to provide students with the best educational experience possible. In addition to policies, student assessment tools should be revisited to guarantee appropriate learning outcomes for science education. Jordan is also well known to have a good technology infrastructure for science education and research, some of which are described in this chapter. Although the Jordanian education system faces many challenges, adaptation mechanisms are highly required to provide a sustainable educational environment that meets international excellence for creating resilient future generations.

S. M. B. Bataineh (✉)
Department of Biotechnology and Genetic Engineering, Jordan University of Science and Technology, 3030, 22110 Irbid, Jordan
e-mail: smbataineh3@just.edu.jo

J. H. M. Ibbini
Department of Land Management and Environment, Hashemite University, P.O. Box 330127, Postal Code 13133 Zarqa, Jordan
e-mail: jhibbini@hu.edu.jo

© The Author(s), under exclusive license to Springer Nature Singapore Pte Ltd. 2022   43
R. Huang et al. (eds.), *Science Education in Countries Along the Belt & Road*,
Lecture Notes in Educational Technology,
https://doi.org/10.1007/978-981-16-6955-2_4

Keywords Jordan · Science education · Online learning · COVID-19 · Digital teaching

4.1 Overview of the Country

4.1.1 Geographical Location, Population, and Political System

The Hashemite Kingdome of Jordan is located in the north of the Arabian Peninsula and West Asia. The country got its name from the Jordan River, located on the western border next to Palestine (West Bank). The Gulf of Aqaba, located in the southwest of Jordan, is the country's only outlet to the Red Sea. Other neighboring countries are Syria to the north, Iraq to the east, Saudi Arabia to the south and south-east. The country includes 12 governorates, and Amman is the capital city. The official language is Arabic, though English is the first foreign language.

Jordan's total population was estimated to be 10.4 million people in 2019; according to the latest national census conducted by the Department of Statistics, there were 4966,000 females and 5588,000 males, with a sex ratio of 112.5, representing the number of males per 100 females. It is worth mentioning that 42% of the total population in Jordan resides in the governorate of Amman (Department of Statistics (DOS), 2021). Following the political instabilities in the region, which drove people to migrate to Jordan from Palestine, Iraq, and Syria, Jordan's population increased significantly. Although most of Jordan's population are Sunni Muslims, other religious minorities are also worth mentioning, such as Christians who mostly follow the Orthodox Church. The total land area of the Kingdome is 88794 Km^2, with a population density of 118.9. The highest point stands at 1854 m above sea level in the summit of Mount Umm Al-Dami, and the lowest point on earth is the surface of the Dead Sea that reaches around 408 m below sea level (DOS).

The political regime in the Hashemite Kingdom of Jordan consists of three authorities: parliament heritable royal. On the throne of the kingdom sits his Majesty King Abdullah II, who leads the three authorities and serves as the supreme commander of the armed forces (Ministry of Foreign Affairs and Expatriates, 2021).

Article 25 of the Jordanian constitution states that: "The Legislative Power shall be vested in the National Assembly and the King. The National Assembly shall consist of a Senate and a Chamber of Deputies" (Jordan Constitution). The senators (Majlis Al-A'yan) are appointed by the king, while the house of Deputies (Majlis Al-Nuwwab) is elected by the general public every four years. Both houses (Al-A'yan and Al-Nuwwab) are responsible for debating and voting on legislation for the law-making process (The Hashemite Kingdom of Jordan Government).

Within the political context, the Ministry of Political Development and Parliamentary Affairs was established to amend the country's political and parliamentary development, focusing on activating the participation of the silent majority in the country. These include women and youth, as well as various civil societies and organizations in the political decisions towards a civilized model of a democratic state (Ministry of Political Development and Parliamentary Affairs, 2021).

Following the Israeli—Arab conflicts in the region between 1948 and 1967 and the occupation of the West Bank, many Palestinians fled their homes and moved to neighboring countries. According to the United Nations Relief and Works Agency for Palestine Refugees in the Near East (UNRWA, 2021), more than 2 million registered Palestinian refugees currently live in Jordan. UNRWA provides humanitarian services, vocational education, primary health camp improvement, and emergency response during armed conflict situations. About 370,000 Palestinian refugees live in ten refugee camps throughout the country. Moreover, UNRWA operates and manages 169 schools with 118,296 students, a Faculty of Science and Educational Arts, and two vocational and technical training centers.

Although Jordan's economy is among the smallest in the Middle East, with insufficient supplies of water, oil, and other natural resources (world fact book), it hosted several migrants and refugees over the years. According to the United Nations High Commissioner for Refugees (UNHCR), Jordan hosted over 655,000 Sirians in 2018. The same resource also mentions that Jordan hosted asylum seekers and refugees from other countries in 2019, including 67,000 Iraqis, 15,000 Yemenis, 6,000 Sudanese, and 1,700 from other countries. This influx created more pressure on several sectors, including education (UNHCR, 2019).

4.1.2   Current Situation of Economic, Technologies, and Cultural Development

Economy is a powerful factor that shapes a nation's development and stability. To describe the current situation in Jordan, we need to address its most important natural resources. Although Jordan is a non-oil-producing country, it contains some mineral resources, such as phosphates, potash, fertilizers, and derivatives. Among the most important industries are phosphates (with a global production ratio of 3.36%), potash, oil refining, cement, photovoltaic industries, pharmaceutical industry, in addition to the tourism industry and remittances from abroad as well as foreign aid (Britannica Encyclopedia website, 2021).

The majority of the country's land is classified as arid and semiarid with a fluctuating average rainfall, making most agricultural activities reliable on irrigated systems (National Environment Strategy for Jordan Report, 1991). The agricultural sector in Jordan contributes to 4.9% of the total Gross Domestic Product (GDP). The most important crops are barley, vegetables, fruits, olives, and grains. Animal production mainly relies on raising poultry, goats, sheep, and cows (Global Economic, 2019).

Recently, the Department of Statistics (DOS) issued some information to the press related to the national economy on their website archives (DOS):

- The deficit in the trade balance has decreased by 20.0% during the first nine months of 2020.
- 23.9% Unemployment Rate during the third Quarter of 2020.
- 3.6% is the decrease of the GDP rate at constant prices in the second quarter of 2020.
- National exports decrease by 2.0%, and Imports decrease by 16.9% during the first seven months of 2020.

The World Bank reported that in 2020, Coronavirus (COVID-19) Pandemic drove the global economy into a recession due to the lockdowns implemented to control the spread of the virus. Not far from the global scenario, the Pandemic exaggerated the Jordanian economy's vulnerability and the related socio-economic issues, such as low labor force participation and high unemployment, particularly among women and youth. The report recommended responding to this crisis through substantial mobilization of resources and short-term liquidity (World Bank Report, 2020).

The telecommunications sector has also developed rapidly in the last decade. Jordan has a well-developed telecommunications infrastructure. The industry is preparing itself for the next wave of developments relating to 5G and IoT/M2M to meet public demand and the need to shift into the digital system due to the current Pandemic. The development of a national broadband network based on fiber optics is currently underway. The largest fixed network operator in the country is Orange Jordan followed by Zain Jordan. Both companies and others invest in fiber-based network infrastructure (World and Middle East Financial news (MENAFN), 2021). In this direction, the Ministry of Communications and Information Technology in Jordan estimated that the prevalence of Internet use in the Kingdom has risen to about 40% of the population by the end of 2010. Not surprisingly, Internet subscriptions have increased about 73% by the end of the third quarter of 2020. This sudden rise was mainly attributed to the increase need to employ digital systems in different life sectors following the lockdown and social distancing practices which accompanied the COVID-19 Pandemic. This situation created a competitive market to enter the fourth generation services, which will add a new concept to using the Internet by entering the user to the global network via cell phone or the so-called "mobile broadband. (Telecommunications Regulatory Commission (TRC), 2021).

Advancement in the telecommunication systems bushed the need for other technological advancements that lay the foundation for the transition to the country's digital system. One is the e-government program, which aims at providing citizens, residents, and investors with electronic services. These services are provided through a platform to ensure excellence, quality, and continuous improvement (Electronic government website).

Jordan was one of the first countries in the region to respond to the COVID-19 crisis by enforcing a lockdown and national closure of all educational institutions. The Ministry of Education has turned to distance learning tools to insure the continuity of the learning process during the Pandemic. Officials were quick to leverage materials from the private sector to develop an education portal called "Darsak", as well as two dedicated TV channels that offer online lectures. These resources cover the curriculum's core subjects of Arabic, English, Math, and Science for grades 1_12. Besides, the country's television sports channel has been repurposed to broadcast educational material tailored to students preparing for grade 12, also known as "Tawjihi," the secondary school graduating examination. The Ministry of Education also supported teaching staff by rolling out new interventions to make it easier to transition to distance learning. A newly launched platform for teacher training offers courses on distance learning tools, blended learning, and educational technology. These timely measures helped contain and mitigate the impact of the outbreak on learning in Jordan (Queen Rania Foundation QRF, 2020). However, progress in school enrollment and outreach did not equally benefit all children; refugees and children with disabilities were not sufficiently included in the process; thus, improving educational quality requires more focus and investment (UNICEF, 2021).

With regards to higher education, Jordanian Universities host about 342 thousand students in both governmental (74%) and private (26%) universities. Students from all levels of studies; Diploma, BCs, Master, and PhD used several E-learning platforms such as Microsoft Teams, Zoom, and Moodle. These global platforms are used to provide students and teachers with proper communication and interaction during the Pandemic. The learning environment was aided by live lectures and video streaming of the lectures, discussion, and online quizzes and assignments (Ministry of Higher Education and Scientific Research).

The Ministry of Culture in Jordan is primarily responsible for organizing cultural manifestations, arts, and other related activities, such as cultural events, festivals and participating in Arab and international cultural events. In Jordan, many festivals and events are held annually, some of which have become "international", such as the Jerash Festival, the Fuheis Festival, the Amman Summer, the Castle Mountain concerts, the "Shabeeb Festival", and the musical nights organized by the Abdul Hameed Shoman Foundation. Those events are often conducted in historical locations to promote archaeological sites through concerts of international talents, which contribute to the promotion of heritage and culture in Jordan at the same time. Art galleries and exhibitions are also important features of the Jordanian culture, as well as handcrafts and mosaics (Ministry of Culture, 2021). Beyond the rich traditional culture, new forms of music and arts are constantly growing and developing, leading new generations to create their identity halfway between the traditions and modernity. The paintings, sculptures, graffiti, and photography in the streets, the numerous galleries, and cafes in Amman represent Jordan's artistic movements.

Jordan has been celebrating the International Day for Cultural Diversity annually to promote the authentic heritage in the hearts of emerging generations by holding a cultural diversity festival in Amman or one of the governorates; the circumstances of the new Coronavirus pandemic have prevented these celebrations from taking place for the current year (Ministry of Culture, 2012).

OVERVIEW OF THE EDUCATION DEVELOPMENT

4.1.3 Education System and Policy

The education system in Jordan is managed by two main ministries; the Ministry of Education (MOE) and the Ministry of Higher Education and Scientific Research (MOHESR). The Ministry of Education supervises the school education stages, which begin with the pre-school (kindergarten) of a maximum of 2-years, followed by 10-years of primary education that is compulsory and free in governmental schools. The final phase of school education is a 2-year secondary education, including year 12, which is the school's final year. The Preschool education phase is optional and offered by private institutions and some governmental schools (Ministry of Education, 2021). Kids are meant to gain good attitudes and the right skills to join the formal school education. Kids at six-year of age are accepted in the first year of primary education. For the following 10-years, students gain attitudes, basic sciences, and different fields of knowledge and learn skills that will enable them to become good citizens capable of contributing to the development of the country, the Arab world, and humanity. The final phase of schooling is the secondary education stage. Students enroll in secondary education according to their interests and potential. This stage consists of two major streams, comprehensive and applied secondary education (MOE, 2018). Upon completing this stage, students are expected to integrate with society as good citizens who are considerate, aware of local, national, and international issues to be part of the solution, and leaders for improvement and change. The primary aim is to create an independent student who is able to develop and improve his confidence with self-learning skills and teamwork attitudes (National education strategic plan (ESP), MOE). By the end of secondary education, students must take a national exam (TAWJIHI). According to their performance and grades in TAWJIHI, they can proceed further with their higher education at university (National education strategic plan (ESP), MOE). It is worth noting that some year 12 students in private schools chose to take the Scholastic Assessment Test (SAT) or International General Certificate of Secondary Education (IGCSE) examinations to compete for international opportunities that allow them to pursue their higher education abroad.

All of the mentioned education levels are taught in both governmental and private schools. Nevertheless, the COVID-19 Pandemic has recently forced a nationwide school shut down, obliging the Ministry of Education to switch to online (distant) teaching. Online teaching has been established using various teaching platforms; for example, TV channels assigned classes for all school levels. Teachers do their best to convey knowledge to students during such difficult times. Online teaching platforms were also initiated to support this process. Even the Higher education ministry has been obligated to shift to online teaching in a short period of time.

Higher education in Jordan is administered by the Higher Education Ministry, which manages the public and private universities. Ten public universities, 17 private universities and 2 Regional, and 51 community colleges, in addition to the world Islamic Sciences and Education University. Nearly 236 thousand students are enrolled in higher education in Jordan (Ministry of Higher Education and Scientific Research, 2021). The Jordanian higher education system's quality and effectiveness make Jordan a country of choice for many overseas students, with nearly 28,000 students from around the world studying in Jordan (MoHE).

4.1.4   Statistics on the National Education

Distribution of **Teachers** by Authority of Supervision, Educational Stage and Gender (MoE), 2018/2019

Authority of supervision and gender	Educational sage			No. of teachers
	Kindergarten	Basic	Secondary	
Ministry of education				
Male	0	24,577	8079	32,656
Female	1826	43,006	10,246	55,078
Total	**1826**	**67,583**	**18,325**	**87,734**
Other government				
Male	0	1061	522	1583
Female	6	276	87	369
Total	**6**	**1337**	**609**	**1952**
UNRWA				
Male	0	2139	0	2139
Female	0	2193	0	2193
Total	**0**	**4332**	**0**	**4332**
Private education				
Male	0	2442	1770	4212
Female	6356	28,599	2877	37,832
Total	**6356**	**31,041**	**4647**	**42,044**
Total				
Male	0	30,219	10,371	40,590
Female	8188	74,074	13,210	95,472
Total	**8188**	**104,293**	**23,581**	**136,062**

Distribution of **Students** by Authority of Supervision, Educational Stage and Gender (MoE), 2018/2019

Authority of supervision and gender	Educational stage			No. of students
	Kindergarten	*Basic*	*Secondary*	
Ministry of Education				
Male	17,956	565,218	85,077	668,251
Female	19,266	636,291	99,680	755,237
Total	**37,222**	**1,201,509**	**184,757**	**1,423,488**
Other Government				
Male	95	11,667	2465	14,227
Female	104	2514	533	3151
Total	**199**	**14,181**	**2998**	**17,378**
UNRWA				
Male	0	62,542	0	62,542
Female	0	58,781	0	58,781
Total	**0**	**121,323**	**0**	**121,323**
Private education				
Male	50,862	256,411	18,042	325,315 ·
Female	46,450	166,166	14,599	227,215
Total	**97,312**	**422,577**	**32,641**	**552,530**
Total				
Male	68,913	895,838	105,584	1,070,335
Female	65,820	863,752	114,812	1,044,384
Total	**134,733**	**1,759,590**	**220,396**	**2,114,719**
Dropout rate	-	**0.38%**	NA*	

*Not Available: This stage is not compulsory. Therefore, the dropout rate is not supported

4.1.5 *Educational Research and International Collaboration*

The vision of his Majesty King Abdullah II, best described in this statement: "Jordanian students are taught the fundamental knowledge and skills to function and formulate their contribution to the human family" (Abdullah bin Hussein II). This vision is translated into educational systems, programs, and plans that provide students with the needed technological, geographic, religious, political, economic, and interpersonal abilities and skills required to carry out their future responsibilities.

The main education research institutions in Jordan include universities in both the governmental and private sectors. In each university, the deanship of scientific research takes the responsibility to encourage faculty and researchers to get involved in scientific research by funding research activities that lie within the national priority needs. Furthermore, each university is equipped with specialized centers that focus on specific domains, and here we mention some examples:

- **The University of Jordan**: Cell Therapy Center, Hamdi Mango Center for Scientific Research, Infectious Disease and Vaccine Center, Water, Energy, and Environment Center, Center for Women's Studies and Nanotechnology Center.
- **Hashemite University**: Risk management Center, Information, Communication and E-Learning Technology Center (ICET), Center for Big Data and Artificial Intelligence, Pharmacy Management and Pharmaceutical Care Innovation Center, Language Center, Center for Studies Consultation and Community Service, Community based Rehabilitation Center, Center for Women Studies in the Community.
- **Jordan University of Science and Technology**: Academic Development and Quality Assurance Center, Center for E-Learning and Open Educational Resources, Center of Excellence for Innovative Projects, Information Technology and Communication Center, Consultative Center for Science and Technology, Pharmaceutical Research Center, Princess Haya Biotechnology Center, Drug Information Center, Dental Teaching Center, Water Diplomacy Center, and Queen Rania Al-Abdullah Center For Environmental Science and Technology.
- **Yarmouk University**: Computer and Information Center, Refugees, Displaced Persons and Forced Migration Studies Center, Princess Basma Center for Woman's Studies, Accreditation and Quality Assurance Center, Entrepreneurship and Innovation Center, and Queen Rania Center for Jordanian Studies and Community Service.

The National Center for Human Resources and Development (NCHRD) was established in 1987 and has played a vital role in capacity building and the initialization of education development initiatives. Moreover, they extended their role to include a broader human resource development framework that connected investment in training, education, and qualifications to labor market needs. Between 1991 and 2018, around 188 documents were conducted and issued by the center focusing on the education system in Jordan; most of these studies were funded by foreign donors. The center has a long record of collaboration with regional and international institutions through its outreach projects, programs, research studies, training workshops, and consultancies. Given its track record of accomplishments in human resource development, NCHRD has been named a Center of Excellence for Educational Planning and Management in 2012. Moreover, a regional and international collaboration between the center and the following organization can be listed: Arab Fund for Economic and Social Development, Canadian International Development Agency (CIDA), GIZ, UNESCO, EU, World Bank, Japan International Development Fund, European Investment Bank, German Development Bank, United States Agency for International Development (USAID), Japan International Cooperation Agency (JICA), Department for International Development (DFID)/ UK Government, European Union and UNICEF (NCHRD, 2021).

International collaboration in Jordan's education system takes different forms and occurs at various levels. International and regional aids and funds drive research in education, develop educational programs, and build an infrastructure that fosters proper education within both the Ministry of Education and the Ministry of Higher Education and Scientific Research. Some universities in Jordan, such as the German Jordan University, has developed educational programs that mandate a joint study program in Jordan and Germany. It is also worth mentioning that the Higher Council of Science and Technology (HCST) has several international collaboration through joint projects with the European Union; some projects can be mentioned, for examples:

- Support to Research, Technological Development and Innovation in Jordan (STRD)
- Euro-Mediterranean Cooperation through ERANET Joint Activities and Beyond (ERANETMED)
- Mediterranean Science Policy Research and Innovation Gateway (MEDSPRING)
- Coordination and Support for the Partnership for Research and Innovation in the Mediterranean area (4PRIMA)

Moreover, the Ministry of Education established the Planning and Educational Research Department. One of the Ministry's objectives is to allow students to think objectively and critically and adopt scientific observation methods, research and problem-solving (Ministry of Education, 2021). Based on personal communication with ministry personnel, the ministry conducted several research to evaluate the online learning experience during the COVID-19 Pandemic. These studies include student assessment evaluation as well as parent satisfaction from health perspectives and from learning perspectives. Other studies targeted schools in private sectors because many small to intermediate-sized schools were challenged during the Pandemic as many students shifted from private to public schools. Results of these studies are either under investigation or under publication approval mechanism.

Recently, the inclusive education principle was introduced in the Jordanian education system at a very low scale and under observation. Its philosophy and concept are based on the inclusion of students with special needs into mainstream schools. A study revealed that teachers' lack of awareness and negative attitude toward students with special needs and the lack of appropriate support from both school administration and families might hinder the effectiveness of this program (Al-Natour et al., 2015).

4.2   Current Situation of Science Education

4.2.1   Policies and Standards

Ministry of Education Strategic Plan 2018–2022.

The Ministry of Education has made tangible efforts in the field of policies and strategic planning, releasing the strategic plan for 2018–2022, and participated in preparing the National Strategy for Human Resource Development in public education. The Ministry has also convened the Educational Development Conference (2015), which witnessed broad participation from all the educational sectors. This Conference worked to develop various recommendations that aimed to improve the quality of education in Jordan and the educational system's performance in general. Nevertheless, the Ministry is still working on building the capacities and the provision of qualified human resources as well as bridging the gaps in performance in these areas. The Ministry believes in the role of elective planning as a means to improve the performance of the educational system. In addition, the Ministry will diagnose the current situation of the education system, analyze its performance through doping the current strategic planning track, and work on developing the programs, the projects, and the operational plans to achieve its vision, mission, and strategic objectives in cooperation with all the supporting parts (MoE Education Strategic Plan 2018–2022).

The Following Educational Policy Principles are Listed by MoE:

- Modifying the Educational System in order to have more sustainability for both individuals and society needs also making a balance between them.
- Clarifying the importance of the education system and promoting participation, justice and democracy practices.
- Promoting the implementation of scientific methodology in planning, conducting and assessment system, as well as research and follow-up systems.
- Extending types of education in the educational institutions to involve them in programs related to special education and other programs designed for gifted learners and special needs.
- Ensuring that teaching is a message and a career which has its own ethical and occupational dimensions.
- Enhancing the social status of teachers as they play a prominent role in building up individuals and societies (Ministry of Education, 2021).

Policies and Standards for Ministry of Higher Education and Scientific Research (MOHESR).

For the Ministry of Higher Education and Scientific Research, the key performance indicators of the strategy of higher education appear clearly through: percentages of (1) males and female's enrollment into regular admission programs and parallel programs; (2) the steady increase in faculty members; (3) financial government support for institutions of higher education; (4) turnout for expansion of private universities (private sector) that aims to participate in shouldering the burden and responsibilities of education with the public sector; (5) the Higher Education Accreditation Commission that supervises on quality assurance at both public and private institutions of higher education to be consistent with the international standards; (6) updating libraries of universities and linking all institutions of higher education to the electronic periodicals and universities networks; (7) the Scientific Research Support Fund that finances projects with national priorities, offering grants for outstanding graduates, granting the outstanding research prize, the outstanding researcher prize and the outstanding student prize; (8) and finally accrediting the TOEFL certificate as an admission certificate for joining master and PhD programs.

Moreover, the ministry worked to bridge the gap between higher education output and labor market demand in order to meet current and future needs for qualified and specialized cadres in various areas of knowledge; and to compensate for the region's lack of natural resources by developing qualified human resources fortified by knowledge and efficiency (Ministry of Higher Education and Scientific Research, 2021).

Strategic Plan for Ministry of Higher Education and Scientific Research Highlights the Following Points:

- Developing programs that promote sustainable learning and motivating task groups to ensure providing high-quality services.
- Activating A blog related to employees' behavior and job ethics during the actual monitoring for ensuring the implementation of these effects and accountability.
- Building an institutional culture based on public service and providing the tools and methods required to activate these cultures and making society aware of it.
- Designing and implementing studies and surveys related to services of the ministry and the internal or external partners.
- Maximizing the utility of innovation and creativity.
- Capacity building of employees, managers, women, and those with special needs and promoting human resources skills. (Ministry of Higher Education and Scientific Research Strategic Plan 2019–2021).

The Higher Council for Science and Technology (HCST) Resealed the Following Policy Documents.

1. **National Innovation Strategy 2013–2017**
   Jordan seeks to initiate the innovation economy through recruiting an efficient workforce, using the available institutional potentials built up and developed over long years through Public–Private Partnership (PPP) and efforts exerted by innovative and pioneering individuals. Due to the difficult international and regional economic conditions, harnessing innovation to serve the economy has become a need that leads to economic and social development.
   Based on the confirmed data related to the role of innovation in the efficacy of the economy, interdependence between technology and initiative increases and leads to creating investment opportunities and new jobs; hence, the emergence of a knowledge-based economy. Innovation and technology dissemination policies and strategies aim at creating a proper environment for turning new thoughts, products, and patterns into economic and social benefits. To achieve this goal, a solid basis of knowledge and tremendous innovative capabilities should be made available while creating favorable conditions for disseminating technology in all parts of the economy (Higher Council for Science & Technology, 2013).

2. **The National Policy and Strategy for Science, Technology, and Innovation (2013–2017)**
   The Higher Council for Science and Technology develops policies that result in forming a national scientific and technological base. These policies are designed to meet the evolving technologies and national needs in order to

become an effective tool in stimulating the Kingdom's economic, social, and cultural development. Since its inception, the Council has worked through the Secretariat to develop a national policy for science and technology with the participation of a wide scope of Jordan's scientific and technological community and to formulate that policy in line with the requirements of the development sectors stated in the national economic and social development plans.

One significant project emerged in response to the Royal visions for promoting the role of scientific research and development in Jordan's economic and social development process. This project aims to identify the Hashemite Kingdom of Jordan's scientific research priorities for the next ten years (2011–2020 AD). The project's objective is to enable Jordanian universities and research and development centers to contribute to long-term, sustainable national developments that keep up with the scientific and technological advances (The National Policy and Strategy for Science, Technology and Innovation 2013–2017).

3. **Defining Scientific Research Priorities in Jordan for the Years 2011–2020**
    In order to substantiate the exalted royal visions in reinforcing the role of scientific research in the process of social and economic development in the Hashemite Kingdom of Jordan, and in the light of the recommendations of the "national agenda" and "we are all Jordan" related to scientific research, also to realize the content of the national policy and strategy for science and technology, and their executive plan for the years (2006–2010), which is prepared by the HCST, and in accordance to the national strategy of the scientific research support fund at the Ministry of Higher Education and Scientific Research in its fourth axis: "tightening and coordinating the cooperation with the HCST in the field of its work", as well as to reinforce the existing cooperation between the Council and the Fund in the field of scientific research. The project "Determining the Scientific Research Priorities in the Hashemite Kingdom of Jordan for the Coming Ten Years (2011–2020)" came to map out the way in front of the national institutions and researchers at Jordanian universities and research and development centers as a contribution in realizing the comprehensive, sustainable national development to cope with scientific and technological development.

    For the sake of activating the role of scientific research in the economic and social development, it was necessary to determine the national priorities focused upon by national policies and strategies concerned with scientific, technological, and innovative activities, including scientific research activities and initiatives in which the concerned institutions are highly interested. Therefore, this important national project came to set the fundamentals of sound and efficient planning for the progress of scientific research and development in Jordan. The project deals with all scientific sectors. Fourteen sectorial committees have been formed, covering all scientific fields and areas. Each of those committees includes a group of experts, researchers, and specialists, who represent different national institutions, such as public and private universities, scientific centers, private sector institutions, civil society institutions, in addition to the members of the steering committee and the members of the technical committee of the project, with a total of (139) researchers and specialists. The aforementioned committees have also called for the assistance of (570) experts who filled the questionnaires for the four rounds (Defining Scientific Research Priorities in Jordan 2011–2020).

4.2.2 Curriculums, Digital Resources, and Teacher Training

In order to promote teachers' training and educational skills, Queen Rania Teacher Academy (QRTA) was launched in 2009 in partnership with the Ministry of Education (MOE) to offer professional development programs for teachers in accordance with the educational needs in Jordan and the Arab World. QRTA aimed to advance the quality of teaching and promote excellence in education in Jordan and the region. Also, to enable every educator to positively influence Jordan and the Arab World's future generation by spearheading teacher professional development (QRTA, 2020).

Most countries agreed on the importance of proceeding with the educational process remotely and referring to technological solutions as much as possible during the lockdown and even after the gradual opening resumed. In Jordan, the closure of schools, universities, and training institutions started in mid-March 2020, during the second semester of the 2019/2020 academic years. The government declared that learning in schools and universities will continue to be online; gradually, teaching and learning became more distant.

QRTA recognized the need for online education. therefore, many communication tools were used by teachers such as: videotaped presentations using the ZOOM application, online tasks and assignments, interactive online sessions, and WhatsApp chats. To maximize the benefit of the online training, beneficiaries were required to be trained on how to use these tools. The academic teams were sharing their experiences of developing online training materials while also reviewing and reflecting on the ongoing feedback from the field through the Monitoring and Evaluation (M&E) Department. These evaluation processe identified effective digital tools to be used, assessments and follow-up techniques that work best for the online training approach. The academic teams were also provided with opportunities to participate in international webinars and conferences, enabling them to reflect on training activities in light of the best practices in the field of online training and learning (QRTA Online Learning Experience Evaluation Report 2020).

The Ministry of Education—Jordan MOE has implemented a set of standards, procedures, and decisions to improve the learning environment; these standards comprise educational institutions of varying forms and with different governing frameworks. While quality standards and procedures vary from institution to institution, they all focus on the same outcome—the quality of student learning. Quality, in the Jordanian MoE, is integral to all aspects of the educational system; however, several aspects will be addressed and emphasized here, including assessments, curricula, information and communication technology in education, school leadership and community participation, accountability, safe and stimulating school environment, and school feeding program. The Queen Rania Award for Excellence in

Education recognizes the best teaching practices to improve the quality of education. The Ministry uses these practices to identify areas for further research and inform the policy discussions and the educational development process. In this way, the investment of individual teachers in education quality informs the general policies and programs of the MoE. (Education Strategic Plan 2018–2022).

Curriculum

The general and specific curriculum and evaluation frameworks and outputs were developed for each subject in 2013. The general consensus is that there is a need to reform the curriculum and assessment system in order to ensure that schools move away from rote learning and develop higher-level thinking skills. A recently conducted review of the lower primary revised textbooks revealed that these textbooks offer limited hands-on activity and group-based learning and do not emphasize critical thinking and problem-solving strategies. The subject material is often significantly outdated to the point that textbooks' examples no longer relate to real-world practices.

As required, the MoE carries out research and survey studies of the curriculum (general framework of curriculum and evaluation, general and special outputs for each subject, student books, and teachers' books) in cooperation with specialized experts to evaluate each subject. These teams evaluate the curriculum and identify opportunities for improvement in order to respond to national and global trends.

The MoE teams also prepare the general framework for curricula, general and specific outcomes, and assess teachers' books and textbooks for all students, including the students with special needs. The author edits (linguistically and technically) and designs textbooks and teachers' books. They also produce diverse learning resources teaching aids that support the national curriculum in the primary and secondary stages of academic education as well in Vocational Education VE, KG, and illiteracy eradication. (Education Strategic Plan 2018–2022).

4.2.3 Student Assessment and Achievement

To reach the international standards in education, Jordan follows quality indicators to measure students' performance as the EMIS system at the national level. The Ministry of Education has adopted the Education Management Information System (EMIS) as the main tool for student assessment. EMIS provides reliable data in a sustainable way to gather large student marks data, process and analyze it for further analysis. The EMIS is considered a planning tool that allows the MoE to make appropriate policies to evaluate and improve the education system. Moreover, Jordan has been one of 17 countries that implemented the OpenEMIS in 2014 as a platform for EMIS to enhance data collection, assessment, and management. OpenEMIS is one of the UNESCO initiatives to assist the educational strategic planning (OpenEMIS in Jordan __December 2020). The student's achievements are further assessed and compared at the international level. Jordan has been participating in both the Trends in International Mathematics and Science Studies (TIMSS) and the Program for International Student Assessments (PISA) since the'90 s. To control the quality of education, the results of the national exam for students in grades 4, 8, and 10 are analyzed to assess students' achievement. Upon analysis of the exams' results, recommendations are provided to the MoE and the related institutions to further improve the education system, including science.

4.2.4   Science and Technology Venues and Centers

Several centers and educational venues are available in Jordan, focusing either on science education or scientific research. In this part, few examples are mentioned.

- **Higher Council for Science and Technology (HCST)**

The Higher Council for Science and Technology was established in 1987 as an independent public institution. It acts as a national umbrella for all science and technology (S&T) activities in Jordan. With the objective to build a national science and technology base to contribute to the achievement of the development goals. These objectives are achieved through increasing awareness of the significance of scientific research and development, granting the necessary funding, and directing scientific and research activities within national priorities, in line with development orientations (HCST, 2010).

- **National Center for Agricultural Research and Extension (NCARE)**

The center is a semi-autonomous institution aiming to conduct applied agricultural research and extension services to transfer improved technologies to farmers. As a research arm of the Ministry of Agriculture, the Center receives funding from the Government of Jordan and other national and international donors. NCARE supports Jordanian agricultural research, focusing on soil and environmental, horticulture, and animal and plant protection sciences.

- **Synchrotron-Light for Experimental Science and Applications in the Middle East (SESAME)**

SESAME is a "third-generation" synchrotron light source that was established in Jordan in 2017. It is the first synchrotron light source in the Middle East and neighboring countries and its first major international center of excellence. It was developed under the auspices of the United Nations Educational, Scientific, and Cultural Organization (UNESCO). The Synchrotron provides radiation from Infrared light to X-rays of unparalleled quality, which offer a unique tool for scientific investigations in living matter and new materials (SESAME, 2021).

- **Regional Center for Space Science and Technology Education for Western Asia-Jordan/UNRCSSTEWA.**

It was officially opened in 2019 under royal patronage. The center's vision is to promote regional and international cooperation in the field of space science by assisting participating countries in developing and enhancing their citizens' knowledge and skills in space science and technology so they would be able to effectively contribute to the national space development programs. In cooperation with the Office for Outer Space Affairs (UNOOSA), the Regional Center grants a master's degree in the following specializations (RCSSTEWA, 2021):

- Remote sensing and geographic information systems.
- Space meteorology.
- Satellite communication.
- Space and astronomy sciences.
- Space Law.
- Global Navigation Systems (Arabic—English).

- **Royal Scientific Society (RSS)**

HRH Princess Sumaya bint El Hassan is the president of the Royal Scientific Society, which was established in 1970. HRH is greatly recognized for her work in promoting various scientific endeavors on a local and international level. The Royal Scientific Society is the largest applied research institution, consultancy, and technical support service provider in Jordan and is a regional leader in the fields of science and technology.

The RSS campus is home to a range of specialized organizations that deploy science and technology to strive for excellence in specific fields. Working both independently and collaboratively, they enrich the RSS ecosystem and drive knowledge-based change in Jordan. Such as MESIS, WANA Institute, IRADA, UN-ESCWA, Arab Science Forum, and Phi Science Institute.

The Abdul Hameed Shoman Foundation (AHSF)

It was established in 1978 as a charitable initiative launched by the Arab Bank. Based on its belief in the importance of developing a common ground for Arab progress, the Arab Bank allocated a percentage of its yearly profits to fund the Foundation's establishment. It aimed to ultimately support the national economy while also pushing forward scientific research, humanities studies, cultural enlightenment, and innovation. The Foundation has since sought to achieve its objectives through three pillars: thought Leadership, literature and arts, and Innovation. The foundation encourages pioneering ideas by supporting various fields of scientific research. It launched the Abdul Hameed Shoman Award for Young Arab Researchers in 1982 and the Abdul Hameed Shoman Fund for Scientific Research in 1999. It also provides opportunities that bring together the public, on the one hand, thinkers, researchers, and scientists on the other, through the activities and events of the Abdul Hameed Shoman Cultural Forum established in 1986 (AHSF website).

Talal Abu Ghazaleh Group (TAG)

The Talal Abu-Ghazaleh Group and Co. Consulting (TAG-Consult) was established in 1972 and has become one of the leading professional consulting firms in the Arab world. TAG-Consult is dedicated to providing the best quality of consulting services, including business and investment advisory services, organization and restructuring services, financial consulting, quality management systems, and privatization services.

With its multi-disciplinary professionals working across the Middle East and many other parts of the world to provide services for private and public sectors, TAG-Consult succeeded in introducing development methodologies for businesses and governmental bodies where advancement became a pressing necessity in an era of mass production and globalization (TAG, 2021).

Children Museum

Her Majesty Queen Rania Al Abdallah launched the Children's Museum in Jordan in 2007 as a non-profit educational institution. The Museum has over 180 indoor and outdoor interactive exhibits and educational facilities, including the Library, Art Studio, Thinkers Lab, and Secret Garden, on an area of 8000 m^2, besides its year-round educational programs, events, and shows. More than 3 million people have visited Jordan's Children's Museum since its inception (Children Museum website, 2021).

Little Genius Program

It is an international program that develops children's mental qualifications and raises their intelligence levels, using mental arithmetic and abacus learning. The Little Genius program consists of eight levels of mental arithmetic and abacus learning, each of which lasts three months and includes (26) hours, provided along two hours per week. The program divides the enrolled students into two groups according to their ages; Junior (5–9 years old) and Senior (10–13 years old) (Little Genius Program, 2021).

Association of Jordanian Women Academics

The association was established in May 2014 for academics from public and private Jordanian universities seeking to reach a prestigious professional level that enables them to bring change and development to serve their society, advance scientific research, and achieve effective employment and investment of energies and capabilities. In order to serve the public interest and achieve comprehensive development. The association's goals are to raise societal awareness, change the stereotypes related to women, enhance career advancement for women academics, build their capabilities, adopt their academic, social, and political issues, and activate partnerships with community institutions. These goals are achieved by building databases for academies in particular and studies of Jordanian women in general (Association of Jordanian Women Academics website).

Jordan Society for Scientific Research, Entrepreneurship, and Creativity (JSSREC)

Scientific research is a key component for development in any nation. It is a tool to acquire knowledge in the development, renewal, and use of materials, resources, and commodities to make them useful for individuals and institutions. Due to the significance of scientific research in Jordan, there is a vital need to link the different institutions involved in scientific research. For these reasons, the Jordan Society for Scientific Research, Entrepreneurship, and Creativity (JSSREC) was established in 1999 to meet the following objectives (JSSREC, 2021):

- To develop a national interest in scientific research and to support research activities and programs in the different sectors in Jordan.
- To establish mechanisms to support scientific research and connect it with the national social and economic development by enhancing collaboration between the scientific institutions, economic forums, and businesspeople.
- To monitor the outputs of scientific research both locally and internationally and maximize the benefits of these outputs in Jordan.
- To bridge better communication between researchers and decision-makers to enhance the contribution of research outcomes in decision making.
- To educate the public about the significance of scientific research in the development process.

Conference and Meetings infrastructure

In Jordan, there are many venues to conduct meetings, seminars, and lectures. These venues are equipped with the latest presentation tools and equipment as necessary infrastructure. These venues are distributed in many locations around the country, mainly in Amman and other tourist attraction sites such as the Dead Sea and Aqaba. These Facilities encourage conducting national, regional, and international meetings and conferences.

4.2.5 Utilizing Emerging Technologies

Jordan's science education system is constantly being revised. The Ministry of Education and the Ministry of Higher Education and Scientific Research are working to improve the students' learning experiences through emerging technologies. Emerging technologies like AI are being taught in some universities as part of the computer information system like the University of Jordan and some of the private schools. Using other types of technologies like AR and VR is still at the planning level; it has gone through pilot trials but has not been implemented yet. The implantation of these technologies requires infrastructure establishment and teacher training to give the students the best teaching experience that matches the international level.

With the spread of the COVID-19 pandemic, online platforms were established in a short period of time to support distant learning. The Ministry of Education has launched the DARSAK platform to provide online educational resources to all education levels. In addition, The Ministry of Higher Education and Scientific Reseach used multiple online education tools to offer students the knowledge they deserve. Online applications like BigBlue Button, ZOOM, and Microsoft TEAMS have been used to provide an interactive live teaching experience to students.

4.3 Requirements for Future Development of Science Education

The science education in Jordan at both school and university levels goes through continuous refinement processes using national and international assessment tools. Jordan considers the investment in education very important for the country's development, and the link between education and industry is slowly evolving. The future development of science education depends on improving multiple parameters including, students learning skills, teacher training, school infrastructure, and digital education tools. Students need to understand the importance of learning and its crucial role in improving the country. This requires teachers trained with the most advanced digital tools to keep up with the international standards to provide students with a better learning experience. To ensure the best science education experience, school infrastructures must be improved to provide facilities with up-to-date digital tools. Overall, science education is the cornerstone to a better future as it prepares future generations to contribute to the international community and respond to human challenges.

4.4 Discussion and Conclusion

The education sector in Jordan has encountered forward steps in many fields. However, performance checks and improvement cycles identified some challenges that require greater focus and investment. According to UNICEF, almost 97% of children go to primary schools. The overall progress in gender parity in Jordan has not benefited all children equally; it mainly profited children with disabilities, refugees, and children with a poor socio-economic background.

In Secondary schools, boys of low-income families often drop out of school and get involved in labor activities to help their families. Assessments revealed poor academic achievement in boys' schools. This can be attributed to teacher quality and the availability of male teachers. Meanwhile, girls in some areas tend to drop out after finishing secondary school to help younger siblings and participate in household activities. Students in public schools consistently perform poorly in international standardized tests, especially in math and science subjects. The year 2020 was exceptionally different as education had to switch to online education to mitigate the effects of the COVID-19 Pandemic. This pandemic situation affected the quality of education since the availability of infrastructure tools and resources such as laptops, internet access other digital resources was not fairly available to all children.

In alignment with Jordan's Education Strategic Plan 2018–2022, some recommendations and future pathways are recommended. These include mobilizing efforts and allocations toward a student-centered education system, which can be achieved by raising the schools' infrastructure, especially equipping schools with science labs that use hands-on learning and digital resources. Moreover, curriculum development and modification are necessary to make the context in which it is presented relevant to the Jordanian students and to meet international excellence. Curricula should be modified after Identifying students' skills and competencies required to improve the economic sector by prioritizing sectors that need to be focused on in the short and long terms. Moreover, necessary investments should be made to deliver the required student education, training, and certification. Along with material development, it is equally important to monitor, evaluate and improve teacher training programs to develop teaching pedagogies, especially in science and math education. Future insights in educational research and international collaboration should continue to orient efforts and invest in enhancing the role of science and technology centers that address science education in the country.

Acknowledgements The authors would like to thank The Ministry of Education, with special sense of gratitude to Queen Rania Center for Education and Information Technology for providing important resources, data and statistics. The authors also would like to thank the Ministry of Higher Education and Scientific Rsearch for providing important documents. In addition, special acknowledgment is addressed to the National Center for Human Recourses Development for providing recent updated reports and studies. Finally, sincere gratitude goes to the Talal Abu Ghazaleh (TAG) group represented by Dr. Talal Abu-Ghazaleh who shared with us important insights about future education in Jordan.

References:

AHSF. (2021). The Abdul Hameed Shoman Foundation. Retrieved April 17, 2021, from https://www.shoman.org/en/.

Al-Natour, M., Amr, M., Al-Zboon, E., & Alkhamra, H. (2015). Examining collaboration and constrains on collaboration between special and general education teachers in mainstream schools in Jordan. *International Journal of Special Education, 30*(1), 64–77. Retrieved from https://files.eric.ed.gov/fulltext/EJ1094801.pdf.

Britannica Encyclopedia. (2021). Resources and power. Retrieved April 27, 2021, from https://www.britannica.com/place/Jordan/Resources-and-power.

Children Museum. (2021). No Title. Retrieved April 17, 2021, from http://www.cmj.jo/.

Department of Statistics. (2021). 0.3% GDP growth rate at constant prices in the first quarter of
    2021. Retrieved April 13, 2021, from DOS Archives website: http://dosweb.dos.gov.jo/archive/.
Higher Council for Science & Technology. (2010). *Defining scientific research priorities in Jordan
    for the Years 2011–2020*. Retrieved from http://www.hcst.gov.jo/sites/default/files/defining_sci
    entific_research_priorities_in.pdf.
Higher Council for Science, & Technology. (2013). *National Innovation Strategy 2013– 2017*.
JSSREC. (2021). Jordan Society for Scientific Research, Entrepreneurship, and Creativity. Retrieved
    April 14, 2021, from https://jssr.jo/en/home/.
Little Genius Center. (2021). The little genius center mental arithmetic & abacus learning. Retrieved
    April 17, 2021, from https://tlg.center/.
MENAFN. (2021). Jordan Telecoms, Mobile and Broadband Market Growth Analysis, Challenges
    and Industry Key Players Jordan Telecom Zain Jordan Batelco Virgin Mobile | 2019–2020.
    Retrieved March 23, 2021, from https://menafn.com/1099595099/Jordan-Telecoms-Mobile-and-
    Broadband-Market-Growth-Analysis-Challenges-and-Industry-Key-Players-Jordan-Telecom-
    Zain-Jordan-Batelco-Virgin-Mobile-2019-2020.
Ministry of Culture. (2021). No Title. Retrieved April 21, 2021, from https://www.culture.gov.jo/
    node/33645.
Ministry of Culture. (2012). *Cultural diversity in Jordan*. Retrieved from http://www.ich.gov.jo/
    sites/default/files/culturehekmat.pdf
Ministry of Education. (2021). The philosophy and objectives of education. Retrieved March 14,
    2021, from https://moe.gov.jo/en/node/19404.
Ministry of Education. (2018). *Education strategic plan 2018–2022*. Retrieved from http://www.
    unesco.org/new/fileadmin/MULTIMEDIA/FIELD/Amman/pdf/ESP_English.pdf.
Ministry of Foreign Affairs and Expatriates. (2021). The Hashemite Kingdom of Jordan. Retrieved
    March 15, 2021, from https://portal.jordan.gov.jo/wps/portal/Home/AboutJordan.
Ministry of Higher Education and Scientific Research. (2021). No Title. Retrieved April 17, 2021,
    from http://www.mohe.gov.jo/ar/pages/Statistics.aspx.
Ministry of Political Development and Parliamentary Affairs. (2021). No Title. Retrieved April 14,
    2021, from https://portal.jordan.gov.jo/wps/wcm/connect/gov/egov/government+ministries+_+
    entities/ministry+of+political+development+and+parliamentary+affairs/ministry+of+political+
    development.
NCHRD. (2021). National Center for Human Resources and Development. Retrieved March 23,
    2021, from http://www.nchrd.gov.jo/HOME_En.aspx.
QRTA. (2020). *Online learning experience evaluation report 2020*. Retrieved from https://qrta.edu.
    jo/uploads/2020/11/QRTA-Online-Learning-Experience-Evaluation-Report-Oct-2020_1.pdf.
Queen Rania Foundation. (2020). Educational response to covid-19 from Jordan and other Arab
    countries. Retrieved March 14, 2021, from https://www.qrf.org/en/latest/blog/educational-res
    ponse-covid-19-jordan-and-other-arab-countries.
RCSSTEWA. (2021). No Title. Retrieved March 15, 2021, from http://rcsstewa.com/en/.
SESAME. (2021). Synchrotron-light for experimental science and applications in the middle east
    (SESAME). Retrieved April 17, 2021, from https://www.sesame.org.jo/.
Talal Abu-Ghazaleh Group (TAG). (2021). No Title. Retrieved March 15, 2021, from https://www.
    tagorg.com/?lang=en.
The Hashemite Kingdome of Jordan. (2021). The Hashemite Kingdom of Jordan government - the
    Legislative Branch. Retrieved June 25, 2021, from http://www.kinghussein.gov.jo/government3.
    html.
The Hashemite Kingdome of Jordan Government. (2021). The official website of the e-Governmen.
    Retrieved March 23, 2021, from https://portal.jordan.gov.jo/wps/portal?lang=ar/.
The World Bank. (2020). Jordan economic monitor – spring 2020: Weathering the Storm (English).
    Retrieved April 13, 2021, from http://documents.worldbank.org/curated/en/895901594653936
    142/Jordan-Economic-Monitor-Spring-2020-Weathering-the-Storm.
TheGlobalEconomy.com. (2019). Jordan: GDP share of agriculture. Retrieved March 17, 2021,
    from https://www.theglobaleconomy.com/Jordan/Share_of_agriculture/.

Telecommunication Regulatory Commission (TRC). (2021). Statistics and Indicators for Telecom Markets. Retrieved April 22, 2021, from https://trc.gov.jo/Pages/viewpage?pageID=171.

UNHCR. (2019). UNHCR continues to support refugees in Jordan throughout 2019. Retrieved April 26, 2021, from https://www.unhcr.org/jo/12449-unhcr-continues-to-support-refugees-in-jordan-throughout-2019.html.

UNICEF. (2021). Education Inclusive and quality education for every child. Retrieved March 12, 2021, from https://www.unicef.org/jordan/education.

UNRWA. (2021). United Nations relief and works agency for Palestine Refugees in the Near East. Retrieved April 17, 2021, from https://www.unrwa.org/.

Sereen M. B. Bataineh earned her Ph.D. from the University of Western Australia in 2012 in Microbiology and Forensic Sciences, BSc and MSc degrees in 2001 and 2004, consequently, in Applied Biology from Jordan University of Science and Technology (JUST). She joined JUST in 2011 as a full-time lecturer in the Forensic Science program, and was promoted to Assistant Professor in 2012. Dr. Bataineh has published her work from postgraduate studies in international Journals and Conferences.

Her passion for all things microbial came from a childhood where she gained an interest in disease and how to find out what causes them. Her curiosity from childhood grew, and she quickly wanted to progress her career through her Ph.D. and research so she can contribute back to humanity to help solve world problems. Dr. Bataineh was Assistant Dean during the Academic year 2013. Her duties involved reconstructing the study plan for the Forensic Science program and supervision of the student training program. She was also a member of the Accreditation Committee for the Higher Education Accreditation Commission for the Forensic Program, organized an open-carrier day for the Forensic Science community as well

Dr. Bataineh is the first Jordanian who designed a Massive Open Online Course (MOOC/2014) and one of a few academics in the Arab world. The course is free and offered with Open2Study Australia. The course is the first online course combining Microbiology and Forensic Science, and over 21,000 students took the course from all over the world (https://www.youtube.com/watch?v=5i_GDL2i3jQ). Dr. Bataineh was appointed as an expert with the Arab League, Educational, Cultural and Scientific Organisation (ALESCO) and a member of the Advisory Committee of the Open Book program. Her research interest is antibacterial activity of plant extracts and recently I combine the extracts with different nanoparticles to investigate their efficacy.

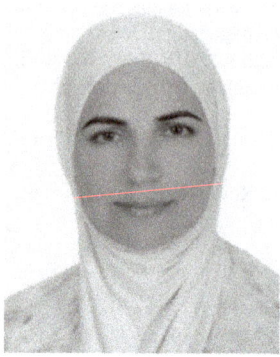

Dr. Jwan Ibbini is an associate professor in Environmental Microbiology. Earned her Ph.D. from the USA (Kansas State University) in 2008. Currently, she is a Faculty at the Department of Land Management and Environment at Hashemite University in Jordan. Her research interest includes biorisk management, bioremediation and exploring antimicrobial agents from natural resources. She was involved in few projects related to environmental education in higher education systems. Over a decade of teaching experience in environmental topics, she is able to explore new teaching methodologies in classrooms that focus on adult learning systems which provoke and facilitate student learning outcome.

Chapter 5
Science Education in Kenya

Edmond Kizito Makoba and John Otieno Odhiambo

Abstract This chapter discusses the situation of science education in Kenya. In this discussion, the geographical and political system of Kenya is given as well as an overview of its educational system and its development. The policies governing science education since independence are also considered including curriculum reforms. The number of schools, students, and teacher profiles is given as well as their gender segregation. Kenya's standing in educational research as well as international collaborations are discussed. The provision of learning resources including digital devices and emerging technologies are covered. Their contribution to improving science education is considered. Student assessment and achievement in international as well as local examinations are considered with a view to gauge Kenya's standing in the attainment of requisite knowledge and skills. The infrastructure to support science education such as science technology venues and centres is discussed. In conclusion suggestions and recommendations for the future development of science education in Kenya are given.

Keywords Emerging technologies · Sciences technology parks · Teacher professional development · Innovation and creativity · Competency based curriculum

5.1 Overview of the Country

5.1.1 The Geography, Population Situation and Political System

The Republic of Kenya is an equatorial country in East Africa with Nairobi as the capital city. The country is named after Mount Kenya, the second highest mountain

E. K. Makoba · J. O. Odhiambo (✉)
Centre for Mathematics Science and Technology Education in Africa (CEMASTEA), P.O
Box 24214-00502, Nairobi, Kenya

© The Author(s), under exclusive license to Springer Nature Singapore Pte Ltd. 2022   67
R. Huang et al. (eds.), *Science Education in Countries Along the Belt & Road*,
Lecture Notes in Educational Technology,
https://doi.org/10.1007/978-981-16-6955-2_5

in Africa. Kenya lies on the equator and borders the Indian Ocean (south east), and Tanzania to the south, Uganda and Lake Victoria to the west, South Sudan and Ethiopia to the north and to the east Somalia.

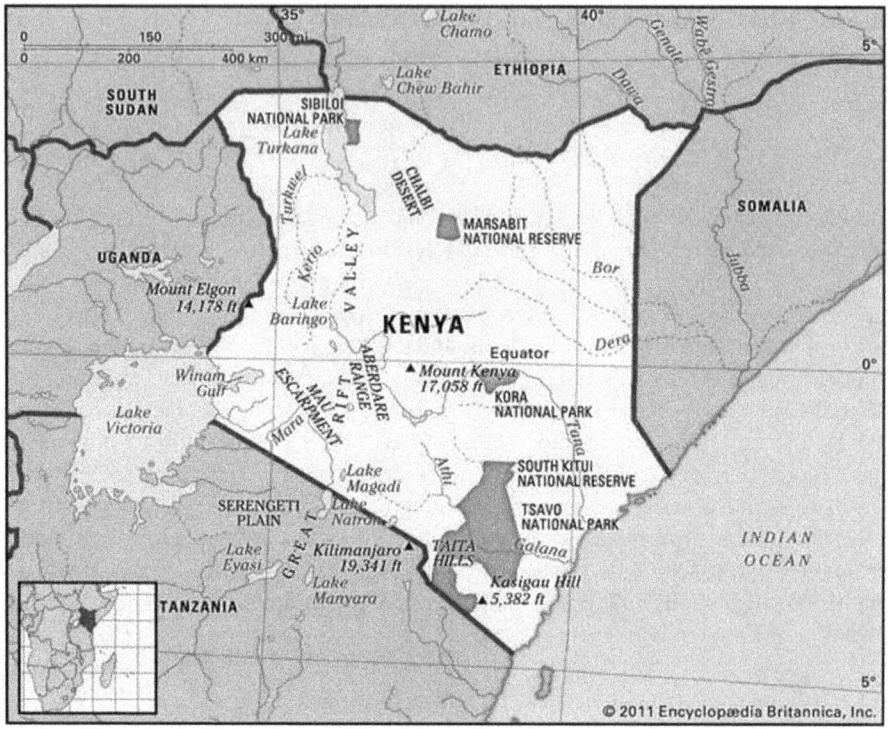

Map of Kenya

It has a land area of 582,647 km^2. Kenya's topography is diverse with features ranging from semi-deserts to snow-capped mountain, sandy coastlines to fresh water lakes, savannah grasslands to fertile agricultural zones, extinct volcanoes to beautiful coral reefs. On the eastern side the land gently slopes towards the Indian Ocean, while on the west the land is hilly with alternate plateaus and the magnificent Great Rift Valley.

The country's population as per the national census 2019 data indicated the population stands at 47,564,296 million people with 24 million more women than men, who stand at 23.6 million.

In 2010 Kenya promulgated a new constitution and this changed the administrative structures with the focus at the county levels. This was a turning point in the county's history as it reconfigured the balance of power by devolving power and responsibility from the national government to 47 elected county governments.

The Political System of Kenya

Kenya gained internal self-rule from the British colonial rule on the 1st of June 1963 when the country was allowed to form its first self-government with Mzee Jomo Kenyatta as the first President. On 20th December, 1963, Kenya attained full independence. It is a unitary State divided into 47 counties. The country is run by the National Government and 47 devolved units. The two levels of government work in close consultations. Generally, it is a multi-party system. Parliament of Kenya is a bicameral house consisting of the National Assembly and the Senate. The members of the National assembly are called Members of Parliament (MP) while the members of the Senate are referred to as Senators. Each serve a five-year term after which general elections are held.

The President is elected directly for a five-year term and is deputized by a Deputy President. Together with the Cabinet Secretaries they comprise the Executive arm of the government. Parliamentary politics in Kenya is open, free, fair and highly competitive.

5.1.2 The Current Situation of Economic, Technologies and Cultural Development

Kenya has made significant political, structural and economic reforms that have largely driven sustained economic growth, social development and political gains over the past decade (WB). Key challenges to its development include poverty, inequality, climate change, continued weak private sector investment and the vulnerability of the economy to internal and external shocks. In 2019, Kenya's economic growth averaged 5.7%, placing Kenya as one of the fastest growing economies in Sub-Saharan Africa. This was boosted by a stable macro-economic environment, positive investor confidence and resilient service sector.

Kenya's economy is being hit hard through supply and demand shocks on external and domestic fronts, interrupting its recent broad-based growth path. Apart from the Covid-19 pandemic, the locust attack which started early 2020, has affected many parts of Kenya especially North East. It has had a negative impact on the food security and growth of the agriculture sector in the country. Real gross domestic product (GDP) growth is projected to decelerate from an annual average of 5.7% (2015 to 2019) to 1.5% in 2020. The downside risks include a protracted global recession undermining Kenya's export, tourism and remittance inflows, further tightening of covid-19 health response measures that disrupt the domestic economic activity, fiscal weather-related shocks.

5.2 Overview of the Education Development

5.2.1 Education System and Policy

The Kenyan education system has changed a number of times since the colonial times, when education was discriminatory against the Africans. In 1902, a school for European children was opened in Kenya, followed by Asians schools. By 1910 35 mission schools had been founded, and emphasized reading, writing and arithmetic. The main idea by the missionaries was to give the locals a superior culture, since the African culture was viewed as inferior. The main objectives of education were not to help the African live in harmony with their local environment but rather to perpetuate their master-servant status quo between the colonizers and the colonized (Sheffield, 1973).

Colonial education for the Africans was inadequate in both quantity and scope, with narrow and restrictive objectives (Sheffied, 1973). This meant that a large majority of children of school age were not going to school. Only a small number of Africans passed through the system, hence creating a small educationally elite group, many assuming leadership positions after independence. When Kenya got independence, the Ministry of Education set up its first national commission (Ominde Commission, 1964) to assess the county's education system, review policy needs and recommend improvements to the government.

Another notable change in the education system was made by President Daniel Moi, whose administration was interested in resolving the challenges of unemployment and underemployment due to the 7-4-2-3 system of education to 8-4-4 as recommended by The Gachathi Report of 1976 (GoK, 1976). The report called for change to increase the technical and vocational aspects of the curriculum.

The 8–4-4 System of education started in 1985, following the Mackay report of 1982. The 8-4-4 policy emanated from the concerns that a basic academic education might lack necessary content to promote widespread sustainable (self) employment. It was therefore thought that it would equip the students with skills to enable the dropouts at all levels to be either self-employed or secure employment in the informal sector (Wanjohi, 2011). Students had practical activities like building, knitting, cooking among others.

The 8–4-4 system is divided into primary, secondary and tertiary institutions. The first 8 years of schooling, out of which first 3 years may be taught in mother tongue depending on the catchment area of the school. English is the language of instruction from class 4 to 8. Most of the schools are public and some are private. Most schools are co-educational though a few are only girls' schools. The primary school ends with the students sitting for the Kenya Certificate of Primary Education (KCPE).

Secondary education which caters for 14–17 year-old pupils, leading to the award of the Kenya Certificate of Secondary Education (KCSE) takes 4 years to complete.

Science education in Kenya is currently broadly structured with 3 years of preschool (preprimary), eight years of primary school science and four years of secondary/high school science four years of university (basic bachelor of science degree). (Okelo, 1997). Science education begins in preschool as a diffuse, amorphous introduction of simple scientific concepts through informal but guided activities. The role of the teacher at this stage is to provide facilities and an environment that promote informal exploration and experimentation during class activities (Okelo, 1997). At the primary school aspects of agriculture and Home Science have been incorporated (KIE, 2002). The syllabus also addresses emerging issues such as the environment, drug abuse and HIV/AIDS. The concepts are organized and presented as units. For Standard One, Two and Three, there are ten units (KIE, 2002). The upper primary classes, Standard four through eight, have nine and eleven units. These units refer to topics covered in the teaching process. As the topics become more advanced and detailed in the upper primary, the pupils are given an indication of the discipline-based structure of the secondary school (Okelo, 1997). At the secondary level, there are three sciences namely, biology, chemistry and physics. These three sciences for the basis of the careers a student would wish to pursue at the university level. The O-Level requirement is that a candidate can take at least two sciences as per the Kenya National Examinations Council (KNEC). The assessment has been mainly summative and places a lot of focus on examination results. This has been viewed as a major drawback to enable the students to cope with the current trends in the world, rather than skills and knowledge acquisition. An evaluation conducted by Kenya Institute of Curriculum Development (KICD) in 2009 pointed at a number of gaps which had to be addressed at the primary and secondary levels. This study indicated that the 8-4-4 system was geared towards passing examinations and did not embrace a holistic education (NESSP Plan for the period 2018–2022). The study also pointed out that there was a lack of capacity among curriculum implementers (teachers and field officers). The study also pointed out that there were inadequacies in the assessment. In the new curriculum more emphasis is placed on practical and vocational education as well as nurturing talent and de-emphasizing academics (NESSP, 2018–2022). It also called for the establishment of special schools for talent in areas such as music, athletics, and sports and in addition to the mainstreaming of ICT at all levels of Basic and Tertiary education.

The new Competency Based Curriculum (CBC) covering preprimary and primary Grades 1, 2 and 3 was rolled out nationally in 2009 and is expected to be rolled out in primary Grade 4 in 2020 (Kenya Economic Survey, 2020).

5.2.2   School, Students and Teachers' Profiles

In this section some statistical data on the schools, students and teachers will be discussed. Enrolment rates and retention will also be looked at pre-school education, primary school education, secondary education and higher education. It is also important to note that the Government has put in place a policy to ensure 100 per cent transition of all learners from primary to secondary education.

Schools

As per the report on the economic report 2020, the total number of schools reduced by 2.5 per cent to 89,337. The number of primary and secondary schools declined by 14.7 per cent and 8.2 per cent to 32,344 and 10,463 respectively in 2019, while pre-primary schools increased by 10.0 per cent to 46,530 in the same period (Economic Survey, 2020). The reduction in the number of schools is attributed partly to stringent measures and inspection by the Ministry of Education staff to ensure they meet the standards set.

Students

At the secondary level the school enrolment as per the Kenya Economic Survey (2020), by class and sex are as shown in the table below.

The total enrolment in public and private secondary schools increased by 10.8 per cent to 3.3 million in 2019 from 2.9 million in 2018.

Enrolment in Form 1 was at 861.4 thousand in 2019. This was a record increase of 4.3 per cent from 826.0 thousand in 2018.

This growth in enrolment was attributed to the Government policy of achieving 100.0 per cent transition from primary to secondary education. In 2019 according to the report, girls accounted for 50.9 per cent and 49.1 per cent of total enrolment in Form 1 and Form 4 respectively.

Number ('000)

Class	2015			2016			2017			2018			2019*		
	Boys	Girls	Total	Boys	Girls	Total	Boys	Girls	Total	Boys	Girls	Total	Boys	Girls	Total
Form 1	380.4	352.2	732.7	382.8	375.1	757.9	405.5	396.1	801.5	414.6	411.4	826.0	423.2	438.2	861.4
Form 2	359.8	331.6	691.4	372.5	357.9	730.4	386.4	372.3	758.7	408.7	392.5	801.2	450.8	457.3	908.1
Form 3	331.1	296.4	627.5	345.0	324.3	669.4	356.0	335.7	691.7	369.4	348.5	717.9	389.0	387.9	776.9
Form 4	277.1	230.3	507.4	296.6	266.3	562.9	303.0	275.9	578.9	312.6	285.0	597.6	363.1	350.5	713.6
TOTAL	1,348.4	1,210.5	2,559.0	1,396.9	1,323.6	2,720.6	1,450.8	1,380.0	2,830.8	1,505.3	1,437.4	2,942.7	1,626.1	1,634.0	3,260.0

Source : Ministry of Education

* Provisional

(Source: Kenya Economic Survey 2020)

Teachers in recent years

The number of teachers in public secondary schools and teacher training colleges by qualification or category and sex from 2015 to 2019 is presented in Table below. The total number of teachers increased by 6.0 per cent from 99,272 in 2018 to 105,234 in 2019. The number of male teachers accounted for 59.0 per cent of the total teachers in 2019.

Number

Qualification/ Category	2015			2016			2017			2018			2019*		
	Male	Female	Total	Male	Female	Total	Male	Female	Total	Male	Female	Total	Male	Female	Total
Masters and Doctorate (PhD) Degrees	1,351	1,114	2,465	1,346	1,107	2,453	1,285	1,075	2,360	1,085	930	2,015	1,024	896	1,920
Bachelors Degree	45,240	29,052	74,292	46,952	30,704	77,656	51,614	34,341	85,955	55,313	37,554	92,867	59,074	40,551	99,625
Post Graduate Diploma in Education	36	16	52	37	18	55	34	18	52	10	6	16	9	6	15
Diploma	4,898	3,728	8,626	4,914	3,732	8,646	3,643	2,814	6,457	2,343	1,969	4,312	1,951	1,663	3,614
Contract Teachers										54	8	62	52	8	60
TOTAL	51,525	33,910	85,435	53,249	35,561	88,810	56,576	38,248	94,824	58,805	40,467	99,272	62,110	43,124	105,234

Source : Teachers Service Commission

* Provisional

† Data excludes teachers on unpaid study leave and those with disciplinary cases

(Source: Kenya Economic Report 2020)

5.2.3 *Educational Research and International Collaboration*

A brief introduction of the main educational research institutions, education research programs, and international collaboration.

5.3 Current Situation of Science Education

5.3.1 *Policies and Standards*

Kenya has various policies that guide the implementation framework in the provision of basic education. In this section, the policies are discussed with a specific focus on provisions aimed at promoting science education. The constitution of Kenya provides for access, equity, and inclusivity in education as well as the right to free and compulsory basic education for every child (GoK, 2010). This is a recognition that education is a critical component of her development. On the other hand, Kenyas' blueprint for development, The Kenya Vision 2030 envisions the repackaging of STEM Education and Training to promote experiential learning, innovation, and creativity. Attraction to STEM-related disciplines will also be enhanced through well-coordinated programmes in education, R&D, and Training in all aspects of ST&I at all levels starting from Early Childhood to Primary and Secondary Education levels up to University (GoK, 2018).

The Basic Education Act 2013 spells out what is required of the Ministry of Education concerning the provision of education in the country. For instance, to ensure that quality basic education conforming to the set standards and norms is provided. It also recommends the provision of required infrastructure including schools, learning and teaching equipment, and appropriate financial resources (GoK, 2013).

The National Education Sector Plan (NESP) is a five-year plan that outlines the education sector reform implementation agenda based on the challenges affecting the sector. According to NESP, 2013–2018, the overall objectives of the education sector are to ensure equitable access, attendance, retention, attainment, and achievement in education. This also includes science, research, and technology by ensuring the affordability of education services. Other objectives include the promotion and improvement of Science and technology culture, improving the quality of all aspects of education and training so that recognized and measurable learning outcomes are achieved, and equipping schools to ensure that all primary and secondary schools meet minimum quality standards of teaching and learning, including equipping all secondary schools with science laboratories and science equipment among others.

The Science, Technology, and Innovation (STI) sector is governed by the Science and Technology and Innovation Act, 2013. The Act provides the legal framework to facilitate the promotion, coordination, and regulation of the progress of science, technology, and innovation of the country; to assign priority to the development of science, technology, and innovation, and to entrench science, technology, and innovation in the national production system. However, there exists a shortage of human resources needed for the development of Science, Technology, and Innovation. The 2014 African Economic Outlook Report shows that Kenya had only 13, 012 research personnel in 2010, which translated to 322 research personnel per million people in the population, which is low compared to established knowledge-driven economies. More so, data shows that there is low participation by women in research. Similarly, enrolment of females in higher education is lower than males at all levels of training, especially in STEM programmes.

5.3.2 Curriculums and Digital Resources

The vision of the basic education curriculum is to enable every Kenyan to become an engaged, empowered, and ethical citizen while its mission is nurturing every learner's potential. This can only be achieved by providing every Kenyan learner with quality education. The provision of quality instruction, school environments including resources is essential for quality human capital (KICD, 2016).

This section will therefore give an overview of the science curriculum at basic education and the digital resources used at those levels. A good curriculum contributes to the development of thinking skills and the acquisition of relevant and acceptable global competencies that learners need in daily life and careers. Since

2009 Kenya has embarked on reviewing its curriculum and is transitioning from the current knowledge-based academic curriculum to the Competency-based Curriculum (CBC). The new curriculum is seen as a solution to the gaps realised from the old curriculum. One of the gaps is about its form of assessment which is largely summative with too much focus on passing examinations rather than skill and knowledge acquisition (KICD, 2016). The new curriculum has therefore shifted focus towards the acquisition of focal competencies in line with the 21st Century skills one of which is digital literacy. In this curriculum, Inquiry-Based Learning has been proposed as the teaching approach for adoption.

The curriculum is divided into primary and secondary level curriculums. At the primary level, the science curriculum focuses on applying creativity and critical thinking skills in problem-solving, exploring the immediate environment for learning and enjoyment. And practicing hygiene, nutrition, sanitation, and safety skills to promote health and wellbeing as learning outcomes in science. At the secondary school level, the science curriculum is offered in distinct subjects such as biology, chemistry, and physics all integrated with ICT.

Digital Resources

The government of Kenya has invested in ICT integration in education to enhance access, quality, and equity in education as well as to enhance digital literacy. Through the Digital Literacy Programme (DLP) at the primary level, schools were provided with digital resources. These resources were tablets and laptops with accessories to receive the internet. This was aimed at ensuring effective curriculum delivery. This initiative is however currently facing some challenges like unreliable electricity supply, unreliable and/or lack of internet connectivity and limited ICT skills among teachers. There is also the unwillingness of teachers to integrate ICT in teaching and learning. These challenges are affecting the sustainability of the programme.

At the secondary level, a programme on Computers for Schools was put in place. Through this programme 3000 schools were provided with ICT learning resources. The resources include personal computers (PC), laptops, tablets, smartboards, and projector(s). The programme is however faced with a lack of adequate ICT learning resources and guidelines for ICT Integration. The challenges of internet connectivity, as well as electricity, have limited access to content. It is also affecting communication and collaboration among teachers and learners hence a low level of ICT use in science education.

5.3.3 Student Assessment and Achievement

Assessment of students is an important component of any educational system as it enables achievement to be monitored. It also provides a basis for student selection and placement in various courses as well as career pathways. In Kenya, the mandate to assess students is vested in the Ministry of Education through Kenya National Examinations Council (KNEC) for the basic education and middle-level colleges.

The universities are autonomous and hence have their system of assessment. This section discusses student assessment in PISA and TIMMS as well as assessment by KNEC in Science Education at the basic level of education.

Student achievement in PISA and TIMSS

The Programme for International Students Assessment (PISA) and Trends in International Mathematics and Science Study (TIMSS) are international frameworks for monitoring student achievement. Concerning these international assessments, Kenya has so far not participated. It has however participated four times in the Southern and Eastern Africa Consortium for Measuring Education Quality (SACMEQ) assessment which is tailored along the TIMMS. KNEC (2020). The SAQMEQ assessment was however limited to grade 6 learners in mathematics and English subjects. Therefore, no results are available for science education. There is however a proposal to develop and implement the framework for participation in the PISA and TIMSS (NESSP 2018). This will help to gauge the skills acquisition of Kenyan children, with those from other countries. It will also provide internationally comparable test scores to inform research and policies on education.

Student assessment in science education

The assessment of students in Kenya happens at two levels. There is an internal level in which every school has a mechanism for continual assessment of students. Then there is the national level in which the national examining body assesses final students at the primary and secondary school levels. This section focuses on national assessment because the results of school-based assessments are not translated into a national report for public access and use. The assessment at the primary school level is called the Kenya Certificate of Primary Education (KCPE) administered in grade 8(Class 8) and the Kenya Certificate of Secondary Education (KCSE) administered in grade 12 (Form 4). The results are used to place students in secondary schools and to join tertiary education respectively.

Overall results from student assessments by the KNEC show that learning achievement in KCPE national mean score is still low. This mean score, based on 500 total marks has remained slightly above 50% for the last 7 years. In KCSE assessment, the entry grade to university which is C + and above has been decreasing over the years. For example, in the 2017–2018 KCSE results, more than half of the candidature scored a mean grade of D and below (MOE, 2018). On the other hand performance in sciences Biology, Chemistry and Physics at the secondary level have been consistently low for the last nine years for the period 2006 to 2014 with a national mean score below 50% in all science subjects.

5.3.4 Science Technology Venues and Centers

Experience worldwide has shown that Science and Technology Parks (STPs) can provide an innovative infrastructure through which to nurture a country's development. They can spur indigenous technology-based economic activity when well understood and properly harnessed (MoE,2018). For instance, STPs have been catalysts for sustainable innovation and economic development of the Asian Tigers like Singapore and South Korea. Kenya in recognition of the significance of such parks has a draft STP policy of 2012. One of the objectives of this policy is to bridge the gap between research and industry.

Although this draft policy has not reached the implementation phase, some milestones have been realised such as the development of a 10-year science and technology park (STP) Master Plan in 2017. The Vision of the Master plan is a "Globally competitive knowledge economy leveraging on a dynamic science, technology, and innovation ecosystem. Besides, incubation centres have been set up in some universities and by public–private partnership sector players. Others such as the Konza Techno city are under construction. These STPs will focus on biotechnology, design, manufacturing, and materials, pharmaceuticals, and ICT hardware and software support systems among others.

On the other hand, Kenya has other institutions that can be regarded as science centres. They include Technical and Vocational Education and Training (TVET) institutions, STEM model schools, the National ICT and Innovation Centre, and the Centre for Mathematics Science and Technology Education in Africa (CEMASTEA). There are therefore just a few such centres in the country.

The science-related activities and competitions associated with these centres vary according to the levels. At the primary and secondary school levels, learners take part in Science and Engineering faires. The faires are held each year and are open to all learners. Each year has a unique focal theme upon which participants are expected to model their innovations to solve a real-life challenge. The competitions commence from the sub-county level and culminate at the national level. The winning innovations some times get opportunities to be presented in continental as well as global Science faires.

Some centers such as the National ICT Innovation and Integration Centre (NI3C) was established to provide support and helpdesk services to teachers and schools in ICT integration, testing and advising senior management about innovative solutions, and hosting the national teachers' portal. Currently, the centre is undergoing some reorganization to make it achieve its full potential. CEMASTEA on the other hand provides In-Service Education and Training to teachers in STEM subjects. It also promotes interest in science by holding Annual Science faires as well as student exhibitions in science education.

5.3.5   Utilizing Emerging Technologies

Technology is critical in transforming education and addressing significant challenges of access, quality, relevance and equity faced by the education system. Furthermore, the rate at which technology is evolving is so rapid which calls for the regular review and adoption of relevant emerging technologies to promote science education. In this section, the use of emerging technologies to impel science education in Kenya is discussed concerning basic education. When curriculums and digital resources were discussed, it was observed that various interventions are in place to promote ICT integration in education. It was also observed that there are challenges faced concerning ICT resources, internet connectivity, electricity, and low teacher skills in ICT. As a result of these factors, the use of emerging technologies in public schools is minimal. However, due to the COVID-19 pandemic, there has been an upsurge in the demand for remote learning. This has increased the use of virtual platforms for teaching and learning. Some of the commonly used platforms by teachers in Kenya include Google classroom and Google meet, Microsoft Teams, Big blue button, and Cisco Webex. The social networking platforms such as Whatsapp and Facebook are now being used by teachers to keep their learners engaged during this pandemic period.

Emerging technologies such as Artificial intelligence, Augmented or Virtual Reality tools are used to a very limited extent in public schools. However, through science and engineering faires some learners are getting exposed to aspects of Ai through robotics and AR/AI. Data on their use is also limited.

5.4   Requirements for Future Development of Science Education

The rapidly increasing youth population in Kenya has the potential to transform the country's economic fortunes. So far Kenya has achieved in terms of access to education since the introduction of free education with a gross enrolment rate of 92.1% in primary and 70.3% in secondary school levels (MoE 2018).). However, increased access has posed the challenge of quality. Concerning this, the government needs to improve the quality and relevance of the skills young people possess when entering the labour market. This requires sustained development in science education at all levels.

Policy declarations on what needs to be done to develop science education in Kenya need to be implemented for them to realise their intended purposes. For instance, the establishment of Science and Technology Parks will go along way in enhancing interest in Science and how it is learned. Such parks will provide learners with space to experience science as well as try out their creative and innovative ideas.

Policies on education recognise the role of technology in education and science in particular. It is for this reason that all of them have provisions for improving ICT in education through the provision of digital resources. They however also recognize the challenges being faced such as limited ICT skills by teachers. To improve the quality of science education, therefore, the professional development of teachers in the use of technology must be addressed. Experience has shown that the provision of digital resources without training teachers makes resources to be underutilised. Besides digital resources, science education will develop through rationalisation of science infrastructure in schools. Although there are guidelines on requirements for school registration, enforcement has been a challenge due to the demand for school brought about by free universal education. There are plans for providing minimum specifications for the science labs and resources across all schools to ensure equity across schools.

The revised curriculum has made provision for learners to pursue various pathways according to their interests and abilities. STEM as one of the pathways is meant to introduce creativity and innovation to students from an early age. In this respect, the Government has designated some schools as STEM schools. These schools should be enhanced to become centres of innovation and creativity. To realise this resourcing them and building teachers' capacity on learner-centered pedagogies and multidisciplinary teaching approaches will be required.

Regarding assessment, the participation of Kenya like other developing countries in international assessments such as PISA and TIMSS is significantly low (Sandefur, 2016). There is a need to increase participation to gauge skills acquisition of Kenyan children using international parameters.

5.5 Discussion and Conclusion

Kenya aspires to be a middle-income economy as per its Vision 2030 blueprint. Science Technology and Innovation is a great catalyst for the realisation of this dream. From its policies on education it is clear that the role of science in her development has been well recognised. It is now a matter of translating policy statements on science, technology, and innovation as well as improving the quality of education into practice. Further, the curricula should be strengthened accompanied by teacher development to promote scientific thinking skills as well as digital literacy. Putting in place strategies to make digital devices especially mobile ones affordable will go along way to increase uptake of technology in education. The efforts being made to increase internet connectivity at affordable rates and connecting all schools to the national grid should be sustained.

The feeling that the curriculum was examination focused informed its revision to a Competency Based Curriculum (CBC). When a curriculum is focused on passing exams it leads to a teaching approach that promotes drilling and recall at the expense of acquiring applicable skills and competencies. It is suggested that emphasis on standardised examinations should be downscaled and more emphasis given to nurturing learners' creativity and innovativeness through practical learning and application to real life situations. Participation in international assessments such as PISA and

TIMMS should be encouraged however, their customisation for regional or local use should also be encouraged and explored.

Development of well-resourced science innovation hubs and parks is a good idea to provide avenues for the youth to explore and apply their creative minds while in school and also while out of school. This calls for the implementation of all policies regarding ICT, Science Technology and Innovation and Science Technology Parks. Besides, there is a need for creating awareness on the availability of such centres and linking them to educational institutions. Such centres will be used to expose learners and educators to emerging technologies for enhanced learning and creativity.

References

Government of Kenya. GoK (2010). Constitution of Kenya, 2010. http://www.kenyalaw.org:8181/exist/kenyalex/actview.xql?actid=const2010.

Government of Kenya (GoK). (2013). *Basic education Act No 14 , 2013.* https://www.education.go.ke/index.php/downloads/file/96-basic-education-act-no-14-of-2013.

Government of Kenya.GoK (2018). *Kenya vision 2030.* https://vision2030.go.ke/publication/third-medium-term-plan-2018-2022/.

Justin Sandefur. (2016). Internationally comparable mathematics test scores for fourteen African countries. In *CGD Working Paper 444.* Washington, DC: Center for Global Development. Retrieved from http://www.cgdev.org/publication/math-scores-fourteen-african-countries.

Kenya Institute of Curriculum Development (KICD). (2016). *Basic education curriculum framework.* KICD. Retrieved from https://kicd.ac.ke/curriculum-reform/basic-education-curriculum-framework/.

Kenya National Examinations Council. KNEC (2020). *Monitoring Learner Achievement at Class 7 Level of primary school education in Kenya.* Retrieved from https://knec.ac.ke/wp-content/uploads/2020/09/FINAL-NASMLA-Class-7-Report-17.08.2020-Copy.pdf.

Kenya National Bureau of Statistics (2020). *Kenya economic survey.*

KIE (2002). Kenya institute of education.

Okelo, O. (1997). Science education in Kenya. *Frontiers: The Interdisciplinary Journal of Study Abroad, 3*(2), 83–90.

Sheffield, J. R. (1973). *Education in Kenya: An historical study.* Teachers College Press.

Wanjohi, A. M. (2011). Development of education system in Kenya since independence. *KENPRO Online Papers Portal.*

Edmond Kizito Makoba is a dean in biology education and chairman innovations committee at the Centre for Mathematics Science and Technology Education in Africa (CEMASTEA) based in Nairobi Kenya. Edmond has 19 years of experience working with the rest of the team on teacher professional development. Edmond is responsible for developing teacher training contents and resources as well as formulating teaching strategies for improved classroom practices in biology. Edmond also conducts training of teachers in biology as well as in the use of technology in teaching and learning including administrative duties in the department. Edmond in his working is driven by the idea of "giving meaning to learning"

John Otieno Odhiambo is a National Trainer in the Department of Biology at the Center of Mathematics, Science and Technology Education in Africa (CEMASTEA). John is a PhD student at the Catholic University of Eastern Africa (CUEA) in Educational Research and Evaluation. As a National Trainer in Biology, John regularly interacts and trains biology teachers at secondary school level to showcase new teaching trends, such as the application of the 5E instructional methods, ICT integration in teaching and learning of biology and lately promotes remote learning with the Covid-19 pandemic at hand in the last one and a half years. Between 2006 and 2008, John was involved in providing technical support on monitoring and evaluation of the National INSET in Uganda (SESEMAT). John is currently the chairperson of Disability mainstreaming Committee in CEMASTEA, had taken a lot of interest to laisse with the National Council for Persons with Disabilities (NCPWD) to mainstream disability in the institution, and was awarded the best institution by DAAR for a starling performance in making CEMASTEA a disability friendly institution. John is also the deputy coordinator of Partnerships and Linkages (P&L) where he was in 2019 the Chair of the COMSTEDA 17 (Conference of Mathematics, Science and in Africa), where the main discussion and paper presentations were on STEM education in Africa.

Chapter 6
Science Education in Libya

**Mohamed Elbeshti, Masauda Elaswed, Fathi Bribesh,
and Mohamed Abushafa**

Abstract This chapter provides a brief introduction to the geographical location of Libya, its population and the current political system. We highlight the levels of the education system in Libya (primary, basic and higher education). This includes the number of enrolled students and the number of dropout rates per year. Training and development in some schools and universities that have established eLearning and emerging technologies to enhance science education teaching or learning are discussed. The assessment methodology is also considered. The chapter also provides some statistics about the economic situation technologies and cultural development in Libya. Challenges, strategies, reflections, issues, and future paths form the latter part of this chapter.

Keywords Level of study · Enrolled students · Opportunities · Issues

M. Elbeshti (✉)
Director of eLearning Center, Zawia, Libya
e-mail: m.elbeshti@zu.edu.ly

M. Elaswed
General Manager of General Centre for Training and Education Development, Tripoli, Libya
e-mail: masauda@moe.gov.ly

F. Bribesh
International Cooperation Office, University of Zawia, Zawia, Libya
e-mail: f.bribesh@zu.edu.ly

M. Abushafa
eLearning Center, University of Zawia, Zawia, Libya
e-mail: m.abushafa@zu.edu.ly

© The Author(s), under exclusive license to Springer Nature Singapore Pte Ltd. 2022   83
R. Huang et al. (eds.), *Science Education in Countries Along the Belt & Road*,
Lecture Notes in Educational Technology,
https://doi.org/10.1007/978-981-16-6955-2_6

6.1   Overview of the Country

6.1.1   Geographical Location, Population and Political System

Libya is a North African country whose northern shores border the Mediterranean Sea which is located between Egypt and Tunisia. Apart from these two countries, Libya also shares borders with Algeria, Chad, Niger and Sudan. Most of its population of around 6.8 million lives along the Mediterranean coastline (CIA Factbook, 2020). The main ethnic group is Arab, the language Arabic and the official religion is Muslim. Although elections were held in 2014, and a government was appointed which gained international recognition, there has since been dispute over the results and the ensuing political crisis has created turmoil and divided the country (Fanack, 2020).

6.1.2   Current Situation of Economy, Technologies and Cultural Development

Due to the instability in Libya, its economy has dropped significantly. The country is dependent on oil and gas exports but these have been sporadic and various factions have tried to control the oil wealth. Consequently, any investment in developing technologies has been very limited as both the poor infrastructure and lack of security have not encouraged sustainable development (Fanack, 2020). It is estimated that currently one third of Libyans live at or below the national poverty level (CIA Factbook, 2020) and the political instability has had a significant impact on everyday life in the country. This has been exacerbated by the presence of terrorist groups and migrants flooding into the country from Sub-Saharan Africa and Asia (CIA Factbook, 2020).

6.2   Overview of Education Development

Libya was granted its independence by the United Nations on 24 December 1951. It was described as one of the poorest and most backward countries of the world. The constitutional declaration issued by the Revolutionary Command Council (RCC) on 11 December 1969 (World Data On Education. 6th edition, 2006/07), clarifies that "education is a right and a duty for all Libyan citizens, it is free and compulsory until the end of the preparatory level, and the State is responsible for building and establishing schools, institutes, universities and educational and cultural foundations". The education system in Libya consists of five stages of education. Libya established its first university based in Benghazi in 1957. Initially, there were only

15 students graduating from university, all with degrees in humanities. Independence brought rapid changes in the education system and during the 1975–76 academic years, about 13,418 students were studying in university. By 2004, an estimated 200,000 students were studying in Libyan universities with another 70,000 studying in higher technical and vocational training colleges. To accommodate the rising interest in learning, more institutions of higher learning have been established throughout Libya. The main factor contributing to this high uptake of higher education programs is due to the increased participation of the Libyan government and the free aspect of education. The government under a public budget program funds most of the universities. A number of private universities have been accredited to give the Libyan Education system new impetus. The education system in Libya has seen rapid development shape its future. It has significantly improved and advanced compared to other developing countries. The free aspect of it has resulted in more students enrolling and taking up studies. However, this rapid structural advancement has come with framework pressures that seem to be weighing heavily on it. Socio-cultural issues have also weighed on the system and in the past saw the ban of foreign influences being imposed, barring universities from teaching foreign languages (Tamtam, 2011). Since the Libyan regime change in 2011, political instability has negatively affected all levels of Libyan education system (lack of budget, infrastructure damage, schools being closed in many conflict areas). These problems have yielded very low outcomes in student performance.

6.2.1 Education System and Policy

The structure of the Libyan regular education system is divided generally into two main parts: schools and University or Higher Vocational Education (Tamtam, 2011). In general, for all stages, the academic year begins in September and ends in June. The school system consists of pre-schools, basic education and secondary education. These are sorted into five main stages of education as follows:
Statistics on national education.

1 **Pre-school:**
    This lasts for two years and is for children aged four and five. This stage is optional. Pre-schools do not have specific curricula or educational programs. The goal is developing children linguistically and to prepare them for schools.
2 **Basic Education:**
    Basic education is for nine years. It is compulsory and free for all Libyan citizens, who have reached the enrolment age of six at registration time, both male and female, according to the Education Law No. (95) of 1975, regarding compulsory education, and as stipulated in Resolution (210) of 2011, regarding the regulation of basic grade education. It includes six years of primary and three years of primary secondary. Each level at primary education has a number

of subjects which pupils have to learn and pass to move on to a higher level, with successful pupils in the ninth year obtaining the Basic Education Certificate. The six-year primary school's aim is to teach students to read, write, and count, as well as the natural sciences, hygiene, arts and crafts, and physical education. During their final year of primary school, students must pass an examination to qualify to enter junior high school. Junior high school, or Intermediate school, adds three more years to primary education, making it last for nine years.

After the basic education is successfully undertaken, pupils can enrol either in secondary school or three years in vocational training centers.

3 **Secondary Education:**

This stage of education has witnessed changes during the last two decades. Recently, it has stabilized as follows:

Now, students who have successfully finished basic education are able to enrol in secondary education. The first year is general and in the rest of the three years period students have to choose one of two special

4. **University and Higher Vocational Education system:**

Students who complete secondary education may enrol in one of the university faculties suited to the specialized secondary certificate they hold and also their average grade certificate. Public higher education is free in Libya, only those who enrol in Open University or private higher education are required to pay a tuition fee. Depending on the faculty and specialism, the university study period extends from four to six years. The university level is controlled by Regulation No. (501) of (2010), regarding the regulation organizing higher education, and faculty internal regulations (Ministry of Education, Libya 2020).

Secondary school specialism	University Faculties and Higher Vocational Training Centers that student can enrol in
Sciences	Faculty of Science, Faculty of Education, Faculty of Engineering, and Faculty of Education, Medicine, Dentistry, Pharmacy, Veterinary, Medical Technology, Faculty of Education, Higher Institutes of Health. Also, students can enrol in higher technical and vocational institutions as well, which includes polytechnics, higher institutes for trainers, higher institutes of technical, industrial and agricultural sciences
Arts	Economy, Social Sciences and Languages

5 **Postgraduate Studies:**

These include studies after a student has graduated or got a higher vocational level such as a postgraduate diploma, Master's Degree, or PhD Degree in several specializations and fields. This stage is also controlled by Regulation No. (501) of (2010), regarding the regulation organizing higher education (Ministry of Education Libya, 2020), as well as faculty internal regulations.

6.2.2   *Educational Research and International Collaboration*

Under the previous regime, vocational education was tailored largely for domestic, public sector jobs. Efforts to shift focus in vocational education towards private sector jobs and improved international relations became an important issue for the new government in Libya. In 2013, the Libyan Board for Technical and Vocational Education signed a memorandum of understanding with UK based TVET UK in order to facilitate efforts to modernize the institutions established in the 1970s and 1980s and overhaul vocational education in Libya. TVET UK agreed to work with UK based suppliers to set up workshops in Libya to prepare the transition of skilled works into new private sector industries.

In May 2013, the General National Congress began a funding initiative to send students to study overseas. The program chose at its inception 2,004 educators with master's degrees and 5,692 students to complete their education abroad, with plans to send another 3,616 top students over the coming years and to provide another 31,000 students with English language training. Initially, the fund gave scholarships to students who fought in militias during the civil war, but was later expanded to allow women and handicapped students to receive scholarships as well. (Education in Libya, n.d.)

Many Libyan Universities have become very active in their international collaboration, particularly under the umbrella of **Erasmus +** . This collaboration covers capacity building projects and international credit mobilities - see for instance National Erasmus Office (2020) and Mediterranean University Union (2020).

6.3   Current Situation of Science Education

6.3.1   *Policies and Standards*

Up until 2011 Libyan education was meeting its goals of universal access to education, although the quality of such education was still a concern, issues including poor outcomes and a lack of standardized planning and management (Carter, 2018). Currently there is a lack of data on policies and standards but two initiatives are ongoing: UNICEF (2017) is monitoring the implementation of education policies and UNESCO (2017) is planning a monitoring and evaluation system.

6.3.2   Curriculums, Digital Resources and Teacher Training

Digital curricula and resources

Curricula in Libya are still based on hardcopy books at all education levels. In 2006 the Libyan state went through the experience of e-learning and increased the level of performance in some educational subjects, especially mathematics and English. Therefore, it was decided to launch a pilot project for e-learning (with RiverDeep GESL company) and extend to include scientific education. However, this project was not completed, despite an active attempt in subsequent years.

Teacher training

Educating teachers in Libya coincides with the launch of the e-learning program, as the General Center for Training and Education was established in 2006. The General Center began its work in educating teachers and raising their professional competence, and developing educational skills, during the launch of the basic and secondary training program for teachers of the basic and secondary education stages. They were also trained in primary education, training and active education in preparation for teacher training in all schools. The efforts of the General Center began to train teachers and inspectors over long periods, which were a problem due to a lack of general conditions in the country until the project to introduce the English language for the first grades in 2018, where the center undertook the training of teachers for the first grades. From all regions of the country, east and west, a total of 110 subjects were created for national education, with local projects and local participation.

Among the projects of the General Center, one has been training 30 trainers in the French language in preparation for its introduction to secondary schools as a third language after Arabic and English. Also, 85 trainers were trained in teaching skills in an English language project.

In order to develop the level of education and keep pace with modern methods of teaching, the center has trained 8,665 teachers in national education, English language, French language and teaching strategies. The Center has also contributed to the completion of some recorded lessons, up to 1131 teaching sessions for the basic education level (Grade 1–9).

6.3.3   Student Assessment and Achievement

School restructuring, improvements, training, and curriculum development contribute greatly to the preparation for assessment. For example, exams are standardized in some years, such as Grade 9 and 12 in Libya. All schools are assumed to have completed the curriculum scheduled for these years. For other years, evaluations are conducted internally in schools, either by the class teacher or by a number of related teachers. Most of the exams are paper-based and are corrected manually or automatically (if they are multiple choice tests).

Universities have a legacy, and there are mid-term and final exams (for each semester). In addition, the practical is also assessed alongside the theoretical side (if the course has a practical component). The questions are comprehensive to the course the student has studied, so each professor chooses the method and approach that is best suited for his test.

6.3.4 Utilizing Emerging Technologies

Some universities have added e-learning centers in their structure, which in turn contributes to the production of electronic lessons (in support of education). Libyan legislation has not adopted electronic courses in Libyan universities, apart from the Open University. In addition, schools and universities have recently used digital education for some teaching curricula (this is more applicable to private schools).

6.3.5 Moving to Diazotizing

Most of the Libyan universities have turned to the eLearning programs in teaching, including the University of Tripoli, the University of Benghazi, University of Zawia, and University of Sabha.. The use blend eLearning methods in most disciplines. But the use of eLearning rate is still weak, according to the study conducted by the Center for E-Learning at University of Zawia in the analytical study on the application of blended e-learning during the years 2020 and 2021. However, the rate of using the eLearning Apps like virtual classroom and using digitizing information are relatively increasing by 20% each year.

6.4 Discussion and Conclusion

This chapter has presented a brief introduction in the Libyan Education system and structure. The fact is that education in Libya is one of the pillars of the renaissance, and it needs more attention. However, the education system does not have a successful strategic plan and a future vision to achieve current and future education goals. In the pre-university stage, for example, many decisions made show the lack of a methodology followed. Moreover, in 2010, the development and construction of building new schools that are commensurate with the number of students has stopped (currently there may be more than 30 students per class). This, in turn, reflects on the level of understanding and the discussions in class and has an impact on low outcomes (especially given class time per subject is 45 min, including teaching, discussing and explaining). Speeding up the course completion may reduce efficiency of the material (this is due to late starting dates for schools, short teaching time, and the number of holidays in the year). Increasing the number of students in class and taking monthly exams per teacher requires a fundamental development study, in order to enhance the absorption rate. Development of the level of education has started, along with keeping pace with modern methods of teaching that has recently been introduced (2019–2020), and this is helping to promote the development of education in Libya for the future.

On the other hand, since 2014 support and development has reduced in universities and this, in turn, reduces providing modern scientific resources such as laboratories, curricula and electronic services. Also, there is a lack of participation in international conferences and research journals, due to the suspension of foreign exchange. This has forced university professors to choose local conferences and journals, due to the cost of paying in foreign currencies or high travelling expenses. At the same time, 2020 has found that most universities now have strong local scientific journals, which are refereed and are scientific assets, where most of the universities are participating in Open Education Resources OER (i.e., University of Zawia, University of Benghazi, University of Tripoli, University of Sabaha).

Acknowledgements We extend our thanks and appreciation to Dr Yousef Zawayd the IT manager - Ministry of Education for contributing to the completion of this chapter

References

Carter, B. (2018). *Girls' educational needs in Libya.* K4D Helpdesk Report. Brighton, UK: Institute of Development Studies.

CIA Factbook (2020). *Africa: Libya.* .Retrieved September 15, 2020, from https://www.cia.gov/library/publications/the-world-factbook/geos/print-ly.html.

Fanack (2020). *Governance and politics.* Retrieved September 15, 2020, from https://fanack.com/libya/governance-and-politcs-of-libya/.

Mediterranean University Union (2020). *Libyan University future cooperation.* Retrieved September 14, 2020, from https://www.uni-med.net/en//?s=Libya.

Ministry of Education Libya. (2020). *Regulations and decisions* Ministry of Education. Retrieved September 16, 2020, from http://moe.gov.ly/اللوائح-القرارات.

National Erasmus Office (2020). *Erasmus+Libya.* Retrieved September 14, 2020, from https://www.erasmusplus.ly/.

UNESCO. (2017). *Country Plan. Libya.* Retrieved September 24, 2020, from http://unesdoc.unesco.org/images/0026/002614/261414e.pdf

UNICEF. (2017). *Libya Quarterly Inter-Sector Reporting: Education, April 2017.* Retrieved September 24, 2020, fromhttps://reliefweb.int/sites/reliefweb.int/files/resources/Education%20%20Libya%20Monthly%20Report%20%28April%202017%29.pdf.

Mohamed Elbeshti Finished my Master degree in computer science from Acadia University, Canada on 2000. Then completed my Ph.D. in IT on 2014 from Murdoch University, Australia. I have worked in several positions, including the Ministry of Libya Education and the Ministry of telecommunications. I also held several positions at universities, including the head of the department and the dean of the College of Information Technology and currently the director of the e-learning center.

Since 1990, I have worked in the field of teaching, learning and curriculum development of the undergraduate or the postgraduate studies. I have leaded small and enterprise projects and taught and trained many academic and industry courses and programs.

Masauda Elaswed is the General Manager of the General Centre for Training and Education.

Development, the centre is an institution of the Libyan Ministry of Education. Its objectives include the training and capacity-building for school teachers, conducting research and analysis on the Libyan educational systems and implementing strategies to improve education in Libya. As part of her role, she implemented a number of key projects aimed out fulfilling the duties of the centre. The most recent project was conducted in collaboration with IFIS, it involved the training of 109 teachers to become trainers for 2000 civic studies teachers in Libya. She was also involved in a project that was conducted in partnership with Garnet. The project was successful in training 120 teachers to become trainers. They are currently training 4000 English language teachers in Libya about the modern English language teaching techniques and strategies. In addition, she is currently taking part in the World Bank's Collaborative Governance Leadership Development Program for Libya. Masauda is a graduate with a PhD in Education from the University of Sheffield in the UK.

Chapter 7
Science Education in Morocco

Khalid Berrada, Khadija El Kharki, and Hana Ait Si Ahmad

Abstract The purpose of this chapter is to give an overview of Science education in Morocco describing the most important aspects that affect practices and challenges our educational system is currently facing. From historical of Science education in Morocco to the new development of Science educational research we will be exploring current reforms in terms of teacher's professional development, social-cultural context, innovative models of learning and teaching in the digital age. This chapter focuses on cooperation programs and the current situation of science education and targets also policies and standards in Morocco in comparison to other emerging countries. We will then highlight a few requirements and dressed some ideas that can make science education in the future an important pillar in producing the human resource to succeed in the industrial revolution. As a result, we will conclude that the rapid development of science education, government, society, and industry should be in co-operation and work together.

Keywords Science education · Future citizenship · Morocco educational reforms · Curriculums · Digital resources · Morocco teacher's training

7.1 Overview of the Kingdom of Morocco

7.1.1 Geographical Location, Population and Political System

The Kingdom of Morocco is located in the northwest of the African continent, bordered on the south by the Mauritanian state, on the east by the Algerian state, on

K. Berrada (✉)
Mohammed V University in Rabat, Faculty of Science, N°4, Avenue Ibn Batouta, B.P. 1014, Rabat, Morocco
e-mail: k.berrada@um5r.ac.ma

K. El Kharki · H. Ait Si Ahmad
Trans ERIE - Faculty of Sciences Semlalia, Cadi Ayyad University, BP.: 2390 Marrakech, Morocco

© The Author(s), under exclusive license to Springer Nature Singapore Pte Ltd. 2022   93
R. Huang et al. (eds.), *Science Education in Countries Along the Belt & Road*,
Lecture Notes in Educational Technology,
https://doi.org/10.1007/978-981-16-6955-2_7

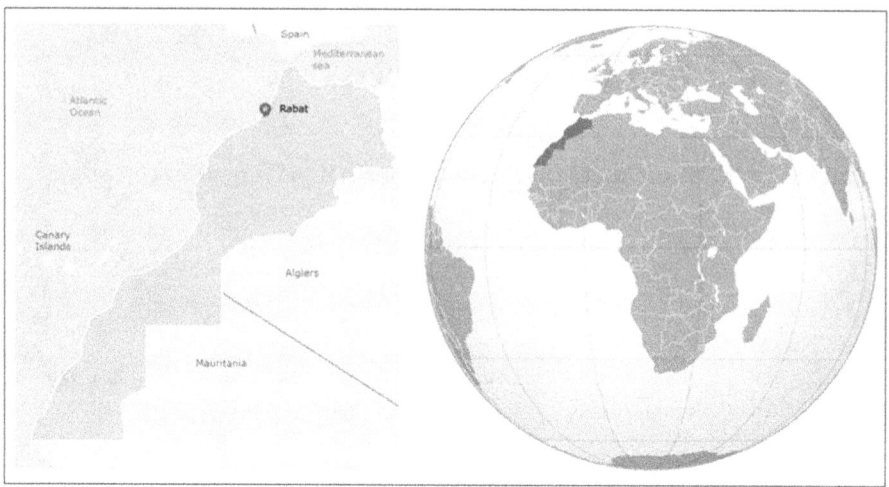

Fig. 7.1 The Morocco kingdom map and location

the north by the Mediterranean Sea, and on the west by the Atlantic Ocean (Fig. 7.1). Morocco spans an area of 712,550 km². It is the only African country and one of only three nations (along with Spain and France) to have both Atlantic and Mediterranean coastlines.

The Moroccan population is about 36 million in 2020 (HCP, 2020), which is equivalent to 0.47% of the total world population. Morocco is ranked number 40 in the list of countries by population, with 49% of the male and 51% of the female. About 27.8% of the population is under 15 years old, 66.1% are between 15 and 64, and 6.1% are above 64. The median age is 29.5 years, the life expectancy is 75.9 years, and almost 63.8% of the population is urban (Worldometer, 2020).

The government system in the Kingdom of Morocco is a constitutional monarchy, parliamentary democracy, and social (Constitutional Court, 1973). The constitutional system of the kingdom is based on the separation, balance, and collaboration of powers, as well as citizenship and participatory democracy and the principles of good governance and the correlation between responsibility and accountability. The King has the highest authority in the state, being the Commander of the Faithful (Amir Al Mouminine) and the highest commander of the Royal Moroccan Armed Forces. The Parliament of Morocco is the bicameral legislature. It is formed of two parliamentary chambers: The House of Representatives and the House of Councilors. The House of Representatives is formed of 395 members, who are elected directly for a five-year term, while the House of Councilors has not less than 90 and not more than 120 members. They are elected indirectly for a six-year term. The Parliament is composed of territorial groups, municipality councils, professional chambers, and elected representatives (Kingdom of Morocco, 2020).

7.1.2 Current Situation of Economic, Technologies and Cultural Development

The Moroccan economy is considered a relatively liberal economy governed by the law of supply and demand. The key sectors of the economy include agriculture, tourism, aerospace, automotive, phosphates, textiles, apparel, and subcomponents. Given the richness of Morocco's soil, the economy is dominated by the agricultural sector, the economic growth relies heavily on this sector, which employs nearly 37.9% of the workforce, and contributes to 12.3% of GDP (gross domestic product), the main crops are barley, wheat, citrus fruits, grapes, vegetables, olives, and livestock. Morocco has a relatively small amount of mineral resources; phosphates is the main source of wealth. The industry contributes 25.9% of the GDP and employs 21.6% of the workforce, the main sectors are textiles, leather goods, food processing, oil refining, and electronic assembly. The services sector accounts for slightly less than half of the GDP (49.5%) and employs 40.5% of the workforce, it is spearheaded by real estate and tourism (Nordea Trade Portal, 2020).

Morocco has increased investment in its port, transportation, and industrial infrastructure to position itself as a center and broker for business throughout Africa. Industrial development strategies and infrastructure improvements most visibly illustrated by a new port and free trade zone near Tangier are improving Morocco's competitiveness (Fanack, 2020; IndexMundi, 2019).

Morocco entered into a bilateral Free Trade Agreement with the United States in 2006 and an Advanced Status agreement with the European Union in 2008. Morocco also seeks to expand its renewable energy capacity to make renewable more than 50% of installed electricity generation capacity by 2030 (IndexMundi, 2019).

Since the launch of the Industrial Acceleration Plan (2014–2020) in 2014 in the Kingdom of Morocco, the industrial sector has created about 405.5 thousand new job opportunities during the period from 2014 to 2018, achieving about 81% of the target set by the industrial strategy, for the Ministry Industry, investment, trade, and the digital economy. According to the study completed by the Ministry, women received 49% of the jobs created, and more than 21% of these opportunities were provided by enterprises created after the launch of the Industrial Acceleration Plan (Fanack, 2020).

Morocco is the 5th largest African economy by GDP. The World Economic Forum placed Morocco as the 1st most competitive economy in North Africa and at the level 75 among the most competitive nation in the world out of 140 countries ranked in the 2018 edition of the Global Competitiveness Report (Trading Economics, 2020).

The information and communications technology (ICT) sector generates between 5 and 6% of GDP, of which telecommunications companies represent about half. However, technological developments coupled with strong political will and economic imperatives are pushing Morocco's ICT sector growth beyond the telecommunications subsector. Incorporating digital solutions in business operations is being a must to keep Moroccan companies at the forefront of regional and international standards.

Morocco aims to position itself as a strategic hub in the Middle East and North Africa by becoming one of the top-performing countries in the region in terms of Datacom infrastructure and Information Technology (IT) business environment. The investment reform plan presented in July 2016 by the Ministry of Industry, Trade, Investment, and the Digital Economy, marked a strategic step in the realization of new reforms for building a competitive and efficient economic model. As an important step to further building the country's international positioning, the Minister of Industry, Trade and Green and Digital Economy (MCINET) launched a new Digital Program for 2020, after the Morocco Numeric 2013 Plan, emphasizing the importance of introducing more diversification to improve the competitiveness of the country. According to the Ministry, in order to "reach an emerging country status and enable all citizens to fulfil their aspirations, Morocco should attempt to create the conditions for a sustainable economic growth, in which investment acts as a catalyst". The Digital Program plans to accelerate Morocco's digital transformation and reinforce the country's status as a regional digital hub. The program consists of a 750 million USD investment in reducing the digital divide by 50% through the digitization of administrative services, improved access to the internet through free Wi-Fi in public spaces, and digital literacy programs, aiming to train over 39,007 ICT professionals by 2020 (Infomineo, 2017).

7.2   Overview of the Education Development

7.2.1   Education System and Policy

The educational system of Morocco is based on the principles of the Islamic faith. It aims at forming a virtuous citizen, open to science and knowledge, and endowed with a spirit of initiative and creativity (MEN, 2020a).

According to the Education Statistics Compendium published by the Ministry of National Education in 2019, the total number of public and private primary, middle, and high schools increased to 16,860 in 2019 from 12,873 units in 2011. And of course, the number of higher education institutions is also steadily increasing over the same period, from 310 to 409 establishments (*Education Statistics Compendium*, 2019; MEN, 2020c). However, there are four levels of education in the Moroccan education system (preschool, primary, secondary, and university).

7.2.1.1   Pre-School Education

Pre-school education lasts two years and is open to children aged four to six. Its objective is to facilitate the child's physical, cognitive and emotional development, and initiation to basic religious, ethical and civic values (MEN, 2020d). Quranic schools are also part of pre-school education. Generally located in rural areas, which focus mainly on the fight against illiteracy.

7.2.1.2   Primary Education

Primary education, (six years), is open to children aged six years or older, whether or not they have completed pre-school education. It is divided into two cycles. The first cycle, which lasts two years, and the second cycle of primary school, lasting four years, is open to children of the first cycle of the same school. The main objectives of this cycle will be the development of children's skills and abilities, particularly through learning to read, write and express themselves in their first foreign language, and a first introduction to modern information, communication and interactive creation technologies. At the end of primary school, children must obtain their primary school certificate to be admitted to secondary school.

7.2.1.3   College Secondary Education

The training at the College, which lasts three years, is intended for young holders of the primary school certificate. The objective of this college cycle is to support the development of young people's formal intelligence (MEN, 2020b). The completion of college education is linked to the students' success in the final exam of the 3rd year for a College Education Certificate (CEC). Holders of the CEC may continue their secondary school studies according to their orientation and learning abilities.

7.2.1.4   Qualifying Secondary Education

Qualifying secondary education includes two types of training:

- The professional qualification cycle.
- A baccalaureate program, consisting of two cycles: The Common Core (1 year), and the General and Professional Baccalaureate cycle (2 years).

The professional qualification cycle (one to two years) aims to train personnel capable of adapting professionally to the industrial and service sectors. It is open to learners who hold an CEC. At the end of this cycle, learners must receive a Diploma of Vocational Qualification (DVQ). The Common Core cycle, which is open to students who hold the CEC. During this cycle, learners have a choice of options: literature, science, mathematics or economics. They will choose the stream of their choice for the next two years.

The Professional Baccalaureate aims to train technicians with scientific and technical skills in all sectors. It is open to students who have completed the Common Core cycle. At the end of this cycle, learners must pass the national examination to obtain their Baccalaureate of Technological and Vocational Education (BTVE). The General

Baccalaureate aims to provide learners with scientific, literary and economic training to pursue higher education, and is open to Common Core learners. At the end of this cycle, learners must pass the national examination to obtain their Baccalaureate of General Education (BGE).

7.2.1.5 Higher Education

Morocco's higher education system consists of three main sectors: universities, higher schools, and engineering schools, under the supervision of the Ministry of National Education, Vocational Training, Higher Education and Scientific Research (MNEVTHESR).

Moroccan universities accept students on condition that they obtain their BGEs. They adopt the LMD system (Bachelor - Master - PhD). The Bachelor's degree is the first year of the higher education cycle corresponding to the first 3 years of studies. It is organized in 6 semesters. The Master is the second stage of the higher education cycle leading to doctoral studies or professional activity. It is organized over two years, i.e. 4 semesters. The PhD represents the last phase of the LMD program, it is intended for holders of a Master's degree who wish to start research studies in three years.

Higher schools are specialized higher education institutions (6 years of study), under the financial supervision of the technical ministries. They accept students, who have high final grades in the baccalaureate.

Engineering schools train professional cadres able of occupying positions requiring qualified skills and in-depth knowledge. They are distinguished by a selection process based on a competitive entrance examination. The studies take place over 6 years after the BGE and lead to a state engineering degree.

7.2.1.6 Government Actions to Improve Education

The objective of the national education department of the MNEVTHESR is to develop education through the integration of a pedagogical approach based on active learning, without forgetting the important role of the education and training reforms that the ministry has developed to improve the education sector in Morocco. Among these major reforms, the National Charter for Education and Training (NCET), a reform based on the democratization of education and aims to promote quality education (Tawil, Cerbelle, & Alama, 2010). The second reform is the National Emergency Education Plan (NEEP) was implemented between 2009 and 2012 in response to the failure of the NCET (Colombo, 2011). The NEEP has been implemented with the aim of increasing enrollment rates and promoting the competitiveness of Moroccan schools through the creation of schools of excellence and preparatory schools. The third reform is the Action Plan for Education (APE) (NAJAH Program) which aims to consolidate the gains of previous reform initiatives (MEN, 2008), and to increase the number of students enrolled in the professional qualifying secondary education.

After one year of implementation, the APE resulted in a 27% increase in student enrollment in vocational education programs (Morchid, 2019). The final reform is the Strategic Vision of the Reform 2015–2030, it was developed to develop four pillars: equity, equality of opportunity, quality of education, and he advancement of society. This Strategic Vision of the Reform succeeds in implementing an affirmative action policy aimed at empowering students from disadvantaged socio-cultural backgrounds (CSEFRS, 2015).

7.2.2 Statistics on the National Education

According to statistics published by the MNEVTHESR, at least 8,948,624 students have entered education for the 2019–2020 school year. In 2010–2011, educational institutions received 1,492,296 new students and in 2019–2020, more than 1,900,383 students. A continuous increase due to population growth and awareness (Fig. 7.2).

Teachers are an essential link in the conduct and success of the teaching process. Their contribution is important in improving the quality and performance of Moroccan schools. Moreover, the total number of teachers in Morocco is 267107 for more than 11,400 schools. primary, middle school, and secondary school teachers with qualifications represent 92%, which corresponds to 246,784 teachers, and university teachers represent 8%, which is equivalent to 20,323 teachers. These numbers remain very low compared to the high number of students.

According to the Superior Council of Education, Training and Scientific Research (CSEFRS), in 2015, the dropout rate in the school system was about 508,300, a rate

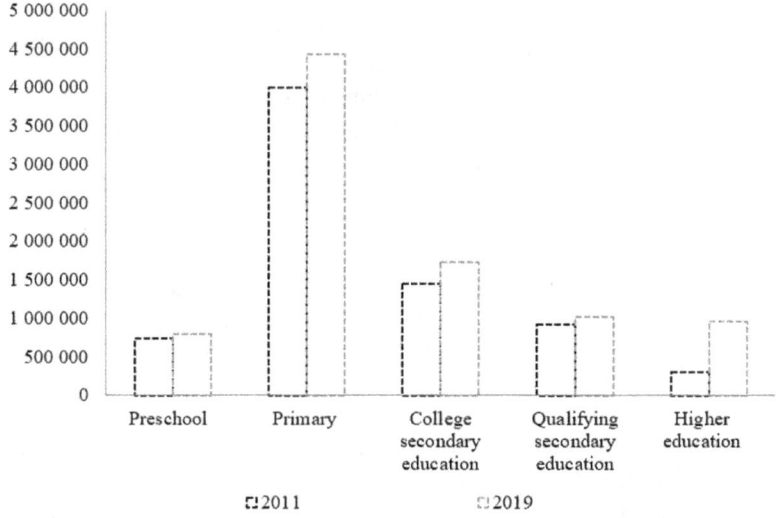

Fig. 7.2 Evolution of the number of students registered from 2011 to 2019

of 8.8% at the national level. The year 2016 saw a significant decrease in the number of dropouts to 407,674 (7.1% of all students enrolled). However, since that date, the trend has been reversed and the number of dropouts has increased in 2017 and 2018 to reach 431,876 students, or a proportion of 7.4% of all students in the 3 cycles (CSEFRS, 2019a). The analysis of dropouts by cycle shows that the majority of these dropouts occur in the basic cycle with 78.3% of the total dropouts, or approximately 338,000 dropouts in the two cycles of elementary and secondary college, cycles of compulsory education that are normally supposed to keep children in school at least until the completion of the college cycle (CSEFRS, 2019a). For the public Higher education sector in Morocco, it has shown in many internal studies that for open access institutions (Faculties of Sciences, Faculties of Law, Economy and Social Sciences, Faculties of Human Sciences and Arts) that the drop out for students enrolled is the first year of their studies can varied around 25% depending on the nature of institutions and disciplinary fields. One of the challenges taken during the NEEP is to reduce the drop out to less than 20%.

7.2.3 Educational Research and International Collaboration

7.2.3.1 Educational Research Institutions

Research in the field of education in Morocco aims to understand how learning takes place, and promotes the development of new tools and methods for the teaching/learning process.

In Morocco, first developed research in the field of education was conducted at the Faculty of Science Education (FSE) that has been created at Mohammed V University in Rabat in 1983. Nowadays, FSE supports and coordinates educational research concerning all levels of education. It brings together and makes available to researchers all the documents relating to teaching programs and methods. In addition, the FSE provides professional training in Communication and Expression Technique that now are deserving all Moroccan Higher Institutions.

The National Center for Scientific and Technical Research (CNRST) that is considered as an operator whose mission is the development and valorization of scientific research. The CNRST contributes to the dissemination of scientific and technical information and the publication of research work, particularly in the field of education. CNRST also encourages research in various fields, including educational sciences, and promotes the best students through a program of research excellence scholarships for doctoral studies to promote research in Moroccan universities (CNRST, 2018).

In parallel to all those valuable contributions many Higher education faculties and institutions in Morocco were developed scientific research around science education inside laboratories and teams. This involvement was done by necessity since we have noticed that to perform a good quality of teaching and learning even at universities we should base our effort on results of empirical studies that need to be developed within our program and on our curricula. So far, we will find now in most universities initiatives round research on science education that are now providing quality to strengthen the way our teaching and learning are developed.

As example at our local level, the Cadi Ayyad University (UCA) as well as for other Moroccan public universities created in 2018 the Center for Pedagogical Studies, Evaluation and Research. This center encourages innovation in pedagogical practices and supports teachers in the production of innovative pedagogical resources (UCA, 2015). This center was created after an enormous hard work developed by the Centre Pedagogical Innovation at CAU since 2013 and other research laboratories developing Science education as main topics of their activities that was it self-supported by the launch in Morocco of UNESCO Chair in 2010 "Teaching physics by doing" (Berrada, Channa, Outzourhit, Azizan, & Oueriagli, 2014). We will also cite locally the experience of Transdisciplinary Research Group on Educational Innovation (Trans ERIE) in developing research on Science education. Trans ERIE has initiated with other local research teams the creation in 2019 of a new program of Training Doctoral entitled "Didactic of Sciences and Pedagogical Engineering at Cadi Ayyad University.

7.2.3.2 Cooperation Program

Taking advantages of his special status as partner of the European Commission, the Kingdom of Morocco has initiated a wide range of cooperation tools and programs throughout many actions which were supported and funded the last fifteen years. In another hand, the ministry plan focused on the development of the international cooperation strategy by promoting a set of actions that have been presented in previous studies (Kaaouachi, 2020):

- *Focusing and strengthening current cooperation programs with foreign countries,*
- *Creating of new partnerships in the fields of higher education and scientific research,*
- *Consolidating the cooperation with the regional environment in the framework of south-south cooperation,*
- *Strengthening the mobility of Moroccan students and professors.*

According to the results obtained by Morocco in previous programs, the Ministry aims to strengthen its position to be able to apply for more projects, be an associated country in order to coordinate programs, and to host a number of research projects. Recently, both parties agreed to explore the potential paths for the implementation of cooperation actions in higher education and scientific research by 2027. In terms of bilateral and multilateral cooperation, the Ministry has also established several agreements with Arabs, Africans, European and Americans countries. More than that, cooperation programs in Morocco have been also extended to international organisms and private sectors. As example, we report below in more details some national and international cooperation programs:

Morocco-Wallone Cooperation

Morocco and Wallonia-Brussels have agreed to set up a cooperation program for the years 2015–2016-2017. Moroccan institutions of higher education and research, in partnership with their counterparts in Wallonia-Brussels, have participated in the implementation of joint projects in various fields, including those of training in educational professions and pedagogical innovation: New pedagogical approaches, use of ICT in teaching, e-learning (Issam, 2015).

Moroccan-French Cooperation

In accordance with the partnership agreement for cooperation and cultural development between the government of the Kingdom of Morocco and the government of the French Republic, joint efforts have been made by the two countries to promote training and research on the basis of joint programs open to the national academic and scientific communities. They agreed to launch a call for projects, which aims to support scientific research through research training for the benefit of young doctoral students and researchers through mobility programs between French and Moroccan universities. Among the different themes of this project, we can mention research in educational sciences with a particular emphasis on thesis projects in the field of massive open online course (MOOC) (MEN, 2015).

Ibn Khaldun Program

As part of the 2015–2030 vision of the Higher Council for Education, Training and Scientific Research (CSEFRS) and the strategy of the Ministry of National Education, the latter is launching, through the National Center for Scientific and Technical Research, a call for projects in the field of human and social sciences. This program called "Ibn Khaldun" aims to promote the development of a research of excellence for a better knowledge of Moroccan society and its evolution on the cultural, social and economic levels, and to encourage researchers to collaborate with each other at the national and international level. In addition, the development of the education system, and research science of education is among the themes on which this project focuses (MEN, 2018).

7.3 Current Situation of Science Education

7.3.1 Policies and Standards

As part of the new reform of teacher training, Morocco proceeded about 20 years ago to overhaul its education system through a new reform which aimed at the implementation of the orientations defined by the National Charter of Education and Training (NCET). This charter advocated a set of guidelines concerning the fundamental principles of the Moroccan education system. As a result, in accordance with these new principles and orientations, the NCET had reserved a strategic place for the training of teachers. To this end, "lever 13" of the charter made the link between "the commit-ment of teachers" and "the renewal of the school", through a set of guidelines such as the quality of teacher training as a condition for academic success; the integra-tion of all training establishments, "in order to mobilize all available potential"; the need to integrate continuing education into the management of teachers' professional careers.

Despite this urgency and these principles, training was a difficult component to master in the reform process launched since 2000, which resulted in a decade of delay in its implementation, which was still in progress when the Emergency plan was launched for the period 2009–2012. This Emergency plan distinguished between four major measures to improve teacher training: a definition of criteria and selection processes for school teachers; the establishment of "University Education License" (UEL); the creation of CRMEF; the implementation of a strategy for in-service teacher training.

The reconciliation between the choices adopted by Morocco and international trends in teacher training, shows, on the one hand, that Morocco is part of these inter-national trends by adopting the principle of "l 'universitarisation' of initial teacher training and that, on the other hand, the country is also committed to the application of the principle of professionalization of training.

From an organizational standpoint, it should be noted the increased involvement of universities with the FUEs created in 2012. In addition to that, initial training was provided in most cases by the ENS in Morocco, which since 2011 have been dependent on the universities. Morocco count till now eight ENS since 1978.

The second phase, called "qualifying training", is provided by the 15 new CRMEFs, which have replaced the old Regional Pedagogical Centers (CPR) and the Centers of Training Teachers (CFE).

In the CSEFRS report concerning the 2015–20130 strategic vision (Chapter II, Lever 9), it was clearly mentioned that the main dysfunction of the school, in its various components, appears in the quality of its services and of its performance. It is for this reason that the work on quality is essential as the first priority of the strategic vision. This aims for total and equitable quality, which concerns the various players in education and training, curricula and programs, school governance and scientific research.

Improving the performance of educational actors: teachers, trainers, supervisors, researchers and directors, appears at the top of the priorities likely to promote the quality of the school, improve its performance and achieve its reform. The council calls for work to improve their training, strengthen their motivation, upgrade their functions, respect their dignity and improve the conditions for exercising their profes-sion. This is all the more necessary as we are witnessing the transformation of the age structure of teachers which tends towards the opening of a demographic window of progressive rejuvenation and which requires an anticipation of the needs for managers and new profiles for the training of future generations, at the same time as the improvement of the qualifications of current managers is required.

According to this strategic vision of CSEFRS, initial training is a condition of access to the teaching profession in preschool and school cycles, in both the public and private sectors. However, the universities through the ENSs remain essential collaborators in the CRMEF. It is then incumbent on the ENS to properly prepare the preparatory cycle for access to CRMEF and to devote more freedom of initiative in the fields of education, research and innovation.

Another factor which reconfirms the position of the ENSs as a main training actor is its integration as a higher education establishment within universities. The ENSs will then be able to strengthen the representativeness of institutions in the field of educational sciences, which unfortunately only represents 1% today compared to other disciplinary fields in Morocco.

Statistical studies carried out by the supervisory ministry in 2017 identified Moroccan higher education institutions by disciplinary field. This study revealed that less than 9% of establishments could be in charge of pre-qualifying training in the form of License and Master curricula, leading to teaching professions in Morocco.

7.3.2 Curriculums, Digital Resources and Teacher Training

7.3.2.1 Curriculum and Initial Teacher Training

In line with its strategy to improve the quality of education in Morocco, the MNEVTHESR has several programs to ensure quality teacher training, hence the creation of the CRMEF, which mainly responds to national needs for training competent teachers (primary, secondary and qualifying). Within these centers, teachers are trained to participate in the education of learners, and transmit knowledge and know-how according to validated contents and methodology, and prepare the student to integrate into professional and social life, and provide him/her with the necessary skills (CSE, 2008).

7.3.2.2 GENIE Program

The new strategies adopted by the government to improve the teaching process aim to promote, facilitate and enhance a pedagogical culture that fosters the integration of ICT in teaching and learning. With this vision in mind, the Moroccan government adopted, starting in 2006, the Generalization of Information and Communication Technologies in Education in Morocco (GENIE) program, which aims to provide IT infrastructure to equip schools, create a national digital resources laboratory and a national ICTE portal to accelerate the development of digital content, and train school directors, inspectors, and teachers in computer science and the use of ICTE in the students' pedagogical environment (Idrissi, 2020).

For the pedagogical integration of ICTs in teaching, an agreement with INTEL to use the "Intel Teach to the Future" program, based on the project's pedagogy, was signed in 2007 to enrich the skills of teachers in their teaching practices. Thus, to train them in the use and integration of ICT in their effective teaching practices in order to improve student learning (Said, M'hammed, Mohamed, & Mohamed, 2009).

7.3.2.3 Strategic Vision for School Reform 2015–2030

The manifestation of the Moroccan government's efforts to ensure the development and innovation of teacher training was concluded with the implementation of the Strategic Vision for School Reform 2015–2030, initiated by the Higher Council for Education, Training and Scientific Research (CSEFRS) (CSEFRS, 2015). It recommended a set of orientations concerning the fundamental principles of the Moroccan education system, and set six areas for renewal, including the curriculum for teacher training. On the basis of this vision, the Moroccan education system has undergone several restructurings and innovations (Lahchimi, 2015), namely:

- The creation of thirteen model courses of study at the various universities in the Kingdom, covering all primary and secondary specialties, in order to simplify the accreditation process.
- The organization of the qualifying training cycle of future teachers in two years instead of one year: a first year of face-to-face training in the CRMEFs; and a second year of training by alternating between the place of practice and the CRMEF, and those with the help of the implementation of a distance learning platform for the benefit of trainee teachers.
- The generalization of continuing education and making it compulsory and decisive in the evolution of the careers of the actors of education, through the elaboration of the national strategy for the continuing education of the actors of education.

7.3.3 Student Assessment and Achievement

As well as for many countries in the region, Morocco is also aligned with international standards and participates in the International program for monitoring student achievement (PISA), Trends in International Mathematics and Science Study (TIMSS), etc. tests. In 2018, PISA placed Morocco 75th out of 79 countries. This international test assesses the abilities of 15-year-old students in reading, mathematics and science. For this test, China dominated the top of the rankings in all fields of study. Singapore is the country which occupied the first place in 2015. China did not even place in this same test in the top five of the ranking in 2015.

Approximately 600,000 students from 79 countries attended the two-hour computerized assessment. However, Morocco ranked ahead of the Philippines, the Dominican Republic, Lebanon and Kosovo. This was Morocco's first participation in the international assessment that the Organization for Economic Co-operation and Development (OECD) has been carrying out for about twenty years. Morocco remains Morocco, an African country which took part in these evaluations alongside five Arab countries.

Carried out every three years, the study took place in April 2018 on a sample of around 7,000 Moroccan students from almost 180 qualifying high schools. Unfortunately, in almost all fields of study, Moroccan students scored below the average PISA mark.

Regarding reading, Moroccan students obtained a score of 359 points, well below the international average for PISA considered at 487. Morocco's national average score in mathematics is 368 points, or more than 100 points in below the international average of 489. In science, Moroccan students achieved a score of 377 points, against an international average of 590 points. This was explained by the fact that 13% of Moroccan students have vulnerable social origins and were still able to obtain the national average. As a result, the PISA assessment is one of the many education-related assessments that have placed Morocco in the wrong position.

In November 2018, the World Bank published a report highlighting the shortcomings of the Moroccan education system. The report, "Expectations and Aspirations: A New Framework for Education," suggests that Moroccan programs rely heavily on rote learning, a memorization technique based on repetition. The World Bank noted that the learning method is inefficient and makes students unable to demonstrate a basic understanding of knowledge translation in everyday life.

7.3.4 Science and Technology Venues and Centers

Science and technology venues and centers in Morocco have significantly developed in recent years. The government has been implementing reforms to encourage scientific research in the Kingdom. scientific research is becoming one of the most national priority in Morocco, the country has a good supply of well-trained high-quality human resources that could transform its research and development sector into a key vehicle for development. The following part lists a number of science and technology centers that exist in Morocco and provides a description of their missions:

The CNRST is an operator whose mission is to promote, develop and enhance scientific research, according to the cultural, economic and social needs of the country and in conjunction with public and private organizations pursuing the same objectives (http://www.cnrst.ma/index.php/fr/).

The Moroccan Institute for Scientific and Technical Information (IMIST) reports to the CNRST in its capacity as the national operator for the promotion of scientific research. The IMIST provides scientific and industrial circles with the information and scientific and technical documentation they need to be at the cutting edge of their activities (https://www.imist.ma/).

The Institut Pasteur of Marocco (IPM) is a public establishment created by Royal Decree No. 176–66 of June 23, 1967, also regulated by Law 17–04 which assimilates it to an industrial pharmaceutical establishment. IPM is a member of the Institut Pasteur International Network and of the International Association Of National Public Health Institutes. The research and teaching department addresses several themes related to human health (bacterial, parasitic, fungal diseases, etc.).

IPM supervises more than 300 interns per year and its researchers have published more than 260 articles in the last five years (http://www.pasteur.ma/).

The Research Institute for Development (IRD), the creation of an IRD representation in Morocco at the beginning of the 2000s gave new impetus with the broadening of the partnership and the implementation of multidisciplinary co-constructed programs which today make it possible to tackle themes that are at the forefront. both strategic with regard to the Sustainable Development Goals (SDGs) and which correspond to the priorities defined by Morocco as part of its "National Strategy for the Development of Research by 2025" (https://maroc.ird.fr/).

7.3.5   Utilizing Emerging Technologies

The cognitive science insists that learning needs to be attentive, engaged, receiving and consolidating information, and giving immediate feedback (Dehaene, 2015). While traditional education does not fulfill these fundamental requirements, it offers the learners an education that does not mesh with a world of exponentially increasing information and changing technology (CSEFRS, 2015). Furthermore, school and university learners no longer want to learn only by reading books and copying texts. Today's learners want to get benefit from the power of technology in their classrooms and to have access to a modernized education without the lack of advanced technologies. In the two last decades, the educational field is very actively impacted by technology and digitalization. Besides, adapting technological solutions to education is becoming increasingly popular. As it raises the quality of education to much higher levels and it increases the learner engagement and interest in learning.

The current educational reforms (CSEFRS, 2015, 2019b) are calling educators and teachers to integrate educational technology into their teaching process, in order to enhance the effectiveness and attractiveness of teaching and learning for learners, which has a high impact on the learner's socialization, development, and academic success. Also, it is considered as one of the most effective ways of providing future generations with the necessary skills and competencies needed to succeed in future labor markets.

More recently, the new emerging technologies, and their implementation and support by specialized e-learning platforms, have led to the apparition of new forms of learning that are dramatically changing the traditional roles of both teachers and learners, where learning is no longer about the content transfer but rather about knowledge construction and skills development (Bouroumi & Fajr, 2014).

For about two decades now, teachers from Moroccan universities are conducting research aimed at the integration of the new emerging technologies into education to develop innovative learning solutions for the educational system in Morocco: MOOC (Idrissi et al., 2020), Open educational resources (Zaatri et al., 2020), Gamification (Lamrani & Abdelwahed, 2019), Virtual laboratory (El Kharki, Bensamka, & Berrada, 2020), Artificial intelligence (Karkouch et al., 2017), Augmented reality (Elmqaddem, 2019), Virtual reality (Fahim et al., 2019), and many other emerging technologies that have a main objective which is enhancing and improving science education teaching and learning.

7.4 Requirements for Future Development of Science Education

As we know, education today plays a very important role in developing the human resources that modern society needs. Talking about the school of tomorrow or the jobs for the future is based on new curricula and the need of innovative training courses adapted to the twenty-first century. This is the reason why science education programs must involve the whole community with a vision integrating social, economic, political, psychological, and cultural dimensions (Science Learning Hub, 2020). We are aware that the development of Science has also contributed to the planetary crisis we are facing today but it helps also in growing our economy. So what kind of education can prepare young students for the best future? and what are their impacts on teaching science?

It is well known that the approach taken in science education is a disciplinary one. Today our society faces many complex problems and any resolution undertaken should be in harmony and meet the United Nations objectives on sustainable development. We need to train a citizen capable of finding solutions to problems whose complexity could be transversal and not unidimensional. This citizen needs to be trained in trans-disciplinarily and to accept the complexity of today's world. Hence the need to reform the curricula and introduce the history and philosophy of science, which would help this citizen to see that science is a way of making sense of the world in which he lives.

Another not insignificant aspect needs to be explored and concerns the ethical dimension of science teaching. Today, we produce knowledge and development through science, but we also need to know that science is not a process devoid of values. Learning ethics in school and analyzing the problems of today's world from several angles could be a way of helping the citizen to develop more complex understandings. One of the successful ethics experiments was conducted by the National Academy of Sciences in its integrated and consolidated MENA-level program called "Teaching Responsible Science" (Husbands, 2014) where several modules are delivered to raise awareness of the role of ethics and professional conduct in the area of Scientific integrity.

We must also create programs to sensitize the citizen on the civil responsibility to take actions and measures in favor of society. Give him hope that any planetary crisis is manageable and it is his duty to be an agent of change. We also need to think about science education programs that put the student with the natural and real world much more than the world of abstract science in order to help him protect the planet, and contribute to the development of a society more truly bicultural.

For African countries rethinking Science education stay a key to solve many problems this continent will be facing in the future. This aspect should be widely supported by government of those countries in the respect of all recommendations as mentioned below:

- Science education should allow independence for citizens.
- Science education should contribute to the development of citizens and society.
- Science education must be more practical, efficient and responsive. He must help the citizen to change.
- Learners should develop a mindset associated with research into any action undertaken and with discovery rather than memorizing facts.
- Science has a duty to inform the public and promote their interests.
- Technological knowledge should be adapted to the society and its needs.

7.5 Discussion and Conclusion

Improving science education is needed and has received considerable attention around the world. The challenges, so great in developed countries, are even greater in developing countries, where there is often a lack of well-trained teachers, high-performance materials and even the most basic science supplies and equipment.

As it has been emphasized above, Science education are very crucial to the development of countries. Some countries spend a lot of money on development although there are various problems that affect the development of science education. As a conclusion we can say that developing countries have to understand well the problems they are facing and aspects that those problems are having behind. To come over that problems it will be needed to accelerate the development of science education as learner-centered, teacher-assisted, action-oriented, project-based education programs according to previous recommendations. Continuing education teacher's training programs must be effectively improved. For the rapid development of science education, government, society and industry should be in a co-operation and work together.

References

Berrada, K., Channa, R., Outzourhit, A., Azizan, M., & Oueriagli, A. (2014). UNESCO active learning approach in optics and photonics leads to significant change in Morocco. In M. Costa & M. Zghal (Eds.), *12th Education and Training in Optics and Photonics Conference* (Vol. 9289, pp. 196–202). https://doi.org/10.1117/12.2070289

Bouroumi, A., & Fajr, R. (2014). Collaborative and cooperative e-learning in higher education in Morocco: A case study. *International Journal of Emerging Technologies in Learning, 9*(1). https://doi.org/10.3991/ijet.v9i1.3065

CNRST. (2018). Programme de Bourses d'Excellence de Recherche PBER. Retrieved from https://www.cnrst.ma/index.php/fr/financement-de-la-recherche/suivi-et-bilan/bourses-excellence.

Colombo, S. (2011). *Morocco at the Crossroads: seizing the window of opportunity for sustainable development*. Retrieved from https://www.jstor.org/stable/pdf/resrep09864.9.pdf.

Constitutional Court. (1973). The constitution of the Kingdom of Morocco. Retrieved from https://www.cour-constitutionnelle.ma/.

CSE. (2008). *Etat et Perspectives du Systéme d'Education et de Formation - Conseil Supérieur de l'Enseignement*. Retrieved from https://www.csefrs.ma/wp-content/uploads/2009/01/vol4-vf1.pdf.

CSEFRS. (2015). *The strategic vision of the reform (2015–2030) - Higher Council for Education, Training and Scientific Research*. Retrieved from https://www.csefrs.ma/wp-content/upl oads/2017/09/Vision_VF_Fr.pdf.

CSEFRS. (2019a). L'Atlas territorial de l'abandon scolaire. Retrieved from https://www.csefrs.ma/publications/latlas-territorial-de-labandon-scolaire/?lang=fr.

CSEFRS. (2019b). *Law Framework No. 17.51 relating to the system of education, training and scientific research - Higher Council for Education, Training and Scientific Research*. Retrieved from https://www.uiz.ac.ma/sites/default/files/doc/loi-cadre-51-17-AR.pdf.

Dehaene, S. (2015). Experimental Cognitive Psychology. Retrieved from https://www.college-de-france.fr/site/en-stanislas-dehaene/course-2014-2015.htm.

Education Statistics Compendium. (2019).

El Kharki, K., Bensamka, F., & Berrada, K. (2020). Enhancing practical work in physics using virtual javascript simulation and LMS platform. In Burgos, D. (Ed.), *Radical solutions and eLearning* (pp. 131–146). https://doi.org/10.1007/978-981-15-4952-6_9

Elmqaddem, N. (2019). Augmented reality and virtual reality in education. Myth or reality? *International Journal of Emerging Technologies in Learning (IJET), 14*(03), 234–242. https://doi.org/10.3991/ijet.v14i03.9289

Fahim, M., Ouchao, B., Jakimi, A., & El Bermi, L. (2019). Application of a non-immersive VR, IoT based approach to help moroccan students carry out practical activities in a personal learning style. *Future Internet, 11*(1), 11. https://doi.org/10.3390/fi11010011

Fanack. (2020). Economy of Morocco. Retrieved from https://fanack.com/morocco/economy/.

HCP. (2020). Population clock. Retrieved from High Commission for Planning of the Kingdom of Morocco website: https://www.hcp.ma/.

Husbands, J. L. (2014). Responsibilities or requirements: Framing dual use issues for scientific engagement. *Frontiers in Public Health, 2*, 107. https://doi.org/10.3389/fpubh.2014.00107

Idrissi, A. J., Berrada, K., Bendaoud, R., Machwate, S., Miraoui, A., & Burgos, D. (2020). Starting MOOCs in African University: The experience of Cadi Ayyad University, process, review, recommendations, and prospects. *IEEE Access, 8*, 17477–17488. https://doi.org/10.1109/ACCESS.2020.2966762

Idrissi, A. N. (2020). *Les TICE au Maroc : entre usage et gestion , c as de l ' e nseignement du français dans le cycle collégial à Agadir ICTE in Morocco : between use and management , case of teaching French in secondary schools in Agadir*, 64–73.

IndexMundi. (2019). Morocco economy profile 2019. Retrieved from https://www.indexmundi.com/morocco/economy_profile.html.

Infomineo. (2017). The Development of Morocco's IT Sector. Retrieved from https://infomineo.com/the-development-of-moroccos-it-sector-2/.

Issam. (2015). Programme de Coopération entre le Maroc et la Communauté Française de Belgique. Retrieved from https://www.enssup.gov.ma/sites/default/files/PAGES/482/Appel_projet_MES_03_02_2015.pdf.

Kaaouachi, A. (2020). *Internationalization of Higher Education in Morocco: Progress and Challenges - Eurasia Higher Education Summit, Turkey*. Retrieved from https://eurieeducationsummit.com/wp-content/uploads/2020/08/Abdelali-Kaaouachi-Eurie-2020.pdf.

Karkouch, A., Mousannif, H., & Al Moatassime, H. (2017). A ubiquitous students responses system for connected classrooms. *Proceedings of the Mediterranean Symposium on Smart City Applications*, 528–537. https://doi.org/10.1007/978-3-319-74500-8_49.

Kingdom of Morocco. (2020). The national portal. Retrieved from http://www.maroc.ma/en.

Lahchimi, M. (2015). La réforme de la formation des enseignants au Maroc. *Revue Internationale D'éducation De Sèvres, 69*, 21–26. https://doi.org/10.4000/ries.4402

Lamrani, R., & Abdelwahed, E. H. (2019). Game-based learning and Gamification to improve skills in early years education. *Computer Science and Information Systems, 00*, 43. https://doi.org/10.2298/CSIS190511043L

MEN. (2008). Présentation du Programme NAJAH. Retrieved from http://www.abhatoo.net.ma/maalama-textuelle/developpement-economique-et-social/developpement-social/education-enseignement/politique-de-l-enseignement/pour-un-nouveau-souffle-de-la-reforme-de-l-education-formation-presentation-du-programme-najah.

MEN. (2015). Cooperation Maroco-Française. Retrieved from https://www.enssup.gov.ma/sites/default/files/PAGES/482/TDR_AP-ES-2015.pdf.

MEN. (2018). *Programme Ibn Khaldoun d'appui à la recherche Scientifique dans le domaine des Sciences Humaines et Sociales Appel à Pré-projets de recherche.*

MEN. (2020a). Charte Nationale d'Education et de Formation. Retrieved from https://www.men.gov.ma/Fr/Pages/CNEF.aspx.

MEN. (2020b). Enseignement secondaire collégial. Retrieved from https://www.men.gov.ma/Fr/Pages/Ens-collegial.aspx.

MEN. (2020c). Higher Education in Numbers. Retrieved from https://www.enssup.gov.ma/sites/default/files/STATISTIQUES/5656/Brochuredesstatistiqus2019-2020VF_16092020.pdf.

MEN. (2020d). L'Enseignement préscolaire et primaire. Retrieved from https://www.men.gov.ma/Fr/Pages/enseignement-presco-prim.aspx.

Morchid, N. (2019). The determinants of use and acceptance of mobile assisted language learning: The case of EFL students in Morocco. *Arab World English Journal (AWEJ) Special Issue on CALL*, (5). https://doi.org/10.2139/ssrn.3431747

Nordea Trade Portal. (2020). The economic context of Morocco - Economic and Political Overview. Retrieved from https://www.nordeatrade.com/en/explore-new-market/morocco/economical-context.

Said, A., M'hammed, D. M., Mohamed, K., & Mohamed, T. (2009). Programme GENIE au Maroc : TICE et développement professionnel. Retrieved from http://www.taalimtice.ma/fr/node/225.

Science Learning Hub. (2020). Thinking about science education for the future. Retrieved from https://www.sciencelearn.org.nz/resources/2890-thinking-about-science-education-for-the-future.

Tawil, S., Cerbelle, S., & Alama, A. (2010). *Éducation au Maroc: analyse du secteur.* Retrieved from https://unesdoc.unesco.org/ark:/48223/pf0000189743.

Trading Economics. (2020). Morocco Competitiveness Rank. Retrieved from https://tradingeconomics.com/morocco/competitiveness-rank.

UCA. (2015). Centre d'Etudes, d'Evaluation et de Recherches Pédagogiques (C.E.E.R.P.). Retrieved from https://www.uca.ma/fr/page/ceerp.

Worldometer. (2020). Morocco population (2020). Retrieved from https://www.worldometers.info/world-population/morocco-population/.

Zaatri, I., Margoum, S., Bendaoud, R., El Malti, I. L., Burgos, D., & Berrada, K. (2020). Open educational resources in Morocco. In *Current State of Open Educational Resources in the "Belt and Road" Countries* (pp. 119–134). https://doi.org/10.1007/978-981-15-3040-1_7

Khalid Berrada is professor of physics at Mohammed V University in Rabat since January 2021. He was professor of physics at UCA since 1996 and director of the Centre for Pedagogical Innovation at Cadi Ayyad University (UCA) and UNESCO Chairholder on "Teaching physics by doing". He has been member of many national and international conference and meetings committees. He is also one of the developers of the successful French program of UNESCO Active Learning in Optics and Photonics. He was coordinating the UC@MOOC project created in 2013 at UCA and was director of Trans ERIE research group on educational innovation at UCA (Trans ERIE, 2016–2020). He is leading the Morocco Declaration on Open Education since 2016.

Khadija El Kharki is a PhD student at Cadi Ayyad University (UCA). She is a holder of a Master's Degree in Engineering and Technology of Education and Training. She is developing research on virtual laboratory based on digital simulation with the JavaScript programming language at Trans ERIE group of research of UCA.

Hana Ait Si Ahmad is a PhD student at Cadi Ayyad University (UCA). She is a holder of a Master's Degree in Multimedia and Pedagogical Engineering from High Normal School at UAE. She is developing research on active learning and teaching methods and tools at Trans ERIE group of research of UCA.

Chapter 8
Science Education in Oman

Taking Oman to Better Future

Abdullah K. Ambusaidi⬤, Mohamed A. Shahat⬤, and Ali S. Al Musawi⬤

Abstract The purpose of the present chapter is to give the reader an overview of the current science education research in Oman. The chapter focuses on three main sections as an overview of the country, an overview of the education development, and the current situation of science education. It addresses the emerging technologies in relation to science education including blended education and e-learning developments. The chapter then discusses the requirements for better implementation of science education along with the future strategies to overcome the challenges and issues that facing science teachers in Oman. It concludes by recommending the importance of making greater efforts to instill the twenty-first-century skills in the future graduates of science to enable them to respond to the requirements of the global economic market.

Keywords Science education · Education system · Research and international collaboration · Oman

8.1 Overview of the Country

8.1.1 Geographical Location, Population and Political System

Oman, officially the Sultanate of Oman is a country located in western Asia and is the third largest in terms of area in the Arabian Peninsula. It occupies the southeastern

A. K. Ambusaidi (✉)
Ministry of Education, Postal Code 132, Al-Khoud, Muscat, Oman

M. A. Shahat · A. S. Al Musawi
College of Education, Sultan Qaboos University, Postal Code 123, Al-Khoud, Muscat, Oman
e-mail: m.shahat@squ.edu.om

A. S. Al Musawi
e-mail: asmusawi@squ.edu.om

M. A. Shahat
Aswan University, Aswan, Egypt

part of the Peninsula with an area of 309,500 square kilometers bordered on the west by the Kingdom of Saudi Arabia, from the southwest by the Republic of Yemen, and from the northwest by the United Arab Emirates. Oman enjoys a strategic location as it controls the Strait of Hormoz, the gate to the oil-rich Persian Gulf region. The country is administratively divided into eleven governorates with sixty-one wilaya (states). The climate is hot dry inland and wet hot along the coast where the northern and central regions of the country have the opportunity to rain during the winter season with strong southwest monsoon in the far south (MOI, 2020).

Oman has a long trade history as an important connection station for commercial caravans with cultural and commercial links with Mesopotamia, China and India. It was a powerful empire in the late seventeenth century, vying with the United Kingdom and Portugal for influence under the 'Al-Bu Said Dynasty' in the Persian Gulf region and Indian Ocean. At its peak in the nineteenth century, Omani influence stretched across the Strait of Hormuz into the modern era, Iran and Pakistan and as far south as east African coasts. Historically, its capital city, Muscat, was the main commercial port in the Gulf region and was also among the most important commercial ports in the Indian Ocean.

Following an isolation period of a hundred years, the late Sultan Qaboos bin Said has accessed to power in 1970 and worked on setting up political and economic reforms in the country witnessing a turning shift. Since then it has become a prosperous, secure, and developed country supported by its commercial and maritime history. Its distinctiveness by picturesque nature attracts visitors and tourists from home and abroad. In addition, oil and gas resources and trade have given Oman global attention as a commercial partner (MOI, 2020).

Oman is an Arab country with a monarchial hereditary system. The current sultan is His Majesty Sultan Haitham bin Tariq Al Said, who assumed power on January 11, 2020, after the death of the late Sultan Qaboos bin Said who is well recognized as the founder of the Omani renaissance. The Omani constitution gives the right to vote for every Omani citizen who has reached the age of twenty-one to choose the members of the Shura Council (the lower house of parliament). The Basic Law of the State its amendments and principles state the rights and duties in a balanced and fair independent judiciary system with the separation of executive and legislative powers.

The Omani population has reached about 4.9 million people in 2020, with a population density of about 10.6 people per square kilometer. Vast majority are Omanis, at a rate of 55%, while expatriates (mostly Indian and Pakistani, as well as Europeans) constitute 45% with and the rate of population growth of 5.8% for the last five years. The Omani people have remained basically Arab Muslims with a great majority of the population are those under the age of thirty years. The people mostly educated and acquired academic qualifications to help replacing the semi-skilled expatriate workforce. In addition to working for public and private sectors' jobs, the Omani society includes citizens who live by fishing, farming, seafaring, and livestock. The Omani citizen enjoys good living standards as the Omani economy is currently classified as (emerging economy) and has strong growth potential (MOI, 2020).

8.1.2 Current Situation of Economic, Technologies and Cultural Development

The Sultanate has developed a reasonable economic status. Article 11 of the Omani Basic Law states that *"the national economy is based on justice and the principles of a free economy."* The commercial export of oil began in 1967 and since then, many advanced oil fields have been found. With the new discoveries, Oman has proven reserves of total oil of about 5.5 billion barrels, which is the 24th largest in the world. Oil is extracted and processed by the Petroleum Development Oman with production is estimated at 1,015 million barrels per day of oil in 2016 ranking 21st in the world; equivalent to 7.0% annually of its total proven reserves. Since the decline in oil prices in 2019, Oman has made effective plans to diversify its economy and put more emphasis on other areas of industry, such as tourism. Other sources of income such as agriculture and industry are small in comparison to oil and represent less than 1% of the country's exports, but diversification is seen as a priority in the Omani government. Within this perspective of economic diversification, the Sultanate offers many investment opportunities especially by supporting the legal framework with a number of policies and new strategies. The country's laws and regulations, the government, its regulatory efficiency, and its open markets reflect a high level of choice, protection and high competition in the market. These efforts obtained a worldwide recognition and the Sultanate was ranked 56th in the world and 6th in the Middle East in terms of freedom of the economy in 2015 (MOI, 2020).

Providing technology within the Omani context contributed to many sociocultural and economic changes. However, the advent of more technologies demands to develop a national framework for technology training and awareness. The adoption of the Digital Oman Strategy in March 2003 was one of the steps towards achieving this vision which intends to transforming Oman into a knowledge-based economy, as the strategy is concerned with aspects of create the Omani knowledge society and transform the state infrastructure into an e-government. The Strategy confirms the need for training the citizens and government employees on the basics of information technology with a view to eradicating digital literacy. It emphasizes that school students should be provided with the basics of information technology. The Information Technology Authority was established to realize this strategy. After completing many projects in terms of technological infrastructure and employees' training, the Ministry of Information Technology was recently created to continue efforts in various fields. The Digital Oman Strategy has been updated so that over the next five years, greater attention and focus will be given to areas of empowering society and individuals through advanced training programs in order to obtain approved certificates. It also provides the community especially schools and universities with computers and Internet access at reduced prices. His Majesty Sultan Haitham also issued directives to the government to facilitate its operations, harness modern technologies to serve its daily activities, and focus on providing its services through electronic means (MTC, 2020).

On the educational level, there are many educational institutions that have already adopted electronic delivery of their courses, which supports traditional lectures based on a blended format. For example, Sultan Qaboos University and the of Higher Education, Scientific Research and Innovation are making efforts such as introducing Moodle®, BlackBorad® and other specialized software to facilitate the design of teaching and learning through the Internet. Training workshops and guides to integrate online technologies and how to adapt for instructional purposes were conducted specifically within the given context of emergency remote teaching during the spread of the Coronavirus (COVID 19). Nowadays, interactive electronic classrooms, multimedia labs., e-portals and online libraries link the buildings of the Omani public and private schools, and universities and colleges (Al-Musawi, 2018).

8.2 Overview of the Education Development

8.2.1 Education System and Policy

Education in Oman is one of the most important priorities. Several international and regional organizations such as UNESCO, the World Bank, the Arab Education Office for the Gulf States, and the Arab Organization for Education, Culture, and Science with those efforts. These have been emphasized in particular on the Sultanate's commitment to education for all, the achievement of gender equality in education, and the introduction of modern subjects into the study plan, such as life skills and the development of post-basic education. The curriculum includes concepts and skills related to peace, tolerance, children's rights, human rights, traffic safety, the environment, etc. In addition to the interest in human development and the preparation of national cadres, able to meet the needs of the labor and life market in the twenty-first century (Council of Education-Oman [CoE], 2014).

The public-school system in Oman has undergone noticeable developments and changes (Ambusaidi & Al-Balushi, 2015). Basic Education is divided into two cycles: In Cycle 1 (grades 1 to 4), students of both genders are taught in the same classes. The staff in these schools are exclusively females. In Cycle 2 (grades 5 to 10) male and female students at this level are taught in separate schools. Accordingly, the staff can be either made up of males or females. From grade 1 to grade 8, there is one single textbook including integrated science subjects (biology, chemistry, and physics). These subjects are taught by one science teacher. From Grades 8–9, the science subject is divided into biology, chemistry, and physics.

Post-Basic Education (grades 11 to 12) represents the education after the completion of the second cycle of basic education in terms of teaching students by either male or female instructors. Students study science as separate subjects such as physics, chemistry, and biology (Ambusaidi & Elzain, 2008). Students can choose to study either all three or two science subjects if they aim to major in science-oriented specializations at the university level. The language of instruction in all public schools is

Arabic (Ambusaidi & Al-Shuaili, 2009). After completing grade 12, students receive a certificate in General Diploma in Education. Then, depending on their marks the post-school options will be followed. Students with good marks guarantee a place in colleges and universities inside or outside Oman. Students with low marks get useful jobs at different public and private sectors in Oman (Ambusaidi, 2016).

During the time its existed, the Council of Education in Oman has been undertaken several initiatives, including building a new national education strategy until 2040, preparing plans and studies to restructure its various stages and processes, developing its legislation, plans and programs. The policy encompassed other initiatives and projects through which the Council hopes to achieve a qualitative shift in education in Oman. Besides, it seeks to prepare generations capable of managing the development and development of society and meeting the challenges of the future and meeting its requirements (Council of Education -Oman [CoE], 2017).

8.2.2 Statistics on the National Education

In the report of 2016/17 and 2017/18 for pre-school, it was found that the enrolment rate for 2017/18 decreased from 52.30% in the previous year to 50.20%. The net enrolment decreased to 42% from 44% during the same period. There was no significant disparity in the enrolment rate between males and females. The number of kindergarten schools increased from 265 in 2017/18 to 327 with a growth rate of 23.40% compared to the previous year. Kindergarten teachers reached 2014 in 2017/18 with an Omani employment rate of 90.40%. The total enrolment rate of students was as fellow: 100.60 in (grades 1–4), 102.60 in grades (5–10), and 94.60 grades (11, 12). There are 1125 public schools (276 schools for Cycle 1; 273 schools for Cycle 2; 462 for Continuing schools; and 89 for schools with grades 10–12; 25 for schools with grades 11–12) with 579,024 students and 309 private schools in Oman with 56,040 students. In the 2017/18 year, the total male students were 291,068 while female students were 287,956. The average class size, in public schools, is about 27 students. The teachers' number is 13765 in Cycle 1, 14257 in Cycle 2, 22,036 in Continuing schools, 5000 in grades (10–12), and 1327 in grades (11, 12). This means 1862 male teachers with 32.40% and 38,125 female teachers with 67.60%. Of these numbers, 1,742 are Omani teachers and 5,849 are non-Omani (CoE, 2018). The number of dropout rates was 0.8% in 2010/11, 0.7 in 2011/2012, 1.1% in 2012/13. It was noted that the dropout rate from 2011–2013, in grade 11, was higher in males than that of females. As reported by the National Centre for Statistics and Information (2013), 95,146 students are studying in different Higher Education Institutes (HEIs) inside and outside the country. Males represent 45.3% and females 54.7%. The number of students studying in different HEIs inside the country is 87,615 among which 38,899 (44.4%) students study in public HEIs (Al-Balushi, 2016).

8.2.3   *Educational Research and International Collaboration*

There are different donor agencies for funding educational research applications in Oman. One of these is the Research Council (TRC) (recently emerged with the Ministry of Higher Education) that was established in 2005. TRC is a national research council that funds research projects submitted by researchers in higher education institutions and ministries in the country. These applications could be applied for open research grants or strategic grants (The Research Council [TRC], 2020). The second funding body is the Sultan Qaboos University (SQU). SQU has three systems as His Majesty Grant (HMG), the Internal Grant (IG), and Joint Grant (JG) for funding research projects applied by its faculty members. The HMG is used to support long-term, multidisciplinary strategic research projects of importance to the Sultanate. Researchers are expected to obtain major results and identify antici-pated direct benefits to prospective users. This type of research should generally go along with the Sultanate's long-term planning goals. The HMG provides an annual research budget that reaches half a million Omani Rials (USD 1,250,000). The IG funds small-scale research projects. Its annual budget exceeds often half a million Omani Rials (USD 1,250,000) (Al-Balushi, 2016). The JG funds joint applications to jointly support a research program. This type of research grant is governed by a signed research agreement with the other institution (Sultan Qaboos University [SQU], 2020).

The Ministry of Education in cooperation with international research institutes conducted several studies aimed at identifying the strengths and weaknesses of the education system. The following are among the most notable studies: "Education in Oman: The Drive for Quality," conducted in 2012, in cooperation with the World Bank. In 2013, the Ministry conducted an extensive study entitled "Evaluation of the education system in Oman for 1–12" in cooperation with the New Zealand Associa-tion for Research in Education. The two studies concluded with several proposals and recommendations which the ministry benefited from as starting points for achieving quality in education. The Ministry also began the drafting of the document of the general framework for education policy development along with its implementation plans (Oman Educational Portal, 2020). Besides, science education researchers have conducted several collaborative research projects with their peers in the US, UK, Australia, New Zealand, and other Arab countries (Al-Balushi & Ambusaidi, 2015).

8.3   Current Situation of Science Education

8.3.1   *Policies and Standards*

Following Oman Vision 2040 education policy is based on comprehensive education, sustainable learning, and scientific research that leads to a community of knowledge and competing national capacities (Oman Vision-2040, 2020). As a Muslim country science education in Oman influence the Arabic context and role of Arab and Muslim scientists in preserving the development of different sciences during the Middle Ages. One of the fundamental principles of science education in Oman is supporting scientific thinking (Al-Balushi, 2016; Al-Balushi & Ambusaidi, 2015). The Ministry is developing a policy for the implementation of comprehensive science education in its broadest sense, where a study has been carried out to identify the challenges facing its implementation (Oman Educational Portal, 2020).

Science education policy and standards are being developed in Oman. The Sultanate seeks, as a policy, to educating science through which the student is prepared for life and work in the knowledge age society. This happens by providing knowledge, skills, and values to students. To this end, the Ministry of Education seeks to build and adopt national science standards of a global nature coming from the cultural and social identity of the Omani human being. Therefore, the standards of science education for grades 1–8 in Oman focus on linking to international standards, and the latest updates of science and technology, focusing on the skills of the twenty-first century (such as creativity, meditation, inquiry, problem solving, teamwork, and appreciation), and basic skills (Information and communication technology, computation, reading, oral communication, writing). Besides, these standards focus on promoting Islamic and Omani values, the role of science and scientists in the service of humanity, interacting with the problems of society, and participating in the sustainable development of Oman.

These standards had their impact on the scope and sequence of science subjects and its related aims for grades 1–10. The scope and sequence in all classes of new science curricula focused mainly on the development of integrated science, the skills of the twenty-first century, and Oman's national and Islamic identity. This is achieved through topics that are logically organized and sequenced, in which theoretical and practical knowledge is combined (MoE, 2020).

8.3.2 Curriculums, Digital Resources and Teacher Training

In Oman, science learning outcomes are designed to help in supporting student gains of the knowledge, skills, and attitudes needed for developing scientific literacy and sustainable development (Mullis, Martin, Goh, & Cotter, 2016). These learning outcomes are based on new science curricula implemented in 2017 in an agreement with Cambridge University Press (Oman Educational Portal, 2020). In addition, Oman has been given considerable attention to the application of Next Generation Science Standards (NGSS), which are linked to the integration of Science, Technology, Engineering, and Mathematics [STEM] education as an interdisciplinary approach based on real-world applications. Considering the Omani context, in 2017 STEM Oman is being implemented, designed by Rolls-Royce UK, as an enrichment education program for students (Oman Educational Portal, 2020). Regarding NGSS (2013), there are eight practices of science and engineering that are essential in K-12 education for all students to learn, and they are: asking questions (for science) and defining problems (for engineering), developing and using models, planning and carrying out investigations, analyzing and interpreting data, using mathematics and computational thinking, constructing explanations (for science) and designing solutions (for engineering), engaging in argument from evidence, obtaining, evaluating, and communicating information (Banko, 2013). The use of science and engineering practices in instruction implies a change in the teacher's role from dispensing content information to encouraging thinking and experimental/problem-solving skills in the student (Ibrahim et al., 2017).

These curricula are directly used in developing scientific inquiry, new ways of thinking, engineering habits of mind, and problem-solving skills especially through hands-on experiments as one of the fundamental principles of education (Al-Balushi, 2016), and assessing the learner's performance in a variety of ways. All these curricula were translated into Arabic. Besides, the reform includes the importance for students to acquire 21st-century skills, which requires the implementation of inquiry and engineering design-based learning to be taught by the teachers (Oman Educational Portal, 2020).

In June 2014, the Specialized Institute for Professional Training of Teachers oversees improving the teachers' quality, promoting their skills, and measuring their performance. It carries out specialized programs aimed at training teachers according to international standards and developing clear mechanisms and plans for training them and monitoring their performance in the field of education (Oman Educational Portal, 2020).

8.3.3   Students' Assessment and Achievement

In the Trend in International Mathematics and Science Study (TIMSS) of 2015, the average science scores of Omani fourth-and eighth-grade students were 431, and 455 respectively, which was significantly lower than the TIMSS average of 500. As a result, Oman ranked 36th and 28th among the 39 and 41 participating countries in both grades (Mullis, Martin, Foy, & Hooper, 2016). One of the weakest points in students' performance in TIMSS was their inability to: (1) apply the knowledge gained to new situations, (2) their failure to understand the shapes and graphs, and the daily natural phenomena (Ambusaidi, Al-Hajri, & Al-Maharooqe, 2020). The Omani Ministry of Education (MOE) attributed this poor result to students' abilities to read and understand the requirements of the questions, and the inability of some science teachers to practice these types of activities that focus on inquiry processes, and instructional pictures and questioning in the classroom (Ambusaidi & Al-Balushi, 2015).

In both grades (4 and 8) the results in TIMSS 2015 and also in the previous test of 2011 showed that only about half of the students got engaged in science lessons and liked science for learning, and more than two-thirds of them valued the importance of science. Besides, in TIMSS 2015, girls' achievement with a mean of 478 was significantly higher than boys with a mean of 433. Besides, in TIMSS 2011, more than seventy percent of science teachers in both grades in Oman felt that they were confident of teaching science, and most of them were either satisfied or somewhat satisfied with their profession. Also, more than seventy percent of science teachers reported that they used science investigations in about half the lessons or more. Most of them related lessons to students' daily lives and engaged students in learning in most lessons. On the other hand, very low emphasis on computer-based instruction was reported. Also, most science teachers felt that their science instruction was limited by students lacking prerequisite knowledge and skills (Al-Balushi, 2016; Martin, Mullis, Foy, & Stanco, 2012).

The Omani MoE (2012) clarified that one of the weakest points in students' performance in TIMSS was their inability to apply the knowledge gained to new situations, as well as their failure to understand the shapes and graphs, as well as the daily natural phenomena (Ambusaidi et al., 2020). To overcome these low TIMSS results, the Ministry of Education has experienced huge changes in its education system, including reforms in curricula and teaching/learning processes. One of these reforms includes developing science and mathematics curricula by signing, in 2017, an agreement with Cambridge University Press (Oman Educational Portal, 2020). These curricula are directly used in developing scientific inquiry, problem-solving skills, and assessing the learner's performance in a variety of ways. Accordingly, teachers can express their knowledge and skills in ways that fit their abilities and learning patterns (Al-Balushi, 2019). All these curricula were translated into Arabic. Besides, the reform includes the importance for students to acquire 21st-century skills, which requires the implementation of inquiry-based learning to be taught by the teachers (Oman Educational Portal, 2020).

Assessment practices have seen revolutionary changes with the introduction of the new curricula of Cambridge. Less emphasis has been put on paper-and-pencil examinations. A combination of formative and summative assessment strategies has been formally enforced by the science curriculum. Formative assessment techniques focus on problem-solving skills and inquiry processes (Al-Balushi, 2016; Alquraan & Al-Shaqsi, 2019).

8.3.4 Outreach Science Education

The Ministry of Education is putting a lot of effort to provide good science education for Omani citizens. At the formal level, the ministry is working to provide schools with all the facilities they need for effective teaching of science. In this regard, well-equipped laboratories have been established in each school. At high schools (grades 11–12), there is a lab for each subject (biology, chemistry, and physics). Some of these labs equipped with new technologies in teaching such as an interactive white-board. In addition, learning resource centers have been established in each of the Sultanate's schools. These centers contain many facilities including computers, software, and various resources in the field of science teaching. In conjunction with the establishment of these centers, a subject of information technology was introduced from the first grade to grade 10 in order to help students acquire technical skills that help them search for various sources of knowledge. Furthermore, the ministry established the Scientific Exploration Centers, which are currently operated in six educational governorates. These centers have different halls that provide an opportunity for students to practice different aspects of science and technology. Examples of these halls are the Geology Hall, which is an integrated geological environment that shows the types of rocks and their locations in different environments. The technology hall which includes various tools in the field of robotics and a group of interactive scientific devices that shows the mechanism of computer work, the Internet, and other scientific devices that simulate virtual reality.

Considering the spread of the Coronavirus (COVID 19), the Sultanate adopted blended learning that provides a practical and useful solution for students in teaching various subjects, including science. In this type of teaching, science teachers teach scientific subjects as face to face on some days and through E-platform on other days from home. These platforms, such as the Google Classroom, have many educational opportunities for an interactive environment with each other and with teachers. Science teachers will be trained to use such platforms. Besides, students will be provided with simulation games, videos, animations, and other electronic materials.

At the informal level, many activities have been run by the ministry in collaboration with public, private, and non-governmental sectors. The first activity is the "Oman Science Festival" which is targeting students at different levels; schools, colleges, and universities and it aims to deliver science to students in particular and society in general by easy means. Moreover, to stimulate methods for creative thinking, finding a positive direction towards science, innovation, and scientific research, and encouraging students to realize the importance of science in life. More than 1,200 participants participated in the Oman Science Festival in its second edition, representing 60 institutions from the government, military, and private sector institutions. The exhibition witnessed wide participation from international organizations such as the International Atomic Energy Agency (IAEA), the European Organization for Nuclear Research (CERN), the World Intellectual Property Organization (WIPO), and the Rolls Royce Foundation.

The second intuitive is "The National Science, Technology, Engineering, and Mathematics Week (STEM)" which aims to simplify science and increasing awareness of technology and innovation. It is also a platform to motivate and support teachers and students, those interested in the field of STEM. There is a wide range of participation from public and private sector institutions to conduct activities related to STEM. Some of these organizations run full training packages for students in STEM during summer time.

Related to STEM education, the ministry in collaboration with the British Rolls Royce Foundation establish an enrichment program called STEM OMAN (Ministry of Education, 2019). The program is implemented in Omani schools for three years. It is targeting grade 10 students. The implementation of the program in these schools is accompanied by equipping the rooms with the equipment and tools necessary for its implementation, as they have been configured to be important incubators for STEM learning (Alkharusi, 2020; Alkharusi, Ambusaidi, & Alkharusi, 2020).

8.3.5 Utilizing Emerging Technologies

Reliance on modern technology which is a must for science education in the twenty-first century, the Ministry of Education adopted various educational initiatives and projects and started to introduce some ideas related to virtual and augmented reality. The introduction of this technology in teaching requires the availability of appropriate equipment and tools. The Ministry has worked to provide all 1186 schools with the interactive whiteboard through its available resources or in cooperation with the private sector. A science teacher can take advantage of these boards to present some of his/her lessons using virtual reality. As for augmented reality, the Ministry hopes in the near future that schools will be provided with tablets that can be used to provide this type of e-learning. A committee is formed recently in the Ministry of Education to move towards more digitalization in education including teaching and learning. Research projects were conducted on methods of integrating three types of a virtual lab in science instruction at basic education schools and proved that this technology was positively received by students and improved their academic achievement (Ambusaidi et al., 2018; Al-Amri et al., 2020).

In light of the spread of the Coronavirus (COVID 19), the Sultanate adopted blended education that provides a practical and useful solution for students in teaching various subjects, including science teaching. In this type of teaching, science teachers teach scientific subjects as face to face in some days and through E-platform on other days from home. These platforms, such as the Google Classroom platform, have many educational opportunities for an interactive environment with each other and with teachers.

8.4 Requirements for Future Development of Science Education

Considering the current status of science education in Oman, we will find the relatively high density of content of science subjects (biology, chemistry, physics) in grades 11 and 12. This density obliges the teacher to focus on transferring the content more than developing students' skills and attitudes for learning. The average class size in some classes especially in grades 11–12 is more than 35 students. Some teachers still do not believe in the feasibility of professional development and scientific research or are reluctant to join them for different reasons such as they see them taking time without any reward and influence on their practices in the classroom.

Given the above-mentioned science education situation in Oman, science educators in Oman are required to adapt new applications and innovations for developing science education. This objective can be attained by:

- Providing the students with scientific presentations, simulation and modeling programs, field trips to natural environments, videos, visits to science museums are important to design the implementation of science teaching methods.

- Employing new methodological applications and innovations through learning based on exploration, inquiry, problem solving, and project-based learning to make the instructional processes more effective, interesting, and sensible for all grades.
- Including supportive technology such as instructional games, e-labs, assistive technologies for the disabled students, and smart robots in addition to personal computers and laptops on the micro and macro level of the science education curricular structure.
- Designing more interactive and communication tools such as discussion forums, mobile/ubiquitous learning, and social media networks in science instruction to explain the scientific concepts through students' imagination, abstract perceptions, or interactions.
- Providing professional development programs/sessions to foster the professional knowledge and skills of science teachers to improve their performance level in the classroom.
- Developing a science teacher's culture as a researcher is a fundamental pillar of any development of the educational process and the current trend internally linking the practice of teaching with evidence.

8.5   Discussion and Conclusion

It seems that science education in Oman needs a close look to restructure its curriculum and develop its tools for future generations. There are a number of challenges facing science education (Ambusaidi, 2016; Ambusaidi & Al-Balushi, 2015). The following is a discussion of these challenges and issues and future pathways to overcome them.

The first challenge is related to the density of content of science subjects (biology, chemistry, physics) in Grades 11 and 12, which hinders the teacher's employment of teaching methods such as inquiry, problem solving, and project-based learning. This is because teacher focus will be on covering the content more than on developing skills and attitudes.

The second challenge is related to the number of students per class which is more than 35 students in some classes especially in grades 11–12. The results of this will let teachers use traditional teaching methods instead of inquiry-based learning or Project-based learning.

The third challenge is related to the achievement level of students in science. This challenge is one of the major challenges facing the teaching of science in Oman, especially among male students. Local and international studies such as TIMSS indicate a decline in the performance of Omani students. In order to overcome this challenge, an understanding of the school's culture and the teacher's practice in the classroom should be done, not just only overcome it by introducing new curricula or teacher training.

The fourth challenge is related to the development of the teachers' skills in assessment. The Teacher is the one who manages the classroom situation, and if this management is successful, then there will be effective learning. The assessment process is related to the teacher's ability to use it in order to improve the learning process. Therefore, besides developing the teaching skills of the science teachers, attention must also be paid to developing their skills in the assessment process which allow them to use the assessment to improve teaching.

The fifth challenge is related to the professional development of science teachers. Professional development is an integral part of the teaching process as it will improve the level of teacher performance in the classroom. Despite the availability of supportive programs for the Omani science teachers, including short- and long-term courses offered by the Specialized Institute for Professional Training for Teachers, some teachers still do not believe in the feasibility of professional development or are reluctant to join it for different reasons such as they see them taking time without any reward.

The sixth and final challenge is developing a culture of the teacher as a researcher. Scientific research is a fundamental pillar of any development of the educational process and the current trend internally linking the practice of teaching with evidence. The Ministry of Education is moving towards empowering teachers as a researcher by involving them in training courses in scientific research especially action type of research. The challenge comes when teachers use it to improve their practices to the extent to which they actually do so even if they are trained, and the extent to which some of them are convinced of the importance of scientific research in improving practice.

To sum, science education is well established in Oman but needs to continue its steady development by facing this century's challenges building culture, awareness, and positive attitudes towards the use of the above-mentioned solutions innovative approaches, and applications by both the teacher and the student. As Oman is ambitiously looking forward to a future based on a knowledge and information society, science educators need to make greater efforts to instill the twenty-first-century skills in the future graduates to enable them to respond to the requirements of the global economic market.

References

Alkharusi, A. (2020). *Perceptions of teachers and students participating in the STEM OMAN program.* (Unpublished Master Thesis). Sultan Qaboos University. Muscat, Sultanate of Oman.

Alkharusi, A., Ambusaidi A. &, Alkharusi, H. (2020). *Perceptions of teachers and students participating in the "STEM OMAN" program in light of some variables.* Paper Accepted for publication at An-Najah University Journal for Humanitarian Research.

Al-Amri, A., Osman, M., & Al Musawi, A. (2020). The effectiveness of a 3D-virtual reality learning environment (3D-VRLE) on the Omani eighth grade students' achievement and motivation towards physics learning. *International Journal of Emerging Technology Learning, 15*(5), 4–16. https://doi.org/10.3991/ijet.v15i05.11890

Al-Balushi, S. M. (2016). Science education research in Oman: Opportunities, trends, and challenges. In M.-H. Chiu (Ed.), *Science Education Research and Practice in Asia* (pp. 129–153). Singapore: Springer Singapore. https://doi.org/10.1007/978-981-10-0847-4_8

Al-Balushi, S. M. (2019). Science and Math teaching and learning in Oman: Opportunities and challenges. *Paper presented to the 3rd Excellence Conference in Teaching and Learning of Science and Mathematics,* King Saud University, Riyadh, KSA.

Al-Balushi, S. M., & Ambusaidi, A. (2015). Science education research in the Sultanate of Oman. In N. Mansour & S. Al-Shamrani (Eds.), *Cultural perspectives in science education. science education in the Arab Gulf States: Visions, sociocultural contexts and challenges* (pp. 23–47). Rotterdam, the Netherland: Sense Publishers.

Al-Musawi A.S. (2018). *Oman.* In: Weber A., Hamlaoui S. (Eds.), E-learning in the Middle East and North Africa (MENA) Cham: Springer. https://doi.org/10.1007/978-3-319-68999-9_13

Alquraan, M. F., & Al-Shaqsi, A. A. (2019). Math and science post-basic education school teachers' use of assessment for learning and assessment of learning practices in Oman. *Journal of Educational and Psychological Studies [JEPS],* *13*(4), 615. https://doi.org/10.24200/jeps.vol13iss4pp6 15-627

Ambusaidi, A. (2016). Teaching science in the Sultanate of Oman. In B. Vlaardingerbroek & N. Taylor (Eds.), *Teacher quality in upper secondary science education: International perspectives* (pp. 119–130). Palgrave Macmillan.

Ambusaidi, A., & Al-Balushi, S. (2015). Science education in the Sultanate of Oman: Current status and reform. In N. Mansour & S. Al-Shamrani (Eds.), *Cultural Perspectives in Science Education. Science education in the Arab Gulf States: Visions, sociocultural contexts and challenges* (pp. 189–204). Rotterdam: SensePublishers.

Ambusaidi, A., Al-Hajri, F., & Al-Maharooqe, M. (2020). The difference between reality and desirability in science teachers' pedagogical content knowledge as perceived by their students. *Cypriot Journal of Educational Science, 15*(1), 1–19.

Ambusaidi, A., Al Musawi, A., & Al-Balushi, S. (2018). The impact of virtual lab learning experiences on 9th grade students' achievement and their attitudes towards science and learning by virtual lab. *Journal of Turkish Science Education, 15,* 13–29. https://doi.org/10.12973/tused. 10227a.

Ambusaidi, A., & Al-Shuaili, A. (2009). Science education development in the Sultanate of Oman. In S. BouJaoude & Z. R. Dagher (Eds.), *Cultural and historical perspectives on science education* (Vol. 3). *The world of science education: Arab states* (pp. 205–219). Rotterdam: Sense Publishers.

Ambusaidi, A., & Elzain, M. (2008). The science curriculum in Omani schools: Past, present and future. In R. K. Coll & N. Taylor (Eds.), *Science education in context: An international examination of the influence of context on science curricula development and implementation* (pp. 85–97). Sense Publishers.

Banko, W. (2013). *Science for the next generation: Preparing for the new standards.* NSTA Press.

Council of Education –Oman. (2017). *Philosophy of education in Oman:* Author.

Council of Education –Oman. (2018). *Oman's annual education report:* Author.

Council of Education-Oman. (2014). *Education in the Sultanate of Oman:* Author.

Ibrahim, A., Aulls, M. W., & Shore, B. M. (2017). Teachers' roles, students' personalities, inquiry learning outcomes, and practices of science and engineering: The development and validation of the McGill attainment value for inquiry engagement survey in STEM disciplines. *International Journal of Science and Mathematics Education, 15*(7), 1195–1215. https://doi.org/10.1007/s10 763-016-9733-y

Martin, M., Mullis, I., Foy, P., & Stanco, G. (2012). *TIMSS 2011 international results in science.* Chestnut Hill, MA: TIMSS & PIRLS International Study Center.

Ministry of Education-Oman (MOE). (2019). *Report on the progress of STEMOMAN program.* Muscat, Ministry of Education.

Ministry of Education-Oman (MOE). (2020). *The range and sequence matrix of science (grades 1–8).* Muscat, Author.

MOI- Ministry of Information. (2020). *Oman.* Retrieved from https://www.omaninfo.om/.

MTC- Ministry of Transport, Communications and Information Technology. (2020). *E-Oman Strategy*. Retrieved from https://www.mtc.gov.om/.

Mullis, I. V. S., Martin, M. O., Goh, S., & Cotter, K. (2016). *TIMSS 2015 encyclopedia: Education policy and curriculum in mathematics and science*. Retrieved from Boston College, TIMSS & PIRLS International Study Center: http://timssandpirls.bc.edu/timss2015/encyclopedia/.

Mullis, I. V. S., Martin, M. O., Foy, P., & Hooper, M. (2016). *TIMSS advanced 2015 international results in advanced mathematics and physics*. Boston College, TIMSS & PIRLS International Study Center.

Oman Educational Portal. (2020). STEM Oman programme. Retrieved from https://home.moe.gov.om/updates/3/show/1057.

Oman Vision-2040. (2020). *Oman vision 2040*. Muscat: Supreme Council of Planning.

Sultan Qaboos University. (2020). *Research grants*: SQU. Retrieved from https://www.squ.edu.om/education/Research.

The Research Council. (2020). *Grants*. Retrieved from https://www.trc.gov.om/trcweb/.

Prof. Abdullah Ambusaidi is a Professor of science education at Sultan Qaboos University previously and currently working as Undersecretary of the Ministry of Education for Education. He earned his PhD in science education from the University of Glasgow, UK, and obtained a master's degree in science education from Warwick University, UK. He taught several courses in the science education field at SQU. He published more than 130 articles in international, regional, and national journals.

Dr. Mohamed A. Shahat is an Assistant Professor of science education at Sultan Qaboos University, and an Associate Professor at Aswan University (AU), Egypt. He earned his PhD in science education from the University of Duisburg-Essen, Germany, and obtained a master's degree in science education from South Valley University, Egypt. He attained a post-doctoral fellowship at the University of Duisburg-Essen. He taught several courses in the science education field at SQU and AU. He published articles in international, regional, and national journals.

Professor Ali S. Al Musawi is a Professor of education technology at Sultan Qaboos University. He earned his PhD in education from the University of Southampton, UK. He taught several courses in the education technology field at SQU. He has conducted and published (184) scientific works such as research, books, articles, and academic reports in which he followed quantitative and qualitative approaches using experimental design, comparative and correlational research, focus groups, case studies, digital ethnography, and meta-analysis methods. He was among the founding members of the Omani Society of Education Technology, the Omani Cultural Club, the Literary Forum, and the Scientific Club.

Chapter 9
Science Education in Palestine

Hope for a Better Future

Jamil Itmazi and Zuheir N. Khlaif

Abstract The aim of this chapter is to give the reader an overview of the real situation, opportunities, trends, and challenges related to science education in Palestine. The Palestinian educational system is struggling from the Israeli occupation and the procedures of the occupation procedures on the ground which negatively influence the development of the Palestinian infrastructure in different aspects of the Palestinian people including the educational system. The Palestinian organizations including the governmental ones and the local community try their best to mitigate the negative influence on education through many initiatives. This chapter was to introduce the actual situation of Science education in the Palestinian territories exploring the potential of emerging technology and utilizing it in the Palestinian educational system especially in science education in both higher education institutions and in K- 12. The Ministry of Education and non-profit organizations as well as the local community are supporting science education in public schools through improving the infrastructure of science labs, providing schools with scientific equipment and teachers' training. The Palestinian policy is to encourage students to learn science through lunching different national competitiveness in science field and participating regional and international exams in science field. Along with these procedures, the Palestinian Ministry of Education encourages teachers to integrate technology in science education especially using new emerging technology such as Artificial Intelligence, Augmented Reality, and Virtual Reality. In addition, training teachers to design technological activities and interactive content.

Keywords Palestinian territories · Science education · Palestinian policies

J. Itmazi (✉)
Head of Information Technology Department, Palestine Ahliya University, West Bank Bethlehem, Palestine

Z. N. Khlaif
Faculty of Educational Sciences and Teachers Training, An Najah National University, Old Street, An Najah, Nablus, Palestine
e-mail: zkhlaif@najah.edu

© The Author(s), under exclusive license to Springer Nature Singapore Pte Ltd. 2022   129
R. Huang et al. (eds.), *Science Education in Countries Along the Belt & Road*,
Lecture Notes in Educational Technology,
https://doi.org/10.1007/978-981-16-6955-2_9

9.1 Overview of the Country

9.1.1 Geographical Location, Population and Political System

Geographical location:

Palestine typically (Called also: historic Palestine) referred to the geographic region located between the Mediterranean Sea and the Jordan River. It was composed of Jewish and Palestinian. It included West Bank (the area west of the Jordan River) and Gaza Strip (along the coast of the Mediterranean Sea) and the West Bank. The people who call this territory home are known as Palestinians.

The occupied territories which include Gaza, the West Bank and East Jerusalem have a surface area of 6,220 km^2. The Israel has a surface area of 21,937.00 km^2. Combined, the areas of West Bank, Gaza strip, Jerusalem and behind the green line make up 28,157 km^2 (CIA, 2020).

The Oslo Accords represented the first direct Palestinian-Israeli peace agreement. This led to the formation of the Palestinian Authority (PA). On the ground, the occupied West Bank was divided into three areas - A, B and C. (Haddad, 2020):

Area A: initially comprised three percent of the West Bank and grew to 18% by 1999. In Area A, the PA controls most affairs.

Area B: represents about 22% of the West Bank. In both areas, while the PA is in charge of education, health and the economy, the Israelis have full control of external security.

Area C: represents 60 percent of the West Bank. Instead, Israel retains total control over all matters, including security, planning and construction.

The territory of historic Palestine is greater than the territory of The PA (the State of Palestine). In order to limit the scope of this chapter, the information and issues are concerning the West Bank (Without Israeli settlements), the Gaza Strip and the city of Jerusalem. See Fig. 9.1: Palestine People and land from 1946–2010 (PNC, 2019).

Population

The country is divided into two physically separated geographic regions, the West Bank and the Gaza Strip. According to the results of the 2017, Population, Housing and Establishments census, the State of Palestine had an estimated population of about 4.78 million people in 2017 with 47% of the population under the age of 18. The data indicates a decline in the total population growth rate between 2007 and 2017 to 27%, with an average annual growth rate of 2.7%. The average size of a Palestinian family is 5.1 persons. Muslims make up 96.5% of the total population. Christians are a minority (46,850), and other religions account for only 1,384 people. (PCPS, 2017).

According to the results of the 2016–2017 Palestinian Strategic Report, issued by the Zaytouna Centre for Studies and Consultations, the number of Palestinians around

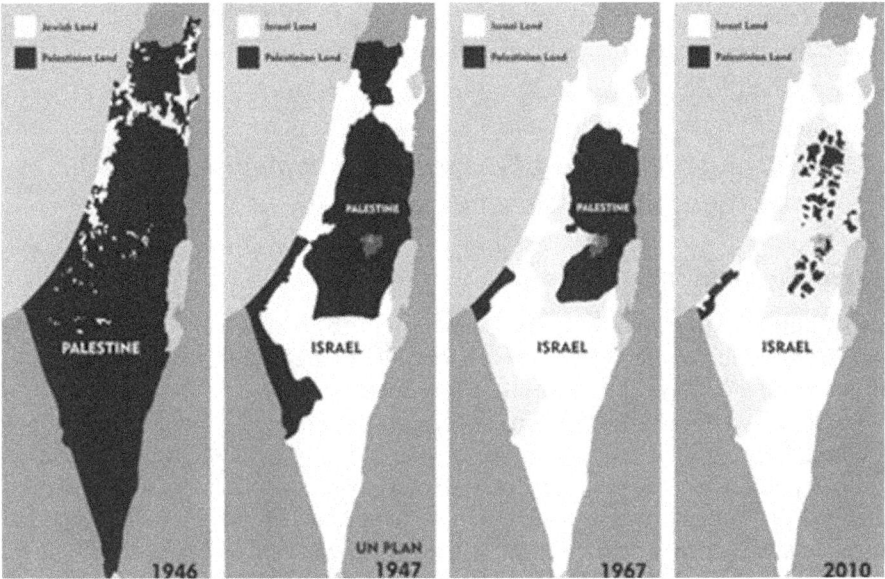

Fig. 9.1 Palestinian loss of land 1946–2010

the world was estimated at 13 million, more than half of them (50.6%) residing in historical Palestine (Israel and Palestine). The rest live abroad, about 75% of them in neighboring countries including Jordan, Lebanon, Syria and Egypt. Palestine is a member of the League of Arab States, and Arabic language is the dominant language. (Zaytouna Centre, 2018).

Political System

The Palestinian Political System is complicated. The Declaration of Principles on Interim Self Government (DoP) signed between the Palestine Liberation Organization (PLO) and the Government of occupation in 1993 led to creation of PA as an interim administration to govern non-contiguous or weakly connected parts of the West Bank and the Gaza Strip.

Nowadays, the political System of PA come theoretically within the framework of a semi-presidential multi-party republic, with an executive president, a prime minister leading the cabinet, and legislative council. Elections for the PA were held in Palestinian Autonomous areas from 1994 until their transition into the State of Palestine in 2013.

The Palestinian general elections of 1996 were the first elections for the President of the PA and for members of the Palestinian Legislative Council (PLC), the legislative arm of the PNA. The second Palestinian presidential election took place on January 2005 while the second Legislative elections were held in the Palestinian territories in January 2006. The third Legislative and presidential elections were scheduled to be held in 2009 but was postponed until our days and never happened.

The local elections (Municipal elections) were held in 2005, 2012 and last one were held in 2017 to elect local councils (Central Elections Commission, 2020).

9.1.2 Current Situation of Economic, Technologies and Cultural Development

Current Situation of Economic

The Palestinian Ministry of National Economy states four main explanations for the poor performance of the Palestinian economy: border closures, the withholding of tax revenues, labor flow reductions to Israel, and dependency.

Border closures increase transaction costs for business. The Revenue Clearance System has led to the occupation government regularly withholding tax revenues, often for politically motivated reasons. Labour flow reductions have led to a smaller number of Palestinians being able to commute to Israel for work due to increased movement and security restrictions. Finally, the dependency of Palestinian trade on Israeli whims has artificially inflated prices and reduced competitiveness (World Bank, 2013).

Table 9.1 show the most significant outputs of economic forecasts based on the assumed scenarios in PA for 2019 (PCBS, 2019):

In the World Bank report; following a fiscal crisis in 2019, the Palestinian economy was projected to slowly recover in 2020. However, the COVID-19 outbreak seems to be largely weighing on economic activity. Living conditions are difficult with a quarter of the labor force unemployed and 24% of Palestinians living below the US$5.5 (2011 Purchasing power parity -PPP) a day even prior to the recent outbreak. A larger than expected decline in aid and a further spread of the COVID-19 virus

Table 9.1 The most significant outputs of economic forecasts based on the assumed scenarios

Macroeconomic indicators	Base line scenario	Optimistic scenario	Pessimistic scenario
Value in Million Dollars			
Gross Domestic Product (GDP)	13,840.3	14,401.4	13,396.4
GDP per capita (Dollars)	2,979.2	3,100.0	2,883.7
Gross National Income	15,447.6	16,283.0	14,839.9
Gross National Disposable Income	16,946.6	18,081.8	15,739.3
Real Average Daily Wage per employee (Dollars)	22.3	22.5	22.2
Unemployment Rate (%)	31.5	29.3	32.7

Data excludes those parts of Jerusalem governorate which were annexed by Israel in 1967. Data at constant prices with 2015 as the base year

pose significant downside risks. Growth in 2019 is estimated to have been weak, reaching 0.9%, as economic activity was disrupted for most of the year due to the fiscal crisis. Going forward, measures put in place by the PA since early March 2020 to halt the spread of the COVID-19 outbreak, while effective in limiting the spread of the virus, seem to have resulted in disruptions in economic activity, especially in the West Bank. As a result, the economy is expected to contract by 2.5% in 2020. There are significant risks to the outlook. If the COVID-19 outbreak is not controlled soon, its impact on economic activity and livelihoods is going to be severe. Under this scenario, drastic measures are expected to be put in place including a complete lock down and a ban on movement between cities. Private sector representatives have also announced plans to cut pay by 50%. The largest impact would be through a decision to stop the entrance of more than 140,000 Palestinian workers to Israel, as those workers and their families account for a third of private consumption given their higher take-home pay. A sharp reduction in the number of tourists is also expected to contribute negatively. The lack of policy tools, such as fiscal stimulus, liquidity injections, or external borrowing could leave the economy vulnerable. The impact on consumption could be severe and the economy could contract by more than 7% (World Bank, 2020).

Current Situation of Technologies

In Palestine, Israel has controlled the Palestinian ICT infrastructure hindering its development and Palestinians' digital rights including recent mass surveillance and monitoring of Palestinian content online. More than a decade after the Palestinians made the initial request for the release of the Third Generation (3G) frequencies; the service became available for Palestinian customers in the West Bank in early 2018. Israel's control of the ICT infrastructure and flow of information has allowed it to limit and violate Palestinians' digital rights, specifically the rights to access the internet, privacy and freedom of expression" (7amleh, 2018).

The Israeli occupation hinders technological progress. It impedes the import of equipment and hardware, controls frequencies and prevents Palestinian ICT companies form offer 4th generation of mobile networks as well as 5G. In general, Palestine needs to improve the ICT infrastructure; they need to have rights to operate 4th generation of mobile networks as well as 5G.

Current Situation of Cultural Development

The archaeological history of Palestine is very rich, with approximately 12,000 sites in the West Bank, including East Jerusalem, and the Gaza Strip. As home to three monotheistic faiths (Islam, Christianity and Judaism) that consider the city of Jerusalem a holy city, Palestine is the site of many holy shrines and a unique architecture (Fanack, 2020).

Palestinians, no doubt, exhibit one of the most resilient and enduring cultures in the world. Despite the incredible pressure of occupation, this society has resolutely refused to let go of tradition, a sense of humor, and a passion for community. Nevertheless, despite this incredible resilience, the threats to Palestinian cultural heritage and identity are ever increasing. The split between Gaza and the West Bank, checkpoints, and the barrier are pushing communities further and further away from each other, dividing and further fragmenting Palestine. For many Palestinians, their cultural treasures are too frequently out of reach. The awe-inspiring Old City of Jerusalem is cut off from both Gaza and the West Bank, and therefore inaccessible for the majority of Palestinians who live in these areas. Many historical and natural wonders are located in Area C, under full Israeli control, preventing many Palestinians the right to access them, also weakening the sustainable development and effective management of infrastructure and of natural and cultural resources. Under the current blockade, Gaza's treasures are not only isolated from most Palestinians and the rest of the world but also under the threat of destruction from war and economic decline (Roberto & Lodovico, 2017).

It is well known that Palestinian culture is most closely related to those of the neighbors from the Levantine countries such as Jordan, Lebanon and Syria as well as the Arab World. Cultural contributions to the fields of art, literature, music, costume and cuisine express the distinctiveness of the Palestinian experience.

9.2   Overview of the Education Development

9.2.1   Education System and Policy

The related Ministry of Education sector in Palestine called (Ministry of Education–MoE), also sometimes, it called (Ministry of Education and Higher Education–MOEHE) and nowadays there are two related Ministries; MoE and (Ministry of Higher Education and Scientific Research–MOHESR).

The MoE holds responsibility for managing public schools (including regulatory overview of schools run by the private sector), overseeing 67.08% of the total number of Palestinian students. United Nations Relief and Works Agency (UNRWA) oversees 24.07%, while the private sector oversees 8.85% of the total general education student population. The educational system in Palestine is a centralized one, in terms of using the same textbooks for all grades and subjects in all schools in the country, and also having the same in-service teacher training programs, and a unified regulations (for public schools) in terms of teacher evaluation forms, teacher recruitment and upgrading procedures, and matriculation examinations. UNRWA schools, and despite that they are using the "centralized" national curriculum. They have their own system administration procedures: UNRWA schools in west bank are controlled by UNRWA education office in Jerusalem, while UNRWA schools in Gaza are controlled by UNRWA education office in Gaza, both west bank and Gaza offices are controlled by the UNRWA regional office in Amman – Jordan. (Mohammed Matar, 2017).

The Palestinian general education system is composed of the following sub-sectors (MOEHE, 2014):

A.   Pre-school education: Pre-school education refers to services for children from 4–6 years for two years. Pre-school services are provided by local and international institutions, with the local private sector services increasing rapidly. Currently, the MoE only indirectly oversees this type of education. However, the MoE is directing its policies towards establishing pre-school education in public schools as part of this education strategy.

B.   Primary Education: It includes grades 1–10. Basic education ranges from Grade 1–10 and is compulsory. It is divided into two levels the lower basic stage of grades 1–4 and upper basic stage (empowerment) including grades 5–10.

C.   Secondary Education: It consists of academic and vocational education for grades 11–12. Vocational education includes four streams: commerce, agriculture, industry and tourism. Academic education includes the science and humanities stream.

D.   Non-formal Education (NFE): The MoE grants licenses for NFE centers according to specific conditions. MoE provides two non-formal education programs: parallel education program provided to dropouts who had completed 5–6 years of basic education, and literacy program and adult education, provided for those over the age of 15 who are not proficient in reading and writing.

The main resource for the governmental policies that were initiated to further enhance education in PA are stated in the "Education Sector Strategic Plan 2017–2022" issued in 2017 from Palestinian MOEHE as well as "Summary of Education Development Strategic Plan EDSP 2014–2019. Palestine 2020: A Learning Nation. Directorate General of Planning" from the same Ministry in 2014.

The three sector goals are as follows (MEHE, 2017; MOEHE, 2014):

1.   Ensuring safe, inclusive, and equitable access to education at all levels.
2.   Developing a student-based teaching and learning pedagogy and environment.
3.   Enhance Accountable Results-Based Leadership, Governance and management.

9.2.2   Statistics on the National Education

The following data do not include the Israeli Municipality and Culture Committee Schools in Jerusalem. The source of data is (PCBS, 2020): The Indicators for education by Level of Education for Scholastic Year: 2017/2018 in PA are listed in Table 9.2.

The number of students, graduates and teaching staff in universities, university colleges and community colleges by sex in 2018/2017 are shown in Table 9.3.

Across all Universities, university colleges and Community Colleges, the number of female students now exceeds the number of males; it is nearly 60% female and 40% male. In contrast with student enrolment rates by gender, academic staff employs significantly fewer female than male professors do.

Table 9.2 The indicators for education by level of education

Pre-school education	Number of kindergartens	1954
	Number of children	150,850
	Average number of students per class	22.7
School education	Number of schools	2998
	Number of students	1,253,238
	Average number of students per class	1,253,238
Higher education	Number of students in universities, university colleges	211,294
	Number of students in community colleges	11,480
	Number of graduates in universities, university colleges	42,773
	Number of graduates in community colleges	2949

Number of Schools Teachers in PA for Scholastic Year: 2018/2019 is 57458 person. The number in Private sector is 7450, in UNRWA is 10004 and in Government sector is 40004 person.

The average number of students per teacher in PA schools for the same year is 22.3

While the Completion rate by Level of Education in PA for year 2018 in Primary level is 97.9, Lower Secondary level is 86.2 and Upper Secondary is 59.3

Note: the indicator was calculated depending on the years of schooling which the person successfully completed. The last grade of primary level is 6th grade, for lower Secondary is 10th, and for upper Secondary is 12th.

Mean Years of Schooling for Persons (15 years and above) in PA for Year 2018 is 10.92; Females is 10.94 and Males is 10.91.

Table 9.4 show the Percentage Distribution of Persons (10 years and Over) by Completed Years of Schooling in 2018:

Table 9.5 show the Literacy Rate of Persons (15 Years and Over) in PA by Age Group in 2018.

Table 9.6 show the Percentage Distribution of Persons (15 Years and Over) by Educational Attainment and Sex in 2018.

9.2.3 Educational Research and International Collaboration

Educational Research

Formally, The Scientific Research Council (SRC) was started in 2002 to formulate research policies within the framework of Palestinian higher education institutions. SRC was re-established in 2013 by a presidential decree restructuring the council's membership. The annual budget allocated for SRC came from the ministry's budget to support scientific research. Nowadays the "Scientific Research" is part of Palestinian

Table 9.3 The number of students, graduates and teaching staff in Palestinian institutions of higher education (HEIs)

Institution	Teaching staff			Graduates			Students		
	All	Females	Males	All	Females	Males	All	Females	Males
Universities and university colleges[*]	6263	1456	4807	42,773	27,033	15,740	211,294	128,417	82,877
Community colleges	389	95	294	2949	1713	1236	11,480	5870	5610

[*]Data include students and graduates of intermediate diploma, bachelor and graduates studies

Table 9.4 Percentage distribution of persons (10 years and Over) by completed years of schooling

Completed Years of schooling	Both sexes	Females	Males
0	2.5	3.7	1.3
1–6	17.5	16.9	18.1
7–9	22.4	20.3	24.6
10–12	32.4	32.0	32.7
13 +	25.2	27.1	23.3
Total	100.0	100.0	100.0

Table 9.5 Literacy rate of persons (15 Years and Over)

Age group	Literacy rate
15–19	99.3
20–24	99.4
25–34	99.2
35–44	99.1
45–54	91.0
55–64	*
65 +	*
Total	97.2

Table 9.6 Percentage distribution of persons (15 Years and Over) by educational attainment

Educational attainment	Both sexes	Females	Males
Illiterate	2.8	4.3	1.3
Can read and write	5.2	5.3	5.2
Elementary	12.8	10.9	14.7
Preparatory	37.3	34.6	39.9
Secondary	21.3	22.8	19.8
Associate diploma	5.6	5.6	5.6
Bachelor and above	15.0	16.5	13.5
Total	100.0	100.0	100.0

Ministry of Higher Education and Scientific Research which may indicate for government interest. Despite the economic and political impediments during the Palestinian recent history and up to date, several academic institutions, non-governmental organizations and independent research centers were established.

The diverse research framework in Palestine includes both public and private universities, along with research centers. Research centers are either university-based, or civil society (registered at the Ministry of Information or at the Ministry of Specialization). In recent years, the number of researches centers in universities increased to 80 centers and laboratories in total. Public research carried out by them often focus on specific fields: water and environment (5 universities), biotechnology,

ICT, and renewable energy (3 universities). In particular, while Palestinian universities show favourable environment for research; research productivities is still inefficient. Academic staff can be promoted according to universities' regulations. The promotion regulations in all universities require academic staff to perform research and publish in acknowledged journals. Although most research done at universities is hence related to academic promotions, research productivity is still weak due to the heavy teaching loads, and the tendency to grant teaching overtimes to teaching staff due to financial constraints, and lack of proper funding for research activities as well. Not to mention the use of part-time contracts has risen in last decade … Furthermore, typical research in Palestine is widely produced either to obtain a degree or promotion, leading universities to look at research and knowledge production as more of a step to upgrade one's rank. (Isaac & others, 2019).

There were 4,533 full time equivalent researchers in Palestine in 2013 of whom 1,023 were females, which translates to 655 researchers per million inhabitants (PCBS, 2014).

Recent MOEHE numbers indicate the total number of academic researchers in Palestine for the year 2017/2018 is 653, which translate to almost 150 researcher per million inhabitants. In North America and Europe there are 3,500 and 2,500 researchers per million inhabitants respectively, while this figure in the Arab world ranges between 200–250 researchers per million inhabitants. Currently, there are more than 150 research centers registered at the Ministry of Information, but most of them are not functional. Only four research centers are registered by MOEHE, namely the Applied Research Institute-Jerusalem (ARIJ), the Palestinian Economic Policy Research Institute (MAS), the Palestinian Center for Policy Research and Strategic studies (MASARAT) and a newly established Palestine Institute for National Security Studies. (Isaac & others, 2019).

The research production by civil society mostly specialize in development, advocacy, and cooperative efforts. In Palestine the research production is rather marginalized when it comes to university affiliated institutions (constituting 10% of research output), while the majority of organizations conducting research are NGOs. Moreover, 80% of social sciences research is produced through research centers or consultative agencies outside the university settings (Qumsiya & Isaac, 2012).

The Center for Applied Research in Education - CARE (www.care-palestine.com) is a registered nonprofit organization, established in 1989, that aims to promote democratic ideals, develop a more robust civil society, foster respect for different cultures, and help to create a more humane world for all through education. CARE has spearheaded numerous programs and projects in cooperation with local partners and international institutions.

The Palestinian International Cooperation Agency (PICA) established by a presidential decree on January 2016, is the main public diplomacy tool of the State of Palestine and works as a national coordinator for South-South and North–South Cooperation. PICA transfers Palestinian know-how to countries of the South, in the form of development and technical assistance.

The creation of PICA "is guided by a national responsibility towards the international community and is an expression of gratitude and appreciation for the support given to the State of Palestine over the years. Guided by a vision of solidarity through development, the State of Palestine seeks to mobilize its skilled human resources to provide technical assistance to support partner countries in responding to crisis and tackling development challenges with the aim of promoting South Cooperation and realizing the 2030 Agenda and its 17 Sustainable Development Goals. To work towards this vision, PICA was created as the development cooperation arm of Palestinian foreign policy with specific attention given to support developing countries across and beyond the Arab region" (PICA, 2020).

According to PCBS, total expenditure on R&D in 2013 was USD 61.4 million, representing USD 24,641 per full-time equivalent researcher. The governmental sector contributed 56.1% of all R&D expenditure, nongovernmental organizations contributed 20.9%, and HEIs contributed 23.0% of total expenditure on Research and Development. The Percentage of External Funds for R&D is 26.9%. The major outputs of R&D in 2013 were as follows: 72 international awards and 116 local awards, 149 international standard book numbers (ISBN), and nine patents. Research was distributed by field of research as follows: 26.7% studies and consultations, 34.4% basic research, 30.6% applied research, and 8.3% experimental research. The Outputs of Research and Development in Palestine by Type in 2013: National Prizes for Researchers and Inventors is 116, International Prizes for Researchers and Inventors is 72, Patents is 9 and International Standard Book Number (ISBN) is 149. (PCBS, 2014).

The expenditures included both bilateral and multilateral projects, studies and consultancies. In 2018, MOEHE allocated $5.6 million for funding research projects in Palestinian HEIs.

9.3 Current Situation of Science Education

9.3.1 Policies and Standards

In the Palestinian educational system, the science education curriculum has been reformed many times towards providing greater opportunities for the Palestinian students to equip them with the 21 century skills and to realize that principals studied in different grades are meeting the daily requirements. For example, the reform was including redesign the activities in the curriculum to be suitable with the infrastructure in schools such as schools laboratories, the activities were student-centered approach asking students to conduct the activities were related more to the real daily life. Some curriculum reformed through integrating more advance technology especially the science curriculum in high school settings. Various science initiatives have been established to encourage students to study science in the higher education institutions from governmental organizations and non-profit organizations including scholarships, science competitions on the country level as well as the ISEF each year. ISEF was an international competitive encouraged student in 9th and 10th grades to participate in the activities through designing a project in different fields one of them was science. The project could be research, a product to be benefits for the local society. Another initiative to encourage students to study science was STEM summer club. This initiative was launched by the Mistry of Education (MOE, 2018) and the Palestinian universities. STEM club was a summer camp for elementary and middle school settings. The MoE launched the cap cooperating with the local universities. The main objective of these camps were training students to learn science by doing and playing through interactive activities. These summer clubs were for two weeks. It involved 3000 students from different background.

According to the strategic plan of the MOE (2017–2022), integrating technology into science activities, the science curriculum integrated local and relevant technologies and their environmental impact on science concepts. Based on our reviewing the science curriculum in different grades, invaluable information in the technological and scientific activities were included whenever suitable.

The majority of Palestinian science teachers have a degree in science and educational sciences from universities in Palestine and other countries. There are different universities and colleges in Palestine offer bachelor's degree in science (i.e. science education, science, Physics, chemistry, and biology).

9.3.2   Curriculums, Digital Resources and Teacher Training

We can conclude that the proposed curriculum of the last version released at the beginning of the academic year 2020–2021 that the curriculum is designed based on learner-centered based, multidisciplinary to some extent with other topics such as math, technology, and engineering, also it is relevant to national development goals and consistent with the ELESCO standards. The curriculum is flexible to suit the separate needs of first year students in higher education as well as the students in K-12. The development of the science curriculum in designed based on the consequences of the grades and student's level. The curriculum includes textbook, which provides students with basic concepts and its relevant applications from real-world problems, supplementary materials and each class should implement different activities related to the concepts in the science laboratory in the school. The majority of the Palestinian schools have a lab for science and science teachers have 90 min per week to work individually in the science lab to prepare for students' activities as well as to clear the tools of the lab. The MoE in Palestine developed professional development programs for science teacher for training them to implement activities and to integrate technology in teaching science and using new trends in science education such as Virtual reality and simulation in science. Along with the professional development programs, the MOE has established the national center of raining teachers on new educational strategies and on using new technological tools in teaching science.

Many technological initiatives have been launched by organizations and individual to design interactive curriculum for science to tack attention of students and to make teaching science fun by using gamification, learning by doing, and learning by playing. The new trend in the policy of the MoE is using open educational resources (OER) as well as open educational practices (OEP) because both of them can mitigate the challenges facing science education in the unique case of Palestine since it is under occupation for more than 60 years and cannot provide schools with the necessity tools for science education especially for chemistry and Physics. As mentioned previously using open educational repository in teaching science is important to reduce the running costs of teaching science though benefit from national, regional, and international experiments in teaching science (Affouneh & Khlaif, 2020). The Israeli occupation has a negative impact on the educational system in Palestine. The MoE cannot provide schools with the important equipment for science education which prevent schools from obtaining essential materials for science experiments (Traxler et al., 2019).

Providing schools in Palestine with a digital content is a big challenge because science teacher in Palestine are not qualify to design interactive content for science education and recruiting firms to design digital content is expensive which cannot achieve in these days because of economic and financial situation. Therefore, we think the best solution to mitigate these challenges and provide schools with digital content is to use OER and have agreement with International organization to support science education in Palestine and to use free platforms to design e-content.

Moreover, the role of the MoE not only to provide curriculum and instructional materials for teachers but also developing the infrastructure of schools to be suitable to do experiments in the lab through providing the schools with different tools and instruments suitable for the curriculum and the school level.

9.3.3 Student Assessment and Achievement

Different tools have been used in schools for student's assessment including the traditional exams which are weekly, monthly, and at the end of the semester. The MoE participated in regional science exams especially in math, physics, and Chemistry in Jordan. The MoE has established Evaluation center to design the exams and assessment based on the international standards. The achievement of the students in the science international exams was low compared with regional countries, especially in the Arab world. For example, the achievement in PISA and TIMS was at the end of the scale which caused the Palestinian decision makers to stop participating in the assessment.

9.3.4 Science and Technology Venues and Centers

Students in Palestine in both higher education institutions and K-12 were participated in many competitive in the Arab world and international. Most of these activities centered on using technology in Augmented Reality and Virtual reality as well as using Artificial Intelligence in describing scientific phenomena. The purpose of these activities was to encourage students towards science education and the MoE supported students financially to participate in these activities. However, comparing Palestinian participation in regional and international activities is low due lack of financial support and lack of budget for these activities from the higher education institutions and the MoE as well as from the non-profit organizations.

The MoE used various approaches to facilitate outreaching of science education programs through launching formal and informal activities in the directorate of education and in the higher education institutions two times each year. Moreover, the MoE encouraged schools to visit the universities to participate in the science activities at the end of the academic year such as the Open days. On the open day, students organize an exhibition to present samples from their work in science education.

9.3.5 Utilizing Emerging Technologies

With the development of technology and adapting of Emergency Remote Teaching and Learning in Palestine, teachers and schools started to integrate different types of Information Technologies and Communications in teaching science and to use open educational resources in teaching science. Based on my experience while working in the educational technology department in the MoE in Palestine which focused on visiting schools and the science lab, I can confirm that the main emergent technologies were used for teaching and learning science including and not limited (1) simulation through using phet.com and other software especially in physics and chemistry. In addition to virtual labs in schools with high Internet speed as well as visualizations of some activities especially in Biology; (2) computational thinking the MoE conducted different workshops fort raining teaching on using CT in teaching STEM concepts including science education; (3) mobile devices; (4) Virtual reality; (5) robotics in computer science education; (6) Artificial Intelligence in specific private schools; and (7) Augmented reality.

Last year the MOE adapted the best rewards for science teachers who can design a learning object by using augmented reality and Virtual reality. The aim of this competition was to encourage teachers to design their technological activities to use in teaching science. Along with the aforementioned competition, students in middle high schools were invited to participate in a competition on the country level. These activities aimed to encourage students to explore the potential of advance technology in science education and how we can use these technologies to mitigate the challenges of science education in the Palestinian context.

Artificial Intelligence was a new trend in specific context and specific environment which was in higher education institutions and in some private schools were adapted AI in science education and encouraged students to participants in the activities supported by local non-profit organization in Palestine. The competition was to create an application to teach some science concepts or to explain a scientific phenomenon related to the local community and the Palestinian context. The competition enables students to utilize technology in different phases of the project and build strong teams. Utilizing technology by students enhances collaboration work among them and reduces the challenges of travelling to meet their partners and to communicate with international experts from the world.

The MoE utilizing technology in science education through building curriculum and designing interactive content for science education to be attractive for students. Moreover, technology has been used to facilitate science education through communicating with international virtual laboratories especially in the higher education institutions. Using advance technology as learning tools including Artificial Intelligence in higher education and creating virtual reality labs for teaching science in different universities and private schools.

In the age of COVID19, science education in Palestine has been affected by hidden factors such as the digital inequity, digital privacy, and technostress on teachers and students because of telework which is teaching and learning from homes (Khlaif & Salha, 2020). However, the MoE utilize different technologies to mitigate the negative impact of COVID-19 on science education through using interactive platforms for science education such as H5P and Phet. Moreover, training teachers to design technological activities for science education before COVID-19 (Khlaif et al., 2019).

9.4   Requirements for Future Development of Science Education

The complex issues that facing our planet today e.g. Corona pandemic is a challenge to our Science education system for help prepare young people for the complex issues we are now facing and could help students to learn to think in different ways and to be able to support multiple perspectives as well as to deal complexity.

Theme like science education for the future is open the door for discussion the inspiring and challenging ideas. These ideas must follow the common human ethics, Otherwise the uncontrolled development of sciences, promiscuous treatment of nature and Overuse usage of nature resources will bring more and more future pandemics and disasters.

Within Palestinian context, the information from formal resources of the future development of science education in PA are rare; one of these resource is "Education Sector Strategic Plan 2017–2022: An Elaboration of the Education Development Strategic Plan III (2014–2019)" from Palestinian Ministry of Education and Higher Education. In fact this plan is discussing the education sector not the science education in specific.

Previously, we mention the three sector goals, we give some details here.

The New Priorities of the Education Sector Strategic Plan 2017–2022 are: (MEHE, 2017).

Goal 1: ensuring safe, inclusive and equitable access to equal education at all levels of the system.

Priorities of the Strategic Goal 1: The priorities of the first goal can be classified in three areas:

a. Improving enrollment in education at all educational levels and maintaining good enrollment rates:
b. Providing protection and safe access to education at all stages and to all premises:
c. Equity and justice

Goal 2: developing student-centered teaching and learning pedagogy and environment.

Priorities of the Strategic Goal 2: priorities of the second goal can be grouped under four categories:

a. Developing programs, curricula and evaluation systems:
b. Training and qualifying human resources working in the educational system:
c. Adapting education, life and labor market needs
d. Developing capacities in the field of scientific research:

Goal 3: promoting accountability, result-based leadership, good governance and management.

Priorities of the Strategic Goal 3: priorities of the third goal can be categorized in four areas:

a. Developing the legal and legislative framework of the Palestinian educational system:
b. Developing the administrative and governing environment:
c. Partnership, Coordination and Cooperation
d. Education in emergencies and difficult situations

May be it is worth mentioning that The A. M. Qattan Foundation, via the Educational Research and Development Programme (ERDP) which looks forward to improving the quality of education in Palestine and Arab world by means of continuing programmes and projects. One ot their project is The Science Studio.

The Science Studio is a product of a long-term research-led experience driven by inquiry-based teaching approaches in science education. It aims to spark people's curiosity about science, and to make scientific knowledge relevant and attainable within the Palestinian and international human contexts. It aspires to communicate science in a way that achieves engagement, emotional involvement and retention of acquired knowledge, and to offer learning opportunities for students, educators, and the general public. Officially launched in October 2017, The Studio is a five-year pilot project (2017–2021) that will lay the foundations for the establishment of an Interactive Science Centre in Ramallah in the near future. It is one of the projects implemented by A. M. Qattan Foundation, in partnership with the Ramallah Municipality. The Studio is also in partnership with the Exploratorium: The Museum of Science, Art and Human Perception in San Francisco (USA). The Science Studio offers a free and creative space for creators, scientists, artists, engineers, educators and technicians to collectively take part in brainstorming, ideation, design, prototyping and creation of interactive science exhibits directed at the Palestinian public for learning and educational purposes (A. M. Qattan, 2020).

It also worth to mention the "Scientific Societies" which play a significant role in the advancement of teaching, learning and research in various disciplines. The Palestine Academy for Science and Technology (PALAST) has undertaken the initiative to catalyze and support the formation of national disciplinary and multidisciplinary scientific societies and clusters in Palestine. This is done in close coordination with all Palestinian universities and expatriate scientists. Within the first phase of this initiative, supported by Friedrich Naumann Foundation, Palestinian scientific societies in mathematics, physics, chemistry, biology, and Agricultural Sciences were formed. Societies in other sectors are in the pipeline. The Founding Council for each society consists of representatives from each of the relevant academic departments from all Palestinian universities in Jerusalem, West Bank and Gaza. In addition, each council will include a number of expatriate scientists in the given discipline. The Academy will provide the nascent societies with management, organizational, legal, and technical support, in addition to pump priming seed funding to help them take off. Each council will convene a general assembly, consisting of university faculty in the given discipline. (PALAST, 2020).

9.5 Discussion and Conclusion

Science education, literacy and scientific research are main indicators of development. In Palestine, science education continues to rely mainly upon teacher-centered learning approach for delivering material and contents as well as administering assessments. Indeed the Science education in Palestine remains to be traditional where the students' role is restricted by classes based on textbook-driven, and the students' performance is evaluated based on their knowledge of these textbooks.

As Battrawi and Muhtaseb (2013) stated that, the students' interest in science may inhibit as they see it as a rigid topic with minimal hands-on experimentation. Furthermore, there is a general lack of science culture in the Palestinian society. Science literacy is very important and achieving it is a challenge; however, developing informal science learning environments could be effective in enhancing science literacy in Palestine. The Walid and Helen Kattan Science Education Project (WHKSEP) is a Palestinian project aiming at "improving the quality of science education in Palestine's schools and effectively transmitting its value into the wider society", (Qattan 2011).

The rigidity of science education in Palestine remains to be a challenge. PA government have not a solid plan about future development of science education in Palestine, we did not found a really good plan means a plan that has a high chance of success.

The results of the study (Ghani & Saifi, 2012) showed that science fiction was not available through science textbook in general science for grades four and five primary.

The Palestinian educational system show many successful stories about Palestinian students and teachers which indicate a somehow good system, but it need more and more. In general, the governmental, local and non-profit organizations in Palestine are working on development of science education in the Palestinian context since 1993 to equip students to the 21-century skills and to provide students with high quality education in science, but due to the economic and political issues, they failed to succeed in their endeavor. However, we can say they did great work in teacher's professional development, improving the infrastructure in schools, and developing the core curriculum. We still need more work on the digital content and have agreement with international organizations to use OER and OEP. New trends in using emerging technology is started before COVID-19 but it was accelerated to use these trends especially Artificial Intelligence, Augmented Reality, and Virtual Reality.

References

7amleh. (2018). The Arab Center for the Advancement of Social Media, Connection Interrupted: Israel's Control of the Palestinian ICT Infrastructure and Its Impact on Digital Rights, Retrieved July 11, 2019 from https://7amleh.org/wp-content/uploads/2019/01/Report_7amleh_English_f inal.pdf.

A. M. Qattan Foundation. (2020). The Science Studio. Retrieved September 29, 2020, from www. qattanfoundation.org/en/ss.

A.M. Qattan Foundation. (2011). The Walid and Helen Kattan Science Education Project Document Ramallah, Palestine: A.M. Qattan Foundation: http://qattanfoundation.org/en.

Abdel Ghani Hamdi Abdullah Saifi. (2012). Science Fiction in Science Textbook for Four and Five Grade in Palestine. Jamea'a ,16,141–168. Retrived September 29, 2020, from www.researchgate.net/publication/260640549_Science_Fiction_in_Science_Text book_for_Four_and_Five_Grade_in_Palestine.

Affouneh, S., & Khlaif, Z. N. (2020). A hidden dream: Open educational resources. In *Radical Solutions and Open Science* (pp. 57–67). Singapore: Springer.

Battrawi, B., & Muhtaseb, R. (2013). The use of social networks as a tool to increase interest in science and science literacy: A case study of 'Creative Minds' Facebook Page. The 2nd Edition of the International Conference "New Perspectives in Science Education. Florence, Italy. 14–15 March 2013. Retrieved September 09, 2020, from https://conference.pixel-online.net/confer ences/npse2013/common/download/Paper_pdf/043-SPI08-FP-Battrawi-NPSE2013.pdf.

Central Elections Commission. (2020). The Palestinian Central Elections Commission. Previous Election Events. Retrieved September 09, from www.elections.ps/tabid/235/language/en-US/Def ault.aspx.

CIA. (2020). "Middle East: WEST BANK"."Middle East :: GAZA STRIP". The World Fact Book. CIA. Retrieved September 04, 2020 from www.cia.gov/library/publications/the-world-factbook.

Fanack. (2020). Fanack.com, 2020. Governance & Politics of Palestine. Retrieved September 09, 2020 from https://fanack.com/palestine/history-past-to-present/palestinian-leadership.

Haddad, M. (2020). Palestine and Israel: Mapping an annexation. Aljazeera.com. 26 Jun 2020. Retrieved September 09, 2020 from www.aljazeera.com/news/2020/06/26/palestine-and-israel-mapping-an-annexation.

Isaac and others. (2019). Study of Higher Education and Research in Palestine. Jad Isaac, Hatem Jemmali, Belal Fallah, Ferdoos Al-Issa, Abeer Istanbuli, Mazen Abu Qamar and Ala Al Azzeh. August 2019. arij.org. Retrieved September 09, 2020, from: www.arij.org/files/arijadmin/2019/ uni_2019.pdf.

Khlaif, Z., Gok, F., & Kouraïchi, B. (2019). How teachers in middle schools design technology integration activities. *Teaching and Teacher Education, 78*, 141–150.

Khlaif, Z. N., & Salha, S. (2020). The Unanticipated educational challenges of developing countries in Covid-19 crisis: A brief report. *Interdisciplinary Journal of Virtual Learning in Medical Sciences, 11*(2), 130–134.

MEHE. (2017): Education Sector Strategic Plan 2017–2022: An Elaboration of the Education Development Strategic Plan III (2014–2019). Palestinian Ministry of Education and Higher Education. Ramallah, Palestine. Retrieved September 09, 2020 from www.lacs.ps/documentsShow.aspx?ATT_ID=34117.%20.

MOEHE. (2014). Ministry of Education and Higher Education. Summary of Education Development Strategic Plan EDSP 2014–2019. Palestine 2020: A Learning Nation. Directorate General of Planning, MoEHE February 2014. Retrieved September 09, 2020 from: https://planipolis.iiep.unesco.org/sites/planipolis/files/ressources/palestine_education_development_strategic_plan_2014_2019_summary.pdf.

Mohammed Matar. (2017). The State of Accountability in Palestine: Educational Planning with Uncertainty: A State under Military Occupation. a paper was commissioned by the Global Education Monitoring Report as background information to assist in drafting the 2017/8 GEM Report, Accountability in education: Meeting our commitments. Retrieved September 09, 2020, from www.un.org/unispal/wp-content/uploads/2017/11/UNESCOCS_021117.pdf.

PALAST. (2020). Scientific Societies .The Palestine Academy for Science and Technology (PALAST). Retrieved September 29, 2020, from http://palast.ps/index.php/en/what-we-do/science-societies.

PCBS. (2014). Palestinian Central Bureau of Statistics, Survey of Research and Development: Main results Ramallah-Palestine. Retrieved September 09, 2020, from www.pcbs.gov.ps/Downloads/book2075.pdf.

PCBS. (2020). Palestinian Central Bureau of Statistics, Ramallah-Palestine. Retrieved September 20, 2020 from www.pcbs.gov.ps/Portals/_Rainbow/Documents/Education-1994-2018-E.html, www.pcbs.gov.ps/Portals/_Rainbow/Documents/Completion_Rate_2009-2018_E.html, www.pcbs.gov.ps/Portals/_Rainbow/Documents/Education-2000-2018-16E.html.

PCBS. (2019): The Palestinian Central Bureau of Statistics (PCBS). Economic Forecasting - Annual Statistics: The most significant outputs of economic forecasts based on the assumed scenarios in Palestine* for 2019. Retrieved September 09, 2020, from www.pcbs.gov.ps/Portals/_Rainbow/Documents/E-themostforecast-2019.html.

PCPS. (2017). The Palestinian Central Bureau of Statistics. Population, Housing and Establishments census – 2017. Retrieved September 09, 2020 from www.pcbs.gov.ps/Downloads/book2364-1.pdf.

PICA. (2020). The Palestinian International Cooperation Agency. Retrieved September 25, 2020. http://pica.pna.ps.

PNC. (2019). Palestine National Council, Palestine People and land (palestinian-loss-of-land-1946–2010). Retrieved September 09, 2020 from www.palestinepnc.org/en/news/item/14-palestine-people-and-land-palestinian-loss-of-land-1946-2010.

Qumsiya, Mazin and Isaac, Jad.)2012(. Research and Development in the occupied Palestinian territories: challenges and opportunities. *Arab Studies Quarterly, 34*(3),158–172.

Valent, R., & Folin-Calabi, L. (2017). Culture, Development, and Identity. United Nations Development Programme: Programme of Assistance to the Palestinian People. Retrieved September 09, 2020 from www.ps.undp.org/content/papp/en/home/presscenter/articles/2017/07/01/culture-development-and-identity-.html.

Traxler, J., Khaif, Z., Nevill, A., Affouneh, S., Salha, S., Zuhd, A., & Trayek, F. (2019). Living under occupation: Palestinian teachers' experiences and their digital responses. *Research in Learning Technology, 27*.

World Bank. (2013). World Development Indicators 2013. Washington, D.C.: World Bank. Retrieved September 09, 2020 from http://data.worldbank.org/data-catalog/world-development-indicators.

World Bank. (2020). Poverty & Equity and Macroeconomics, Trade & Investment Global Practices. Retrieved September 09, 2020 from http://pubdocs.worldbank.org/en/394981554825501
   362/mpo-pse.pdf.
Zaytouna Centre for Studies and Consultations. (2018). Palestinian Strategic Report 2016–2017.
   Beirut – Lebanese. Retrieved September 09, 2020 from www.alzaytouna.net//arabic/data/attach
   ments/PlsStrRep/STR2016-17/PSR-16-17_Summary.pdf.

Jamil Itmazi is Associate Professor; Dean of Scientific Research at Palestine Ahliya University in Bethlehem - Palestinian. He holds Ph.D. in Computer Sciences, specifically in the development of e-Learning software, from the Granada University, Spain, 2005. Dr. Itmazi is an active
researcher; He has published more than 50 papers in international journals and conferences, in
addition to 13 books, and numerous articles in newspapers. Media have also interviewed him
many times. His research interests include but are not limited to the following areas: e-Learning
systems, Open Educational Resources (OER), Learning Management Systems, Recommendation
Systems and Open-Source Systems. He also enjoys participating in a wide-range of IT-related
projects.

Zuheir Khlaif is a full time lecturer at An Najah National University. Before coming to ANNU,
Khlaif was working on designing an interactive content as well as an instructional designer. Khlaif
published more than 20 articles in international journals in technology integration field, computational thinking, mobile technology, and STEM. These days he is working on different research
projects including factors stressed female in online learning during COVID and Techno-stress
impact on the productivity of professors at the Palestinian universities.

Chapter 10
Science Education in the State of Qatar

Amal Rida Malkawi

Abstract The State of Qatar has taken pioneering steps to reform and develop its educational system to match the education systems in developed countries, and to fulfill the Qatar National Vision 2030. Qatar has made great efforts to elevate science education due to its close relationship with the nation's development. In this regard, it has built schools in accordance with the latest international standards, and provided a rich learning environment for interactive learning. It has developed modern national standards that imitate international standards, provided modern curricula for motivating learning, and teachers training, and created an evaluation system based on modern grounds that supports science teaching according to modern global trends. The State of Qatar has three forces that lead to rapid access to the reform and development processes, namely: an ambitious vision, adequate resources, and political will. The "general framework for the national curriculum" in Qatar reflects a competency-based curriculum. In fact, the science curriculum requires developing five core competencies: research and inquiry, communication, creative and critical thinking, collaboration and participation, and problem solving. Qatar is keen on making its students particpate in international exams, namely: (TIMSS, PISA, PERLS). The findings showed that students' performance in science has improved during their participation from 2007 until 2019. However, their performance level and their global ranking failed to reach the required international level. The most important pillars that Qatar adopts in its plans are both education and scientific research in order to become a modern state with a knowledge-based economy, global competitiveness, and sustainable development. Qatar Foundation for Education, Science and Social Development undertakes many reforms in education, training, occupation, educational aspects such as science and technology. Qatar Science and Technology Park (QSCP) made significant contributions in this field. Qatar University has made major contribution also. Qatar has taken significant steps in the field of digital transformation and using information and communication technology, to provide integrated technology with the highest quality. The State of Qatar has

A. R. Malkawi (✉)
College of Education, Department of Educational Sciences, Qatar University, Doha, Qatar
e-mail: a.malkawi@qu.edu.qa

© The Author(s), under exclusive license to Springer Nature Singapore Pte Ltd. 2022   151
R. Huang et al. (eds.), *Science Education in Countries Along the Belt & Road*,
Lecture Notes in Educational Technology,
https://doi.org/10.1007/978-981-16-6955-2_10

advantages that qualify it to benefit from artificial intelligence data, which is why a comprehensive national strategy for artificial intelligence has been prepared.

Keywords Qatar · Education · Science education · Qatar's national vision 2030 · Science standards · TIMSS

10.1 Introduction

In view of the great importance of the role that education plays in achieving sustainable human development, security, stability, prosperity and progress, as well as ensuring the development of human societies and their joining the ranks of developed countries, the Qatari constitution stipulates that "**Education is one of the basic pillars of social progress. The State shall ensure, foster and promote education.**" Thus, the Ministry of Education and Higher Education supports a large number of projects and initiatives to achieve these goals.

In this chapter we will shed light on the most important projects and initiatives undertaken by the State of Qatar, as an attempt to understand the current situation of science education in the State of Qatar. The chapter included a brief description of the geographical location of the State of Qatar and some statistics related to the population, the political system, and the economic, technological and cultural development of the State of Qatar. In addition, a general overview of the education system in the State of Qatar was given in terms of policies, educational stages, and evaluation methods and methods used in the various stages of education. It also shed light on the State of Qatar's participation in international tests, and the emerging technologies used in the State of Qatar, such as: artificial intelligence, as well as a summary of the most prominent challenges facing the science education system, and future trends.

10.2 Overview of the Country

10.2.1 Geographical Location, Population and Political System

The State of Qatar is located in the Middle East, which is a peninsula located in the center of the western coast of the Arabian Gulf, with an area of 11,521 m^2, and it contains many islands. Qatar is dominated by a desert climate which is hot in summer, mild winters, and rare rains.

The State of Qatar is an Arab, sovereign and independent state. It is an Islamic country that derives its legislative laws from Islamic law. Its official language is Arabic, and English is widely spoken.

The Amir is the Head of State. His person shall be inviolable and he must be respected by all. The Amir is the Commander-in-Chief of the armed forces. He shall supervise the same with the assistance of the Defense Council, which is set under his direct authority. The Amir shall represent the State internally and externally and in all international relations.

The total population of the State of Qatar until the end of October 2020 is estimated to be 2,717,360. The number of males is estimated at 1,960,876 and females are 756,484. This data includes the population of all ages, both Qatari and non-Qatari, who are within the borders of the State of Qatar.

10.2.2   Current Situation of Economic Technologies and Cultural Development

Qatar National Vision 2030 is a development plan to transform Qatar into an advanced society that can achieve sustainable development. This plan's development goals are divided into four central basic elements: economic, social, human and environmental development. The government seeks to meet development goals by developing a strong bureaucratic framework and implementing strategies to address the challenges presented in human development reports.

Effective long-term management of Qatar's hydrocarbon resources is considered a primary component to achieve the sustainable economic development. So, the oil and gas sector plays an essential role in the development of the Qatari economy. Meanwhile, the Qatar National Vision looking to support the growth of non-oil sectors and achieve diversification, in order to transform Qatar's economy into a knowledge-based economy.

The Ministry of Transport and Communications has been working restlessly to transform Qatar to an information society achieving public prosperity and well-being on both social and economic levels. Several projects have been established in this regard, such as the Science and Technology Park in Education City. In this regard, the opening of the Qatar Financial Center is one of the new economic initiatives aimed at attracting international institutions and multinational companies to Qatar. Currently, the Qatari economy is considered as one of the strongest regional economies, despite global and regional challenges, the country is maintaining the rates of economic growth in a balanced way, thus expanding the state's GDP by more than 5% in the first half of 2018 compared to the same period in 2017. Consequently, the State of Qatar is achieving record numbers towards economic sufficiency, through the support it provides to national production, in addition to establishing major projects that support sustainable energy, such as establishing the largest water storage station in the world with a storage capacity of 1500 million gallons, in addition to establishing the largest electricity production station of solar energy in the world.

The Communications and Information Technology sector in Qatar is witnessing a great expansion that seeks to transform Qatar into an advanced society at the social and economic levels, in order to be in line with the national objectives of the Qatar National Vision 2030. For this purpose, the Ministry of Transport and Communications has been working restlessly on improving data access through offering high-speed Internet networks, as well as developing attractive business opportunities for the digital Arabic content.

All residents of Qatar, along with Qatari citizens, contribute to the realization of their transformative vision for the state. The State of Qatar is always seeking to develop and reach the ranks of the developed countries in the world, such as hosting international projects such as the FIFA World Cup Qatar 2022.

Qatar won the FIFA World Cup bid for the football championship that will be held in 2022, thus becoming the first Arab country in the Middle East to host this big event. The matches will be held at international environment-friendly Stadiums that operate with solar powered cooling technology, designed by professional architects from around the world. Its design is characterized by modernity and sophistication, while at the same time reflecting the history and culture of the State of Qatar. Some of them were in the shape of the tent and its name was "Al Bayt", including "Al Rayyan", which reflects the Qatari culture through its wavy exterior, and the "Education City" stadium, which is located in Education City among several world-class universities, and takes the form of a serrated diamond.' Qatar sees this stadium as a symbol of innovation, sustainability and progress over the decades, in line with the Qatar National Vision 2030 and the objectives of both Qatar Foundation and the Supreme Committee of the projects and legacy.

10.3   Overview of the Education Development

10.3.1   Education System and Policy

The State of Qatar follows a policy of compulsory education until the end of middle stage of school. The basic education Qatar consists of three stages which are elementary, preparatory, and secondary. The country provides free education for all Qatari students from pre-school stages till university level. Education in Qatar is compulsory for children between the ages of (6–15) years.

Schools in Qatar can be classified into many types; governmental schools that teach the national curriculum, and international schools that teach international curricula with international standards, such as the American, British, French, German, and others curriculum, as well as private Arab schools that apply the standards of the Qatari curriculum, schools for communities that are partly affiliated with the Ministry of Education and Higher Education, private schools that provide special education, and integration schools and specialized schools for people with special needs and talents.

Educational stages

- Preschool or kindergarten stage, which is an optional stage for children under the age of six.
- The primary stage (elementary stage), which lasts for 6 years, in separate coeducational primary schools
- The preparatory stage (middle school), which lasts 3 years, from the seventh to the ninth grade.
- The secondary stage, which starts from the tenth to the twelfth grade.

Government scholarship program

The State of Qatar has provided a government scholarship program at its own expense for students who want to study for pre-university "diplomas", undergraduate and postgraduate studies. This scholarship is of two types; either within or outside Qatar. The Ministry of Education and Higher Education adopts scholarships for students in cooperation with the Ministry of Administrative Development, Labor and Social Affairs, which coordinates with government agencies for scholarships to prepare the necessary plans and then provide the Ministry of Education with them. The ministry grants scholarship students many privileges, such as guaranteeing the material cost of studying for his children, in addition to that the person on scholarship can accompany his family with him, and a monthly salary is paid to the scholarship student and his family if he is married, and other privileges.

10.3.2 Statistics on the National Education

The State of Qatar is developing a strong educational system, in order to be online with various international standards and achieves the Qatar National Vision 2030. According to the World Bank, education spending in Qatar for the year 2019, which constitutes about 3.3% of GDP, is one of the highest rates in the Middle East region and North Africa. Where bout (5.3 billion dollars) from the state budget was allocated to the education sector, which constitutes about 9.3% of the total expenditures. As a result of the great interest in the education sector in the State of Qatar, it has obtained several international rankings in several areas, including: It was ranked sixth in the world in the field of critical thinking 2019 according to the report issued by the World Economic Forum in Davos, in addition to that it ranked sixth in the world according to the indicators of workforce skills in the future, and ranked eighth according to the Graduate Skills indicator, and 16th place in the Quality of Professional Training indicator.

According to education statistics for the year 2018/2019, the number of government schools were 312, including 71 kindergartens, 115 primaries, 64 preparatory, 58 secondary and 4 specialized high schools, which could accommodate more than 121,552 students, while the number of teachers in public schools has reached 14,103 male and female teachers.

The number of private schools were 800, including all stages. These schools were divided into 58 Arab schools (following the national standards Qatar) and 742 foreign schools (following the standards of Qatar or its countries). Private schools received more than 200,017 students, and the number of teachers in these schools is 13,601 male and female teachers. Whereas, the Ministry compels all private schools to teach Arabic, Islamic education and Qatari history. Economic reports predict that investments in private schools in Qatar will threefold that in 2020, while that the total number of students who are enrolled will increase by about 392,000 students by 2022, with a growth rate of 3.9%. Thus, the total number of public and private schools in the State of Qatar becomes 1112 schools that can receive approximately 321,569 students. While the total number of teachers in these schools were 27,704 as it shown in the following Table 10.1.

Institutions of higher education in Qatar are divided into three types: universities, government colleges, and private sector institutions. There are 19 public higher education institutions, 3 of which are governmental which are (Qatar University, Community College, and Ras Laffan College), and 9 of them are associated with the Qatar Foundation for Education and Science. The number of students in higher education institutions reached 352,444, including 26,737 students in government universities and colleges and 8,507 in private institutions.

, The number of Qatari scholarships increased significantly from 2,000 In the year 2012/3013 In the year 2012/3013 to 5,000 in 2017/2018, where the annual increase rate was 18%. It was noticed that the percentage of scholarships for females is greater than for males, reaching 74%. While the percentage of scholarships inside Qatar was the highest, reaching 78% of the total number of scholarships, and the high percentage of internal scholarships for females, which was 86% of the total number

Table 10.1 Students and teachers by level of education, gender and type of education 2018/2019

Education Type		Total		Private schools[a]		Government schools	
Education Level and Gender		Teachers	Students	Teachers	Students	Teachers	Students
Pre-primary[b]	Males	456	28,171	456	23,912	0	4,259
	Females	4,187	26,296	3,142	21,615	1,045	4,681
Primary	Males	2,677	80,306	1,855	52,997	822	27,309
	Females	10,416	76,797	4,542	47,429	5,874	29,368
Preparatory	Males	2,219	29,491	680	17,094	1,449	13,493
	Females	3,055		1,394	14,664	1,661	14,827
Secondary[c]	Males	2,214	25,489	554	12,081	1,660	13,408
	Females	2,570	24,432	978	10,225	1,592	14,207
otal	Males	7,476	164,553	3,545	106,084	3,931	58,469
	Females	20,228	157,016	10,056	93,933	10,172	63,083
	Total	27,704	321,569	13,601	200,017	14,103	121,552

[a]Include Qatar Foundation Schools
[b]Include nurseries
[c]Include Specialized Secondary

of Qataris, is due. In 2018–2019, the number of scholarships outside Qatar reached 1048 students, including 362 female students, while the number of delegates inside the State of Qatar reached 4,748, of whom 3546 were female.

As for the training program in Qatar, based on the statistics issued by the Planning and Statistics Authority in the State of Qatar 2018/2019, the number of trainees was 471,491 males and females in the various sectors of the State of Qatar. While the number of trainees among teachers was 11,250, including 7,637 female teachers and 3613 teachers. A significant increase was observed in the number of trainees in the various training centers, and the annual rate of training.

10.3.3 Educational Research and International Collaboration

The Qatar Science and Technology Park is one of the most important places to support research, development, innovation and cooperation in the State of Qatar. It aims to meet national needs in scientific research and technological development. This can be achieved by working according to four key elements: energy, environment, health sciences, and communication and information technology.

The Education City, which is built on an area of more than 12 km^2, is a unique environment designed in a way that enhances integration and cooperation between different disciplines and sectors, thus contributing to attracting scientific competencies. It includes a group of leading international and local universities, primary and secondary schools, research centers, policy institutes, and various educational institutions.

The Research, Development and Innovation sector at Qatar Foundation is responsible for planning research, development, innovation, cooperation, local and international partnerships, and seizing new opportunities and confronting challenges. These initiatives have played a major role in promoting the growth of the economy, transferring and exchanging knowledge and experiences, and building local and international partnerships in the fields of research, development and innovation to meet the needs of the State of Qatar in the fields of science, technology and health care.

In addition, In addition, this sector is responsible for transforming national priorities into applicable initiatives and projects, development and cooperation, in order to reach the highest levels of national benefit and global impact. Examples of such projects and initiatives are: Business Acceleration and Incubation Programs, Accelerate Technology Project Development Acceleration Program, Business Incubation Center, Dojo MENA Program, Innovation Coupon, and Qatar Genome Program.

10.4 Current Situation of Science Education

10.4.1 Policies and Standards

The Ministry of Education and Higher Education issued the so-called "The General Framework for the National Education Curriculum for the State of Qatar" in 2016 which aims to achieve the goals of the Qatar National Vision 2030. This decision provides a clear future vision for education, in order to develop curricula and to make the educational process more effective and comprehensive.

This educational Framework has been used as a main reference for a comprehensive review of curriculum standards, their revision and publication in 2018 (as they were first published in 2004). In addition, it is a comprehensive reference framework that covers all levels from Kindergarten to twelfth Grade.

Consequently, curriculum standards are considered the basis for the education reform process in Qatar, as they have been prepared in accordance with international expectations of what the student should learn in terms of skills in order to be able to employ them efficiently, in addition to that they set a clear vision of what educational practices should be in the school.

10.4.2 Curriculums, Digital Resources, and Teacher Training

In 2018, Qatar developed the science curriculum, in line with the objectives of the "National Educational Curriculum Framework" in order to ensure that students are able to acquire knowledge, skills and develop their attitudes. Major changes in the science curriculum in the State of Qatar included integrating skills and processes of scientific inquiry into other branches of science; Creating learning experiences through the use of models to facilitate teaching and learning processes, in addition to identifying five basic competencies for science learning: Inquiry, research, communication, critical and creative thinking, as well as collaboration, participation and problem solving.

Due to the importance of the technological dimension in the educational process, the State of Qatar has been intense on supply all educational institutions (K-12) with information and communication technology infrastructure, Teachers have been trained in high-quality intensive training on how to blend technology into the learning and teaching process, as well as for school principals, in addition to that, the Supreme Education Council in Qatar has developed an integrated strategy, which includes many projects that ensure the use of information and communication technology in all stages of education, starting from kindergarten and ending with university education. These projects focus on three main elements: student education and learning, preparation and qualification of teachers and trainers, and education management. The Supreme Education Council created the "Electronic Bag Project" initiative to activate the process of communication, interaction and creative participation for all schools by providing them with a laptop computer (Tablet PC). This bag contains an electronic library, e-mail, personal file, homework, agenda and personal conversations, and interactive content related to the basic materials, adjustable by the teacher to suit the needs of students. It also contains a set of tools and programs that enable the teacher to manage the classroom environment interactively. As well as organizing the learning process and following up on what students are doing on laptops through the main computer controlled by the teacher.

In honor of the efforts made by the Ministry of Education and Higher Education in the field of digital transformation and modern communication and information technology and harnessing them in providing the best services and digital solutions to the educational community, it won the Gold Award for the best digital service "Registration and Transfer Services for Students", as part of the Excellence Awards of Qatar Digital Government For the year 2020. The Ministry received this award as the best government agency providing integrated public high quality digital services.

The State of Qatar pays special attention to the training and development process in order to keep pace with the development of educational systems in developed countries, as it provides high-quality vocational training that improves the performance of the teacher and all members of the educational community to achieve professional development and sustainable development that the Qatar Vision 2030 aims to. In this regard, it has been indicated that:

The Training and Educational Development Center at the Ministry of Education and Higher Education offers different and integrated set of training programs, which reached 107 in 2019. These programs cover 19 specializations which directing to all components of the educational community, while the number of trainees was 5000 during the period 2016–2019. In addition, The Ministry provides trainees with a set of incentives and necessary facilities to be able to reconcile between their jobs and training.

The Training and Educational Development Center at the Ministry of Education holds workshops and training courses in mathematics and science for teachers in all stages (K-12) to improve thier performance, Which will contribute significantly in improving the performance of Qatar's students in international assessments in science and mathematics, as well as keeping teachers updated with global strategies, technologies and developments in education.

There is an ongoing cooperation between the Ministry of Education and Qatar University in this field, specifically with the College of Education, which provides advanced training programs and workshops for teachers. Whereas, the National Center at Qatar University prepares and designs training programs for public school teachers to meet their training needs. The design of the training programs are based on the national professional standards for teachers in the State of Qatar, the skills of the twenty-first century, and the job description of teacher roles.

10.4.3   Student Assessment and Achievement

The assessment process is considered as a primary element of the educational system, as it is not possible to develop the educational system without assessment development. Therefore, it must be done according to scientific standards, foundations and principles.

Students in the State of Qatar are subjected to various types of assessment systems during their academic life (formative class assessment, semester examinations, high school diploma exams, national assessments and international assessments) in order to assess their level of academic achievement, and their shortcomings and develop appropriate treatment plans according to specific academic determinants.

The evaluation in the science curriculum takes many types. Teachers also perform realistic performance-based assessments using different techniques for example: practical application, projects, teacher notes, checklists, self-reflections/diary, model making, wall charts, games, debates, field trips, student achievement files, and others.

The Ministry of Education also applies what is known as a "comprehensive educational survey", whereby it takes the opinions of students, parents, teachers and school administrators about the educational process in Qatar's schools, in order to improve the performance of the educational system, in order to be in line with the best education systems in the world.

Assessment of the first three grades

The students are assessed at this stage based on the observations of their teachers through the use of standardized assessment tools. It aims to measure the student's performance in basic skills and knowledge according to specific criteria in addition to achievement tests.

Assessment for grades (4–12)

The assessment in these grades differs according to the national examination policy, as the sixth and ninth grades are subject to the national examination system. Evening school students and home students are evaluated every semester at 100% of the subject per semester for all grades.

National exams

In this stage which is third-grade students take the national examination in the subjects of Arabic, English and mathematics, while students of the sixth and ninth grades take the national test in the subjects of Arabic, English, mathematics and science.

Secondary school certificate exam

Students'achievement is measured in subjects in Arabic, English, mathematics, physics, chemistry, biology, social sciences, Islamic education, and the optional courses in the field. The two-semester system is approved for these exams; where the total score is calculated at 50% for each semester. This confirms the student's success in the test and completing the secondary stage.

Qatar's participation in international assessments

In order to compare the performance of Qatar's students with their counterparts in the world in science and mathematics; Qatar is interested in the continued participation of its students in the three international assessments (PISA, TIMMS, and PIRLS).

Arab participation in the PISA International Program began in 2006, when 57 countries participated in the PISA 2006 course, and Qatar was one of the three participating Arab countries. (65) Countries participated in the "PISA 2009" session, including three Arab countries, led by Qatar. Qatar continued participation in 2012, 2015, and 2018.

Results of Qatari students in the international tests

The State of Qatar has participated in Trends in International Mathematics and Science Study (TIMSS) with a sample of eighth and fourth grade students in four types of schools which are: national independent schools, private schools, community schools, and international schools. Qatar's first participation was in 2007. The second participation was in TIMSS-2011, followed by TIMSS-2015, and its last participation was in TIMSS-2019.

As for the results of the eighth grade science in TIMSS-2019, four Asian countries won the first ranks: Singapore with an average of 608, followed by Chinese Taipei with 574, Japan 570, and Korea 561. The average performance of Qatar's students was 475 points, which is ranked 25 out of 39 participating countries, and at the second place among Arab countries out of 10 participating countries.

As for the results of fourth grade in science in 2019- TIMSS, Singapore ranked first internationally with an average (595), followed by Korea (588), then Russia (567), Japan (562), and Chinese Taipei (558) in fifth place. Qatar ranked 49th out of the 64 participating countries, and the third Arab country with an average performance of (449).

By reviewing the trends of Qatar's students in all TIMSS assessments that they participated in, starting from 2007 until the end of 2019, it is noted that the fourth grade science results were consistent with the eighth grade results. It is also observed that there is continuous improvement in the level of performance in science during all intervals in which the State of Qatar has participated. Although Qatar has achieved advanced positions in the Arab world, the average performance of students in science in all intervals was less than the internationally accepted average (500). As shown in Figs. 10.1 and 10.2.

The results of the international study TIMSS-2019 also showed the superiority of female students over males in science in all the assessments they participated in as shown in Figs. 10.3 and 10.4.

The results also indicated that the international schools were better than the community schools, while the community schools were better than private schools that follow the Qatari curriculum standards. On the other hand private schools were better than independent national schools. The International schools got the highest rank while the lowest were the independent schools. These results as indicators must

Fig. 10.1 Average science achievement across assessment years (8th grade). *Source* IEA's Trends in International Mathematics and Science Study—TIMSS 2019. Downloaded from http://timss2019.org/dow nload

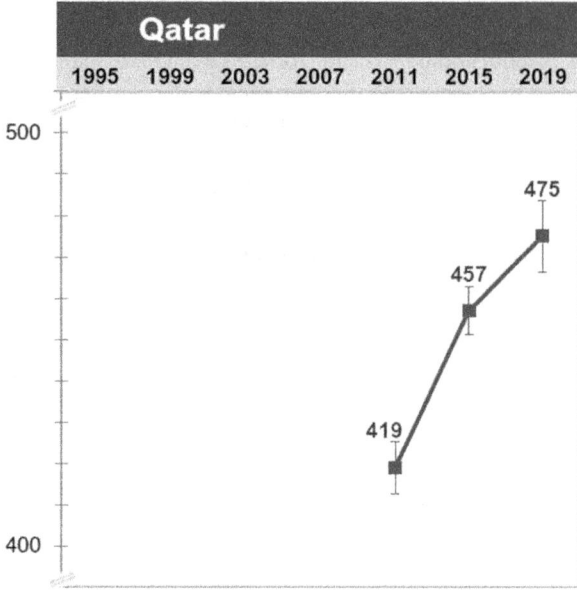

Fig. 10.2 Average science achievement across assessment years (4th grade). *Source* IEA's Trends in International Mathematics and Science Study—TIMSS 2019. Downloaded from http://timss2019.org/dow nload

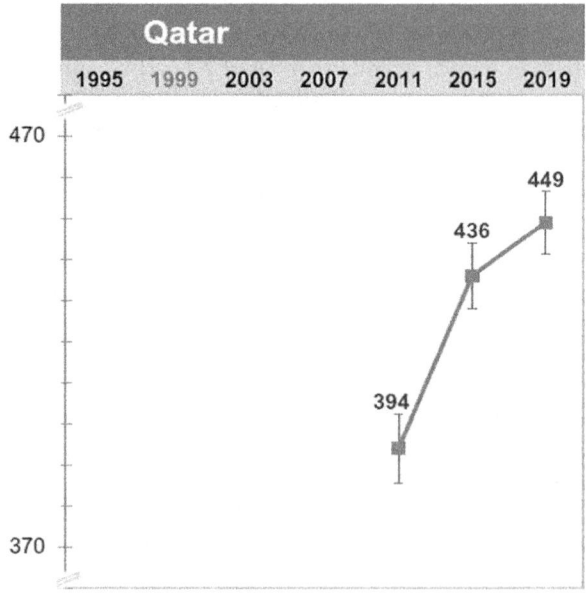

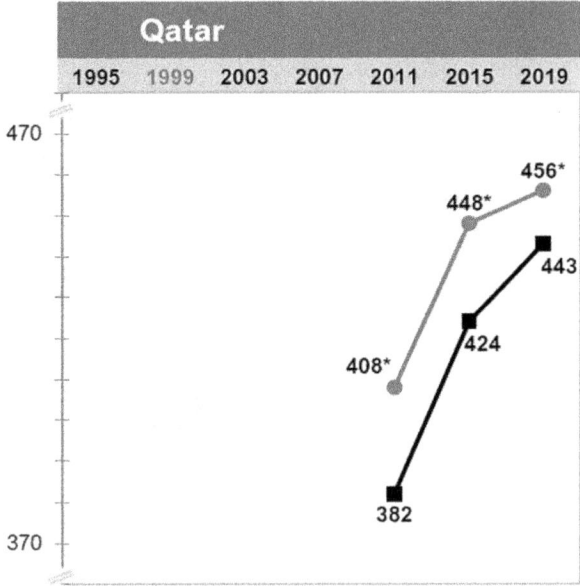

Fig. 10.3 Average science achievement across assessment years by gender (8th grade). Girls ━●━ Boys ━■━ * Average significantly higher than other gender   *Source* IEA's Trends in International Mathematics and Science Study–TIMSS 2019. Downloaded from http://timss2019.org/download

be reviewed and more studies conducted to find out the reasons behind the superiority of the international schools over the independent schools, and to use them in improving the performance of students of the independent national schools.

To improve and develop the performance of Qatar's students in international exams, the Ministry has designated a special department for international examinations affiliated to the Evaluation Authority, whose mission is to prepare, follow up and implement plans and strategies. The improvement policies that the Ministry must undertake to ensure the continuous improvement of the performance of its students in the next decade include the following priorities:

- Increase enrollment in kindergarten.
- Enhancing students' motivation towards learning.
- Continuing to develop standards to keep pace with global standards, individual capabilities and meet needs.
- Continuity of training for educational staff.
- Promote interactive learning.
- Achieving the ideal use of information and communication technology in the learning process.
- Increasing and diversifying learning resources.
- Involving parents in developing education strategies and decision-making.
- Encouraging the culture of self-learning and its continuity for life.

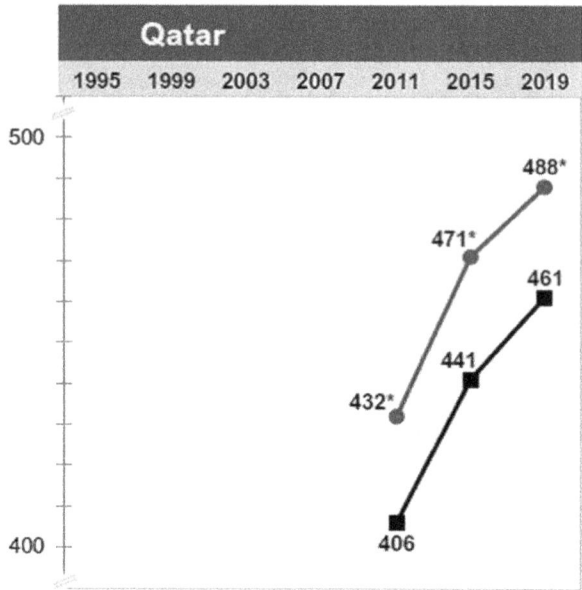

Fig. 10.4 Average science achievement across assessment years by gender (4th grade). Girls ━●━ Boys ━■━ * Average significantly higher than other gender *Source* IEA's Trends in International Mathematics and Science Study—TIMSS 2019. Downloaded from http://timss2019.org/download

10.4.4 Science and Technology Venues and Centers

There are many research and technology centers in Qatar, in addition to many scientific activities and competitions, such as:

Qatar Science and Technology Park: It is an integrated platform that provides a free zone and business park that hosts leading global tech companies, mentor and support a network of startups and rising tech ventures, and have a value chain of acceleration, incubation, and funding programs. In addition it plays a key role in developing technology, innovation, and entrepreneurship ecosystem in Qatar.

The Arab Academy for Innovation: In partnership between the European Academy of Innovation and Qatar Science and Technology Park, it includes the first and largest technology entrepreneurship program in the Arab region, as it supports participants and challenges its ability to establish a new technology company in just 10 days.

STEM Adventure Camp: A collaborative project between Qatar University and Exxon Mobil, camping aims to engage middle school students in learning about STEM.

The Virtual Private Lab for Education Project/Qatar University: It was established by Qatar University in July 2014 with the aim of training teachers and adapting virtual reality in the field of education.

The World Innovation Summit for Education (WISE): It is an international initiative aimed at transforming education through innovation. It was established by Qatar Foundation in 2009, and WISE's mission revolves around promoting new approaches to education and investigating new ways to address pressing global education challenges.

Scientific Excellence Award: It is devoted to scientific excellence, in which Qatari students from different educational levels and university graduates compete, in addition to teachers, schools and researchers.

10.4.5 Utilizing Emerging Technologies

The State of Qatar has many characteristics that qualify it to benefit from artificial intelligence. As the statistics indicated, about 94% of its people use the Internet, which is one of the highest rates in the world. This indicates direct interaction with technology based on Artificial Intelligence (AI).

Qatar's vision is to spread artificial intelligence in all aspects of life and government business in Qatar. To achieve this, Qatar has developed a comprehensive national AI strategy. This strategy aims to invest in artificial intelligence in achieving the Qatar National Vision 2030, and to prepare a society that effectively adopts artificial intelligence technology in a manner consistent with its needs, culture and traditions. The AI strategy is based on six pillars: education, access to data, employment, business, research, and ethics, and will work together to guide Qatar towards the future of artificial intelligence. Qatar Foundation includes many institutes dealing with artificial intelligence such as: Qatar Computing Research Institute, Qatar Environmental and Energy Research Institute, Qatar Biomedical Research Institute, and Qatar Science and Technology Park, Qatar Genome programme.

10.5 Requirements for Future Development of Science Education

As the twenty-first century presents opportunities, it also presents challenges. This makes it imperative that education systems respond flexibly to these global challenges and changing national priorities. For this Qatar has:

– Developed science curriculum standards to simulate the international curricula of the most advanced countries in education, and at the same time have a cultural national character.

- Adapted the science curriculum to contribute to the development of core competencies and an understanding of common issues in Qatar and the world.
- Directed employers in Qatar to pay attention to the skills that their employees possess, including initiative and communication.
- Increased the appreciation of Qatari universities for students who possess scientific inquiry and scientific research skills, to enable them to communicate their ideas and manage their own learning.

The State of Qatar is preparing plans to teach artificial intelligence to all students, and to integrate it into the curriculum at all educational levels, starting from kindergarten.

In this regard, Qatar's National Artificial Intelligence Strategy recommended the following:

- Preparing a strong educational curriculum according to the AI approach for (K-12).
- Providing an active environment for research to reach innovative applications compatible with the national interest in the field of artificial intelligence.
- Adopting strategies to develop local talent and attract skilled talent in the field of AI from all over the world.

10.6   Discussion and Conclusion

The Qatar National Vision 2030 aims at human development through the establishment of an education system at a global level that provides opportunities for enrollment in high-quality education and training, so that all learners acquire the necessary skills and competencies that are consistent with their potentials, capacities, abilities and respect for other cultures. This is why Qatar has given priority to developing the education sector, as it has developed the science curriculum in light of international standards for the most developed and advanced countries in their educational systems, in order to move towards a new direction in science education in the State of Qatar to prepare learners to respond positively to the challenges facing them in the twenty-first century, and to be successful learners and producers of knowledge for life.

In addition, the study showed that the most important pillars that Qatar adopts in its plans are both education and scientific research in order to become a modern state moving towards a knowledge-based economy, global competitiveness, and sustainable development. So, Qatar is keen to participate its students in international assessments, namely: (TIMSS, PISA, PERLS). The findings showed that students' performance in science has improved during their participation from 2007 until 2019. Although Qatar has achieved advanced positions in the Arab world, their performance level and their global ranking failed to reach the required international level which meets Qatar National Vision 2030, to reach an advanced global educational system. Qatar has taken significant steps in the field of digital transformation and using information and communication technology, in providing integrated technology with the highest quality. So, it has prepared a comprehensive national strategy for artificial intelligence.

References

Al-Raya. (2020). School Performance Assessment for the Year 2019–2020. Retrieved From: https://www.raya.com/2020/11/16/%D9%88%D8%B2%D8%A7%D8%B1%D8A9-%D8%A7%D9%84%D8%AA%D8%B9%D9%84%D9%8A%D9%85.

Al-Watan Newspaper. (2017). *Qatar's first education in the Arab world*. Retrieved from: https://www.al-watan.com/news-details/id/81928.

Ministry of education and higher education (Evaluation Authority). (2014). *National Report (TIMSS AND PIRLS 2011).* Ministry of Education Press.

Ministry of Transportation and Communications. Qatar's national strategy for artificial intelligence. Retrieved from: https://www.motc.gov.qa/en/documents/document/national-artificial-intelligence-strategy-qatar.

Ministry of Development Planning and Statistics. (2016). *Concepts and terminology used in education, training, research and development.* Retrieved from: https://www.psa.gov.qa/en/statistics/Statistical%20Releases/Classification/Terminologies_used_in_education_training_research.pdf.

Mullis, I. V. S., Martin, M. O., Foy, P., Kelly, D. L., & Fishbein, B. (2020a). *TIMSS 2019 International results in mathematics and science.* Retrieved from Boston College, TIMSS & PIRLS International Study Center website: https://timssandpirls.bc.edu/timss2019/international-results/.

Planning and Statistics Authority. (2020). *Qatar bulletin: Education statistics.* Retrieved from: https://www.psa.gov.qa/en/statistics1/statisticssite/pages/statisticalabstract.aspx.

Planning and Statistics Authority. (2020). *Qatar second national development strategy* 2018–2022. Retrieved from: https://www.psa.gov.qa/ar/knowledge/Documents/NDS2Final.pdf.

Qatari Legal Portal (Al-Meezan). (2020). *Legislations.* Retrieved from: https://www.almeezan.qa/LawPage.aspx?id=2284&language=en.

Qatar e-government is "Hukoomi". (2020a). *Education and training.* Retrieved from: https://portal.www.gov.qa/wps/portal/about-qatar?changeLanguage=ar.

Qatar e-government is "Hukoomi". (2020b). *Information about Qatar.* Retrieved from: https://portal.www.gov.qa/wps/portal/about-qatar?changeLanguage=ar.

Qatar Foundation. (2020). *Education city in Qatar.* Retrieved from: https://www.qf.org.qa.

Qatar Science and Technology Park. (2020). *Technology transforms into business.* Retrieved from: https://qstp.org.qa/.

Secretariat-General for Educational Planning. (2008). *Qatar national vision 2030.* Retrieved from: https://www.psa.gov.qa/ar/qnv1/Documents/QNV2030_Arabic_v2.pdf.

Supreme Education Council. (2020). *Science standards for the State of Qatar.* Retrieved from: https://www.academia.edu/16786555/%D9%85%D8%B9%D8%A7%D9%8A%D9%8A%D8%B1_%D8%A7%D9%84%D9%85%D9%86%D8%A7%D9%87%D8%.

The Ministry of Education and Higher Education. (2020). Annual statistics about education in the State of Qatar. Retrieved from: https://www.edu.gov.qa/Documents/AnnualStatistics/EdAnnualStat/%D8%A7%D9%84%D8%A5%D8%AD%D8%B5%D8%A7%D8%A6%D9%8A%D8%A7%D

The Ministry of Education and Higher Education. (2020). The strategic plan of the ministry of education and higher education (2018–2022). Retrieved from: https://www.edu.gov.qa/Documents/StrategicPlan/MOEAHE%20-20Strategy%20Booklet%20-%20A4%20landscape%20

Chapter 11
Science Education in Saudi Arabia

Bader A. Alyoubi and Fathi Essalmi

Abstract This chapter presents the current state of the educational system in the Kingdom of Saudi Arabia. The educational system is strongly important for a vibrant society, ambitious nation and power economy. In this context, the Kingdom is investing in the education in order to diversify its sources of income in the field of business, industry and services. In fact, The Kingdom gives a major importance to the knowledge-based economy. In the Kingdom, the investment and budget allowed for education is sufficient. Furthermore, the Kingdom is recruiting the required human capital for the mission of education and conducting research in the related field. From another side, the development and the use of technology are dramatically in continuous evolution in the Kingdom. So, combination of the organized educational system with the evolution of Information and Communication technologies (ICT) is well established in the Kingdom. Further projects could accelerate the development of the educational system such as the project about enhancing the class competition and self-learning, advance in attractive evaluation methods, design of interactive methods and tools, directing the learning by teaching, and personalizing the learning.

11.1 Overview of the Country

The Kingdom of Saudi Arabia (KSA) is the heart of the Arab and Islamic Worlds. In particular, according the vision 2030, the Kingdom has the objective to serve 30 Million Umrah visitors each year. The Kingdom is also characterized by a vibrant society and an ambitious nation. The vision of the Saudi Arabia fosters values of moderation and tolerance, values of excellence and discipline, values of equity and transparency, values of determination and perseverance and upholds the Arabic

B. A. Alyoubi · F. Essalmi (✉)
Management Information System Department, College of Business, University of Jeddah, Jeddah, Saudi Arabia
e-mail: feessalmi@uj.edu.sa

B. A. Alyoubi
e-mail: balyoubi@uj.edu.sa

R. Huang et al. (eds.), *Science Education in Countries Along the Belt & Road*,
Lecture Notes in Educational Technology,
https://doi.org/10.1007/978-981-16-6955-2_11

169

language (KSA, 2020). In particular, the Education Evaluation Commission of the Kingdom is working continuously for enhancing the quality of education, improving the equity of access to education, ameliorating the ranking of educational institutions and ensuring the alignment of educational outputs with labor market needs.

11.1.1 Geographical Location, Population and Political System

The Kingdom of Saudi Arabia occupies four fifths of the Arabian Peninsula, with an area estimated at about 2,000,000 square kilometers (General Authority for Statistics, 2019a). The Kingdom is located in the far southwest of the continent of Asia, where it is bordered to the west by the Red Sea and to the east by the Arabian Gulf, the United Arab Emirates, Qatar and north to Kuwait, Iraq, Jordan, and to the south of Yemen and the Sultanate of Oman (General Authority for Statistics, 2019a).It has a biodiversity in its natural environment (Fanack, 2020). The population of Saudi Arabia is about 35 Millions (WorldoMeter, 2020). Among them, eight million and half aged 14 years and younger and about five million aged 15–24 years (Unesco, 2020).

11.1.2 Current Situation of Economic, Technologies and Cultural Development

The Kingdom of Saudi Arabia is tending to diversify its sources of income in the field of business, industry and services. In 2019, the Gross Domestic Product in the Kingdom is 718,543 million riyals with a growth rate of 2.77 (General Authority for Statistics, 2019a). The Kingdom's economy is among the world's 20 largest economies in terms of economic stability and the attractiveness for investment (Esmail, 2018). Beside the oil sectors, the Kingdom is developing its economy on the non-oil sectors such as non-oil exports, tourism and human resources (Alodadi & Benhin, 2015). In particular, the Kingdom has 35 industrial cities, 28 airports and 9 ports (General Authority for Statistics, 2019a). These sectors influence the economy of the Kingdom (Hatem et al., 2016). This increases the productivity, efficiency and employment and then increases the production of high-quality products and enhances the competition in the markets (Hatem et al., 2016). Furthermore, projects' funding has a positive impact in the development of various sectors in Saudi Arabia (Al Mahish, 2016). In addition, Saudi women are now contributing in the economic growth (Saqib et al., 2016). In this context, the rate of Saudi youth participation in the labor force is increased by 4.4% points over the past four years, and this increase came as a result of an increase in the female participation rate, which reached 6.3% points compared to 2% points for male (Center for Statistical Analysis and Decision Support, 2020).

The 2030 Vision aims to diversify the sources of income and enhance the partnership between the public and the private sectors and the continued fairness and transparency (Esmail, 2018). The development and the use of technology are dramatically in continuous evolution in the Kingdom. In particular, the revenue of the domain of computer and electronic goods in Saudi Arabia will be about 256,7 million U.S. Dollars by 2023 (Huhn, 2019a). In this context, the most online activities selected by Saudi youth (15–34 years) is the participation in social or professional networks and the use of social media (Center for Statistical Analysis and Decision Support, 2020). Furthermore, the income of repair of computers and related goods in Saudi Arabia will be about 1.595,1 million U.S. Dollars by 2023 (Huhn, 2019b).

The vision of the Saudi Arabia promotes the culture and entertainment and grows the Saudi contribution to arts and culture. For example, the Kingdom has 33 museums (General Authority for Statistics, 2019a). Furthermore, the vision develops positive attitude, resilience and hard-work culture among children, empower citizens through the welfare system and improve effectiveness and efficiency of welfare system (KSA, 2020). The culture of voluntary work is well established in the Kingdom. During 12 months, the young (15 years and over) has a participation with 14.7% in the volunteer work. The average volunteer work hours (15 years and more) during this period was 52.2 h (General Authority for Statistics, 2019a).

11.2 Overview of the Education Development

The Kingdom of Saudi Arabia uses a large amount of its revenue in education, modern schools, transport systems and communication networks (Alqarni, 2015). The Ministry of Education is established in 1953 and from that date the Kingdom invests in enhancing the quality of education. The number of students is in continuous increase. In 2019, the number of students of all academic levels in general education has reached 6,257,784. The number of students in all institutions of higher education has reached 1,982,722. The total number of students has reached 8,240,506. The number of teachers in public education has reached 502,050. The number of faculty members in public and private education has reached 85,409 (General Authority for Statistics, 2019a). Furthermore, female education became one of the fastest emergent areas of social development in the Kingdom of Saudi Arabia (Mahboob & Elyas, 2014). The illiteracy rate has been decreased significantly among Saudi youth (15–34 years) when comparing 2007 and 2017, and this decrease is due to the decrease in the female illiteracy rate (Center for Statistical Analysis and Decision Support, 2020). In fact, The Kingdom vision aims to enhance the family involvement in preparing for their children's future and to improve equity of access to education (KSA, 2020).

11.2.1   Education System and Policy

In the Kingdom of Saudi Arabia, the public education has three levels before the access to the University: primary, middle, and secondary levels. Primary school takes 6 years; middle and secondary levels takes 3 years each (Pariona, 2017). The public education in the Kingdom is accessible for all the society without any cost. The Kingdom gives a high importance to the education since it prepares the new generations who are the future human capital (Al-Mousa, 2010). In 2019, the number of general education schools for all grades has reached 36,693. The total number of classes in schools has reached 264,183. The number of universities and colleges has reached 66. The number of students abroad has reached 110,834 (General Authority for Statistics, 2019a). The Kingdom has an adaptive strategy for the motivation and the evaluation of teachers. Since 2011, the Ministry of Education announced its strategy that put 80% of the weight of evaluating teachers' performance based on their students' performance. The results of this strategy encourage teachers to give their maximum in the educational process (Almadina Newspaper). Furthermore, the students' transition rates are increased in the Kingdom. In particular, the transition rates in secondary education for females are 100% (Center for Statistical Analysis and Decision Support, 2020). In fact, the quality of teaching is fundamental in any educational reform (Hargreaves & Shirley, 2009).

The concept of standardization of education is applied in the Kingdom. This is done in order to define the important competencies and knowledge needed by the students in different domains. In this context, the Kingdom's ministry of education underlines the difficulties of other educational systems and transforms them to recommendations for enhancing the education (Alnahdi, 2014). In particular, the ministry of education develops programs and learning outcomes consistent with the economic and social developments (Saudi Arabian Cultural Mission, 2006). This has increased the employment of youth. In the year 2019, the number of employed Saudi youth between the ages of 15–34 years reached 1,489,520 workers, as they accounted for (47%) of the total number of Saudi employees. Center for Statistical Analysis and Decision Support (2020). The ministry of education formed committees to review the subjects and the outcomes that are taught in all levels of school including measurement taken for special education (Saudi Arabian Cultural Mission, 2006). The Numbers of school classes devoted to children with difficulties under the age of 15 for the year 2017–2018 were 423 classes for deaf, 4009 for mental disability, 166 for blind and 897 classes for autistic students (General Authority for statistics, 2019b).

The Kingdom has several projects for enhancing the quality of education. One of them is the "Tatweer" program. The budget of Tatweer program is about $2.4 billion (Kamal, 2012). The goal of Tatweer is to decentralize the educational system by giving more responsibility to schools and educational departments (Hakami, 2010). Tatweer aims to enhance the implementation of the constructivist approach which is based on the learner-centered education. As another important project, The King Abdulaziz and His Companions Foundation for Giftedness and Creativity

"Mawhiba" is a national non-profit organization that aims to discover and nurture talented and creative people in scientific fields of national priority (Center for Statistical Analysis and Decision Support, 2020). The foundation has supported 8 Talented Saudi Students in 2019. In the International and regional competitions, 47 awards were taken. In the National Olympiad for Scientific Creativity, there are 84,418 participants, 6,352 projects, 102 candidate projects and 27 winners (Center for Statistical Analysis and Decision Support, 2020).

.

11.2.2 Statistics on the National Education

In the academic year 2017–2018, the number of teachers in the schools of education in Saudi Arabia was 533.98 thousand. In the same academic year, the number of schools in the Kingdom was 30.75 thousand (Puri-Mirza, 2020). The number of schools' students enrolled in the Kingdom was 6.1 million (Statista Research Department, 2020). The majority of those students, around 3.2 million, study at a primary school level. The primary school participation, and net enrolment ratio from 2008 to 2012 is about 96.6% (Unicef, 2013). In 2019, the Pupil-teacher ratio in primary schools has reached 12. The percentage of primary school teachers trained for teaching has reached 100%. The proportion of primary schools with internet access has reached 100%. The proportion of secondary schools with internet access has reached 100% (General Administration for Innovation and Statistical Development, 2019). In the same year, the number of students of all academic levels in general education has reached 6,257,784. The number of students in all institutions of higher education has reached 1,982,722. The total number of students has reached 8,240,506. The number of teachers in public education has reached 502,050. The number of faculty members in public and private education has reached 85,409 (General Authority for Statistics, 2019a).

11.2.3 Educational Research and International Collaboration

The Kingdom established several Universities which promote research programs aligned with the objectives of the higher education and the economy of the country. Among the important Saudi Universities, we mention the King Abdul Aziz University (KAU) which is ranked as the first one in the Arabic World according to QS Global World Ranking in 2020. King Fahd University for Petroleum and Minerals (KFUPM), King Saud University (KSU), King Khalid University and King Faisal University have also good ranking according to the same classification. These universities have achieved a good reputation in the world.

The Saudi government gives an important role to the educational research and international collaboration. For example, the Kingdom of Saudi Arabia has established educational cultural bureaus in consulate offices in different countries. They focus on the postgraduate education of Saudi students in science fields. Through the educational cultural bureaus, the Saudi government supports and sponsors the students for postgraduate studies. In this way, the students of the Kingdom have access to more than 500 Universities in the world (Shin, 2012). The number of students abroad has reached 110,834 in 2019 (General Authority for Statistics, 2019a).

11.3 Current Situation of Science Education

11.3.1 Policies and Standards

The Kingdom of Saudi Arabia offers education with no cost for the six years of primary level, the three years of the intermediate level, and the three years of the secondary level. Similarly, there is no cost for the higher education in the Kingdom. In addition, higher education is basically free for all Saudi, and a monthly stipend is given to them. Also, non-Saudi students living in Saudi and having high ranks have scholarships from Saudi Arabia based on international agreements between the Kingdom of Saudi Arabia and the countries of the students. The school year includes two semesters. Each semester is 15 weeks followed by two weeks of the final exams. The number of students enrolled in the primary stage has reached 3,076,394 in 2019 and the number of primary schools is 12,687. Then, in average, there is 242 students for each school. The number of students enrolled in the middle school has reached 1,352,460 and the number of middle schools is 5,196. Then, in average, there is 260 students for each school (General Authority for statistics, 2019b). Saudi Arabia has continuous curricular reforms which are implemented to enable the educational system to be aligned with the most advanced educational systems in the world. In particular, the Ministry of Education in the Kingdom has implemented new science and mathematics curricula in the academic year 2009 (Alghamdi & Al-Salouli, 2013). To align the educational system with the most advanced educational systems, Saudi Arabia implements several projects integrating science, technology, engineering, and mathematics in a unified curriculum that enhances the students' competencies. K12 science education in Saudi Arabia is in continuous evolution for improving the quality of learning and to enhance the level of outcomes (Al Shannag et al, 2016). From the grades 1 to 6, general concepts of science are learned. Then, from the grade 7 to the grade 9, more details are given for science education and about 4 h per week are allowed for science learning. After that, from the grade 10 to the grade 12, specific concepts of Physics, Chemistry, and Biology will have more importance (Alsslloom, 1991). It is important, to note that in 2019, the number of expected years of study has reached 17 (General Administration for Innovation and Statistical Development, 2019).

11.3.2 Curriculums, Digital Resources and Teacher Training

In the Kingdom of Saudi Arabia, the Ministry Of Education (MOE) develops digital resources, digital knowledge libraries and smart schools (MOE, 2020a) for all grades of the three levels (primary, intermediate and secondary). In particular, relevant material in the science field such as mathematics and biology are continuously offered and updated since several years (Al-Sulaimani, 2010). Furthermore, the schools collaborate with the MOE in different projects with pedagogical perspective such as developing educational resources (Alyami, 2014). The Saudi digital library includes more than 310,000 educational resources covering the different academic disciplines (MOE, 2020a). In addition, there are several regional projects for enhancing the scientific educational systems and keep up with the rapid development in Saudi Arabia (Madani & Forawi, 2019). Educational resources including assessment materials are developed for the three school levels (primary, intermediate and secondary). Also, these educational resources and the assessment materials are complementary (Alshaya & Abdulhameed, 2011). Interactive books are developed including digital content for the secondary education (MoE, 2020b). In addition, the percentage of primary school teachers trained for teaching has reached 100% (General Administration for Innovation and Statistical Development, 2019).

Teaching of Science, Technology, Engineering and Mathematics is a growing area (UNESCO, 2010). In the Kingdom, the Ministry Of Education aims to enhance the scientific progress, knowledge sharing and supporting the postgraduate programs (MOE, 2020a). In particular, in the primary and middle levels, science is taught as one course. It includes biology, chemistry and physics concepts. Then, in the secondary level more detailed contents about biology, chemistry, physics and geology are done (Ministry of Education, 2014).Teachers of these fields have direct dialogue about the possible applications of the combined concepts. This dialogue is more constructive when there are more collaboration between the teachers of Science, Technology, Engineering and Mathematics (El-Deghaidy, & Mansour, 2015).

11.3.3 Student Assessment and Achievement

Teachers are using continuous assessment methods for evaluating the students as well as for evaluating their teaching methods (Rana & Zubair, 2019). This allows students to observe their achievement (Ghahariand Farokhnia, 2017). In this way, the educational system of the Saudi Arabia is enhanced considerably (Rana & Zubair, 2019). The educational system includes different evaluation methods and learning activities for all grades (Iqbal, 2017). These activities play an important role for ameliorating the quality of learning and form a source of feedback on the performance of the students (Onuka & Onabamiro, 2010). In addition, the ministry of education supports the students' constructive education with digital books allowing the self evaluation (MoE, 2020b).Furthermore, these activities include home assignments and summarized exercises (Mwebaza, 2010). These formative assessments are also considered as tools for evaluating the students' understanding and the misconceptions (Rosas, 2014).

11.3.4 Science and Technology Venues and Centers

The Kingdom of Saudi Arabia gives a major importance to the knowledge-based economy. For that reason, the use of technology plays an important role for the economic development of the country (King Abdulaziz City for Science and Technology, 2020). Then, the innovation and the use of science and technology in Saudi Arabia are growing the economy (King Abdulaziz City for Science and Technology, 2020).

According to Behairy et al (2013), science and technology are contributing in the development of the Kingdom. In particular, King Abdulaziz City for Science and Technology includes several incubators and technology parks. As an example, we mention the Sultan bin Abdul Aziz Center for Science and Technology disseminates the principles of science and technological innovations and presents them in interesting methods (Scitech, 2020). As others examples, we cite the Riyadh Techno valley, Dhahran Techno valley and BADIR ICT incubator which are among the most important incubators and technology parks (Khorsheed et al., 2012).

The Human capital is another aspect of the development of the Kingdom. Human capital training includes professional training and the encouragement of the teachers and the researchers (King Abdulaziz City for Science and Technology, 2020). The professional training of teachers and administrators for designing and implementing new projects is one of the aims of the Kingdom (Kamal, 2009). In this context, the ministry of education has developed several programs of teacher preparation and professional training (Ministry Of Education, 2020d). This is fundamental for the success of the science and technology development process (Behairy et al, 2013). In addition, the Kingdom has several implemented projects in the direction of preparing classrooms with smart equipments, video camera and internet (Tatweer, 2008). Furthermore, schools allow access to virtual laboratories, digital libraries, and advanced e-learning activities such as project-based learning (Tatweer, 2008).

11.3.5 Utilizing Emerging Technologies

The Kingdom of Saudi Arabia has the potential for leading the digital economy through the deployment and the use of emerging technologies such as Internet of Things, Artificial Intelligence, Data science, Cyber security, Cloud, Blockchain, Augmented Reality, Virtual Reality, Robotics and Deep learning (Ministry of communications & information technology, 2019). In particular, The Social Development Bank supports emerging technology enterprises for facilitating new digital enterprises development (The Social Development Bank, 2020). By 2025, revenues from artificial intelligence are expected to raise $90 billion in the Kingdom. Artificial intelligence will participate with 14% of the Gross domestic product by 2030 (Hassan, 2020). For all these reasons, artificial intelligence represents an important business opportunity in the Kingdom of Saudi Arabia.

The Kingdom of Saudi Arabia has an integrated information system including the 'Digital ID', the businesses' service, the systems of the ministry of justice and so on. In particular, the 'Digital ID' allows citizens and residents to connect to the public services. It includes facilitated functionalities for increasing the efficiency of transactions. Furthermore, the businesses' service provides a unique 'ID number' for each private entity in the Kingdom of Saudi Arabia (National Digital Transformation, 2019).

The emerging technologies are implemented in different sectors. For example, information technology is fully used for enhancing the efficiency of the healthcare sector. In particular, tele-medicine, and tele-consultations are starting to be exploited. Furthermore, the access and the use of the e-health service are facilitated (Emerging Technologies in Saudi Arabia, 2020). In addition, the Kingdom gives high importance for the application of ICT in education. The incubators and the technology parks implemented in the King Abdulaziz City for Science and Technology are examples showing this direction.

In the Kingdom, there are several technologies used for science education. (Alqarni, 2015). In particular, Blackboard Collaborate is used officially in higher education. Furthermore, the national center for e-learning offers the needed e-learning tools and develops several useful educational resources (Alqarni, 2015). In addition, schools and Universities have continuous programs for teachers training about using educational technology. This facilitates the development of electronic educational resources including the assessment resources. Emerging technologies such as educational games, virtual reality, and augmented reality are exploited in some pilot studies. This is done based on the volunteer teachers and researchers who design and implement these technologies before their use in classrooms.

11.4 Requirements for Future Development of Science Education

The Kingdom of Saudi Arabia gives high priority to the education and knowledge development. The investment and budget allowed for education is largely sufficient. Furthermore, the Kingdom is recruiting the required human capital for the mission of education and conducting research in the related field. In addition, the Kingdom has the legislation rules for encouraging the development of the quality of education. Also, the Kingdom has a distinguished infrastructure of schools and University. For all these reasons, the educational system could be in continuous development. The following recommendations could be applied for accelerating the development of science education:

- Development of class competition and self-learning: competition between learners in the same class challenges them in different ways. Some of them will do their best to have the best rank. Some others will try to have acceptable rank compared to their peers. Some others will try to avoid the bad ranks. The class competition, if it is well designed, encourages the self-learning which is a way for each learner to prove his/her enhanced competencies. In this case, the concurrence between the learners can encourage their families to help them during their learning process. There are several methods to encourage the class competition and self-learning. For example, the teacher can mention in the beginning of the semester that he/she will select the best students who will help their colleagues in the library (e.g. 2 h each week) and the others students can ask them (selected students) for some concepts learned in the classroom which are not well understood. The selected students will have a special recompense from the school in the case of students in the primary, intermediate or secondary level. They will have the recompense from the University in the case of higher education. This will encourage the students, from the beginning of the semester, to be among the bests.
- Development of attractive evaluation methods: formative evaluation and final evaluation are important for having feedbacks about the learners' progress and achievement. The evaluation instruments range from the knowledge evaluation to the competency evaluation. The multiple-choice questions are the most used techniques for evaluating the knowledge achievement of the learners. However, writing and problem resolution could be used for evaluating more advanced level of achievement. However, several students are not very motivated for participating in formative evaluation which includes writing and problem resolution. In this case, the use of advanced technology such as educational games and virtual reality as evaluation method is highly recommended. This will motivate the learners in the process of learning combined with formative evaluation.
- Development of interactive methods and tools: pedagogy of teaching is a kind of communication and its success depends on the participation of learners. This bidirectional communication between the teacher and the learners could have several forms such as between a teacher and a learner, between a learner and a learner, between groups of learners and so on. Furthermore, there are several tools that could encourage the participation of learners such as the brainstorming sessions and the formative evaluation. From the other side, the innovation in the information and communication technologies has to support high interaction and participation of the learners. In particular, new tools such as simulation, virtual reality and augmented reality could support this pedagogical interactivity.
- Learning by teaching: learner could play the role of teacher when he/she is explaining some concepts for his/her colleagues. What is interesting is that not only the colleagues who are playing the role of learner profit from the situation. The learner who is playing the role of teacher will have more achievement since he/she is encouraged to provide more efforts to explain for his/her colleagues. The

learning by teaching could be supported by ICT tools for more generalization of the approach. In particular, the use of artificial intelligence could provide virtual learners which support the learning by teaching.

- Development of personalized learning: the learners have different backgrounds, different preference, different learning styles and different motivations. These differences have to be considered for more efficiently of the learning scenarios. The personalized learning is a solution for the issue of 'one size does not fit all' in the context of education. Assume for example, that we have two students in ICT course: one of them has a good background about programming language since he/she is participating in a club of programming during the last 5 year. Assume that the other student has very limited knowledge about programming. In this case, giving the same course with the same way for both students could decrease their motivation. The learner who has important knowledge about programming language may consider that the course is for beginner and not appropriate for his/her level. The other student may consider that the course is so difficult and not appropriate for him/her. In this context, artificial intelligence could be used for developing tools allowing personalizing the learning. In this case, the learners profile will be considered for generating appropriate learning scenarios for each course.

11.5   Discussion and Conclusion

The Kingdom of Saudi Arabia invests in the education during more than 50 years. These investments resulted to high quality infrastructure in the schools and Universities. Furthermore, the Kingdom has recruited the human resources for educating the learners who are Saudi or residents. In addition, the Kingdom has implemented several projects in the development of educational resources needed by the instructors in the different levels of education.

The use of ICT for the service of education is accelerated in the last years. In particular, digital educational resources are becoming fundamental for education. This opens the door for more investment in the use of more advanced technology for education. For example, the use of augmented reality for education will be habitual in schools and Universities. As another example, the use of educational games for enhancing the formative evaluation is becoming possible. In this context, the design and the implementation of new projects for implementing advanced technologies supporting education is becoming a priority.

The vision of the Saudi Arabia and in particular the National Transformation Program includes a plan for enhancing the quality of learning from different perspectives. In particular, the professional training of teachers, the design of innovative projects, and the development of modern pedagogy are among the objectives of the National Transformation Program (Patalong, 2016). In this direction, teachers could be trained for the use of modern pedagogy including the use of augmented reality and educational games. This will have added values for the life of classrooms and enhance the skills of the learners. Furthermore, this is synchronized with the increase of the applications of ICT in education.

Acknowledgements Our special thanks go to the Saudi Ministry of Education and the University of Jeddah for their valuable supports.

If you wish to acknowledge persons who contributed or sponsoring agencies, do so here in this optional section.

References

Alghamdi, A.K.H., Al-Salouli, M.S. (2013). Saudi elementary school science teachers' beliefs: teaching science in the new millennium. *International Journal of Science and Mathematics Education, 11*(2), 501–525.

Al Mahish, M. (2016). The impact of financing on economic growth in Saudi Arabia. *International Journal of Economic and Finance*. Retrieved from https://doi.org/10.5539/ijef.v8n8p1

Al-Mousa, N. (2010). The Experience of the Kingdom of Saudi Arabia in mainstreaming students with special educational needs in public schools. Riyadh: The Arab Bureau of Education for the Gulf States. http://unesdoc.unesco.org/images/0019/001916/191663e.pdf

Al Shannag, G., Schreier, H., Abdel Fattah, F., & Alshaya, F. S. (2016). Lost in memorization: The case of science instruction in ksa schools and the argument for a promotion of science literacy based on formative assessment practices. *Journal of Emerging Trends in Educational Research and Policy Studies (JETERAPS), 7*(4), 292–301

Almadina Newspaper. (2011). Education: 80% degrees of the teachers incentives based on students performance. Retrieved from http://al-madina.com/node/306304

Alnahdi, G.H. (2014). Educational change In Saudi Arabia. *Journal of International Education Research—First Quarter 10*(1).

Alodadi, A., & Benhin, J. (2015). Religious Tourism and Economic Growth in Oil-Rich Countries: Evidence from Saudi Arabia. Retrieved from https://doi.org/10.3727/108354215X14464845877995

Alqarni, A. A. (2015). Educational technology in Saudi Arabia: A historical overview. *International Journal of Education, Learning and Development., 3*(8), 62–69.

Alshaya, F., & Abdulhameed, A. (2011). Project of mathematics and natural sciences: Hops and ambitions [In Arabic]. In *A paper presented in the 15the annual conference for the Association of Science Education.* Egypt.

Alsslloom, H. (1991). Education in Saudi Arabia [In Arabic]. The Saudi Arabian Culture Mission (USA).

Al-Sulaimani. (2010). The importance of teachers in integrating ICT into science teaching in intermediate schools in Saudi Arabia: A mixed methods study. Ed.D Dissertation, RMIT University.

Alyami, R. (2014). Educational reform in the kingdom of Saudi Arabia: Tatweer schools as a unit of development. *Literacy Information and Computer Education Journal, 5*(2), 1424–1442. Retrieved from https://doi.org/10.20533/licej.2040.2589.2014.0202

Behairy. H. M., Obeid. A. M., Bensaleh, M. S., Hennache. A. S., Haj Bakry, S., & Qasim. S. M. (2013). The role of national research centers and Universities in the economic development of Saudi Arabia. In *22nd annual International Conference on Management of Technology, IAMOT 2013.* Porto Alegre, Brazil. International Association for Management of Technology.

Center for Statistical Analysis and Decision Support. (2020). Saudi youth in numbers: a special report on the occasion of the International Youth Day 2020. General Authority of statistics

Retrieved in Octobre 4, 2020 from https://www.stats.gov.sa/sites/default/files/saudi_youth_in_numbers_report_2020ar_0.pdf

El-Deghaidy, H., Mansour, N. (2015). Science teachers' perceptions of STEM education: Possibilities and challenges. *International Journal of Learning and Teaching*, *1*(1), 51–54. https://doi.org/10.18178/ijlt.1.1.51-54

Emerging Technologies in Saudi Arabia. (2020). Vision 2030 Healthcare Initiatives Which Will Need Digitalization. National Transformation program 2020. Retreived from http://www.emergingtechksa.com/2019/vision2030/vision-2030-healthcare-initiatives-which-will-need-digitalization/

Esmail, H. (2018). Economic growth of Saudi Arabia between present and future according to 2030 vision. *Asian Social Science*, *14*(12).

Fanack. (2020). Geography of Saudi Arabia. Retrieved fromhttps://fanack.com/saudi-arabia/geography/?gclid=EAIaIQobChMIh9m588PZ6wIVztvVCh0K1wdREAAYASAAEgJDlvD_BwE

General Authority of statistics. (2020). Statistical Yearbook. 51. Retreived from https://www.stats.gov.sa/en/413-0

General Authority for Statistics. (2019a). My country in numbers: a special report on the occasion of the 89th National Day. General Authority for Statistics. Retrieved in Octobre 4, 2020 from https://www.stats.gov.sa/sites/default/files/national_day_report_0.pdf

General Authority for statistics (2019b). Our Children: A special report on in the World Children's Day 2019. General Authority for statistics. Retrieved in Octobre 4, 2020 from https://www.stats.gov.sa/sites/default/files/tfln_tqryr_bmnsb_lywm_llmy_lltfl_2019.pdf

General Administration for Innovation and Statistical Development. (2019). International Indicators Management. Human Development Report. General Authority for statistics Retreived in octobre 4, 2020 from https://www.stats.gov.sa/sites/default/files/human_development_index_report_ar.pdf

Ghahari, S., & Farokhnia, F. (2017). Triangulation of language assessment modes: Learning benefits and socio-cognitive prospects. *Pedagogies: An International Journal*, *12*(3), 275–294. Retrieved from https://doi.org/10.1080/1554480X.2017.1342540

Hakami, A. (2010). King Abdullah bin Abdulaziz public education development project: Tatweer. Retrieved in 28 May 2014, from http://www.tatweer.edu.sa/En/MediaCenter/Documents/Presentations%20of%20Educational%20Standards%20and%20National%20Testing%20Workshop/

Hargreaves, A., & Shirley, D. (2009). *The fourth way: The inspiring future for educational change*. Lynch School of Education. Boston College.

Hassan, O. (2020). Artificial intelligence, Neom and Saudi Arabia's economic diversification from oil and gas. *The Political Quarterly, 91*(1), 222–227.

Hatem et al. (2016). Determinants of economic growth in the kingdom of Saudi Arabia: An application of autoregressive distributed lag model. *Applied Economic and Finance*.

Huhn, Ph (2019a). Industry revenue of manufacture of computer, electronic and optical products in Saudi Arabia. Statista, 2011–2023.

Huhn, Ph (2019b). Industry revenue of repair of computers and personal and household goods in Saudi Arabia. Statista, 2011–2023.

Iqbal, M. (2017). Effect of continuous assessment techniques on students' performance at elementary level. *Bulletin of Education and Research, 39*(1), 91–100.

Kamal, A. (2009). Exemplary use of technology in K-12 education in Saudi Arabia: Dar Al-Fikr Private School. In *Proposal submitted to the conference of the association of educational communication and technology*.

Kamal, A. (2012). Enabling factors and teacher practices in using technology–assisted projectbased learning in Tatweer schools in Jeddah, Saudi Arabia. (PhD thesis). Kansas State University, Kansas.

Khorsheed, M. S., Alhargan, A., & Qasim, S. M. (2012). A Three-Tier service model for national ICT incubator in Saudi Arabia. IEEE. On Management and Service Science (MASS 2012). Shanghai, China.,1–4

King Abdulaziz City for Science and Technology. (2020). Strategic Technology Program Summary Document. Ministry of Economy and Planning. Kingdom of Saudi Arabia. Retreived in September 2020 from http://admission.kau.edu.sa/GetFile.aspx?id=52481

Kingdom of Saudi Arabia. KSA Vision. (2020). Strategic Objectives and Vision Realization Programs. Kingdom of Saudi Arabia. Retrieved in September 23, 2020 from https://vision2030.gov.sa/sites/default/files/report/Vision%20Realization%20Programs%20Overview.pdf.

Madani, R. A., Forawi, S. (2019). Teacher perceptions of the new mathematics and science curriculum: A Step toward STEM implementation in Saudi Arabia. *Journal of Education and Learning, 8*(3). Canadian Center of Science and Education.

Mahboob, A., & Elyas, T. (2014). English in Saudi Arabia: A historical and linguistics analysis. *World Englishes, 33*(1), 12–35.

Ministry of Education (2020a). Ministry of Education—Initiatives and Projects. Retreived in Octobre 9, 2020 from https://www.moe.gov.sa/en/TheMinistry/AboutMinistry/Pages/Initiativesa ndProjectsofTheMinistryofEducation.aspx

Ministry of Education (2020b). Curriculum Digital Content Portal. Retrieved in Octobre 9, 2020 from https://www.moe.gov.sa/en/eproducts/e-Services/Pages/e-Books.aspx

Ministry of Education (2020c). Noor Program. Retrieved in Octobre 9, 2020 from https://www.moe.gov.sa/en/eproducts/e-Services/Pages/NoorProgram.aspx

Ministry of Education (2020d). Education and Vision 2030. Retrieved in Octobre 9, 2020 from https://www.moe.gov.sa/en/Pages/vision2030.aspx

Ministry of communications and information technology. (2019). Saudi Emerging Technologies Forum. Ministry of communications and information technology. Retrieved from https://www.mcit.gov.sa/en/media-center/events/185776#:~:text=The%20Kingdom%20of%20Saudi%20A rabia,4.0%20across%20public%20and%20private

Ministry of Education. (2006). Educational System in Saudi Arabia. Retrieved from HYPERLINK "http://www.sacm.org/Publications/58285_Edu_comple te.pdf".

Ministry of Education. (2014). Ministry of education school study plans, Directorate of Curricula, Saudi Arabia, Riyadh, 2014.

Mwebaza, M. (2010). Continuous assessment and students' performance in 'A' level secondary schools in Masaka district. Ed dissertation, Makerere University, Kampala. https://www.mak.ac.ug/documents/Makfiles/theses/Mwebaza_Michael.pdf

National Digital Transformation. (2019). National digital transformation report 2019. Saudi Arabia's National Digital Transformation Unit. https://ndu.gov.sa/en/news/saudi-arabias-nat ional-digital-transformation-unit-releases-2019-report

Onuka, A., & Onabamiro, A. (2010). Challenges of and possible solutions to conducting educational research and evaluation in Nigeria. In 4th *annual conference of the west african research and innovation management association.* Morovia, Liberia.

Pariona, A. (2017). What Type Of Education System Does Saudi Arabia Have? Overview Of The Educational System In Saudi Arabia. WorldAtlas. Retrieved from https://www.worldatlas.com

Patalong, F. (2016). Vision 2030 and the Transformation of Education in Saudi Arabia. Altamimi & Co. Retrieved in 28 September 2020 from https://www.tamimi.com/law-update-articles/vision-2030-and-the-transformation-of-education-in-saudi-arabia.

Puri-Mirza, A. (2020). Number of teachers in school education in Saudi Arabia 2017–2018. Statista. Retrieved in September 2020 from https://www.statista.com/statistics/1097372/saudi-arabia-number-of-teachers-in-school-education/#:~:text=In%20the%202017%2D2018%20acad emic,Arabia%20amounted%20to%2030.75%20thousand.

QS Top Universities. (2020). QS world University rankings. Retrieved in Octobre 3, 2020 from https://www.topuniversities.com/university-rankings/world-university-rankings/2021

Rana, S., & Zubair, R. (2019). The reality of continuous assessment strategies on Saudi students' performance at university level. *English Language Teaching, 12*(12). Canadian Center of Science and Education. https://doi.org/10.5539/elt.v12n12p132.

Rosas, R. (2014). Elementary school teachers,, and principals,, formative assessment beliefs, practices, and assessment literacy. Unpublished dissertation. California State University, Fresno.

Saqib, N., et al. (2016). women empowerment and economic growth: Empirical evidence from Saudi Arabia. *Advances in Management and Applied Economics*, *6*(5).

Saudi Arabian Cultural Mission. (2006). Educational system in Saudi Arabia. Washington DC, Saudi Cultural Mission. Retrieved from: http://www.sacm.org/Publications/58285_Edu_complete.pdf

Scitech. (2020).Sultan bin Abdul Aziz Center for Science and Technology. Retreived in Octobre 9, 2020 from https://www.scitech.sa/en/page/About/79/Mission

Shin, J. C., Lee, S., & Kim, Y. (2012). Knowledge-based innovation and collaboration: A triple-helix approach in Saudi Arabia. *Scientometrics, 2012*(90), 311–326.

Statista Research Department. (2020). School students in Saudi Arabia 2013–2017. Retrieved in September 24, 2020 from https://www.statista.com/statistics/628665/saudi-arabia-total-num ber-of-students/#:~:text=As%20of%202017%2C%20the%20number,at%20a%20primary%20s chool%20level.

Sultan bin Abdul Aziz Center for Science and Technology disseminates the principles of science and technological innovations and presents them in interesting methods (Scitech, 2020).

The Social Development Bank. (2020). Emerging technology product. Retreived in Octobre 9, 2020 from https://www.sdb.gov.sa/en-us/our-products/projects/productive-loans/products/new technology

Tatweer. (2008). Retrieved from http://www.tatweer.edu.sa/Pages/home.aspx

Unesco. (2020). Education and literacy. Retreived from http://uis.unesco.org/en/country/sa

Unesco.(2010). Annual Report. UNESCO Office Jakarta. [Online]. Retrieved from http://unesdoc. unesco.org/images/0019/001921/192108e.pdf

Unicef. (2013). Saudi Arabia Statistics. Unicef Retrieved in December 24, 2020 from https://www. unicef.org/infobycountry/saudiarabia_statistics.html

WorldoMeter. (2020). Retreived from https://www.worldometers.info/world-population/saudi-ara bia-population/

Bader A. Alyoubi is a PhD in Knowledge Management since 2014. He has more than 35 scientific papers in many filed which are: Knowledge Management, Information Systems, Information Technology, Computer Sciences and Artificial Intelligence and so on which are published in peer reviewed journals and conferences. Furthermore, Awarded from university of Jeddah for excellent IMPACT FACTOR research, Eager to Published a lot of papers in different journals and Attendee conferences. Dr. Bader has obtained the Certified KPI Associate in 2017 and also obtained PMI_PMP Training Certificate in 2018. In addition, he has 6 years of academic experience.

Fathi Essalmi is a PhD in computer science since 2011. He has more than 70 scientific papers published in peer reviewed journals and conferences. Furthermore, he supervised international students in the research master degree and in their PhD thesis. Dr. Fathi has obtained the habilitation degree to direct research in 2017. In addition, he has 15 years of academic experience.

Chapter 12
Science Education in South Africa

Jako Olivier and Donnavan Kruger

Abstract This chapter provided an overview of science education in South Africa. From a broad introduction to the country and its current economic, technological and cultural development it is evident that the country's colonial and apartheid past has made a definite impact on the wider education system. This chapter, furthermore, presented a summary of the country's education in terms of school and post-school sectors. In addition, the relevant governmental structures and national qualification framework was explained. Some statistics were presented in terms of education in general as well as with regard to science education. This chapter also considered the wider research landscape in terms of the focus and outputs for the different levels of education. The current situation in the country regarding science education and policy, curriculum, teacher training and student assessment was also discussed. The chapter also provided an overview of science and technology venues and centers as well as the use of emerging technologies for science education in South Africa. Finally, some recommendations are proposed for future development of science education in the country.

Keywords South Africa · Science education · Primary school · High school · Higher education

J. Olivier (✉)
Research Unit Self-Directed Learning, Faculty of Education, North-West University, Private Bag X2046, Box 575, Mmabatho 2735, South Africa
e-mail: Jako.Olivier@nwu.ac.za

D. Kruger
Research Unit Self-Directed Learning, Faculty of Education, North-West University, 11 Hoffman Street, Potchefstroom 2520, South Africa
e-mail: Donnavan.Kruger@nwu.ac.za

R. Huang et al. (eds.), *Science Education in Countries Along the Belt & Road*,
Lecture Notes in Educational Technology,
https://doi.org/10.1007/978-981-16-6955-2_12

12.1   Overview of South Africa

12.1.1   Geographical Location, Population and Political System

To contextualize this chapter on science education in South Africa a short overview of South Africa as a country is provided. The history of science education and education in general in South Africa can be traced back to the arrival of the San and Khoi tribes in this geographical area at least 150 000 years ago. These earliest inhabitants and other later arrivals from elsewhere in Africa brought with them a rich oral history through which science has been conveyed for centuries.

The Republic of South Africa is situated at the southern tip of the African continent and covers 1 219 602 km^2. In 2019 it was estimated that the country has around 58.78 million people. Furthermore, the country is fairly diverse with the recognition of 11 official languages. However, despite some use of African languages in the early years of schooling, English and to a lesser extent Afrikaans are the main languages of learning and teaching. In terms of its political profile the country can be described as a parliamentary republic with an executive presidency (Tibane et al., 2019).

South Africa's colonial history is linked to Portuguese explorers passing by the coast in 1488 and the establishment of a Dutch victualling presence by the Dutch East India Company in 1652. This was followed by the establishment of a permanent Dutch colony which was followed by two occupations by Great Britain in 1795 and 1803. Most of the country remained under British control, despite opposition from descendants from Dutch, French and German colonists as well as many indigenous groups such as the Xhosa in the southeast and Zulu along the east coast.

The country gained a measure of independence after the formation of the Union of South Africa in 1910 as a dominion under the British Crown and ultimately becoming an independent republic in 1961. However, this history was not only affected by colonial practices of slavery and westernization but also a political strategy based on segregation based on race. This movement lead to discriminatory legislation and ultimately apartheid. Apartheid referred to a system through which racial segregation was institutionalized and this had a definite impact on education in the country (cf. Freund, 2019; Worden, 2012). Consequently, schools and universities were segregated based on race and there were stark differences in the quality between different systems as well as the range of opportunities that were available for students at all levels. Schooling in South Africa was initially linked to missionaries from Europe, but gradually a public-school system evolved.

Only after the fall of apartheid in 1992 and the subsequent first democratic elections in 1994 schools and universities were opened up for all races. A number of national school departments had to be merged into the Department of Basic Education (DBE). Furthermore, the newly formed Department of Higher Education and Training (DHET) focused on technical and vocational colleges, universities of technology as well as the so-called traditional and comprehensive universities (cf. Tibane et al., 2019).

12.1.2 Current Situation of Economic, Technologies and Cultural Development

During the colonial period of South Africa's history the immense mineral wealth of the country was discovered and hence mining for diamonds and gold resulted in many developments in terms of the economy, industry and even technology. However, after independence from Britain 1961 policies promoting racial segregation ultimately lead to sanctions and the country being cut off from many countries. During this period the country especially focused on developing in a self-sufficient manner and many state-owned companies for example used in the production of petroleum, steel and weaponry were extended (Freund, 2019).

After the fall of apartheid in the 1990s some of these companies were privatized and many mines either sold off or abandoned. Yet, South Africa remained one of the most prominent industrial and economic powers in Africa in terms of the economy and technology.

The current gross domestic product is estimated at US$369 billion while the gross domestic product in terms of purchasing power parity is estimated at US$833 billion (IMF, 2020). Despite some ongoing fluctuations, the South African economy showed a third consecutive quarter of economic decline at the beginning of 2020 falling by 2% (Statistics South Africa, 2020).

In terms of cultural development, the recognition of unity in diversity is emphasized. However, there are still stark differences between wealth along racial and cultural lines. Culturally, the country incorporates various African cultures with strong western influences. The concept of *ubuntu* is key to the South African society and is described by Letseka (2012) as an African approach where "a human being is a human being because of other human beings" (p. 57). Governmental initiatives regarding culture are driven by the national Department of Sport, Arts and Culture. In addition, there are also constitutional bodies such as the Pan South African Language Board and various statutory bodies and councils associated with this department.

The importance of indigenous knowledge (cf. De Beer, 2019) in South African education, prior to the arrival of Europeans is clear. Furthermore, there has been a trend towards increasingly recognizing the importance of such knowledge which was marginalized in the past. The inclusion of indigenous knowledge has also become very relevant recently as university students have demanded the decolonization of the curriculum as part of student protests. The protests were driven by social media campaigns associated with the #RhodesMustFall and #FeesMustFall hashtags (Cf. Jansen, 2019; Le Grange, 2019).

12.2   Overview of Education Development in South Africa

12.2.1   Education System and Policy

The South African education system can be divided into three sectors: early childhood development, schooling and post-schooling. Different governmental departments are responsible for these foci. According to the Constitution of South Africa (1996) everyone has the right to a basic education which also includes adult basic education as well as to further education, which the government should make available and accessible through reasonable measures (Constitution of South Africa, 1996).

Recently there has been a renewed focus on expanding early childhood development which covers children from 0 to 4 years (Government of South Africa, 2020). Educational activities for this group are handled by the Department for Basic Education in conjunction with the Department of Social Development.

The primary focus of the DBE is to maintain and support schooling from grade R to grade 12. Although financing and management of schools happen at provincial department level—South Africa has nine different provinces. According to law, school attendance from grade 1 to 9 is compulsory. The different bands, phases, grades and approximate ages as well as science or related subjects for children from grade R to grade 12 are summarized in Table 12.1.

Table 12.1   Bands, phases, grades, approximate ages and science subjects for South African schools

Band	Phase	Grade	Approximate age	Science subjects
General Education and Training (GET) band	Foundation phase	R	5–6 years	Life skills
		1	7 years	Life skills
		2	8 years	Life skills
		3	9 years	Life skills
	Intermediate phase	4	10 years	Natural sciences and technology
		5	11 years	Natural sciences and technology
		6	12 years	Natural sciences and technology
	Senior phase	7	13 years	Natural Science
		8	14 years	Natural science
		9	15 years	Natural science
Further Education and Training (FET) band		10	16 years	Life sciences and physical sciences
		11	17 years	Life sciences and physical sciences
		12	18 years	Life sciences and physical sciences

The Department for Higher Education and Training handles all post-school training. This department was established in 2009 after the then Department of Education was divided into two separate departments as was noted above. Broadly this department has a focus on universities as well as vocation and community education institutions. This includes public universities and around Technical and Vocational Education and Training (TVET) Colleges.

12.2.2 Statistics on the National Education

The total number of students, teachers, and schools in public and independent schools in South Africa for 2019 are summarized in Table 12.2.

An important indicator of school student success is the percentage of students that pass the final school National Certificate examination at the end of grade 12. The following table summarizes the pass percentage for grade 12 s for the past 10 years (Table 12.3).

According to the DBE (2020b) the number of students passing Physical Science are increasing with the following percentage of students achieving 30% or above in Physical Science and Life Sciences for the past five years (Table 12.4).

In terms of general school drop-out rates South Africa does not have a good track record. Despite a high rate of participation for grade 1 to 9 there seems to be quite high

Table 12.2 Number of students, teachers and schools in 2019

	Public	Independent	Total
Students	12 408 755	632 443	13 041 198
Teachers	407 001	37 856	444 857
Schools	23 076	1 922	24 988

DBE (2020a)

Table 12.3 Pass percentage for grade 12 s 2010–2019

Year	2010	2011	2012	2013	2014	2015	2016	2017	2018	2019
%	67.8	70.2	73.9	78.2	75.8	70.7	72.5	75.1	78.2	81.3

DBE (2020b)

Table 12.4 Percentage of students achieving 30% and above for physical science and life sciences

	2015 (%)	2016 (%)	2017 (%)	2018 (%)	2019 (%)
Physical Science	58.6	62.0	65.1	74.2	75.5
Life Sciences	70.4	70.5	74.4	76.3	72.3

DBE (2020b)

Table 12.5 Number of institutions, students and lecturers in the post-schooling sector in 2016

	Public universities	Private universities	TVET colleges	Community education and training colleges	Private colleges
Institutions	26	123	50	9	279
Students	975 837	167 408	705 397	273 431	168 911
Lecturers	19 214	8 188	10 792	14 259	3 090

DHET (2018), (2019)

drop-out among students in grade 10 to 12 with students starting 12 years ago and were supposed to finish in 2013 only 40% and in 2014 only 36% actually completed their schooling (Hartnack, 2017; Weybright et al., 2017).

South Africa's post-schooling sector involves a combination of public and private institutions focusing on vocational training through colleges as well as a wide variety of universities. Tertiary education enrollment in South Africa per 100 000 people is currently at 1 901 which exceeds the sub-Saharan Africa average of 765 (DHET, 2019).

The institution, student and lecturer numbers for the different post-school sectors in 2016 are summarized in Table 12.5.

Regarding science education, it is notable that for the enrolment of universities 295 374 students (30.3%) were doing degrees in science, engineering and technology in 2016 (DHET, 2019).

In terms of graduation rate, the rate was 20.8% in 2016 (DHET, 2019). When success rates, which relates to the proportion of students completing a specific year of university study, is considered then there is a distinct difference between contact and distance education with 83% and 67.6% success rates respectively (DHET, 2019). In 2016 about 65.4% of students studied through contact classes at university and 34.6% through a distance modality (DHET, 2009).

12.2.3 Educational Research and International Collaboration

Educational research for all levels is conducted through private institutions, non-governmental organizations, universities and different governmental departments. In a review of educational research done in South Africa from 1995 to 2006, Deacon, Osman and Buchler (2009) found that most of the research was small scale with 48% relating to schools, 32% on higher education and 9% on teacher education with the rest focusing on other aspects such as early childhood development, adult education, special education etc. The lack of large-scale studies in this context is ascribed to the pressures for increased publication by university academics (Deacon et al., 2009).

This issue is particularly relevant as universities receive financial incentives from government for research outputs.

In terms of school-related research, the DBE has set out the Research Agenda 2019–2023 (DBE, 2019b). This research agenda covers the following thematic areas: teacher development, teacher support, learning and teaching support material, teaching language/reading, teaching numeracy/mathematics, pre-grade R and grade R, heterogeneous classrooms and learning, information and communications technology, sector planning and budgeting, accountability and infrastructure (DBE, 2019b). Internally, the DBE's Education Management Information Systems unit is responsible for the management of education data within this sector.

Research and international collaboration for higher education is generally handled by universities and colleges themselves. In terms of research output, there has been a steady increase in outputs since 2010 (DHET, 2019). The Higher Education Management Information System (HEMIS) directorate within the DHET manages and publishes data on higher education institutions.

The journal publications by South African researchers in 2018 were mainly within the Social Sciences and Humanities (32%), followed by Health Professions and Related Clinical Sciences (18%), Economic and Management Sciences (11%), Physical Sciences (11%), Life Sciences (10%), Engineering & the Built Environment (7%) and another 11% shared among other disciplines (DHET, 2020). In terms of book publications, the Social Sciences dominates with 81.2% of the publications followed by Economics and Management Sciences (5.5%), Engineering and the Built Environment (3.5%), Physical Sciences (3.3%) and Life Sciences (2.2%) in the top five.

There is a move towards increasing publications in international journals and journals that are internationally indexed. But despite increases in education research in South Africa, in 2011 it was already evident that local research has little international impact (Wolhuter, 2011).

A key governmental agency in terms of research is the National Research Foundation (NRF). According to the National Research Foundation Act 23 of, 1998 this body's purpose is to support and fund research activities. In addition to managing funding and scholarship opportunities, the National Research Foundation also has a rating system for South African scientists. Furthermore, the Human Sciences Research Council (HSRC), which falls under the national Department of Science and Technology, also conducts and supports a lot of research in South Africa and the region.

International collaboration is handled through government departments and institutions like universities individually. In addition, South Africa is also part of the BRICS group of countries and consequently also works closely with Brazil, Russia, India and China. Furthermore, the DBE also has ties with many international bodies and similar departments in other countries including China, Japan, South Korea and the USA (DBE, 2019a). Due to the prominence of English as main academic and administrative language in the country, South African institutions also tend to have ties especially with the English-speaking world.

12.3 Current Situation of Science Education

12.3.1 Policies and Standards

The curriculum of the schooling system is determined nationally and currently subject content, standards and planning is determined by the Curriculum Assessment Policy Statements (CAPS) as was published in 2011. At school level, the CAPS document forms parts of the National Curriculum statement which 'gives expression to the knowledge, skills and values worth learning in South African schools' (DBE, 2011, p. 4). Furthermore, the CAPS documents for the different subjects (cf. Table 12.1) provide information about the general aims of the curriculum, time allocation as well as subject-specific information. In terms of Physical Sciences (Grade 10 to 12) for example, the CAPS provides an introduction to the subject, list of aims, time allocation and overview by topic and grade, overview of practical work, weighting and summary of formal and informal assessments (DBE, 2011). The following main knowledge areas are identified in the CAPS (DBE, 2011):

- Matter and Materials
- Chemical Systems
- Chemical Change
- Mechanics
- Waves, Sound and Light.

However, most of the content of the CAPS relate to the unpacking of the content in terms of time; topic; content, concepts and skills; activities; resources; and guidelines for teachers (DBE, 2011). The same type of approach is followed for other science-related subjects and grades.

At school level, *Schooling 2025* is the national long-term planning initiative for schools in South Africa (Government of South Africa, 2020). In this document, specifically sets increasing the number of students who pass Physical Science as a target. The South African Council for Educators (SACE) regulates the teaching profession while Umalusi, or the Council for Quality Assurance in General and Further Education and Training developments and manages the framework of qualifications at school level.

In terms of post-schooling the Council on Higher Education (CHE) is an independent statutory body assuring quality in higher education while the Quality Council for Trades and Occupations (QCTO) has a similar function in terms of occupational standards (cf. DHET, 2019). In South Africa there are also several Sector Education and Training Authorities (SETAs) that regulate training in different fields. The National Skills Authority is another statutory body which advises the Minister for Higher Education and Training regarding skills development. Furthermore, South African Qualifications Authority (SAQA) is the oversight body regarding the National Qualifications Framework (NQF). The NQF is summarized in Table 12.6.

Another important governmental department in terms of science education is the Department for Science and Innovation which has been combined with the DHET

Table 12.6 National Qualifications Framework (NQF) levels

NQF Level	Qualification
1	Grade 9 / GET Certificate / Adult Basic Education and Training level 4
2	Grade 10 / national (vocational) certificates level 2
3	Grade 11 / national (vocational) certificates level 3
4	Grade 12 (National Senior Certificate) / National (vocational) certificates level 4
5	Higher Certificates / Advanced National (vocational) certificates
6	National Diplomas / Advanced certificates
7	Bachelor's degrees, Advanced Diplomas, Post-graduate Certificates or BTech
8	Honours degrees, Post Graduate diplomas and Professional Qualifications
9	Master's degrees
10	Doctoral degrees

under the Ministry of Higher Education, Science and Technology. The Department for Science and Innovation's mission is to provide leadership, ensure an enabling environment as well as resources for science, technology and innovation in support of the country's development. The South African government's National Development Plan 2030 (National Planning Commission, 2012) sets increasing the number of students who are eligible to study science-based degrees as a target.

South Africa also has two prominent science education organizations: the Southern African Association for Research in Mathematics, Science and Technology (SAARMSTE) and the South African Association of Science and Technology Educators (SAASTE) (Rollnick, 2017).

12.3.2 Curriculums and Teacher Training

As noted before, the CAPS provides guidelines per grade in terms of the curriculum and assessment at school level. Within the GET band, the focus is on the introduction of scientific principles through the subjects of mathematics, natural sciences and technology and social sciences (Geography forms a component of social sciences) (grades 4–6). In the senior phase (grades 7–9) the subjects are restructured and technology is taken separately from natural sciences. In the FET band (grades 10–12), specialisation of science subjects are now possible as electives.

Mathematics is compulsory, but students have three electives to choose from. These include pure mathematics (algebra, calculus, trigonometry and geometry), technical mathematics (mathematics that support technical subjects typically presented in technical high schools that specialise in trade industries), and mathematics literacy (mathematics for everyday use in real-world scenarios, e.g. in budgeting interest calculations). To be able to qualify for post-secondary education

Table 12.7 Science subjects offered in the Further Education and Training (FET) band

Formal sciences	Pure sciences	Applied sciences (not offered at all schools)
Mathematics	Physical Sciences, Life Sciences	Agricultural Science, Engineering Graphics and Design, Equine Studies, Geography, Information Technology, Nautical Science, Sport and Exercise Science, Technical Mathematics, Mathematics Literacy

admission to fields in mathematics, science, engineering, medicine and accounting, students will need to take pure mathematics.

As for specialization in science subjects at school level (from Grade 10), students may then elect life sciences, physical sciences and several other applied sciences in the science and technology domains (Table 12.7). As determined by CAPS published in 2011, the science subjects available to the FET band includes representatives from formal-, pure-, and applied sciences (Table 12.7).

Teacher training is mainly done through various faculties of education at universities in South Africa after the closure of many teacher training colleges that were historically structured along racial lines (Chisholm, 2019). Teaching qualifications are structured and measured against the Minimum Requirements for Teacher Education Qualifications (MRTEQ).

In order to be able to teach in South African schools teachers need to be registered with the South African Council for Educators (SACE) and have appropriate qualifications. Teachers generally either complete bachelor degrees in education (BEd) or complete school subject-related degrees and then do postgraduate certificates in education (PGCE) or other similarly recognized postgraduate qualification. Lecturer training at college and university level is less regulated and is determined by the various policies and practices implemented at the different institutions.

Teacher training is usually done along the lines of the specific school phase (Table 12.1) and in the BEd degree would typically involve specializing on at least two science-related school subjects with additional subjects such as mathematics, language and various general didactical or education subjects. Generally, subject didactics is infused in the science-related subjects. Qualifications such as the internationally recognized PGCE is regarded as a professional capping qualification and implies that a student has already completed a Bachelor of Science degree with at least two university subjects recognized as being relevant for two school subjects. Here the emphasis is less on content, but more on pedagogics and practice. However, the way in which teaching qualifications are handled differ from different universities.

12.3.3 Student Assessment and Achievement

In South Africa, examinations take place at the end of twelve years of schooling (cf. Tables 12.1 and 12.2). When the National Curriculum Statement (NCS) was introduced in 2008, it was the duty of the Department of Education to set all Grade

12 examination papers and document and standardize the process to be able to provide enhanced reliability, validity and made assessment procedures unbiased, which also allows for comparison of performance across provinces.

The DBE replaced the Annual National Assessments (ANAs) with a new model of national assessments named the National Integrated Assessment Framework (NIAF), which commenced with pilot schools in 2017. The new model is sample based, externally moderated and administered in grades 3, 6 and 9 once every three years. The NIAF is a three-tier model structured on presenting components of (1) diagnostic tools to support classroom assessment, (2) summative end-of-year examinations that aligns to the CAPS, and independently administered systemic evaluations that will be used to monitor student trends and report on the quality of learning outcomes. This assessment framework will better inform government and the education sector, be more transparent if externally moderated, and monitor indicators on learning outcomes (Mlachila & Moeletsi, 2019).

The Human Sciences Research Council (HSRC) conducts the Trends in International Mathematics and Science Study (TIMSS) research in partnership with the DBE and the International Association for the Evaluation of Educational Achievement. South Africa has taken part in the TIMSS since 1995 and has been evaluating mathematics and science achievement of grade 8 or 9 level in 1995, 1999, 2003, 2011, 2015 and 2019 and in the grade 5 mathematics study, for the first time in 2015 (Reddy, 2018; TIMSS SA, 2019).

Trends in International Mathematics and Science Study results were practically unchanged between the assessment periods of 1995 and 2003 due to high educational inequalities in post-apartheid society. In 2011 trend analysis showed a 64 points improvement for grade 9 science and 67 points for mathematics from 1995 and slightly reduced educational inequalities from 2003 to 2011. In 2015 South African achievement improved from "very low" to "low" but are overall still one of the lower-performing countries (Reddy, 2018). Analysis of the TIMSS results for the 2019 trends are not available at time of publication. Furthermore, the Southern and Eastern Africa Consortium for Monitoring Educational Quality (SACMEQ) IV assessment of literacy and numeracy skills for South Africa followed the release of the TIMSS 2015 study and the2016 National Senior Certificate (NSC, cf. Table 12.2) results, jointly reporting advances in student achievement at the Grade 5, Grade 9 and Grade 12 levels (Chetty et al., 2017: 51).

South Africa is also in partnership with the Organisation for Economic Co-operation and Development (OECD) initiatives. Although the country is not included in the Programme for International Student Assessment (PISA) assessment reports, it adheres to 23 OECD instruments and is integrated in half of the OECD datasets (OECD, 2018, 2020). With the new NIAF for tiered assessment currently being implemented, South Africa's sample based assessments of mathematics, science and language literacy will also be more transparent when moderated externally, and ultimately be included in international assessment instruments like PISA.

In the FET band, subject specific assessments are prescribed by the respective CAPS documents. School-Based Assessment (SBA) is a continuous planned process of identifying, gathering, and interpreting information about the performance of students, using various forms of evaluation. These processes and assessments are coordinated by the district education offices. The SBA for science subjects generally consists of formal assessments of experiments for the practical component, two control tests as well as two examinations in grades 10 and 11. The only difference in grade 12 is that they write one control test, one midyear examination, one trial examination and one final examination as formal assessment. Informal formative assessments are at the teachers' discretion.

12.3.4 Outreach Science Education

Numerous science teachers have confidence in the value of partaking in science fairs to support students and develop their attitudes, skills, and knowledge to motivate them to pursue a career in science and be successful in the scientific and technological society (Czerniak, 1996: 21; Mupezeni & Kriek, 2018: 2). South Africa has a 40-year history of science fairs to look back at, with the first science fair, the Expo for Young Scientists, held in 1980 at Pretoria Boys High School (Gray, 2014). Since, more than 35 science fairs, festivals, expos and competitions are taking place across South Africa. From the literature (Mbowane, De Villiers and Braun (2017) it is evident that these science fairs have contributed to teachers' pedagogical knowledge, content knowledge and even pedagogical content knowledge. There are currently two national science festivals taking place in South Africa, namely SciFest Africa and Eskom Expo for Young Scientists.

To further support and promote community engagement, science centers and other venues are often used for school excursions. The South African Agency for Science and Technology Advancement (SAASTA) is a business unit of the National Research Foundation (NRF), the governmental funding body of research institutions in South Africa. The directive of SAASTA is to advance public awareness, appreciation and engagement of science, engineering, innovation and technology in South Africa, as well as to address capacity building within science centers. They offer programs like focus weeks, talent nurturing, public engagement, competitions, olympiads and festivals. Furthermore, the South African Network of Science Centres (SANSC) is an online management system for the Science Centre Network profiling and online monitoring which is coordinated by SAASTA.

There are more than 40 science centers in South Africa, spanning over all nine provinces. Twenty-six (60%) of these centers are located in only three of the South African provinces, with 23% (10) Gauteng, and an equal distribution between Western Cape and KwaZulu-Natal (19% [8] each). Statistics provided here only accounts for education-, science- and discovery centers, planetariums, observatories and science museums. Awareness is also provided by other science awareness spaces, e.g. nature reserves or game reserves, zoos, aquariums, animal sanctuaries, botanical gardens, industries, libraries, national research facilities, media, non-governmental organizations, not-for-profit organizations, community groups, malls, showgrounds and community halls, in addition to the science festivals and events (Hannan et al., 2016: 6).

Due to the importance of capacity building at science centers, the SAASTA commissioned the Southern African Association of Science and Technology Centres (SAASTEC) conference in 2016 (Hannan et al., 2018). The impact of training on the capacity of science centers can improve the promotion of STEM fields. In a survey Hannan et al. (2018) conducted at this conference which were attended by the participants of the Science Center Capacity Building Programme, 70% of the respondents indicated that they implemented new activities, programs or exhibits at their centers, and 68% of the respondents indicated that they collaborated with other staff members from the science center from which they attended job shadowing.

There is an extensive selection of science awareness spaces in South Africa and students show interest in the utilization of scientific methods of research. However, Mupezeni and Kriek (2018) found that students' understanding is very limited because of the absence of readiness and engagement. The South African government invests in capacity building of science centers and aids in the collaboration and growth of the science center network.

12.3.5 Utilizing Emerging Technologies

To be a globally competitive country, technology integration in science education will play an increasingly pivotal role in South Africa. For this to realize, South Africa faces inherent infrastructure challenges and budget constraints. The DBE grasps this and has announced that they have, in partnership with the Department of Communications and Digital Technologies, identified 152 sites in 76 education districts, to equip them with Virtual Classroom infrastructure. The department stated that through "this initiative, the Sector will fully embrace the digital revolution of remote learning" (DBE, 2020c). Furthermore, the DBE assigned a team that has developed the Coding and Robotics Curriculum for the GET band (Grades R–9). The department has finished the repackaging of this curriculum in July 2020 and presented it to Umalusi, the council who sets and monitors standards for general and further education and training in South Africa, for approval, after which we will soon see a phased introduction in schools (DBE, 2020c).

Unfortunately, the inequalities between rich and poor manifests itself in the digital divide. Higher resources schools will ultimately adapt easier and faster to the new requirements of the changing curriculum. The key is to make the technology available to the schools who cannot afford it and provide the necessary infrastructure to support it. That is why the DBE also announced that they will "provide every school with ICT devices, loaded with digital content" and have "finalised the implementation plan with the Mobile Network Operators, to provide all these schools with devices, connectivity, digital content, as well as ICT integration training for teachers" (DBE, 2020c). This statement from the department is evident of their commitment to realize Vision 2030, which expects that "by 2030, South Africans should have access to education and training of the highest quality, leading to significantly improved learning outcomes" (DBE, 2020c).

Virtual Reality (VR), Augmented Reality and Artificial Intelligence (AI) applications and integration still has some distance to clear in the South African schooling environment. However, VR in mathematics and science communities of practice (CoP's) are being explored in the classrooms, e.g. to experience being in the three-dimensional beating of the heart and watching the planets revolve around the sun in the middle of the classroom (Bridge, 2018). The utilization of VR in teaching and learning in South Africa is currently being investigated at research institutions (Ndayizigamiye, 2018). However, Extended Reality (XR), which includes VR, Augmented Reality (AR) and Mixed Reality (MR) is primarily viewed as a recreational activity, and research in the STEM educational context is still in its infancy (Solomon et al. 2019).

12.4 Requirements for Future Development of Science Education

Development of science education takes on a multi-faceted approach that will need to be considered. One of the most important directions needed to be taken is to better comprehend how science awareness spaces are interacting with the promotion thereof, in addition to what the potential contribution of science awareness from these spaces are (Hannan et al., 2019).

Teacher development and the improvement of resources remain an issue within the South African context. To this end, teacher professional development for science teachers is supported. In terms of resources, extending the use of open educational resources might be useful.

Science fairs can also be extended at more levels and in various places including in rural areas. Such events can also become events for teacher and community engagement and spaces where indigenous knowledge can be embraced.

The use of emerging technologies is also a possible aspect of development and hence for the future of education the use of e-learning and other digital elements in science education will become increasingly relevant. With the lack of science teachers in the country, technology could become a possible way through which science teaching can be expanded. To this end, appropriate mobile technologies could be useful.

As stated at the start of the chapter, English is the main language of education in South Africa at all levels. In order to address the classroom language needs of the multilingual South African student population, science resources in languages other than English and Afrikaans should be developed for all levels of education.

However, research has shown positive results in using existing linguistic resources in classrooms and incorporating African languages in this regard (Msimanga et al., 2017) as language has an influence on achievement in science education (Prinsloo, Rogers & Harvey, 2018).

It is acknowledged that in South Africa science is regarded as a "scarce skill subject" (DBE, 2016, p. 27) and that there is a need to improve not only numbers of school students completing Natural Science up to grade 12 level but also improve the teaching of sciences throughout the wider school context (DBE, 2016).

Keeping well-trained and quality science teachers in the schooling system is imperative. An alarming issue is also the attrition of science teachers and departmental officials who opt to do jobs in various industries and companies where their scientific background can be used as, according to the DBE (2016), they tend to "leave the sector to work and specialise in science-related professions outside education" (p. 35).

As with other South African teachers, science teachers and lecturers "are expected increasingly to enable future generations to interact intelligently and creatively with a technologically changing, digital and increasingly robotised world revolutionizing the world of work and education" (Chisholm, 2019, p. 174). In this regard, science teachers and lecturers will also have to adapt to the challenges posed not only by digital divide in South Africa, but also the needs of the emerging Fourth Industrial Revolution where the lines between human and machine are blurring (Schwab, 2016).

12.5   Discussion and Conclusion

From the overview provided it is evident that South African science education functions within a distinct historical and cultural context. There are also clear divides in the quality and availability of science education in the country. However, there is also evidence of a well-structured policy and regulatory framework for education at all levels.

It is evident that the DBE is committed to the promotion and development of science education. In the Department's Revised Five-Year Strategic Plan 2015/16–2019/20 (DBE, 2016) the following statement is made: "Our focus over the next five years will also be the improvement and progression of especially Mathematics, Science and Technology (MST). Our ultimate goal is to have MST and Reading offices in all provinces as part of strengthening support for improved curriculum delivery." (p. 3). In this document, in addition to improvement of access to information and communications technology, supply of workbooks, student performance in terms of Mathematics, Science and Technology is specifically mentioned as strategic objectives (DBE, 2016). According to the DBE (2019a) a Mathematics, Science and Technology Strategy was started in 2018 with training and resource support.

In terms of higher education, it is clear that since 2013 there has been rapid growth in the area of science, engineering and technology as it has become the most popular field of study (DHET, 2019). Despite some decline in research in the fields of sciences, engineering and technology in the period of 2011 to 2016, this area has been the source of most research outputs in the country (54.8% in 2016) as compared to business and management, education and humanities (DHET, 2019).

There is also a need to situate science education in South Africa within the local context and work towards decolonizing the science education curriculum. In this regard, the inclusion of indigenous knowledge can play an integral role. This approach should extend beyond schools and as Waghid and Manthalu (2019) state, universities should be "meaningfully and not tokenistically open to indigenous epistemologies and pedagogies without firstly transforming such epistemologies and pedagogies and benchmarking them with 'intelligible' Eurocentric paradigms" (p. 55).

In conclusion, it is evident that science education is a key priority for the wider South African education sector. However, within the policy and structure context explained in this chapter there is still a need for pedagogically-sound support and an increase in quality teachers and resources in order to foster effective science education.

References

Bridge. (2018). Virtual reality—possibilities for education. Retrieved from https://thevirtualspace.co.za/

Chetty, M., Moloi, M. Q., Poliah, R. R. & Tshikororo, J. (2017). The SACMEQ IV Project in South Africa: A Study of the Conditions of Schooling and the Quality of Education, Short Report. Department of Basic Education.

Chisholm, L. (2019). *Teacher preparation in South Africa: History, policy and future directions.* Emerald Publishing.

Constitution of the Republic of South Africa 1996. Retrieved from https://www.gov.za/documents/constitution-republic-south-africa-1996

Czerniak, C. (1996). Predictors of success in a district science fair competition: An exploratory study. *School Science and Mathematics, 96,* 21–28.

Deacon, R., Osman, R. & Buchler, M. (2009). *Audit and Interpretative Analysis of Education Research in South Africa: What Have We Learnt?* A research report submitted to the National Research Foundation March 2009. Retrieved from: http://www.academia.edu/download/498 5498/FINAL_REPORT_NRF_Education_Audit_31_March_2009.doc

De Beer, J. (2019). *The decolonisation of the curriculum project: The affordances of indigenous knowledge for self-directed learning.* AOSIS.

Department of Basic Education (DBE). (2011). *Curriculum and assessment policy statement grades 10–12: Physical Science.* Retrieved from https://www.sahistory.org.za/archive/caps-grades-10-12-physical-sciences

Department of Basic Education (DBE). (2016). *Revised Five-Year Strategic Plan 2015/16–2019/20.* Retrieved from https://www.education.gov.za/Portals/0/Documents/Reports/Revised%20DBE%20Strategic%20Plan%20March%202016.pdf?ver=2016-03-31-122004-950

Department of Basic Education (DBE). (2019a). Annual Report 2018/2019. Retrieved from: https://www.education.gov.za/Portals/0/Documents/Reports/Annual%20Report%20%20Print%20Version%2027%20Sep%2009H27.pdf?ver=2019-11-14-121458-073

Department of Basic Education (DBE). (2019b). Research Agenda 2019 2023. Retrieved from: https://www.education.gov.za/Portals/0/Documents/Publications/Researchrepository/ResearchAgenda_V1.pdf

Department of Basic Education (DBE). (2020a). 2019 School Realities. Retrieved from: https://www.education.gov.za/Portals/0/Documents/Reports/School%20Realities%202019%20Final%20.pdf

Department of Basic Education (DBE). (2020b). *Report on the 2019 National Senior Certificate Examination.* Retrieved from: https://www.education.gov.za/Portals/0/Documents/Reports/2019 NSC Examination Report.pdf

Department of Basic Education. (DBE). (2020c). Minister Angie Motshekga: Basic Education Dept Budget Vote 2020/21. https://www.gov.za/speeches/minister-angie-motshekga-basic-education-dept-budget-vote-202021-22-jul-2020-0000

Department of Higher Education and Training (DHET). (2018). Statistics on Post-School Education and Training in South Africa: 2016. DHET: Pretoria. Retrieved from: https://www.dhet.gov.za/DHET%20Statistics%20Publication/Statistics%20on%20Post-School%20Education%20and%20Training%20in%20South%20Africa%202016.pdf

Department of Higher Education and Training (DHET). (2019). Post-school Education and Training Monitor: Macro-indicator Trends. Pretoria: DHET. Retrieved from: https://www.dhet.gov.za/SiteAssets/Post-School%20Education%20and%20Training%20Monitor%20Report_March%202019.pdf

Department of Higher Education and Training (DHET). (2020). Report on the evaluation of the 2018 universities' research output. Pretoria: DHET. Retrieved from https://www.dhet.gov.za/Policy%20and%20Development%20Support/REPORT%20ON%20THE%20EVALUATION%20OF%20THE%202018%20UNIVERSITIES%20RESEARCH%20OUTPUT.pdf

Freund, B. (2019). *Twentieth-century South Africa: A developmental history.* Cambridge University Press.

Government of South Africa. (2020). *Education.* Retrieved from https://www.gov.za/about-sa/education

Gray, R. (2014). *Light comes out of the darkness: The history of eskom expo for young scientists.* Oxford University Press.

Hannan, S., Reddy, V. & Juan, A. (2016). Science engagement framework and youth into science strategy: science awareness spaces in South Africa. (Commissioned by the Department of Science and Technology, March).

Hannan, S., Reddy, V., Juan, A., & Arends, F. (2018). Science engagement framework and youth into science strategy: science centre capacity building project evaluation 2016.

Hartnack, A. (2017). *Background document and review of key South African and international literature on school dropout.* DGMT. Retrieved from: https://dgmt.co.za/wp-content/uploads/2017/08/School-Dropout-Background-Paper-Final.pdf

International Monetary Fund (IMF). (2020). Report for Selected Countries and Subjects. Retrieved from https://www.imf.org/external/pubs/ft/weo/2019/02/weodata/weorept.aspx?pr.x=29&pr.y=10&sy=2017&ey=2021&scsm=1&ssd=1&sort=country&ds=.&br=1&c=199&s=NGDPD%2CPPPGDP%2CNGDPDPC%2CPPPPC%2CPCPIPCH&grp=0&a=

Jansen, J. (2019). Making Sense of Decolonisation in Universities. In J. Jansen (Ed.), *Decolonisation in universities: The politics of knowledge* (pp. 12–27). Wits University Press.

Le Grange, L. (2019). Different voices on decolonising of the curriculum. In J. De Beer (Ed.), *The decolonisation of the curriculum project: The affordances of indigenous knowledge for self-directed learning.* (pp. 25–47). AOSIS.

Letseka, M. (2012). In defence of Ubuntu. *Studies in Philosophy and Education, 31*(1), 47–60.

Mbowane, C. K., De Villiers, J. J. R, & Braun, M. W. H. (2017). Teacher participation in science fairs as professional development in South Africa. *South African Journal of Science, 113*(7–8), 1–7. https://dx.doi.org/https://doi.org/10.17159/sajs.2017/20160364

Msimanga, A., Denley, P., & Gumede, N. (2017). The pedagogical role of language in science teaching and learning in South Africa: A review of research 1990–2015. *African Journal of Research in Mathematics, Science and Technology Education, 21*(3), 245–255.

Mupezeni, S., & Kriek, J. (2018). Out-of-school activity: A comparison of the experiences of rural and urban participants in science fairs in the Limpopo Province, South Africa. *EURASIA Journal of Mathematics, Science and Technology Education, 14*(8), em1577.

National Planning Commission. (2012). National development plan 2030: Our future—make it work. Retrieved from: https://www.gov.za/sites/default/files/gcis_document/201409/ndp-2030-our-future-make-it-workr.pdf

National Research Foundation Act 23 of 1998, Government Gazette, 397, 19017, 3 July. Retrieved from: https://www.gov.za/sites/default/files/gcis_document/201409/a23-98.pdf

Ndayizigamiye, P. (2018). Lecturers' perceptions of virtual reality as a teaching and learning platform. In *ICT Education: 47th Annual Conference of the Southern African Computer Lecturers' Association, SACLA 2018*, Gordon's Bay, South Africa. Revised Selected Papers (Vol. 963, p. 299). Springer.

Prinsloo, C. H., Rogers, S. C., & Harvey, J. C. (2018). The impact of language factors on learner achievement in Science. *South African Journal of Education, 38*(1), a1438.

Reddy, V. (2018). TIMSS in South Africa: Making global research locally meaningful. The Human Sciences Research Council. Retrieved from: http://www.hsrc.ac.za/en/review/hsrc-review-april-june-2018/timss-in-sa

Rollnick, M. (2017). The growth of science education organisations in the old and new South Africa. In F. S. Otulaja, & M. B. Ogunniyi (Eds.), *The World of Science Education* (pp. 27–41). Sense.

Schwab, K. (2016). *The fourth industrial revolution*. World Economic Forum.

Solomon, Z., Ajayi, N., Raghavjee, R., & Ndayizigamiye, P. (2019). Lecturers' perceptions of virtual reality as a teaching and learning platform. In: Kabanda, S., Suleman, H., Gruner, S. (Eds.), *ICT Education. SACLA 2018. Communications in Computer and Information Science*, (vol. 963). Springer.

Statistics South Africa (2020).Key findings: P0441 - Gross Domestic Product (GDP), 1st Quarter 2020. Retrieved from: http://www.statssa.gov.za/?page_id=1856&PPN=P0441&SCH=7746

Tibane, E., Mokoena, M., Honwane, M. (Eds.). (2019). South Africa Yearbook 2018/2019. Pretoria: GCIS. Retrieved September 8, 2020 from https://www.gcis.gov.za/south-africa-yearbook-201819

Timss SA. (2019). Exploring educational progress in South Africa. Human Sciences Research Council. Retrieved from: http://www.timss-sa.org.za/download/TIMSS-SA-Newsletter-October-2019.pdf

Waghid, Y., & Manthalu, C. H. (2019). Decoloniality as democratic change within higher education. In C. H. Manthalu & Y. Waghid (Eds.), *Education for decoloniality and decolonisation in Africa* (pp. 47–68). Palgrave Macmillan.

Weybright, E. H., Caldwell, L. L., Xie, H., Wegner, L., & Smith, E. A. (2017). Predicting secondary school dropout among South African adolescents: A survival analysis approach. *South African Journal of Education, 37*(2), 1353.

Wolhuter, C. (2011). The international impact of education research done and published in South Africa. *South African Journal of Education, 31*(4), 603–616.

Worden, N. (2012). *The making of modern South Africa: Conquest, apartheid, democracy*. Wiley.

Jako Olivier is a professor in Multimodal Learning at the North-West University (NWU), South Africa and he holds the UNESCO Chair on Multimodal Learning and Open Educational Resources. He obtained his PhD in 2011 in which he researched the accommodation and promotion of multilingualism in schools through blended learning. Before he joined the NWU as lecturer in 2010, he was involved in teaching information technology and languages in schools in the United Kingdom and in South Africa. From 2010 to 2015 he was a lecturer in the Faculty of Arts of the NWU after being appointed as associate professor in the Faculty of Education in 2015. During 2012 he was a guest lecturer at the University of Antwerp, Belgium. In 2018 he was promoted to full professor at the NWU. In 2018 he received the Emerging Researcher Medal from the Education Association of South Africa. His research, located within the NWU's Research Unit for Self-directed Learning, is focused on self-directed learning, multimodal learning, open educational resources, multiliteracies, individualized and contextualized blended learning, e-learning in language classrooms and online multilingualism.

Donnavan Kruger is a senior lecturer in the Natural Science Education subject group at the Faculty of Education and a member of the Research Unit Self-Directed Learning at North-West University. His research interests include Life Science Education, inquiry-based learning, contextualized education, adaptive learning technology, blended learning and self-directed learning. He was also awarded a research fellowship under the guidance of the UNESCO Chair for Personalised and Adaptive Distance Education.

Chapter 13
Science Education in Republic of the Sudan

Adam Tairab, Omsalam Alameen, and Omer Babiker

Abstract This paper sheds light on science education in the Sudan where educational progress still varies between change and progress based on the political and economic situations. The paper showed the overall situation of science education in the country and discussed the barriers hindering science education progress. The paper used the government official reports, publications and websites to get the data, and then showed the situation of science education. During the study, the situation of the policies and plans, teacher training, digital resources and assessment methods still need more work and financial support from the government and stakeholders to make comprehensive science education reform. Therefore, the Sudanese community needs to wait for a time to see the required progress in science education. As a result, the paper recommends the regional and international community to support education in Sudan to develop and keep-up with the international progress with focus on science education and providing schools with science laboratories, materials and devices combined with massive training for teachers and improve school's infrastructure. Accordingly, the schools and universities may lead the learners to required progress in Science education. Nevertheless, the study faced difficulties to get up to date data due to the less time since the new government took place after June 2019. Hence, the new educational including science education efforts are still in the process to be officially shown and published for the researchers and followers.

A. Tairab (✉)
Smart Learning Institute, Beijing Normal University, Beijing, China

O. Alameen
Department of Educational Planning, Federal Ministry of General Education, Khartoum, Sudan

O. Babiker
Minister Office, Federal Ministry of General Education, Khartoum, Sudan
e-mail: Omer@moe.gov.sd

© The Author(s), under exclusive license to Springer Nature Singapore Pte Ltd. 2022   205
R. Huang et al. (eds.), *Science Education in Countries Along the Belt & Road*,
Lecture Notes in Educational Technology,
https://doi.org/10.1007/978-981-16-6955-2_13

13.1 Overview of the Country

13.1.1 Geographical Location, Population and Political System

Republic of the Sudan this is the official name of the country. The Sudan located geographically in North-East Africa continent with sharing borders with eight African-Arab countries namely Libya and Egypt in the northern side, Ethiopia and Eritrea in the eastern side, South Sudan in the southern side, Central Africa, Chad in the western side (World fact book; Cia.gov; Worldplays.com). The Sudan is third-largest African country with total area 1.9 million square kilometers after separation of the southern side of the country which is called republic of South Sudan in 2011 (UNESCO, 2018; Elmstba.com, 2019). Based on the last estimation the population of the country estimation for the year 2020 shown as follows; Population estimation (43,849,260), world ranking; 35th after Algeria and Iraq, number of males; 21,907,295, with percentage 49.96%, and the number of females 21,941,974, with rate 50.04%. Furthermore, the average age estimated with 66.09 years, and infant mortality rate; 38.0% (Elmstba.com, 2019).

The political system of the Sudan was military system for 30 years from 1989 to 2019 when the situation of the country started to change in December 2018 when Forces for Freedom and Change (FFC) led revolution of September as a biggest and peaceful revolution in the Sudan to enforce the government of Al-Bashir to step down after governance the country for three decades. After very hard negotiation between the Military transition council and the leaders of FCC under help of African union and united nation with special support of the republic of Ethiopia and Arab University the Transitional Military Council (TMC) and the Forces for Freedom and Change (FFC) signed constitutional document in August 2019 to lead the country for 3 years as a transition period (Sudan Constitutional Declaration, 2019). In September 2019 a new government took office leaded by Prime Minister Abdalla Hamdok as part of power-sharing agreement between the military and civilian representatives of the protest groups to transit the country to democracy system (BBC, 2019). The negotiating parties signed a constitutional document to be relied upon to govern Sudan during the transitional period, and negotiations are still ongoing between the transitional government and the groups of armed struggle movements to sign a comprehensive peace agreement that puts the country on the right path for a bright future. As a result, the government revenues will decrease during the current year by not less than 30% due to the decline in work in number of sectors. Therefore, the government is working on developing short and long-term plans if the government resorts to closing Sudan completely, especially since this step requires huge amounts of money (coface.com, 2020). The short-term plan deals with emergency solutions and responding to the Corona pandemic, whilst the long-term focuses on helping the economy to grow up. The government's efforts currently within the national plan focus on assisting the health system and avoiding its breakdown. An amount of 30 million pounds has been allocated from the

Ministry of Finance, other than what has been approved in the current budget. Also, 20 million pounds were approved as part of the approved plan to help vulnerable segments of society and families affected by the ban. Addition to call from the government for concerted efforts by the government and society, as well as the international community, to prevent the breakdown of the health system of the country (sudantribune.net, 2020).

13.1.2   Current Situation of Economic, Technologies and Cultural Development

The economic situation of the Sudan is facing challenges, since 2011 when the Separation of the Sudan and south Sudan has choke hit the economic growth severely, although a bit increases in fields of education, health in terms of employment salaries, however, the standards of living still not stable level where people struggling to get the drinking water, electricity and transport as basic daily life needs (Akol & Okech, 2020). Based on the complexities of the political situation in the country nowadays, the economic situation highly affected, as the prices of the US dollar showing very prevalent. Furthermore, between calls by economic experts to change the currency and other calls to float the exchange rate, the Sudanese economic reality remains complex and difficult to predict its future indicators, which citizens hope to improve through real production and reflection of that on their lives as soon as possible. As a result of the previous governance, the Sudan is in debt distress, reducing its capacity to mobilize the domestic resources or to lent from global markets or receive financial support from the donors. The economics of the country is in unstable situation due to political situation and health situation of pandemic COVID-19. These situations hardly affected the daily life needs of the majority of Sudanese people. The economic growth is facing real challenges for instance in April 2020, based on the International Monetary Fund forecast for the rate of economic growth in Sudan during 2020 decreases from -1.2% to -7.2%—based on the outbreak of the COVID-19 pandemic. By looking at performance of the microeconomics of the Sudan, considering the statistics of the African Development Bank, the GDP of Sudan is estimated at 2.4% driven by contraction in sectors of services and real estate investment and Inflation in business services expected at 61.5% for 2020 and 65.7% for 2021. This is mainly driven by the monetization of the fiscal deficit, which reduced from 7.7% of GDP in 2018 to 5.7% in 2019 nonetheless is projected at 9.9% in 2020 and 10.9% in 2021 (AFDB.Org, 2020). Addition to these statistics, the exchange rate of the Sudanese Pound vs US $ is 55 according to the official data of the Central Bank of the Sudan (CBOS, 2020). Consequently, the transition government working with the international community and the American government to remove the Sudan from the list of countries supporting terrorism. This step is expected to enable Sudan to return to his normal role among the world and make progress in limited time taking advantage of the enormous resources that country has.

The challenges facing Sudan to overcome the economic status also include dealing with massive population movement (IDPs, refugees and migrants) caused by instability of the neighboring African countries (Eretria, South Sudan, central Africa, and Chad) and Arab countries Palestine and Syrian refugees and the continuation of internal conflicts. Addition to the serious environmental challenges that Sudan faces rising mainly from the current extractive livelihood systems, conflicts and climate changes (National Population Council, 2019).

The observer to the current Sudanese scene and its details feels confused about what is happening in a country rich with enormous natural resources for instance agricultural land of the Sudan estimated with 63% of country, however, only 15–20% of this land is under cultivation, the country also rich with other natural resources such as natural gas, gold, silver, chromite, asbestos, manganese, gypsum, mica, zinc, iron, lead, uranium, copper, kaolin, cobalt, granite, nickel, aluminum and tin as well as water and animal resources (AFDB.Org, 2020; Raphaëlle Chevrillon-Guibert, 2017). Consequently, the authors suppose that Sudan lacks who and how to extract these resources and build a strong economy for the country.

Regarding technologies development in the Sudan, there is concern of technology officially and from the private sectors, as example there are many institutions for science and technology part of it is public and the other part private, addition to the youth societies to promote culture of technology and make the Sudanese community more interactive with technology and take advantages of the change in the life and enable technology to best serve the people. There is a national information center (NIC) working as an e-government portal. It is a comprehensive center for all e-government services and the main channel which provides all types of services to individuals, business sectors, government and visitors, from information to conducting transactions, etc. This, in turn, leads to highlighting the positive effects of using information technology to achieve high levels of competitiveness and transparency in government work. This shift enhances the role of information and communication technology in all aspects of life and works to motivate the public to integrate with the new world of information technology and make the advanced uses of information and communication technology an integral part of their daily needs. The portal allows its visitors to know lots of information, services, systems and laws related to the Sudan. Through multiple access channels such as the web or various means of communication technology, the e-government portal site is in line with the strategic vision of the Republic of the Sudan by providing services and interactive transactions, including online payment, and also through the portal site, an ideal platform will be provided for customers to submit their suggestions and opinions and participate in formulating government policies and improving service delivery methods (NIC, 2020 http://nic.gov.sd/structure.php#). Also, as part of the requirement to implement the project of e-government, there are some steps approved such as specifications and standards, telecommunication infrastructure, fiber optics cables, infrastructure of information technology, internet accessibility and expanding the network over the country states. The figure below shows what has been done in these aspects. Number of standards were approved and issued in 2016 in all e-government axes as follows; electronic services estimated with 4 standards, Networking and gear; 5, information security; 1, software and system; 16 addition to 7 designed websites. Notable, the overall standards are 33standards to implement the e-government.

Lengths of optical fiber cables in Sudan reached 31,425 km for sectors of electricity, petroleum and telecommunication companies. Also, the services have been launched. Till January 2020, the number of mobile subscribers in Sudan 2020 reached 32.83 million mobile connections in the country. This number increased by 2.3 million (+7.4%) from January 2019 to January 2020, this statistics equivalent to 76% of the total population of the country. Number of internet users 13.38 million, increased 316 000 during the period 2019–2020 with ratio (+2.4%), with number 1.30 million user for social media information (Kemp, 2020). Moreover, the National Information Center and the National Communications Authority, Bank of Sudan designed a SADAD platform for electronic payment service.

Politically, the expansion of the geographical area has led to the adoption of the federal system as the ideal form of government in Sudan, and according to the Interim Constitution of (2005) Sudan is divided into (18) states, each state governs a government headed by the governor and a number of ministers in addition to the state legislature. But after the December 2018 revolution that toppled the previous regime and resulted in a return to the regional system, new regions were added to the previous regions which were four regions.

The religions in Sudan include Islam, Christianity. Majority of the population are Muslims, there are minorities embracing the Christian religion. The strengthening of the emergence of Christianity is attributed to the movements European Missionary.

Concerning the cultural development, the Sudan is a rich with diverse components such as music, cinema, the stage, fine art, and folklore, Sudan is distinguished by its cultural, ethnic and religious diversity, as it has many tribes, languages and local dialects, and the Arabic language is the language spoken by the majority of the people of Sudan, as the Sudan interim constitution of 2005 and the Comprehensive Peace Agreement (CPA) of 2005 stipulated that English is an official language for dealing Higher education and state bureaus. Moreover, the Sudan extends over a wide area between southern Egypt and the tropical regions in Central Africa, which has led to the existence of various ethnicities, cultures and religions, with Islamic cultural background for the majority of the population of Sudan mixing with African cultures. The site occupied by Sudan represents a meeting point between the civilizations of the Mediterranean region and Asia, and a catalyst for many migrations that have affected modern Sudan with this huge number of ethnic groups inhabiting it and whose cultural roots are diverse. The presence of the Nile River, which crosses his country from top to bottom, had a great impact on this country's civilization and cultural diversity. This has resulted in over a hundred different dialects being used to communicate in different parts of Sudan. There are also over fifty ethnic groups divided into over nine hundred smaller groups. This advantages and the wide and old culture history which is a civilization dated back to 3000 B.C.

(http://mininfo.gov.sd/en/) this characteristic makes the Sudan rich in literature and arts in terms of poetry, prose and music, despite the multiplicity of languages and cultures in Sudan. Culture has a great influence on the majority of the people of Sudan, and Sudanese are intertwined in most of the daily and social practices, marriage, events and social customs. In rural societies, interest in issues related to tribal customs is slightly higher than in cities and societies that are more aware, overlapping and mixed. Sudan abounds in a diversity of literature and arts in terms of poetry, prose and music, despite the multiplicity of languages and cultures.

Regarding Music, Art folklore and heritage, Sudan has distinguished music based on the five-point scale, which is the musical scale to which the music of China, Scotland, Puerto Rico, Mauritania, southern Morocco, Ethiopia, Eritrea and Somalia belong. The roots of modern Sudanese music go back to what is known in Sudan as bag music, which in turn goes back to the religious chants of praise that were widespread among Sufi groups since the kingdoms of Sudan in the Middle Ages. Bag music blended with ancient African and Nubian musical heritage. Percussion instruments were used with clapping, then stringed instruments, most notably the tambour or the rababa, were used in addition to flutes and brass (Ministry of information, 2020). In 1940, British colonial administration in Sudan established Radio of Omdurman, with the purpose of propagating its war against the armies of the Axis powers in North and East Africa, the music received for the first-time official attention. Since that period, Sudanese music has made tremendous steps in its development to keep pace with global development. Sudanese cinema is based on heritage stories and the production of short films and documentaries. Several Sudanese short films have won international awards such as "But the Earth Spins," which won the gold medal at the eleventh Moscow Festival in the 1970 Documentary Film Competition. "The Shrine", which won the gold in the Cairo Short Film Festival in 1970, and "The Camel", which won the Critics' Prize in Cannes in 1986.

Nevertheless, the stage, theatrical activity began in schools and clubs as a means of education, awareness and guidance since the beginning of the first millennium, fine arts and folklore Sudan has known the art of drawing, engraving and sculpture since ancient times, and the walls of the temples of the Nubian kingdoms in northern Sudan still bear the effects of these works and the situation continued until the eras of the Christian kingdoms. In the modern era, fine arts have been closely linked to the national heritage, the local environment, social development in Sudan, and global transformations in the fields of arts. Folklore: This heritage is represented in the handicrafts that arose in a natural way from within the community using local materials such as pumpkin plants, leather, fronds and wool. Notable folk dancing arts such as Jeri, Hissis, Tuya, Agila, Kembla, Mardum, Hakama singing, Umm Ady, Duria, Kabsa and Al Bastha. Each tribe has its famous dances on various occasions such as harvest, marriage, circumcision, etc. Yet written literature is almost limited to the Standard Arabic language and the Sudanese Arabic African dialects, where a literary movement emerged and developed similar to its counterparts in the Arab and African world, especially in writing poetry and story. Short, criticism and translation. Luminous names emerged such as Tayeb Salih, who was recently called the genius of the Arabic novel for his world-famous novel Season of Migration to the North, Professor Abdullah al-Tayyib, author of The Guide to Understanding and Making Arab Poems (five volumes), Laila Abu Al-Ola, and the poet Muhammad Miftah Al-Fitori and Al-Tajani Youssef Bashir and others (Ministry of information, 2020).

13.2 Overview of the Education Development

13.2.1 Education System and Policy

The educational system in Sudan consists was 2-8-3 (Tairab & Huang, 2017). But after recommendations of educational conferences since 2012 to change this system, the ministry of general education (MFoGE) and after the revolution of September 2018, the change has been approved and will start in 2022 to be 6-3-3 as follows; 2 years for pre-school stage, 6 years for primary school, 3 years for middle school and 3 years for secondary school to complete the general education to grade 12 (K-12 education). Nonetheless, by the end of grade 12 the students have to set for an exam to get the secondary school certificate or as it called locally by Sudanese certificate which is enable the students to join the higher education level. Pre-school level includes khalawi and kindergartens, khalawi is for teach Holi Quran for kids. By the end of this stage children can join the primary education stage. So, this stage considered as preparation for kids to join the official education which started from the primary school or grade one. The primary schools stage consists of 6 years from the age of 6–11-year-old. By the end of primary schools, the students have to take an exam to join the middle school which students age from 12 to 14-years old. By the end of middle school, the students have to set for national exam to join secondary school. Notable, the general education in Sudan consists of two sectors; first one is academic education, and the second one is technical/ vocational education. Secondary schools encompass of three years. The Students age from 15 to 17 years old, by the end of secondary school student have to take a secondary school exam to join higher education. The universities or higher education institutions under the responsibility of the National Council of Higher Education and the Ministry of Higher Education and Scientific Research, which determines the policies and direction of the country's higher educational system that consists of public and private universities and institutions. The study duration in higher education ranges from 2 to 3 years for diploma degree, 4–6 years for bachelor's degree and 2–4 years for graduate studies (Mayada et al., 2020). Notably, the official language in primary and secondary levels is Arabic but in universities and higher education institutions are Arabic and English languages. Regarding the last update of the educational policies and plans to enhance education in the Sudan, the Ministry of general education held on August 15–16, 2020 the December Revolution Conference for the Advancement of Education, the aim of the conference was to provide opportunities for expanded dialogues between the stakeholders in education, to participate in making and approving public education policies and to prepare a national document for the future of education (December Revolution Conference for the Advancement of Education, 2020).

13.2.2 Statistics on the National Education

In this space the study shows numbers of the teachers, enrolled students of dropout rates per year.

No	Sector	Pre-school education	Basic schools	Secondary Schools
1	Number of teachers/ supervisors	42,780	192,022	71,881
2	Number students	1,099,653	6,226,516	1,107,886
3	Number of schools	21,172	19,398	4567
4	Enrolled students' rates	46.1%	73.5%	39.9%
5	Dropout rates per year	Not accounted	1.7%	Affected by the movement of the students between private and public schools

Source Ministry of general education-department of policies and planning and research 2018

There is no dropout rate per year for basic schools, because of the transition of students from private and public school or the movement of the students between the country states. However, the ministry of general education is working on establishing database for this issue.

13.2.3 Educational Research and International Collaboration

Ministries of education of Sudan, care of educational research through projects to be implemented and also in terms of published literature. For instance, there is a department in the ministry of higher education, there is a corporation of scientific research and innovation. This department works to implement the goals of promoting international cooperation in higher education and scientific research, whether through adopting national policies or setting up special international arrangements regarding facilitating the return of highly trained scientists and researchers to the country, whether permanently or temporarily through the process of recovering competencies through international cooperative programs that would strengthen institutions and facilitate employment Full of local capabilities as part of goals according to the quarter century plan (2003–2027) (http://www.mohe.gov.sd/index.php/ar/pages/details/23/1). The research is continuously working to come out with new and effective solutions to lead the country's progress. Ministry of higher education and the national center for the research and the national center for curriculum and educational research (NCCR) which belong to the ministry of general education working in cooperation and collaboration to provide solutions for the country's educational issues (Mohe; moe, 2020). The international collaboration focuses basically on training and funding projects. For instance, the ministry of higher education and ministry of general education cooperate with countries such as China, Eu, Korea, Japan, African countries such as South Africa and Kenya, and Arab countries in teacher training. Many people study in these countries for postgraduate studies and go back to Sudan to contribute to the universities and schools progress. Furthermore, many Agencies and organizations supporting projects implementation such as World Bank; African development Bank, Jeda Islamic Bank, Japanese agency (JICA), UNESCO, UNICEF, UNDP, etc. however, the cooperation of all these bodies still not in satisfy level due to the status of the Sudan and the need for comprehensive development especially in education and more focus in Science education. Where science labs and teaching still face challenges to be applied in a wide range in the whole or majority of the country regions (Mohe; moe, 2020).

Not so far, technology also involved especially from the people in the cultural field where forums and celebrations and diverse cultural events organized online specially under the current situation epidemic of COVID-19 (unicef.org, 2020). For instance, Sudanese youth used technology to create a better world! using creativity and skills to create change. During fall of 2019, a UNICEF-sponsored hackathon event which took place in three states over the countryside starting with the capital state Khartoum, and part of the western states include South Kordofan and East Darfur. The event is called a hackathon, it is an event that brings together persons interested in using technology to innovate or come up with novel solutions to the world's problems. Two months after the hackathon ended, 30 young guys with skills in information technology, web and application design, participated in a UNICEF-supported workshop to create mobile applications that would find solutions to these challenges. The participants organized themselves into four teams and worked to find out solutions to the following challenges: (a) The unemployment. (b) Bullying. (c) Violence against students. (d) Refugee education (unicef.org, 2019). Furthermore, there is a governmental body called Africa City of Technology (ATC) works to provides the necessary facilities and infrastructure for international companies in the field of technology, so that they are localized and integrated in the local and regional markets, as well as attract owners of new ideas to implement their projects where the environment is prepared to receive and finance ideas and projects. Africa City of Technology was approved by a decision issued by former president Omar Al-Bashir in 2008. In June 2012, Khartoum State and the Ministry of Science and technology agreed to establish a partnership to establish a city that would work in the field of technology incubators for the implementation of scientific research results, and to translate them into economic projects. It drives economic, banking and investment development in the country. Moreover, ATC organized workshops and forums, etc., to implement diverse projects, for instance, in November 2019 an event for implementation arrangements for the reverse linking project between the Sudan and Turkish to promote the localization of technology in Sudan, and in particular, the capacity of Africa Technology City in the field of technology incubation. After completing the project after 18 months, Africa Tech City will be able to provide effective support to research and emerging companies, and in this way contribute to the human and economic development of Sudan and as a hub for other African countries.

To overcome the obstacle of government funding and building a global partnership for education, the ministries of education are working on activating effective cooperation with donors and the international community to provide more funds for educational projects. For instance, UNESCO, UNICEF, World Bank, African development Bank, EU-funded national program and Private sector, in addition to the regional foundations like Bank of Jeddah for development loans, UN agencies and their partners contributed significantly to the provision of education services particularly for the vulnerable in remote and war affected areas. Nonetheless, in 2012 Sector partners together provided access to education for over 120,000 children, nearly 50% of whom were girls. Moreover, over 220 classrooms were constructed or rehabilitated, and some 1,250 teachers trained, contributing to a total of nearly 240,000 children receiving an improved quality of learning against a target of 350,00 children. Sudan: (UN and Partners Work Plan 2013). Moreover, the world bank in 2017 supported the project of Basic Education in Sudan: The Long Road to Stability. UNESCO in 2018 has supported the Sudan Education Policy Review. Based on UNESCO, the government of Sudan spends 2.22% of GDP on education and 10.8 of total government expenditure. The Ministry of Education is working to implement educational and educational cooperation agreements in the field of education quality with similar bodies at the local, regional and global levels. Furthermore, the ministry of education is working to enable the culture of quality and disseminate it to all elements of the educational learning process, and to control it with standards and indicators to improve education outputs to keep up with the local and global labor market. The international collaboration includes a number of agreements, memoranda of understanding and implementation programs were signed with friendly and brotherly countries, and also a number of activities part of it has implemented and part of it will be implemented. For instance; an agreement with the international partners to improve the basic education level in September 2019, an agreement with classira company in March 2020 about e-learning platform, memorandum of understanding with Malaysian ministry of education in July 2020,training courses and attendance at its workshops in (Egypt-Kuwait-Saudi Arabia-Qatar-Tunisia-Jordan), addition executive program with Turkey approved during the year 2018, as well as local agreements and memorandum of understanding with local universities and institutions.

The collaboration also includes Eu and japan, China, Korea, as well as collaboration with UNESCO, UNECIF, and ALECSO (General Administration of Cultural Relations: National Center for teacher training, 2020).

13.3 Current Situation of Science Education in the Sudan

13.3.1 Policies and Standards

Developing science education is one of the priorities for the countries and societies development in order to bring about long-term economic and social development. This comes through developing approaches for learning skills to enable the learners to engage in learning tasks to achieve quality science learning as required goal (Bustamante et al., 2018). In the Sudan, there is a project for Science education development aimed to develop science education in Sudan from all relevant aspects. The focus of all the limited initiatives in this regard in the past was limited in the field of developing science curricula in public education. The details of implementing these curricula with their various components did not find the necessary attention (developing science education project in Sudan, 2019). In the next section, the study shows the details of the science education project in the Sudan.

Within the framework of science education project in the Sudan, important issues in the field of science and technology were identified, part of these issues are (developing science education project in Sudan, 2019):

1. The deficiency in the basic competencies of the science teacher in terms of mastering knowledge in him/her field of specialization, weakness of competencies in research and investigation skills, acquiring thinking skills, lack of familiarity with teaching and learning strategies, good choice and use of teaching methods and teaching techniques in line with the nature of the subject matter and the characteristics of learners, and in a manner that achieves the creation of learning environments It promotes initiative, innovation, critical thinking, decision-making, problem-making and solving.
2. Teachers were not well prepared to become familiar with the goals and contents of the developed curricula and the strategies and methods of teaching of these curricula.
3. Lack of opportunities for science teachers to benefit from professional development programs effectively.
4. Weak professional competencies of educational departments that make them unable to face the related obstacles to the development of science education.
5. Insufficient support and capabilities to teach science in an educational environment.
6. As it is known, competitiveness in education is based on quality and excellence in knowledge acquisition and production. Consequently, science education in

the shadow of globalization is an education for excellence and universality, especially if it meets quality standards and their characteristics.

For the basic purpose of science education development project, it focuses on developing science education and its complementary and supportive materials, represented in mathematics and technology through:

1. Modernizing science curricula to support the development of science education in Sudan.
2. Good preparation for science and mathematics teachers through advanced preparation programs aim to the following: The objectives of the project to develop science education in Sudan: A/Providing the teacher with the main necessary competencies to prepare, represented in professional, performance, and evaluation planning competencies. B/Taking care of the teacher's tasks when using modern technologies and teaching methods. C/Taking care of recent developments in the field of specialization. D/Preparing the teacher for the changes in his roles imposed by the challenges of the times, as his mission towards the learners is no longer merely transferring knowledge and indoctrinating information, but will focus on raising the motivations of learners towards learning, identifying their needs for experiences, following their growth, diagnosing their weaknesses and drawing plans to treat them.
3. In-service teacher development and training: One of the main factors in increasing the teacher's efficiency is his desire to learn continuously, which may ensure him to accept new developments and maintain a high level of competence.
4. Developing and enriching the school environment in support of experimental (practical) aspects enriching science education.
5. Encouraging and motivating students, increasing their motivation to learn, and enabling them to think independently and learn by doing, by adopting modern teaching and learning methods that focus only on imparting knowledge and indoctrinating information.
6. The challenge posed by information requires a comprehensive review of science education in terms of repetition, curricula, teacher preparation and training, methods and tools, and evaluation of learning outcomes; since the goal of education is no longer merely the revival of knowledge per se. Nonetheless, the science education project aims to investigate the following:

   1. The status of preparing the science teachers in faculties of education in universities as preparation institutions.
   2. Science and mathematics curricula in basic and secondary stages which is prepared by the national council of curricula and education research (NCCR)
   3. School environment
   4. Reality of in-service science teacher training
   5. Meeting the government institutions
   6. Developing a constitution document to determine and assess the needs of different levels related to the development of science education in the Sudan.

To conclude this paragraph, federal ministry of general education (FMoGE) is entitled to make considerable progress in science education as part of investment in human capital at all education and training levels. As an appeal from the revolution of September to build the country in a satisfactory form for the Sudanese nation. Therefore, the policies include assessment for the current situation of education, developing the curricula, the infrastructure, teachers training, et., Providing quality, equitable and comprehensive education that takes into account the rights of all groups of children and youth in accordance with the (Sustainable Development Goal (SDG) 2030). Regarding the standards, the ministry of education believes that it is an urgent necessity to apply the basic concepts of quality assurance in the educational system and to use the national educational standards and indicators to measure performance effectively and with high quality. Ensuring the competencies and skills that students receive in the stages of general education, measuring the level of achievement in the different grades, measuring the effectiveness of these methods and conforming to the specifications of achieving goals, as well as measuring the performance of departments, professional, teaching and supervisory (http://www.moe.gov.sd/jooda.html#).

13.3.2 Curriculums, Digital Resources and Teacher Training

Science curricula wait for the teaching subjects in the final exam for the basic schools' exam which prepare the students to join the secondary school. Which means in the wait of the teaching subjects reflected in the wait of the detected degrees for each subject, therefore, the science subject does not show as a separated subject till grade seven in basic schools. The rate of teaching science curricula in basic schools estimated with only 20% form the teaching time during the year. In the curricula of the basic education, the subjects related to science taught with estimation to only 20% from the time of the teaching time during the teaching years. Moreover, in the current education system 3–8 still going on but it will be changed from the year 2021, there is no teaching for science as independent subject till grade seven, and the subjects related to the science knowledge taught from grade 3th to grade 6th, and the related subjects called description and applied arts. Below the study provides details for the science subjects curricula; the concepts and terminologies of the books for grade 3–5 designed based on the criteria method in the new curricula and distributed to the aspects of biology, chemistry, physics and earth science and space. While the books of grades 6–8 were designed based on the central and activities method. Science books included many terminologies and concepts graduated from simplicity in lower grades to extend in the higher grades matching with students' abilities. The teaching and learning activates from grade 6–8 focused on the diverse activities and teaching methods stand on the teacher, whiles the activities and teaching methods for grade 3–5 focused on brainstorming, interrogation, experimentation and dialogue, which indicates that the education in these books entitled to be based on the learner, according to the scope and sequence document prepared by the National Center for Curricula and Educational Research (NCCR). Regarding digital resources there is project IP Schools provide information and assessment of the students. The system also aimed to quickly get the information from schools and statistics of teachers, students, etc., as well as a platform called mint for digital curricula. The platform also includes a system for the school management with all the details related to the teachers and students. The platform does diverse jobs such as showing the educational content in a focused manner at specific points providing help for the teacher in the teaching process with efficient and creative ways. Improving the student's intelligence, stimulating students' thinking, activating intellectual abilities, and supporting modern learning strategies, develop the learner self-confidence, by providing learners with immediate feedback (https://www.manahgsudan.edu.sd). The features of the platform also include adding excitement and interest for the learners and adding creativity to the teaching using the motor, visual, auditory learning patterns. Using different learning strategies to support self-learning processes. Promoting the curricula behavioral values and decreasing the cost of education. Provide the student with enriching information that adds to his knowledge and emotional balance and helps him to carry out the classroom and extra-curricular activities. As part of digitizing the learning resources, the ministry of education created virtual lab experience.

13.3.3   Student Assessment and Achievement

The assessment methods focus on types of objective questions, and there are no tools for realistic evaluation. The teaching units lack continuous evaluation methods such as writing reports and monitoring laboratory learning outcomes. The assessment methods in the curricula books all the books referred to include the pre-constructive and final evaluation, and each unit also included final evaluation questions, but the dominant feature of the evaluation in all its stages and patterns focused on measuring cognitive achievement only and at the lower levels according to Bloom's classification of cognitive goals with a clear neglect of evaluating knowledge at its higher levels neglected as well as in evaluating the student's performance in the skillful and emotional fields. Generally, it can be said that modern criteria and standards for the textbook which responds to the needs of the learners, emotional, psychological, social, and cognitive, and the rapid economic, social, cultural, and informational changes do not seem to be fulfilled in science books at the basic education stage. The books also lack the built-in activities that aid students in their investigation, research and content enrichment skills to develop research skills and critical and creative thinking. There are no research activities to enrich knowledge and expand contemporary issues in order for students to participate in solving real life problems. Nevertheless, to assess student achievement and provide, the ministry of general education applied an education management information system (EMIS) to calculate population census data, enrolment rates, giving key system performance indicators and identifying locations which would benefit from new or continuing targeted interventions. Using EMIS has also helped in collecting relevant data from schools by creating a standardized questionnaire, raising the awareness of trained teachers to fill the questionnaire about key characteristics of their schools and students. The training opportunities for officials in data entry has enabled them to know and practice new technology (UNICEF, 2013). In general education, the student's assessment in science curricula is a part of general assessment and developing many skills include thinking skills through comparison, inference, analysis, problem solving, student's follow-up records, tests and final exams. While in the universities where each university has a college for science use different assessment methods for instance using practical tests in the science laboratories, theoretical tests, seminars and written exams. Some similarities between the schools and universities in using written exams as the main tool for student's assessment. In general, it can be said that modern specifications and standards for the textbook which responds to the needs of the learners emotional, psychological, social, and cognitive, and the rapid economic, social, cultural, and informational changes seem not fulfilled in science books of the basic education stage.

13.3.4 Science and Technology Venues and Centers

Science education is a road to the future in the knowledge field (Levrini, et al. 2019). Regarding science education activities, there is the national center for information, the information center belongs to the ministry of general education and the information center belongs to the ministry of higher education. Universities also have faculties of science with diverse majors and disciplines include natural science, agriculture, geology, energy, medicine etc. also there are many universities and institutions with name science and technology such as Sudan University of science and technology, University of technology, STU, Al-Bayan college for science and technology, Al Salamah college for science and technology, Medical university of science and technology, Abdul Latif Al-Hamad University of Technology, University of Managil for Science and Technology, and Africa Technology City (Mayada et al., 2020). Moreover, there are youth organizations-non government bodies manage completions in science and technology such as NEXT EINSTEIN FORUM (NEF). NEF is an organization and member of international events organizing a competition called falling walls lab every year and the final events of the competition are organized in Germany under German government sponsorship. This event and another science event called Africa science week are activities for a group called Africa Science Week under Sudan NEF organization. Africa science week organizes yearly competitions in many African countries and the winners of each country compete in the final events at African countries. For instance, in 2020, the final will be hosted in Kigali Rwanda by the end of the year 2020 under support of African Union and many other sponsorships bodies (https://nef.org/sudan/).

13.3.5   Utilizing Emerging Technologies

Using emerging technologies still in the planning level more than implementing level. However, there is a small center for technology in Khartoum state belonging to the national center for curricula and educational research (NCCER), in the technology center there is an experimental use for VR technology, virtual labs. Moreover, there are some initiatives for online platforms to offer the digital educational resources, such as the educational management information system (EMIS) established by the cooperation between Sudan and the European Union through UNICEF provided technical support to the project at federal level and in fifteen states in Sudan (UNICE, 2011). There are some other projects established by cooperation between some ministries, and part of it by some universities. As an example for the ministry's cooperation, the Ministry of Science and Communication have implemented Universal Access Projects, States Networking. The networking project includes software and applications such as educational information management system (EIMS), ministries and state education sites. Ministry's website to announce the result of secondary school results, questions bank programs, provide services through the ministry's website, for example: the deployment of student's desk numbers—Previous exams numbers, addition to internal mail and computing curricula. There is range of areas that reflect established and emerging industry best practice including graphic design, digital audio and video production, motion graphics, 2D and 3D digital animation, digital media technology, web and mobile application development, casual game development, social media, marketing, web advertising, web analytics and communications (MoHESR, 2019). Notably, there is a committee for using information and communication technology (ICT) in education, this committee was set up by the federal ministry of general education (FMoGE) to put a comprehensive plan for employing ICT in general school. The committee work is still going on, so the authors hope to publish the outcomes of this committee in the future.

13.4   Requirements for Future Development of Science Education

The requirements for the Sudanese science education include infrastructure: providing schools with interactive boards, white boards, establishing and developing science labs and micro labs in secondary school and basic education schools to conduct experiments and acquire practical and scientific skills. Teachers training in terms of technical programs in Science, English, Mathematics and computer, collaboration in scientific research, support curricula digitalization, offering online platforms, digital learning resources, Support extending educational access due to the big number of out of the school children. Supporting technical and vocational education to establish labs, technical workshops, extend the technical schools, etc. Furthermore, science education in universities also needs more support from the government to make required progress. However, offering all these requirements may not make progress if it is not well financially supported.

13.5   Discussion and Conclusion

Science education in Sudan facing diverse challenges (A general view of education as a service to be provided rather than a long-term human investment. of the academic curricula for science subjects in terms of their objectives, content, strategies and methods of teaching science, patterns, methods, and tools for evaluating the performance of learners in science. Science teacher: in terms of selection, preparation, training, professionalism and attitudes towards science education. The learner who does not find motivation and encouragement to study science. School environment: scarcity and lack of specific resources for teaching science. An educational administration that lacks qualified competencies to lead and manage the educational process in educational institutions, and which is unable to face the obstacles facing the challenges of science education in its various aspects.). Consequently, the need has become urgent for a fundamental reform in the educational system in order to provide students with basic knowledge and skills in science education to meet the challenges of the twenty-first century, which are difficult to confront except by striving to keep pace with modern trends in education, which are as follows based on the ministry of education and the national committee for education and culture: To shift from teaching to learning, focusing on the quality in providing knowledge (how well) rather than the quantity (how much). Shifting from a focus on teacher education to focusing on what the learner should learn. The shift from focus on academic achievement as a final goal to focus on acquiring the skills of self-learning, research, investigation and critical thinking. The main challenge is the political and economic situation. Since 1983–2019, the country has been under an unstable political situation for more than three decades, affecting the economy negatively and this economic situation reflected on the other part of the whole life of people not only on education. Therefore, the implementation of the educational policies and strategies under limited financial resources led to the education system producing the current outcomes may not represent the high expectation of the stakeholders particularly in science and mathematics. This point pushes the authors to appeal to stakeholders to support schools to implement the planned policies with concern for science education as one of the future progress of the nation's projects. Addition to the technical issues such as availability of financial support for printing the new curricula which was changed after the September revolution, and teachers training on the new curricula and using technology in the classroom. The statistics on the national education showed the enrolment rate of students is not high compared to the total population in the school age. For example 41% of Sudanese people are aged 14 years or younger, and 20% are between 15 and 24 years old (UNESCO, 2018). This means the ministry of education needs to find an effective strategy to expand the enrollment rate to not allow any child to be left behind. Besides, the educational collaboration somehow showed good relationships with regional and international countries. Therefore, activating the international collaboration to develop the curricula, train the teachers and offer technical support to improve the infrastructure with focus on digital learning resources and online platforms are necessary aspects for science education in Sudan. In the student assessments matter still the schools use traditional methods (final exam) as the main assessment method, as a result, the methods supporting twenty-first century skills need to be used effectively to encourage the students to develop more critical thinking to invent innovations and creativity. The study also showed existence of less science and technology centers and venous in public and private sectors. Hence, it is important for the government of Sudan to turn to this issue which helps youth and community sectors to be active and pay more attention to the role of science education to keep up with countries in the African continent and middle East. Nevertheless, utilizing emerging technology still in the infant level, where a small number of schools are equipped with computers and connected to the Internet services, and a large number of teachers need to be trained on how to use technology in the classroom. Thus, the issue of school digital infrastructure, digital learning resources, online platforms and learning management system (LMS), real-time feedback and skilled teachers and school principal's knowledge and motivation to employ technology in education will draw the future pathways of science education. Finally, the authors of this chapter believe in the existence of limitations faced in this study. The limitations represent challenges of getting the official data in published form since the existence of change in the ministry of education after the revolution of September still finished 1 year since the new government took place. Notable, the beginning of the science curriculum starts from grade three to the grade 8 in general education, the wait of these curricula assessed mainly by final exam. Consequently, the ministry of education is working on many sides to make changes in education, and this situation has an effect on the progress of education as a whole, not only science education. Regarding assessment; it is important for Sudanese educational system to assess the student learning outcomes and achievements by using standards such as PISA and TIMSS to keep up with regional and international progress in science education.

Acknowledgements The acknowledgment of this chapter to the federal ministry of general education (FMoGE) for the cooperation and providing the authors with the information in a critical time and in specific conditions of the Sudan recently and also under COVID-19 pandemic where country breakdown for long time. FMoGE also working on new plans and strategies for education will be published ASAP and the authors hope to see that in reality.

References

Akol, C. D., & Okech, T. (2020). The impact of financial development on economic growth in sudan. *Journal of Finance and Accounting, 4*(1), 54–71.

Bustamante, A. S., White, L. J., & Greenfield, D. B. (2018). Approaches to learning and science education in head start: Examining bidirectionality. *Early Childhood Research Quarterly, 44*, 34–42.

Raphaëlle, C.-G. (2017). The gold boom in Sudan. Challenges and opportunities for national players. *Open Edition Journal.* https://doi.org/10.4000/poldev.2318.

Ministry of Communications and Information, Community & culture. http://mininfo.gov.sd/-ثقافة-السودان/.

December Revolution Conference for the Advancement of Education. (2020). http://www.moe.gov.sd/#.

Daily Rates for forging Currency. Central Bank of Sudan. (2020). https://cbos.gov.sd/en/exchange-rates.

Education information center. http://www.moe.gov.sd/infocenter.html.

General Administration of Cultural Relations, Ministry of education. (2020). http://www.moe.gov.sd/thagafia.html.

Governmental expectations for negative growth of the Sudanese economy due to (Corona), Sudantribune newspaper. https://www.sudantribune.net.

Kemp, S. (2020). Digital 2020; Sudan. https://datareportal.com/reports/digital-2020-sudan.

Levrini, O., Tasquier, G., Branchetti, L., & Barelli, E. (2019). Developing future-scaffolding skills through science education. *International Journal of Science Education, 41*(18), 2647–2674.

Mayada, B. M., Ahmed, N. E., & Mohamed, M. E. (2020). Higher Education and Scientific Research in Sudan: Current status and future direction. *African Journal of Rural Development, 5*(1), 115–146.Major Macro Economic Indicators-Sudan (2020). https://www.coface.com/Economic-Studies-and-Country-Risks/Sudan.

Ministry of Communications and Information Technology. National Information Center. The position of implementing the liabilities of the e-government project 2016. Ministry of Welfare and Social Security National Population council. Sudan National.

Ministry of higher education & scientific research. (2019). http://mohesr.gov.iq/ar/.

Ministry of information. An Essay on Letters and Arts. (2020). http://mininfo.gov.sd/en/.

Sudan Mint platform for curricula. https://www.manahgsudan.edu.sd.

National information center (NIC). http://www.nic.gov.sd.

Sudan Economic Outlook. Macroeconomic performance and outlook. Report for African Development Bank. (2020). https://www.afdb.org/en/countries/east-africa/sudan/sudan-economic-outlook.

Sudan Education Policy Review Paving the road to 2030. UNESCO. (2018).

224

A. Tairab et al.

Sudan after Bashir: The revolution is on the curriculum. https://www.bbc.com/news/world-africa-50835344.

Sudan Constitutional Declaration August 2019. http://constitutionnet.org/sites/default/files/2019-08/Sudan%20Constitutional%20Declaration%20%28English%29.pdf.

Sudan population for the year 2020 | Sudan's world ranking in terms of population. https://www.elmstba.com/sudan-population/.

Sudanese youth and technology to create a better world/stories. UNICEF. (2020). https://www.unicef.org/sudan/ar/.

Tairab, A., & Ronghuai, H. (2017). Analyzing ICT policy in K-12 education in Sudan (1990–2016). *World Journal of Education, 7*(1), 71–82.

Chapter 14
Science Education in the United Arab Emirates

Towards an Inclusive Quality Education

Samia Kouki and Mouza Al Shemaili

Abstract This chapter addresses the current situation of education in the United Arab Emirates (UAE) and the strategy developed by the ministry of education in 2020, to achieve a significant qualitative improvement in the education system. In this chapter, the educational system of the UAE, with its diverse curriculums is presented. The economic, technological and cultural situation of the UAE is also discussed in this chapter. Since, the education is one of the pillar in the development of the country, this chapter highlights, the policies, the standards and the digital resources available to enhance the long-life learning. This chapter presents as well, the engagement of the UAE in targeting very high international assessment exams, such as PISA and TIMSS, in order to realize the UAE's National Agenda goal of an excellent quality education and also to prepare students with the necessary skills for the twenty-first century to become global citizens. The last section of this chapter presents the different facilities the UAE has established toward an inclusive quality education, by improving the teachers' skills, enhancing the use of emerging technologies in education, reviewing the curriculums and providing smart learning programs.

Keywords Twenty-first century skills · Smart learning · Educational strategy · Technology based education · Curriculum · Assessments strategies · Quality education

S. Kouki (✉) · M. Al Shemaili
Higher Colleges of Technology, Ras Al Khaimah Campuses, Abu Dhabi, UAE
e-mail: skouki@hct.ac.ae

M. Al Shemaili
e-mail: malshemaili@hct.ac.ae

14.1 Overview of the Country

14.1.1 Geographical Location, Population and Political System

On the 2nd of December 1971, the United Arab Emirates got its independence from the United Kingdom and was declared as an independent, sovereign and federal state. It is a constitutional federation of seven emirates (Abu Dhabi, Dubai, Ajman, Fujairah, Ras Al Khaimah, Sharjah and Umm al Quwain), where Abu Dhabi city serves as the capital. Its three largest emirates are Abu Dhabi, Dubai, and Sharjah and have about 85% of the total population (About the UAE, 2020).

The UAE is situated in the Middle East, bordering Oman to the east, and Saudi Arabia to the south and west, as well as sharing maritime borders with Qatar to the west and Iran to the north (About the UAE, 2020).

The total area of the UAE is about 83,600 km^2. The UAE maintains a desert climate. It is warm and sunny in winter, humid and hot, in summer, but it is generally cooler in eastern mountains.

Based on the World meter elaboration of the latest United Nations data, retrieved in September 2020, the current population of the UAE is estimated to 9,918,074. The UAE hosts more than 200 nationalities coming from all over the world. The UAE population is equivalent to 0.13% of the total world population, and ranks number 93 in the list of countries by population (United Arab Emirates Population (1950–2020), 2020).

The UAE is governed by a Supreme Council of Rulers made up of the seven emirs, who appoint the prime minister and the cabinet. Each emirate is governed by a ruler and together, they jointly form the Federal Supreme Council. The Constitution of the UAE provides five federal authorities, the Supreme Council, the President and the Vice-President, The Cabinet, The Federal National Council and the Federal Judiciary (About the UAE, 2020).

14.1.2 Overview of the Education Development

14.1.3 Education System and Policy

14.1.3.1 Overview of Education System

The education has been one of the highest priorities in the UAE since the independence. Based on the latest statistics, the literacy rate for both genders, in the UAE, is becoming very low (less than 1%), which is reflecting the efficiency of the education strategy built by the UAE to eradicate illiteracy (UAE Government, 2020).

The United Arab Emirates offer to citizens a free public education for both genders. The public educational goes from the kindergarten level trough the higher education. A private education system is also provided in the UAE and offered through three different models, national private schools, foreign private schools, and foreign community schools (Al Ghfeli M. 2015).

National public schools are government-funded (public schools are free only for citizens), they are following the federal ministry of education curriculum and using its textbooks. The official language of teaching in public schools is the Arabic language and English as a second language is emphasized (MoE, 2020).

Foreign private schools and foreign community schools are all private and many of them are internationally accredited where fees vary from a curriculum to another. All private schools are implementing their own curricula and textbooks, which should be approved by the ministry of education (Al Ghfeli, 2015).

The kindergarten education is offered for two years in mixed-gender classes, and it is noncompulsory for kids from age 4. However the basic stage of education, comprising cycle 1 and cycle 2, is compulsory for all children. Cycle 1 is covering grades from grade 1 to grade 5, and cycle 2 covering grades 6–8. The secondary level of education is offered at general schools, religious schools or at the Institute of Applied Technology. Thus, after completing grade 10, students are offered the possibility to join a public track or an advanced one, and will be receiving the Secondary School Leaving Certificate upon completion of grade 12 (MOE, 2020).

After grade 12 graduation examinations, students studying at public and private schools following the national curriculum, must pass the Emirates Standardized Test (EmSAT) to be able to join most of public universities and colleges (Education in the United Arab Emirates, WEST, 2018).

Table 14.1 summarizes the k-12 education system in the government schools of the UAE at different levels.

In 2009, children from a variety of special needs centers have joined public schools. A lot of efforts have been implemented to ensure the smooth integration of these children in public schools while providing the necessary support during the learning process (MOE, 2020).

Table 14.1 The public education system in the UAE

Cycle	Kindergarten	School cycles		
Level	Kindergarten (KG1 and KG2)	Basic level: cycle 1	Intermediate level: cycle 2	Secondary level: cycle 3
Age	4–5 years-old	6–11 years-old	12 and 14 years-old	15–17 years-old
Grades	KG1 and KG2	Grades 1–5	Grades 6–8	Grades 9–12
System	Noncompulsory	Compulsory	Compulsory	Compulsory

14.1.3.2   Statistics on the National Education

As per the latest statistics (2019), published by the UAE' Ministry of Education, the total number of public schools in the UAE is 619, where 21,153 teachers are assigned. The total number of students enrolled in public schools has been estimated to 288,794.

- There is a large number of private schools in the UAE which was estimated to 643 in 2019. The total number of teachers in public schools equals 50,869. The number of students enrolled in private schools is much more important than those enrolled in public schools, and it has been estimated in 2019 to total number of 810,537 students (MOEOpenData, 2020b).

In Table 14.2, we present the statists related to the total number of teachers working in private schools in the UAE based on the data published by the ministry of Education, in 2019.

We can notice from Table 14.2, that more than 80% of public school teachers are female and less than 20% are male, with a lower percentage of male teachers at the elementary school (KG, cycle 1 and cycle2) than at the secondary school level.

- In the 2019–2020 school year, there were 21,153 public school teachers in all the UAE public schools. The table below (Table 14.3) is presenting the number of teachers in public schools by gender and by level (MOEOpenData, 2020b).

In 2019–2020, a higher percentage of public school teachers are female, where male teachers represent only 28% of the total number of teachers in all public schools (MOEOpenData, 2020b).

- Based on the data published by the UAE' Ministry of Education (MOE), in 2019, the number of students private schools reached 288,794, however the total number of enrolled students has been estimated to 810,537, which represents three times the number of enrolled students in public schools. The table below (Table 14.4),

Table 14.2 Number of teachers in private schools by gender in each level

	Kindergarten	Cycle 1	Cycle 2	Secondary schools	Mixed classes
Female	3527	10,171	3507	4974	18,529
Male	131	1636	1305	2769	4321
Grand total	3736	12,115	5047	8102	25,363

Table 14.3 Number of teachers in public schools by gender in each level

	Kindergarten	Cycle 1	Cycle 2	Secondary schools	Mixed classes
Female	2191	5836	2892	2945	1352
Male	0	537	2356	2619	425
Grand total	2191	6373	5247	5565	1777

Table 14.4 Number of enrolled K-12 students in private schools by gender for each level

	KG1	KG2	G01	G02	G03	G04	G05	G06	G07	G08	G09	G10	G11	G12
Female	35,247	39,480	38,207	35,225	32,251	31,716	30,226	27,501	25,206	23,281	21,213	19,855	16,684	14,672
Male	37,422	42,094	40,326	37,571	34,704	34,604	32,491	30,316	27,899	25,178	22,919	20,773	17,716	15,760
Grand total	72,669	81,574	78,533	72,796	66,955	66,320	62,717	57,817	53,105	48,459	44,132	40,628	34,400	30,432

summarizes the distribution of student number by gender in different level of education (MOEOpenData, 2020b).

We can notice from Table 14.4, that the total number of enrolled boys in 2019, reached 419,773, where the number of girls has been estimated to 390,764, which gives a total number equals to 810,537.

- As per the statistics published by the UAE Ministry of education, in 2019, among the 288,794 students enrolled in the public K12 education, 138,004, are boys and 150,790 are girls. The table below (Table 14.5) showcases the distribution of students by gender in different levels of K-12 education in public UAE schools.

The most recent available statistics about the dropout rate, are related to 2017. Where it was declare by the ministry of education, that the dropout rate at UAE federal educational institutions is 14 per cent. Globally the rate of 4 out of 10, is not considered as a huge rate but still higher than the average. No new statistic have been published so far about the rate of dropout for 2018 or 2019.

14.1.3.3 Educational Research and International Collaboration

The development and the enhancement of any educational system can be achieved only by focusing more on the educational research. Many researchers have been interested on the educational research topic, since many years, where they have processed systematic collections of data and conducted interesting analysis related to education in the UAE. Many research papers have been published in peer reviewed and ranked international and national journals and conferences.

In order to consolidate the UAE education strategy and the requirements needed for the Furth Industrial Revolution (Industry 4.0), many research centers have been created in the UAE, and are offering a valuable contributions to many research communities around the world.

In alignment with the UAE vision 2021, the Emirates Center for Strategic Studies and Research (ECSSR), one of the most important research centers in UAE, established on 1994, is holding an annual education conference, where strategies, plans, and initiatives that aim to enhance the quality of education, are deeply discussed by researchers. Most of the published educational researches are stressing on the use of innovation and technology standards, to move toward a knowledge and technology based economy (ECSSR, 2020).

The ECSSR, is continuously emphasizing the vital role of education to achieve the UAE goals of sustainable development, by encouraging the use of the new Technologies of Information and communication (TICs) in developing education systems that fits the requirements of the future (ECSSR, 2020). The ECSSR is in charge of publishing distinguished scholarly works in both Arabic and English language (ECSSR, 2020).

Another important research institute, the Sheikh Saud bin Saqr Al Qasimi Foundation for Policy Research, part of Al Al Qasimi Foundation, was established in

Table 14.5 Number of enrolled K-12 students in public schools by gender for each level

	KG1	KG2	G01	G02	G03	G04	G05	G06	G07	G08	G09	G10	G11	G12
Female	9894	10,114	10,369	11,197	9748	11,044	11,143	11,711	11,099	11,431	11,391	10,777	10,648	10,224
Male	9587	9830	9871	10,361	8957	10,184	10,475	10,636	9745	10,550	10,058	9653	9169	8928
Grand total	19,481	19,944	20,240	21,558	18,705	21,228	21,618	22,347	20,844	21,981	21,449	20,430	19,817	19,152

2009 to enhance the economic, social and cultural development in the city of Ras Al Khaimah, a northern emirate in the UAE, (SSSAQFPR, 2020).

In 2019, Al Qasimi Foundation launched a peer-reviewed, bilingual (English/Arabic) academic journal, called *Gulf Education and Social Policy Review* (GESPR). This journal, aims to promote the open access for educational content (Open Educational Resources) and to support scholarship on education and social policy issues in the Gulf region (GESPR, 2020).

14.1.4 Current situation of Science Education

14.1.4.1 Policies and Standards

The UAE Ministry of Education (MoE), is the regulatory authority of all levels of education in both public and private sectors, including schools, colleges, universities and post-graduation programs.

Even it is not directing the private schools curriculum (except social studies and Islamic studies) the MoE is setting up all the guidelines that private schools should follow (MoE, 2020).

Recently in 2019, The UAE' Ministry of education, established a new set of standards related to academic and administrative aspects to be used in higher education institutions in accrediting their academic programs (MoE, 2020).

The educational policy have evolved in UAE, over the recent years, and is continually debated and revised. More focus has been given to some goals in relation with the cultural and economic global change.

The MoE is investing a lot of efforts to incorporate the information and communication technology (ICT) in the national curriculum and to prepare students with the necessary learning skills to participate in the global economy. To achieve these goals, UAE is following well-known international standards that include, (UAE K-12 Computer Science and Technology Standards, 2015).

1. "The Digital Literacy and Competence",
2. "Computational Thinking",
3. "Computer Practice and Programming",
4. "Cyber Security, Safety, and Ethics".

In higher education, and as part of the UAE's National Strategy for Higher Education, the MoE is focusing on the competitively and the global reputation of its higher education institutions. New accreditation standards have been announced by the UAE' Ministry of education, covering eleven academic and administrative aspects, namely: "governance and administration; quality assurance; scientific and research activities; students; faculty; educational facilities; health, safety, and environment; legal compliance and public disclosure."

In their paper "Fact sheet: Education in the United Arab Emirates and Ras Al Khaimah ", published in 2017, Ridge (2017a) and al. proposed a set of recommendations and initiatives to be followed by policymakers in UAE, such as: "offering more constructive teacher training, expanding the scope of curriculum content, implementing effective evaluation strategies, and investing in long-term local capacity to develop curriculum".

14.1.4.2 Curriculums, Digital Resources and Teacher Training

In addition to UAE's national curriculum, a wide variety of international curricula are taught in UAE schools, which include but not limited to International Baccalaureate (IB), UAE's Ministry of Education, British, American, French, German, Indian, Pakistani and Filipino curriculums (Ridge, 2017b).

Governmental schools follow the standard national curriculum of the UAE. The language of instruction is Arabic for all subjects. English is taught as a second language and used for teaching technical and scientific subjects (MoE, 2020).

Private schools follow their own curriculum, but must include specific courses such as Islamic education, social studies and Arabic as a second-language, in their curriculum. Special classes for non-Arabic speakers are also provided in most of privet schools in the UAE (MoE, 2020).

In 2015, H. H. Sheikh Mohammed bin Rashid Al Maktoum, vice President and Prime Minister of the United Arab Emirates, and ruler of the Emirate of Dubai, launched a national initiate which aims to include the innovation and entrepreneurship programs in both public and private curriculums (UAE government, 2020).

In 2016, His Highness Shaikh Mohammad Bin Zayed Al Nahyan, Crown Prince of Abu Dhabi and Deputy Supreme Commander of the UAE Armed Forces, launched an initiative in coordination with the Ministry of Education, which aims to include a subject entitled "Moral Education" in school curriculum (MoE, 2020).

In October 2016, the National e-Library for Research to provide a dedicated archive of all scientific research papers for graduate students and researchers in the UAE. It hosts currently more than 50,000 research papers and additional research papers (MoE, 2020).

Hamdan Bin Mohammed Smart University (HBMSU), is one of the leader universities in e-learning education in the UAE. It was established in 2002 and accredited by the Ministry of Higher Education and Scientific Research. HBMSU aims to support the paradigm of lifelong learning, by providing opportunities to all kind of learners to study online. Now, HBMSU is part of many recognized and international organizations such as, Global Education Coalition (UNESCO), Peter Drucker Society Europe, The Open Educational Resources University (OERu), Association of Arab Universities (AARU), European Distance and E-Learning Network (EDEN), (HBMSU, 2020).

In the same context of promoting the open educational resources, the UAE' Ministry of Education has launched the biggest public digital library of open educational resources, called "Manara" (MoE, 2020). "Manara" is accessed freely and openly licensed text, media and other digital assets used to create, and collaborate with people around the world to improve knowledge gathering.

Another interesting initiative was launched in Ras Al Khaimah, by the Al Qasimi Foundation in 2010, "The Ras Al Khaimah Teachers Network (RAKTN)", which is a social and professional networking site. Since its establishment, RAKTN is playing a crucial in the professional development of teachers and it is acting as a virtual leaning environment to share open educational resources and to debate toe-ward to improvement of the quality education (RAKTN, 2020).

The TRA academy is one of the successful initiates launched by the Telecommunications Regulatory Authority, the federal telecommunications regulatory agency of the United Arab Emirates. Based on the latest statistics available through the official web site of the TRA, 42, 000 trainees from different 83 countries had taken advantage of the free training course available on the TRA academy during the last week of March 2020 (TRA Academy, 2020).

In the same context and for promoting, lifelong learning, the UAE' Ministry, is constantly investing in creating specialized training centers and launching initiatives to maintain a good level of education for all categories of people. As part of its strategy, here are some of the offered programs: One Million Arab Coders (OMAC, 2017), OMAC, Betha programme (Betha Scholarship Program, 2020).

To support e-Learning, m-Learning and distance learning in UAE, and especially to ensure the continuation of teaching and learning during the COVID-19 Pandemic, the UAE' Ministry of Education, in co-operation with Hamdan bin Mohammed Smart University (HBMSU), trained during March 2020, more than 42,000 teachers and academic staff through a free e-training platform (HBMSU, 2020). A program called *How to be an online tutor in 24 h*", was implemented to teach both local and international teachers and academic staff the necessary skills to use online platforms and to monitor remotely their classes (HBMSU, 2020).

To promote a good quality of education, in 2017, the UAE introduced the phase of the UAE Teachers' Licensing System (TLS). By the end of 2020, The TLS will become a requirement for education professionals to work legally at the UAE. Principals, vice-principals, managers and teachers, working in all public and private schools in the UAE must hold a UAE teacher's license.

At the higher education level, almost all universities in the UAE provide professional development related to classroom management skills, incorporating information and communications technology (ICT) into their classrooms, students' assessments strategies, and pedagogical competencies. The Higher Colleges of Technology (HCT), the largest public university in the UAE, is providing through a hybrid education model, Professional Certifications and Qualifications into all HCT academic programs, and at all Exit Points of study to almost all HCT students (HCT, 2020).

14.1.4.3   Student Assessment and Achievement

PISA is a three-yearly international assessment that rank participating countries based on the ability of students in using their cognitive skills in the areas of Reading

Table 14.6 UAE PISA scores (2009–2018)

	Year	2009 (only Dubai)	2012	2015	2018
Subjects	Mathematics	421	434	427	435
	Science	438	448	437	434
	Reading	431	442	434	432
	Creative problem solving	Unavailable	411	435	Unavailable

Table 14.7 UAE TIMSS scores in (2007–2011–2015)

	Year	2007 (Dubai)		2011 (only Dubai)		2015 (UAE)	
	Level	Grade 4	Grade 8	Grade 4	Grade 8	Grade 4	Grade 8
Subjects	Mathematics	444	461	468	478	452	465
	Science	460	489	461	485	451	477

Literacy, Mathematics Literacy and Science Literacy. Mainly grades ranging from 7 to 12, are assessment in PISA.

The UAE has participated in PISA assessment for the first time in its fourth cycle in 2009. The table below presents the scores the UAE recorded starting from the 2009 PISA edition to 2018 edition. The latest participation of the UAE in PISA was in 2018, were the assessments were conducted in all public schools (government schools) along with a sample of private schools (UAE PISA, 2018).

We can notice from Table 14.6, that since 2009, UAE pupils have improved only in mathematics by 14 points. However for Science and reading there is no remarkable progress in the rank and further efforts must be invested by the Ministry of Education and the education stakeholders.

The Trends in International Mathematics and Science Studies (TIMSS) is a global assessment conducted every 4 years to assess the performance of students in Mathematics and Science for grades Four and Eight. Only Dubai city has participated in 2007. However, in 2011, the whole UAE participated for the first time in the fifth cycle of TIMSS. The table below (see Table 14.7) summarizes the scores achieved by Dubai in 2007 and 2011 and the scores achieved by the UAE students in 2015 (TIMSS & PIRLS, 2015).

14.1.4.4   Science and Technology Venues and Centers

The UAE has many specialized institutions and centers that support the use of Science and technology and promote the development of the education of various sciences. We will be selecting some of the most important science and technology center cross the UAE, grouped by categories (MoE, 2020).

1. **Space science and technology**:

   – Sharjah Center for Astronomy and Space Sciences,

- – Center for Space Science
- – New York University, Abu Dhabi,
- – Aerospace Research and Innovation Center,
- – Dubai Astronomy Group (not for profit).

2.   Information and Communication technology:

- – Center for Cyber Security,
- – Information Security Research Center,
- – Khalifa University Robotics Institute (KURI),

3.   Life science and technology:

- – Center for Genomics and Systems Biology
- – Centre for Advanced Biomedical Research and Innovation (CABRI)
- – Al Jalila Foundation
- – Research Institute for Medical and Health Sciences
- – Khalifa University Center of Excellence in Biotechnology

4.   Nuclear science and technology

- – Gulf Nuclear Energy Infrastructure Institute
- – Department of Nuclear Engineering.

Most of the specialized institutions and centers in UAE, mentioned in the previous section of this chapter, are contributing a lot on the development of the informal education in the UAE. Many projects initiated by governmental institutions are engaging public audiences in the large ecosystem of informal science, technology, engineering and math (STEM) learning. Research institutions are also conducting peer-reviewed educational research conferences, reports, and supervising thesis that explore the impact of the informal STEM learning, its impact and its infusion with the existing formal learning systems.

In the same context, the Department of Culture and Tourism—Abu Dhabi regulates, is promoting the emirate of Abu Dhabi as an extraordinary global destination, rich in cultural authenticity, diverse natural offerings, world-class hospitality and unparalleled leisure and entertainment attractions for every type of traveler. It has launched initiated a range number of programs and outreach initiatives to enhance the accessibility of Abu Dhabi's culture and heritage (DCT-Abu Dhabi, 2018).

Some of the most important projects initiated by the Department of Culture and Tourism—Abu Dhabi, such as Louvre Abu Dhabi–Student Ambassadors, Proud Expressions, Wahat Al Karama, The Culture Summit Abu Dhabi, Mamsha Al Khair, Bait Al Gahwa, aim to engage students, teachers and all education stakeholders with the next generation through academic programs and outreach projects (DCT-Abu Dhabi, 2018).

On 2014, Masdar Institute of Science and Technology in Abu Dhabi, launched three outreach initiatives offering a unique experience to aspiring scientists in the UAE regarding the learning process university students and young professionals can expect from the research and academic.

The three programs—Young Future Energy Leaders (YFEL), Summer Research Internships, and 'Ektashif'—aim to reinforce students' knowledge, interest and academic skills in science, engineering and technology as they relate to the graduate, doctoral and post-doctoral research projects being conducted at the Masdar Institute. These programs are part of Masdar Institute's contribution to knowledge and human capital development in the UAE, as well as regionally and globally (MI, 2020).

In the same context related to science education projects and initiatives, and in order to achieve the desired objectives of the volunteerism, the UAE Ministry of Education, has launched some initiatives and programs targeting the engagement of students in volunteering activities. Some of the these programs are, Good Deeds Points, Specialized Volunteerism Tracks, Student's Volunteerism Award, The Volunteerism Council, The Volunteerism Clubs, Volunteerism Forum, Ramadan Initiatives (MoE, 2020).

14.1.4.5 Utilizing Emerging Technologies

The UAE, is aware about the important role the emergent technologies such as AI, Internet of things, blockchain, etc., have in reshaping the economies around the world. The city of Dubai as one of the most advanced digital economies worldwide, has initiate recently, in 2020, two landmark digital initiatives, the Dubai Internet of Things (IoT) Strategy and the Digital Wealth Initiative (UAE government, 2020). These two initiatives aim to support the strategy of the UAE in developing a sustainable ecosystem that is 100% paperless, by 2021. As a long term objective, "the Dubai IoT Strategy seeks to build the world's most advanced IoT ecosystem in the world's smartest city to improve people's lives" (Dubai Internet of Things (IoT), 2020).

In January 2019, the city of Dubai, organized the Middle East's Largest VR (Virtual Reality) Education Program for public schools, where over a million students and thousands of teachers across the Middle East have experienced the most immersive technologies. Many schools in Dubai will be introducing the VR technology to modernize their learning methodologies and pedagogies (Munfarid, 2020).

The UAE is a regional pioneer in incorporating the use of Artificial Intelligence (AI) technology in all fields. In 2017, The UAE government created the UAE AI Council, and started promoting the use of AI, by organizing many events and initiatives related. The UAE believes that the use AI in various domains such as healthcare, energy, space and education is the key component of the advancement of the country (UAE' AI Ministry, 2020).

In the same context, and in collaboration with technology companies like Alef education, the UAE has established new ways of using AI to educate future generations. Educators have experienced the use of AI in everyday activities within their classrooms, such as Predicting student behavior, facilitating in-class automation, Customizing course curriculum, Personalized Learning, etc. … (ALEF Education, 2020).

14.2 Requirements for Future Development of Science Education

The UAE is diving a big attention to the education and place it at the top of its priorities. As part of its vision, the UAE' Ministry of Education, has set ambitious goals to promote smart, sustainable and inclusive growth of science education (MoE, 2020).

In 2020, the UAE' Ministry of Education, developed the Education, 2020 strategy, which is a series of ambitious five-year plans designed to bring significant qualitative improvement in the education system, especially in the way teachers teach and students learn (MoE, 2020).

In higher education, the UAE' MoE has developed a set of potential mechanisms to encourage students to go toward entrepreneurship opportunities and enhance the quality of education toward the SDGs by infusing Information and Communication Technologies (ICT) (such as Artificial Intelligence) in the education system through a smart learning program and a specialized data center (MoE, 2020).

The job market is perpetually evolving decade, and new requirements are needed to strengthen young people knowledge, innovation capacity, and creative capabilities. In Science education policy makers should invest more efforts to make emerging technologies, such as artificial intelligence, blockchain, internet of things, 3D printing, virtual reality, augmented reality, Robotics automating systems, closer to the classroom and help young people (K-12 level education) use their imagination skills. This engagement with new technologies should be done at a very early age and must be able to fulfil the gap between the job market and people skills.

All jobs of tomorrow are almost related to emergent technologies and require skills from the 21st-century skills, where the passion and the creativity are deeply engaged. In fact, an emphasis must be placed on the four C's: communication, collaboration, creativity, and critical thinking, where science education will be focusing more on a collaborative environment based on a collective effort conducted through class discussions, group-based activities, and creative work implemented by teams.

In teaching with the STEM (Science, Technology, Engineering and Mathematics), more efforts should be aligned with the UAE strategy and vision to enhance the connection between schools, community, educational stakeholders and industry. Curriculum developers need to establish standards that aim to promote more the interdisciplinary skills of students by focusing rigorously on the hands-on activities that allow students to understand and solve real-world problems with innovative ideas and trough a critical thinking method.

As a recommendation, adaptive learning based environments be used in the upcoming years to support the STEM educational system. With such adaptive learning environments, students can figure out where they most need support, and can follow their customized learning path to feel confident and to be able to master the course content. This will lead to a more responsive environment that aims to develop student's abilities and skills and to be ready for their professional career.

14.2.1   Discussion and Conclusion

Many examples illustrate the potential role of the UAE in leading the use of artificial intelligence within the education sector. Although, the UAE is investing in education at its different levels (K-12 and Higher education), many challenges related to the continuous changes in the industry sector, the job market and even in the socioeconomic sector, should be addressed in order to achieve the UAE goal of Emiratising.

Achieving an inclusive quality education involving all stakeholders still remain a challenge for different nations and cultures. New generation of policy strategies in education must be implemented, where policymakers must take in consideration the new economic and technological changes to fit the education requirement in the twenty-first century. Science education must be reoriented toward a teaching and a learning paradigm less rigid than the traditional science education.

References

About the UAE [Fact sheet]. UAE Government. (2020). https://u.ae/en/about-the-uae/fact-sheet. Accessed 18 Sep 2020.

Al Ghfeli, M. (2015). Overview of Education System. TIMSS Encyclopedia. http://timssandpirls. bc.edu/timss2015/encyclopedia/countries/united-arab-emirates/. Accessed 15 Sep 2020.

ALEF Education. (2020). https://www.alefeducation.com/. Accessed 15 Sep 2020.

Ashour, S. (2020). Quality higher education is the foundation of a knowledge society: Where does the UAE stand?. *Journal: Quality in Higher Education, 26*(2).

Betha Scholarship Program, 2020. https://www.tra.gov.ae/en/betha-scholarship-program.aspx. Accessed 22 Sep 2020.

The Department of Culture and Tourism—Abu Dhabi. (DCT-Abu Dhabi, 2018). https://tcaabudhabi. ae/en/default.aspx. Accessed 22 Sep 2020.

Dubai Internet of Things (IoT) Strategy and the Digital Wealth Initiative (DITS&DWI, 2020). https://u.ae/en/about-the-uae/strategies-initiatives-and-awards/local-governments-strate gies-and-plans/dubai-internet-of-things-strategy. Accessed 20 Oct 2020.

Education in the United Arab Emirates, WEST. (2018). https://wenr.wes.org/2018/08/education-in-the-united-arab-emirates#:~:text=There%20were%20624%20international%20schools,with% 20545%2C074%20students%20in%202016).&text=By%20some%20estimates%2C%20the% 20number,4%20percent%20until%202020%20alone. Accessed 22 Sep 2020.

Emirates Center For Strategic Studies and Research (ECSSR). (2020). https://www.ecssr.ae/en/. Accessed 20 Sep 2020.

Farah, S., & Ridge, N. (2009). Challenges to curriculum development in the UAE. *Dubai School of Government Policy Brief, 16*, 1–10.

FT Reporter, Top 3 Technological Advancements In The United Arab Emirates. (2017). http://ftr eporter.com/top-3-technological-advancements-in-the-united-arab-emirates/.

GITEX. (2020). https://www.gitex.com/welcome. Accessed 25 Sep 2020.

Global Competitiveness Report (GCR) [Report], Federal Competitiveness and Statistics Authority (FCSA). (2019). https://fcsa.gov.ae/en-us/Pages/Competitiveness/Reports/Global-Competitiven ess-Report-by-WEF.aspx?rid=15. Accessed 16 Sep 2020.

Gulf Education and Social Policy Review (GESPR). (2020). http://www.alqasimifoundation.com/ en/gespr-journal. Accessed 23 Sep 2020.

Hamdan Bin Mohammed Smart University (HBMSU, 2020). https://www.hbmsu.ac.ae/. Accessed 22 Sep 2020.

Higher Colleges of Technology, (HCT 2020). https://hctcatalog.hct.ac.ae/professionalcertifica tions/. Accessed 19 Oct 2020.

Innovation Month. (2020). https://www.uaeinnovates.gov.ae/. Accessed 25 Sep 2020.

The Masdar Institute of Science and Technology (MI, 2020). https://www.ku.ac.ae/institute/mas dar-institute. Accessed 20 Oct 2020.

McFarlane, D. A. (2013). Understanding the challenges of science education in the 21st century: New opportunities for scientific literacy. *International Letters of Social and Humanistic Sciences, 4,* 35–44.

Ministry Of Education, (MoE, 2020). https://www.moe.gov.ae/En/Pages/home.aspx.

Ministry Of Education, Open Data (MOEOpenData, 2020). https://www.moe.gov.ae/En/OpenData/pages/home.aspx. Accessed 10 Sep 2020.

Munfarid, 2020. https://munfarid.org/arvr-ai-in-classroom-largest-vr-education-program-public-schools. Accessed 25 Sep 2020.

One Million Arab Coders, (OMAC). (2017). https://www.arabcoders.ae/. Accessed 8 Sep 2020.

Open Educational Resources University (OERu). (2020). https://oeru.org/. Accessed 25 Sep 2020.

Ridge, N., Kippels, S., & ElAsad, S. (2017a). [Fact sheet]. *Education in the United Arab Emirates and Ras Al Khaimah.* Sheikh Saud bin Saqr Al Qasimi Foundation for Policy Research.

Ridge, N., Kippels, S., & Farah, S. (2017b). *Curriculum Development in the United Arab Emirates.* Sheikh Saud bin Saqr Al Qasimi Foundation for Policy Research Policy Paper No. 18.

Schvaneveldt, P. L., Kerpelman, J. L., & Schvaneveldt, J. (2005). Generational and cultural changes in family life in the United Arab Emirates: A comparison of mothers and daughters. *Journal of Comparative Family Studies, 36*(1), 77–91.

Sheikh Saud bin Saqr Al Qasimi Foundation for Policy Research, SSSAQFPR. (2020). http://www.alqasimifoundation.com/en/home. Accessed 16 Sep 2020.

Strategic AI Initiatives, Minister of State for Artificial Intelligence, 2020. UAE (MSAI, 2020). https://ai.gov.ae/uae-ai-initiatives/. Accessed 10 Sep 2020.

Telecommunications Regulatory Authority Academy, (TRA Academy). (2020). https://academy.tra.gov.ae/. Accessed 12 Sep 2020.

The Centre for the Fourth Industrial Revolution, (C4IR UAE). (2020). https://c4ir.ae/. Accessed 25 Sep 2020.

The Ras Al Khaimah Teachers Network (RAKTN). (2020). https://raktn.org/about-us/. Accessed 10 Sep 2020.

The United Arab Emirates' Government portal. https://u.ae. Accessed 19 Sep 2020.

Think Science. (2020). http://www.thinkscience.ae/. Accessed 10 Sep 2020.

TIMSS & PIRLS International Study Center. (2015). TIMSS 2015. http://timss2015.org/download-center/. Accessed 10 Sep 2020.

TIMSS & PIRLS International Study Center. (2011). TIMSS 2011. https://www.moe.gov.ae/En/Imp ortantLinks/InternationalAssessments/Documents/TIMSS/Brochure.pdf. Accessed 10 Sep 2020.

United Arab Emirates Population (1950–2020) [Fact sheet]. Worldmeter. (2020). https://www.wor ldometers.info/world-population/united-arab-emirates-population/#:~:text=the%20United%20Arab%20Emirates%202020,(and%20dependencies)%20by%20population. Accessed 17 Sep 2020.

UAE' AI Ministry. (2020). https://ai.gov.ae/. Accessed 15 Sep 2020.

United Arab Emirates GDP, Trading Economics. (2020). [Statistics]. https://tradingeconomics.com/united-arab-emirates/gdp. Accessed 25 Sep 2020.

UAE K-12 Computer Science and Technology Standards, Ministry of Education. (2015). https://www.moe.gov.ae/Arabic/Docs/Curriculum/Learning%20Standard/UAE%20CST%20Fram ework.pdf. Accessed 10 Sep 2020.

UAE PISA. (2018) [Brochure]. https://www.moe.gov.ae/En/ImportantLinks/InternationalAsses sments/Documents/PISA/Brochure.pdf. Accessed 15 Sep 2020.

Warner, R.S., & Burton, G. J. S. (2017). *A fertile Oasis: The current state of education in the UAE.* academia.edu.

Samia Kouki is an Assistant Professor at the higher colleges of technology, UAE, Ras Al Khaimah campuses. She received her Ph.D. in Computer Science from the University of Sfax, in Tunisia. She has more than 20 years of teaching experience in higher education for undergraduate and graduate levels in both Tunisia and UAE. She has an expertise in High Performance Computing (HPC), parallel and distributed systems, mobile learning and Artificial Intelligence. She managed to publish several journals and conference papers. She has been involved in many research and industrial projects. She is a consultant since 2015 in the Arab League Educational, Cultural and Scientific Organization (ALECSO).

Mouza Al Shemaili is an Assistant Professor and Dear of Academic Operations at the Higher colleges of Technology (HCT), UAE, Ras Al Khaimah. She obtained her M.Sc. and Ph.D. in computer engineering from Khalifa University. She got more than ten successful years in higher education teaching at the graduate and undergraduate levels at Khalifa University and the higher colleges of Technology. Her expertise is in security, IoT security, Penetration test, and AI. She managed to publish several journals and conference papers. She won a different type of research grants at the UAE level.

Chapter 15
Science Education in Tunisia

**Mouna Denden, Ahmed Tlili, Mohamed Koutheair Khribi,
and Mohamed Jemni**

Abstract This chapter gives an overview about the current situation of science education in Tunisia. Specifically, it starts by giving an overview about the country and its education development. It then discusses the current situation of science education in Tunisia, where it presented the implemented policies for science education, students' achievement in this field and the used emerging technologies related to science education. While science education outcomes are rapidly progressing in the country, it is seen that more efforts are further needed, for instance, to adopt emerging technologies, such as VR and AI in public schools. Besides, based on the current situation of education science, this chapter presents some requirements for future development in this field. For example, it is very important to encourage science learning in many contexts and throughout the lifespan, as well as involving different stakeholders (e.g., researchers) in the development and design of governmental education policies.

Keywords Education · Science education · Education system · Tunisia

M. Denden (✉)
Research Laboratory of Technologies of Information and Communication and Electrical Engineering (LaTICE), University of Tunis, Tunis, Tunisia

A. Tlili
Smart Learning Institute of Beijing Normal University, Beijing, China
e-mail: ahmedtlili@ieee.org

M. Koutheair Khribi
ICT Access Specialist—Educational Programs, MADA Center, Doha, Qatar

M. Jemni
Arab League Educational, Cultural and Scientific Organization, Tunis, Tunisia
e-mail: mohamed.jemni@alecso.org.tn

15.1 Overview of the Country

15.1.1 Geographical Location, Population and Political System

Tunisia is officially called the Republic of Tunisia. It is located in the Maghreb region of North Africa. The word Tunisia is derived from Tunis, which is the capital city. The country shares borders with the Mediterranean Sea on the north and east, Algeria on the west and southwest, and Libya on the east and southeast. It is a relatively small African nation as it covers an area of just 163,610 km². Jebel ech Chambi is the highest point in Tunisia at 5,065 ft and it is located in Kasserine city, whereas Chott el Djerid is the lowest point at −55,8 ft and it is located in the Tunisian part of the Sahara (Desert). The estimated Tunisian population in 2020 is 11,818,619 people at mid-year, which presents 0.15% of the total world population (Worldometers, 2020). Particularly, 70.1% of the population is urban (Worldometers, 2020). The population growth rate is approximately 1.06% in 2020 and the population density is 76/km² (Worldometers, 2020). The country has been influenced throughout its history by waves of immigration including Phoenician, Arab, Berber, African, Roman, and European, but the most notable immigration was of the Spanich Moors (Muslims), where 200,000 Spanish Muslims settled in the area of Tunis. Therefore, Tunisia's predominant religion is Islam, in its Mālikī Sunni form. Its official language is Arabic, although the Tunisian Arabic dialect (Darija) is used in everyday communications. However, only about 1% of the Tunisian people speak Berber. Previously, only modern Standard Arabic is taught in schools, but after achieving independence from French colonization in 1956, French also came to be taught and it is officially declared as the second language of the country. Since that, the French language continues to play an important role in the press, government, large companies, commerce and governmental institutions (Barbour et al., 2020).

Between 1956 and 2011, the Tunisian political system was a strong presidential system dominated by a single political party, which is the secular Constitutional Democratic Rally (RCD). However, after the Jasmine Revolution and the dissolution of the RCD in 2011, the political system changed and it becomes a unitary semi-presidential representative democratic republic. Tunisia is considered to be the only fully democratic sovereign state in the Arab world and the most liberal nations in the Islamic world, especially in terms of the rights accorded to women (Hamdy, 2007). However, Tunisia still closely connected to France economically and politically even after gaining its independence. The next subsequent section presents the current situation of economy in Tunisia.

15.1.2 Current Situation of Economic, Technologies and Cultural Development

Tunisia has a growing and well-diversified economy that is focused on agriculture, mineral exports, especially petroleum and phosphates, tourism and manufacturing. The main industries in the country are petroleum, the mining of phosphate and iron ore, textiles, footwear, agribusiness, and beverage (Barbour et al., 2020). Additionally, since tourism is a major industry in the country, touristic services are as well. In agriculture sector, the main products are olives and olive oil, grain, tomatoes, citrus fruit, sugar beets, dates, almonds, beef, and dairy products. Tunisia is going through a stifling economic and social crisis that has exacerbated the deterioration of living conditions and unemployment rates and expanded the circle of poverty. According to the board of directors of the Central Bank of Tunisia, a negative GDP growth rate of about 6.5% is expected in 2020, which will cause a severe economic contraction (Argoubi, 2020). Additionally, the Central Bank of Tunisia believes that the main growth engines will witness a significant decrease in the pace of their development, especially investment, due to the conditions resulting from the Covid-19 pandemic and the reduction in the volume of trade exchanges with abroad (Argoubi, 2020). During the first half of 2020, the country's budget deficit also worsened by 56% (1.4 billion dollars) and it is expected that the debit reaches 85%. Furthermore, tourism activity declined by about 80% and it is expected that the unemployment rate increased by about 20% (Argoubi, 2020).

In order to open new opportunities for Tunisia's economic growth, the ministry of Communication Technologies and Digital Economy puts a lot of efforts in order to improve the information and communication technologies (ICT) sector, which contributes 7.5% of the GDP and employs around 86,000 people in 2018, according to the National Institute of Statistics (Cherni, 2020). Based on an ambitious digital plan, Tunisia is gaining new commercial opportunities. Specifically, the Tunisian market is considered as a favorable destination for foreign ICT companies wanting to establish themselves and carry out privileged exchanges (Cherni, 2020). Additionally, the ministry tried to improve the telecoms infrastructure to be accessible for everyone, therefore the number of internet users increased by 48 thousand between 2019 and 2020 in Tunisia (Kemp, 2020). Additionally, the number of mobile networks increased by 219 thousand between January 2019 and January 2020 (Kemp, 2020).

On the other hand, the cultural sector was also one of the strategic sectors that Tunisia focused on it for economic development in addition to the importance of its role in achieving comprehensive and sustainable development and interaction with benchmarking and comparative experiences in developed countries. In this context, the minister of culture claimed that the Ministry of Cultural Affairs fixed a list of goals to work on it in 2020, including translating the principle of the right to culture into a general and comprehensive daily reality, building a citizenship culture through participatory democracy, and the contribution of culture to development and employment through the launch of two new programs, namely the National

Program for Cultural Initiative and Creative Industries and Technological Renewal and making it a tributary, directly and indirectly, of the economy (Kerimi, 2019). Additionally, in light of the openness to other cultures, Tunisia was very influenced by the European culture, even in the education sector. Therefore, there are around a dozen active European cultural institutes/operators in Tunisia, for example the British Council and the Goethe-Institut (Helly, 2014). The next section presents an overview about the education development in Tunisia.

15.2 Overview of the Education Development

15.2.1 Education System and Policy

Education is ranked number one in the priority of Tunisia where about 6,2% of the GDP of the country is spent on education (UNESCO, 2016). It is free to all school-age children, and is compulsory until the age of 16. Additionally, it is a basic right guaranteed to all Tunisians without discrimination based on gender, social origin, color or religion. The Tunisian education is very influenced by the French system and it is delivered in both public and private institutions. Specifically, Tunisian's education system is organized as follows:

- *Pre-school education* is optional for children between 3 and 6 years old. It aims to prepare children for school. There are three types of preschools in Tunisia, namely: (1) Kouttabs which are religious educational institutions supervised by the Ministry of Religious Affairs; (2) Kindergartens which are private schools supervised by the Ministry of Women, Family, Children and the Elderly; and (3) some public schools also provide preparatory studies.
- *Basic education* is compulsory and lasts 9 years. Tunisia has both private and public basic education. Specifically, it is divided into two stages: primary education lasts 6 years and it is oriented toward students aged 6–12 years old, and preparatory education lasts 3 years and is oriented to students aged 12–15 years old. Basic education guarantees sufficient level of knowledge and training to enable students either to continue learning in the next phase or to enroll in training Career or social integration. Upon successful completion of the basic education, students are awarded the Diplome de Fin d'Etudes de l'Enseignement.
- *Secondary education* is not compulsory and lasts 4 years. Tunisia has both private and public secondary education. It is divided into two stages: one year of general education plus one year of pre-orientation, and two years of specialized education, where each student selects one particular field: experimental science, math, letters, economics and business management, or technical studies. At the end of 4-year secondary education, students are required to pass the baccalaureate exam to be able to enter into university.
- *Tertiary education* is under the supervision of the Ministry of Higher Education and Scientific Research (MHESR). Tunisia has both private and public higher

education institutions. Higher education degrees are based on the European education structure known as L.M.D: Licence comes in the first cycle (three years) to obtain the Bachelor degree, Master comes in secondary cycle (two years) to obtain a Master degree, and Doctorat comes in tertiary cycle (three to five years) to obtain a Doctoral degree.

The Tunisian government also provided a non-university level post-secondary studies (technical/vocational training) which take two and a half years and is under the responsibility of the Ministry of Employment. Many efforts were put by the government to improve the education situation in the country and to further encourage people to study, instead of dropping-out at an early age. The next subsequent section presents statistics about students' education enrollment in Tunisia.

15.2.2 Statistics on the National Education

According to the last statistics of the Tunisian Ministry of education, the number of students enrolled in the first cycle of basic education increased in 2020 to 1,171,569 compared to 1,122,693 in 2018, in public schools (Ministry of education, 2020). Particularly, 48% of students are females. Also, the number of primary school teachers increased from 63,642 in 2018 to 65,981 in 2020. The majority of students finished the first cycle of basic education while only 1% of them drop-out their classes. The majority of these students are from rural area and the main reason behind leaving schools at an early age is the difficulties faced while accessing to education in these areas (Canavan, 2016). For instance, many students lack means of transport to school, as well as school supplies.

The number of students enrolled in second cycle, basic education (public and technical) and secondary education, also increased in 2020 to 937,005 compared to 887,615 in 2018, distributed at 1520 school. 54,4% of students are females (Ministry of Education, 2020). Additionally, the number of students enrolled in private schools increased in 2020 to rich 87,936, distributed at 445 schools. Like in the first cycle of basic education, the number of teachers in basic and secondary education increased in 2020 to 75,470 compared to 73,665 in 2018. The number of teachers is increasing in private schools from 11,839 in 2018 to 12,789, in 2020. The dropout rate in second cycle basic education and secondary education, however, is higher than the first cycle of basic education. Specifically, in basic education, 8.9% of students dropped-out classes in 2019, where 13% of them are males. Whereas, in secondary education, 10% of students dropped-out their classes in 2019, where 13,1% of them are also males.

Finally, according to the last statistics of the MHESR, the number of students enrolled in tertiary education decreased from 241,084 in 2018 to 233,692 in 2019, distributed at 203 public institutions (MHESR, 2019). Whereas, the number of students enrolled in private institution increased from 31,177 in 2018 to 33,462 in

2019, distributed at 74 institution. Specifically, the number of students enrolled in License decreased from 205,345 in 2014 to 148,066 in 2019, in public institutions, but increased from 12,198 in 2014 to 15,034 in 2019, in private institutions. This can be explained with the majority of students started having more confidence over private education as well. For the number of teachers, it also decreased from 22,343 in 2018 to 21,823 in 2019. However, the dropout rate in tertiary education is not too large compared to second cycle basic education and secondary education (2.2%). The next subsequent section presents the main education research programs and international collaboration made to improve the quality of education in Tunisia.

15.2.3 Educational Research and International Collaboration

Tunisia has made large efforts to improve the scientific research situation in the country. In this context, the number of research students in Tunisia has increased from 11,408 in 2014 to 11,629 in 2019. Additionally, the number of teachers-researchers increased from 2257 in 2014 to 2609 in 2019. The number of laboratories also increased from 329 in 2017 to 392 in 2018, whereas the number of research units decreased from 301 in 2017 to 215 in 2018 (MESRS, 2019). Particularly, the research centers are the basic structure for carrying out scientific research and technological development activities in all fields of science and technology, within the framework of the national priorities identified in the development plans and national programs determined by the relevant parties. Specifically, research centers are distributed over several ministries, such as the Ministry of Higher education and scientific research, Ministry of health and Ministry of Communication Technologies and the Digital Economy. However, these centers still facing many problems, including the lack of public funding for scientific research and the very basic knowledge and technology infrastructure allocated to it.

In order to improve the scientific research in Tunisia, the MHESR signed several international collaborations with external parties from Tunisia. For example, in 2015, Tunisia signed a collaboration agreement with a Canadian non-profit research and training organization (Mitacs) to promote research collaboration and facilitate student mobility between the two countries through the Mitacs Globalink program (Mitacs, 2015). Additionally, MHESR launched a collaboration with many European Union universities as well as a UK university through the Erasmus Mundus program, which started in 2009 and aimed to enhance the quality of higher education and promote dialogue and understanding between people and cultures through mobility and academic cooperation. It was replaced in 2014 by the Erasmus + program (Britishcouncil, 2014). The Horizon 2020 program (H2020) and MOBIDOC Post-doc H2020-Univ are also international programs financed by the European Union for research and innovation (ANPR, 2020). Moreover, the

MHESR launched the PAQ (Programme d'Appui à la Qualité) project in collaboration with world bank to enhance the quality of teaching.[1] The next section presents the current situation of science education in Tunisia.

15.3 Current Situation of Science Education

15.3.1 Policies and Standards

Science education is the teaching and learning of science to non-scientists, such as school children, college students, or adults within the general public. The field of science education includes work in science content, science process (the scientific method), some social science, and some teaching pedagogy. The standards for science education provide expectations for the development of understanding for students through the entire course of their K-12 education and beyond. The subjects include in the science education are physical, life, earth, space, and human sciences.

The Tunisian government focused in the educational system reform of 1988 to provide new policy to improve science education in the country (Bouhouch & Akrout, 2016). For instance, the development of the educational programs and pedagogical tools on which the application of these programs is based, making it permanently adapted to new scientific data and the needs of the environment. Additionally, training the trainers and promote scientific research as a tool for national growth were one of the main focuses. In 2002, the government announced the opening of the Tunis Science City, an institution in charge of making science an inspiring and exciting experience for people of all ages. Its mission is twofold: on the one hand, it is responsible for establishing social dialogue on science, and on the other hand, it offers a wide range of enrichment programs for students aimed at complementing the school's formal science education. In this context, the previous Minister of Higher Education and Scientific Research, highlighted the importance of working on the development of the services of Science City and the mechanisms that it adopts and to enhance its openness to universities and scientific research centers in order to promote scientific tourism in Tunisia. In 2015, the Ministry of communication technologies implemented the "Tunisia Digital 2020" strategy which aims to make the country an international benchmark in digital technology. Furthermore, in order to encourage scientific research, in 2020, the MHESR allocated an amount of 10.000 million Tunisian dinars for each university to offer teachers grants for their newly published research.

---

[1] Ministry of higher education. (2006). Available on: https://apprendre.auf.org/wp-content/opera/13-BF-References-et-biblio-RPT-2014/Programme%20d%E2%80%99Appui%20%C3%A0%20la%20Qualit%C3%A9%20pour%20l%E2%80%99Enseignant%20Sup%C3%A9rieur_Manuel%20de%20Proc%C3%A9dures%20Op%C3%A9rationnelles.pdf.

With regards to schools, according to the article 52 of the Education Directive Act no. 80 issued on July 23, 2002, mathematics and science are taught until students become familiar with different forms of scientific thinking and practicing the types of inference and evidence, and equipped with the skills needed to solve problems and interpret natural and human phenomena. Additionally, technology is taught in schools in order to help students understand the technological environment in which they live in and to realize the importance of using technologies in economic and social activities. However, recent studies showed the low interest of secondary education students in Arab countries, including Tunisia, in the exact sciences (physics, chemistry, biological sciences, earth sciences, astronomy and mathematics), and the decline of their participation in various scientific activities. According to experts, the main factor for this decline is closely related to the traditional methods of science teaching, which have proven to be ineffective, and necessitates developing a new methodology to increase interest in science and scientific options among students. Specifically, the traditional approach to science education focuses on the "what we know" aspect of knowledge and science, while the required approach should be concerned with the exploration of science aspect, i.e., "how we know/learn". The next subsequent section presents the curriculums, digital resources and teacher training.

15.3.2 Curriculums, Digital Resources and Teacher Training

Since its independence, several reforms were provided by the Tunisian government to improve the education systems. For instance, in 2000, Tunisia focused on improving pedagogical research and teacher training programs and developing curricula that reflect scientific and technological advances. Particularly, the updated school science curriculum aimed at nurturing students' interest in science and related disciplines, strengthening students' ability to integrate and apply knowledge and skills, and fostering students' sense of making informed judgments based on scientific evidence. All students from basic to secondary education are entitled to science education. At the basic education level, several subjects of science such as math and earth science, are included in the curriculum and they are taught in Arabic. Students develop their interest in science and basic scientific knowledge to facilitate their progression to learning in secondary education. At the secondary education level, students can extend their learning by specializing in one science subject including math, science, and computer science. At this level, science subjects, however, are taught in French. Additionally, in order to improve the quality of teaching and research, Tunisia is investing in digital technology to support all school levels, especially higher education. Therefore, one of the main goals of the Ministry of education in 2020 is the digitalization of schools by the integration of ICT in the official curriculum in order to improve the quality of education, develop digital learning and the acquisition of twenty-first century skills that guarantee equity and equal opportunities for learners and their integration into society and the digital economy. In this context, the Ministry of Communication Technologies and Digital Transformation announced the national strategic plan for digital development "Tunisie Digitale 2020".[2] Specifically, the integration of ICT in education is reinforced through the Tunisian Virtual School (école virtuelle), the National Center for Technologies in Education (CNTE) and the Tunisian Virtual University (UVT) established in 2002. Students become able to learn using digital resources through online platforms, such as Moodle. In particularly, Moodle, which is one of the most appreciated free online learning management systems, is included in the Tunisian higher educational system, to teach "Certificat Informatique et Internet" (C2i), to attest the mastery of skills in the use of digital technologies (Khribi & Chebli, 2009). Additionally, UVT offers a free access to a resource library for teachers in order to ensure their educational training and create new open educational resources (OER).

The Tunisian educational system lacks science education teacher training of teachers. This is the result of the absence of academic training for teachers in specialized sciences directed to teaching. Therefore, as part of the strategic plan for the education reforms 2016–2020, Tunisia focused on enhancing teacher training to improve the quality standards of education and reduce dropout rates (OxfordBusinessGroup, n.d). For instance, in 2016, a new major "Education sciences" was added to be taught in university in order to form specialized and successful teachers. Specifically, among the taught subjects, we found communication and activation techniques, interpersonal skills and self-management, psychology. Additionally, in collaboration with UVT, the Higher Institute of Education and Continuing Training (ISEFC) provides a number of educational activities both for the benefit of the teaching staff and for professionals of the Ministry of Education.[3] The government also provided six regional centers of education and continuous training for teachers of primary and secondary school. Furthermore, with the support of international organizations, such as World Bank, UNESCO, Microsoft, and The Arab League Education, Culture and Science Organization (ALECSO), they offered different ICT training programs for teachers to acquire more competences in ICT and in the use of technological resources (ALECSO, n.d., Jemni & Khribi, 2017; Hamdy, 2007). The next subsequent section presents the learning assessment method in Tunisia and the students' achievements in science.

---

[2] Ministry of Communication Technologies and Digital Transformation. https://www.mtcen.gov.tn/index.php?id=14.

[3] Higher Institute of Education and Continuing Training. https://www.isefc.rnu.tn/html/francais/presntation/organigra.htm.

15.3.3   Student Assessment and Achievement

According to the Program for International Student Assessment (PISA) 2015 survey, Tunisia is ranked 65 out of 70 countries, on the learning performance axis (BIAT, 2017). Specifically, the PISA ranking was based on a total sample of 540.000 15-year-old students enrolled in 72 OECD countries (the Organization for Economic Co-operation and Development), where these students were required to take tests in science, reading, mathematics, and collaborative problem-solving.

In sciences, 66% of Tunisian students have not reached the Level 2, which is the minimum level to be reached in order to participate in the life of modern society (see Fig. 15.1). Additionally, no Tunisian student has the skills characterizing Level 6 or even Level 5. Specifically, the percentage of high achievers in science is higher among males than females. It is also found that 34% of Tunisian students are planning to exercise a scientific profession when they grow up, where 39% of them are males and 28% are females. Additionally, the average index of students enjoying learning sciences is 0.52. Specifically, according to PISA survey, students' performance in science and their aspiration to pursue a scientific profession are more correlated with the amount of time spent learning science and the way science is taught than with other factors, such as the equipment and personnel assigned to the science section, the nature of the extra-curricular science activities offered in schools, and the qualifications of science teachers.

On the other hand, according to the Trends in International Mathematics and Science Study (TIMSS) in 2011, which is a series of international assessments of mathematics and science knowledge of students at the fourth and eighth grades, Tunisia ranked 47 out of 50 countries, based on fourth grade students' performance in the mathematics assessment, which indicates the presence of serious problems in the early years of basic education (Mhirsi et al., 2013). Whereas, it is ranked 30 out of 42 countries, on the eighth-grade students' performance in the mathematics

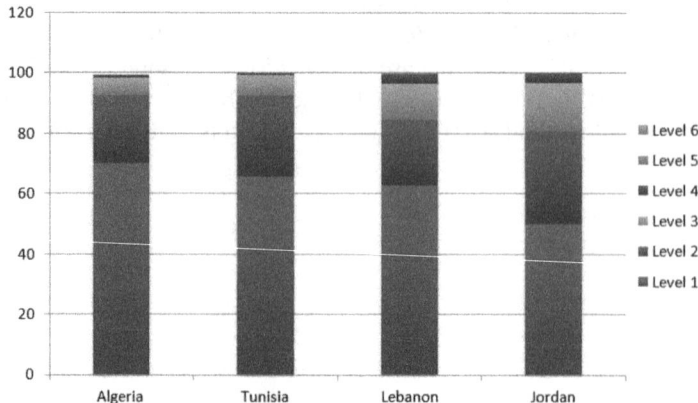

Fig. 15.1 Students' performance in science

assessment. Specifically, the mathematics assessment method at fourth and eighth grades is based on two main dimensions:

- *The Content Domains in Mathematics* which contains four domains, namely: numbers, algebra, geometry, and statistics.
- *The Cognitive Domains in Mathematics* which contains three domains, namely: Knowing, applying, and reasoning. This can reflect levels of mathematics skills acquired by students.

Additionally, in the science field, Tunisia is ranked 48 (out of 50) based on fourth grade students' performance in the science assessment, according to TIMSS, whereas, it ranked 29 (out of 42), based on eighth grade students' performance in science assessment (Mhirsi et al., 2013). Specifically, the science assessment method is similar to the mathematic assessment method (based on two dimensions: The content domains and the cognitive domains), except, in science, the content domains of Fourth Grade are different from those of Eighth Grade. Particularly, at grade 4, students' knowledge in three content domains were tested, namely: Life science, physical science, and Earth science, and in four content domains for eighth grade students' knowledge, namely: Biology, Chemistry, Physics, and Earth science.

15.3.4 Science and Technology Venues and Centers

According to the last statistics of the MHESR, in Tunisia, the total number of research centers in the country is 39, distributed over several ministries. Particularly, 21 research centers are specialized in science and technologies. Table 15.1 presents the distribution of the 21 research centers by ministry.

Furthermore, Tunisia today has three ICT-oriented techno-parks, the largest "Technopark El Ghazala" as well as 18 cyberparks entirely dedicated to training and scientific and technological research. For instance, in 2020, they have organized several scientific days, including "Journée International des Femmes et des Filles de Science", which aims to show the importance of females within the scientific and technological community, "Journée International des Femmes et des Filles de Science", which aims to propose sports and play activities to raise public awareness of the issues related to rare diseases, "Journée des Maths", which aims to

Table 15.1 Distribution of research centers by ministry

Ministry	Number of research centers
Ministry of higher education and scientific research	15
Ministry of health	5
Ministry of communication technologies and digital economy	1
Total	21

raise public awareness of the mathematical cultural aspect, and "Journée des technologies", which provides a robotics competition for children from 8 to 17 years (CST, 2020). Furthermore, in order to encourage scientific research, the ministry launched the national prize for scientific research and technology. Specifically, it includes Best Research Laboratory Award, Prize for the best Tunisian Researcher or Inventor, and Prize for the best national doctoral thesis.

15.3.5 Utilizing Emerging Technologies

To improve the quality of teaching and learning in science education, several new technologies, such as Artificial Intelligence (AI), Virtual Reality (VR) and Augmented Reality (AR) were integrated in teaching strategies of many developed countries. For instance, in Averett University in Danville, Virginia, they used VR system to "virtually" explore the inner workings of the human body (Shenoy, 2020). Additionally, medical students used VR to simulate real-life surgery where errors are allowed to be made without the catastrophic consequences of real procedures. However, the cost of these devices is very high (e.g., the cost of VR consumer devices is about $1.000). Therefore, Tunisia had low adoption of these advanced technologies in their public and private education. However, recently, Esprit, which is a private university in Tunisia, claimed that they planned to introduce VR, AR, and 4K video to deliver superior teaching experiences over the next three to five years (Huawei, 2020).

15.4 Requirements for Future Development of Science Education

In order to enhance the position of Tunisia in science and mathematics compared to other countries, several correction and improvement should be introduced to the education system. Some requirements for future development of science education in Tunisia are presented as follows:

- Promote a system that stimulates innovation by paying attention to education and raising cultural awareness of the importance of science and technology.
- Encourage science learning in many contexts and throughout the lifespan.
- Encourage people to learn via informal science education, which means learning outside the formal school curriculum in places, such as museums, the media, and community-based programs.

- Integrating new technologies, such as Artificial Intelligence (AI) and Augmented Reality (AR) in teaching strategies to improve students' engagement and success rate.
- Involve researchers in the development and design of government policies and propose scientific solutions.
- Allocate a percentage of the national income to scientific research and adhere to it.
- Provide more research centers and teacher trainings.
- Governance of scientific research and the need to encourage the private sector to invest in scientific research.
- Promoting science tourism in the Tunis Science City by modernizing its working methods and mechanisms, developing its infrastructure, and developing the competencies of its employees.
- Promote the design and use of inclusive technology to fulfill accessible and inclusive education.

15.5   Discussion and Conclusion

Although the great efforts already made to enhance the in Tunisian education system, science education is still facing many challenges. The first challenge is related to the lack of policies in implementing science education in Tunisia. Therefore, the government should provide new policies.

The second challenge is related to students learning performance in science. Based on the latest international studies such as PISA and TIMSS, Tunisia still does not occupy an advanced rank in science and mathematics fields. This result is explained, according to the National Center for Pedagogical Innovation and Education Research (CNIPRE), in science, by the weakness of the schedule allocated to science education, which represents only 5% of the overall schedule, in the 2nd basic cycle, compared to 12% worldwide. Whereas, in mathematics, they were explained by lack of student training related to problem solving. In addition to the non-existence in Tunisian curricula of a number of concepts and subjects included in most of the curricula of the participating countries. In order to overcome this challenge, it is very important to allocate more time to teach science and to develop the Tunisian curricula.

The third challenge is the lack of teacher training in the science field. The teacher is the one who manages the classroom situation and chooses the teaching method. Specifically, according to experts, the science teaching method can influence students' interest in science education. Therefore, it is very important to train teachers in this field to further provide new and effective teaching methods.

The fourth challenge is related to the lack of use of new technologies in the science field such as AI and VR. Specifically, according to many studies, the use of these technologies can greatly affect students' achievement in science education. This challenge may be explained by the limitation of the budget allocated by the relevant ministries to integrate new technologies in education.

References

ALECSO (n.d.). تنمية كفاءات المعلمين في مجال تكنولوجيا المعلومات والإتصال. Retrieved from https://ictcft.alecso.org/

ANPR-Agence Nationale de la Promotion de la Recherche Scientifique. (2020). MOBIDOC Post - doc H2020-Univ. http://www.anpr.tn/mobidoc-post-doc-h2020-univ/.

Argoubi, A. (August, 2020). الاقتصاد التونسي.. أزمة حادة في انتظار حلول الحكومة الجديدة. https://www.noonpost.com/content/37886.

Barbour, N., Murphy, E., Brown, L. C., Clarck, J. I., & Talbi, M. (2020). Tunisia. Encyclopædia Britannica. https://www.britannica.com/place/Tunisia/Languages.

BIAT. (2017). Classement PISA et Système Educatif Tunisien. https://www.biat.com.tn/communication-financiere/document/2017-mai-classement-pisa-et-systeme-educatif-tunisien.

Bouhouch, H., & Akrout, M. (2016). الاتّجاهات الكبرى لإصلاح النظام التربويّ، جانفي 1988. http://akroutbouhouch.blogspot.com/2016/03/1988.html.

Britishcouncil. (2014). Mapping UK-Tunisia Higher Education Cooperation: Challenges and Opportunities. https://www.britishcouncil.tn/.

Canavan, K. (2016). Education reform 2016–2020: Building a better future for students in Tunisia?. https://nawaat.org/2016/09/19/education-reform-2016-2020-building-a-better-future-for-students-in-tunisia/.

Cherni, E. (2020). Market study: Information and communications technology. https://www.tradecommissioner.gc.ca/tunisia-tunisie/market-reports-etudes-de-marches/0002801.aspx?lang=eng.

CST-Tunis Science City. (2020). Programme Scientifique 2020. http://www.cst.rnu.tn/fr/events/programme-scientifique-2020?id=590.

Hamdy, A. (2007). ICT in Education in Tunisia. Survey of ICT and education in Africa: Tunisia Country Report.

Helly, D. (2014). Tunisia Country Report. Preparatory action culture in EU external relations.

Huawei. (2020). Tunisia's Esprit University Embraces Smart Education with AirEngine Wi-Fi 6. https://e.huawei.com/ua/case-studies/leading-new-ict/2020/esprit-university.

Jemni, M., & Khribi, M. K. (2017). Toward empowering open and online education in the Arab world through OER and MOOCs. In *Open education: from OERs to MOOCs* (pp. 73–100). Berlin, Heidelberg: Springer.

Kemp, S. (2020). Digital 2020: Tunisia. https://datareportal.com/reports/digital-2020-tunisia#: ~ : text=There%20were%207.55%20million%20internet,at%2064%25%20in%20January%202020.

Kerimi, M.(2019). إصدارات: كتاب «السياسات الثقافية في البلادالتونسية: البناء.. التطور.. الاشكاليات» د. حاتم زير يبحث ثقافة-. Retrieved from https://ar.lemaghreb.tn/. في ذاكرة الثقافةالتونسية بي ن سنوات البناء واشكالياتها التاريخية وـفنون.

Khribi, M. K., & Chebli, H. (2009). Université Virtuelle de Tunis: état des lieux. In C.inaugurale & E. N. O. P. (Eds.), *Espace Numérique Ouvert Pour la Méditerranée –EOMEDConférence inaugurale d'Agadir.* Agadir, Maroc.

MHESR—Ministry of Higher Education and Scientific Research. (2019). L'enseignement supérieur et la recherche scientifique en chiffres http://www.mes.tn/page.php?code_menu=13.

Mhirsi, C., Abdelwahed, S., Helali, H., & Ben Khalifa, S. (2013). Participation of Tunisia in the international assessment "Trends in international mathematics and science study". https://www.iea.nl/publications/study-reports/national-reports-iea-studies/participation-tunisia-international.

Ministry of education. (2020). Statistics of Education. http://www.education.gov.tn/?p=688&lang=en.

Mitacs. (2015). Mitacs and Tunisia work together to advance international research collaborations. https://www.mitacs.ca/en/newsroom/news-release/mitacs-and-tunisia-work-together-advance-international-research-collaborations.

OxfordBusinessGroup. (n.d). Tunisia's education sector to be overhauled. https://oxfordbusinessgroup.com/overview/track-series-reforms-are-set-overhaul-sector.

Shenoy, R. (2020). VR, AR and AI will Transform Universities. Here's How. https://unbound.upcea.edu/online-2/online-education/vr-ar-and-ai-will-transform-universities-heres-how/.

UNESCO. (2016). Tunisia. http://www.unesco.org/new/en/gefi/partnerships/gefi-champion-countries/tunisia/.

Worldmeters. (2020). Tunisia Population. Retrieved from https://www.worldometers.info/world-population/tunisia-population/.

تكنولوجيا المعلومات والاتصال. Retrieved from http://ictcft.alecso.org/.

Mouna Denden is a former assistant professor at the University of Tunis. She received the Ph.D. degree in computer science from the University of Sfax, Tunisia in 2020. She has published several academic articles in international refereed journals and conferences. Dr. Denden has served as a local organizing and program committee member in various international conferences, and as a reviewer in several refereed journals. Her research interest includes providing personalized educational gamification systems according to learners' individual differences.

Ahmed Tlili is a former assistant professor at the University of Tunis, and currently the Co-Director of the OER Lab at Beijing Normal University, China. Dr. Tlili has been awarded the outstanding research award from the Smart Learning Institute of Beijing Normal University for 2019, and the IEEE TCLT Early Career Researcher Award in Learning Technologies for 2020. He has published several books, as well as academic papers in international referred journals and conferences. He has also served as a local organizing and program committee member in various international conferences, and as a reviewer in several refereed journals. Dr. Tlili is the Co-Chair of IEEE special interest group on "Artificial Intelligence and Smart Learning Environments" and APSCE's Special Interest Group on "Educational Gamification and Game-based Learning (EGG)".

Mohamed Koutheir Khribi is ICT expert in education. He has over 20 years' experience in ICT and Technology Enhanced Learning field. Currently, Dr. Khribi is ICT Access expert and researcher at Mada center in Qatar. Dr. Khribi is former assistant professor and researcher at the University of Tunis in Tunisia. He served as programme specialist in educational technology at the ALECSO ICT department (The Arab League Educational, Cultural and Scientific Organization). He'd been leading several projects at large scale in Arab countries, related mainly to the ICT in education field, namely, Smart Learning, OER, MOOCs, ICT-CFT, Mobile Applications and Cloud Computing. He'd been collaborating with regional and international organizations (ALECSO, UNESCO, ISKME, CoL, OIF, etc.), towards promoting the effective use of ICT in education in Arab and African countries. Dr. Khribi participated in the establishment of the Virtual University of Tunis in 2002 where he served as head of the ICT department. He also headed the Online Education Department at the University of Kairouan in Tunisia. Dr. Khribi is a Fulbright Alumnus, He attended the Knowledge Discovery and Web Mining Lab at the University of Louisville in the USA as a Fulbright visiting scholar. Dr. Khribi is member of Latice laboratory at the University of Tunis, and IEEE member. His research interests include Technology Enhanced

Learning; Educational Recommender Systems; Open Educational Resources; Artificial Intelligence, Learning Analytics; Machine Learning, Mobile and Ubiquitous Learning. He has authored several well-cited e-learning related publications in scientific books, journals, and conferences.

Mohamed Jemni is currently the Director of ICT at Arab League Educational, Cultural and Scientific Organization (ALECSO). He received the Engineer Diploma and Ph.D degree in computer science from the University of Tunis, in 1991 and 1997, respectively. He obtained the HDR (Habilitation to Supervise Research) in Computer Science from University of Versailles, France in 2004. He was a professor of Computer Science and Educational Technologies at the University of Tunis. He is leading several projects toward promoting the effective use of ICT in education in the Arab world, namely, OER, MOOCs, mobile applications, and cloud computing. He published more than 250 papers in international journals and conferences.

Part III
Asian Countries

Chapter 16
Science Education in China

Xiangling Zhang, Yao Song, Zhenyu Cai, Lixin Zhu, Tianyue Sun, Ronghuai Huang, and Ahmed Tlili

Abstract In recent years, China attaches great importance to the development of science education and the promotion of scientific literacy of the whole country, and has issued a series of policies to support such national strategies. Meanwhile, it has achieved very good performance in PISA and TIMSS. The development of informal science education, especially the construction of science and technology museums has been growing rapidly in recent years. At the same time, emerging technologies, such as AI and AR/VR, have also been studied and applied in science education. However, there are many challenges for further promotion of science education in China, such as the imbalanced distribution of high-quality resources, the shortage of professional science teachers as well as the lack of sense of fulfillment for science teachers. In order to fully cultivate students' innovative spirit and problem-solving ability, the in-depth reform of science education in compulsory education should be carried out.

X. Zhang (✉)
Beijing Institute of Education, ShiFang Street No 2, Huangsi Street, Xicheng District, Beijing 100120, China
e-mail: zhangxiangling@bjie.ac.cn

Y. Song · Z. Cai · L. Zhu · R. Huang · A. Tlili
Beijing Normal University, No. 19 Xinjiekouwai Street, HaiDian District, Beijing 100875, China
e-mail: songyao@mail.bnu.edu.cn

Z. Cai
e-mail: zhenyu_cai@163.com

L. Zhu
e-mail: 11112018113@bnu.edu.cn

R. Huang
e-mail: huangrh@bnu.edu.cn

A. Tlili
e-mail: ahmedtlili@ieee.org

T. Sun
Columbia University, 525 West 120th Street, New York, NY 10027, USA
e-mail: ts3407@tc.columbia.edu

© The Author(s), under exclusive license to Springer Nature Singapore Pte Ltd. 2022   261
R. Huang et al. (eds.), *Science Education in Countries Along the Belt & Road*,
Lecture Notes in Educational Technology,
https://doi.org/10.1007/978-981-16-6955-2_16

Keywords China · Science education · Inquiry-based learning · Informal learning · Scientific literacy

16.1 Overview of the Country

16.1.1 Geographical Location, Population and Political System

The People's Republic of China is located in the eastern part of Asia, along the western Pacific rim. It is a vast land, covering about 9.6 million square kilometers. The eastern and southern continental coastlines are more than 18,000 km, and the inland and border waters are about 4.7 million square kilometers. There are more than 7600 islands in the sea area, of which Taiwan Island is the largest with an area of 35,798 square kilometers. China borders 14 countries and is adjacent to 8 countries by sea. The provincial administrative divisions are divided into 4 municipalities directly under the Central Government, 23 provinces, 5 autonomous regions, and 2 special administrative regions. The capital is Beijing (State Council, 2021).

China is a multiethnic country consisting of 56 ethnic groups, with the Han ethnic group accounting for 91.11% of the total population, and 55 ethnic minorities accounting for 8.89%. As of November 1, 2020, the total population of China was 1,443.50 million, of which: The population of the Chinese mainland was 1,411.79 million; The population of Hong Kong Special Administrative Region was 7.47 million; The population of Macau Special Administrative Region was 683.22 thousand; The population of Taiwan was 23.56 million. Among 31 provinces, there were 2 provinces with a population of more than 100 million, and most provinces with a population between 10 and 100 million. Of the national population, 51.24% were males, while 48.76% were females. 63.89% of the total population lived in urban areas (National Bureau of Statistics, 2021a). There are more than 80 languages in China. Mandarin is the official language.

China is a socialist country under the people's democratic dictatorship led by the working class and based on the alliance of workers and peasants. The socialist system is the fundamental system of the People's Republic of China. The National People's Congress (NPC) is the highest and powerful organization of the state. The NPC and its Standing Committee exercise the state power to enact laws. The State Council formulates administrative regulations in accordance with the Constitution and other laws and reports them to the NPC for records. The Communist Party of China (CPC) is the only ruling party in China. The eight democratic parties have the status of participating parties under the premise of accepting the leadership of the CPC.

16.1.2 Current Situation of Economic, Technologies and Cultural Development

China is the world's second largest economic entity, as well as the largest manufacturing, trading, foreign exchange reserve, steel producer, and agricultural country in the world. Facing multiple severe shocks such as the outburst of COVID-19 and the recession of global economy, China's GDP in 2020 still reached 101.6 trillion yuan, or about 14.7 trillion dollars, ranked second in the world. The GDP increased 2.3% over the previous year. The proportion accounting for the global economy reached about 17%. The total grain output reached a high record in 2020, and the productions of industrial sector increased. The employment situation was significantly higher than expected. The income growth of residents and economic growth were basically synchronized, and the gap between the income of urban and rural residents has narrowed. By the end of 2020, the per capita GDP has exceeded 10,000 dollars in two consecutive years. Besides, the development gap with high-income countries has continued to narrow (National Bureau of Statistics, 2021b). In 2021, all of the 98.99 million rural poor people and 832 national-level poverty-stricken regions under the current standards have been lifted out of poverty (Xi, 2021). China has realized the first centenary goal, building a moderately prosperous society in all respects.

China has vigorously promoted innovation in science and technology, and has accelerated industrial transformation and upgrading. Innovation is crucial in China's modernization. More than two-thirds of the world's high-speed rail tracks are laid in China, and the total mileage of highways is ranked first in the world. China has over one-third of the 5G standard-essential patents. In 2020, a stream of scientific and technological breakthroughs was achieved, like the Tianwen-1 Mars mission, the Chang'e-5 lunar mission, and the Striver deep-sea manned submersible. China has intensified efforts to make major breakthroughs in core technologies in key fields. Intellectual property protection has been strengthened. More efforts have been done to upgrade the industrial sector with digital and smart technologies (Li, 2021). During the 13th Five-Year Plan, China's scientific and technological strength has reached a new level. The research and development (R&D) expenditure of the whole society has increased from 1.42 trillion yuan to 2.44 trillion yuan, and the R&D investment intensity has risen from 2.06% to 2.40% (National Bureau of Statistics of China, 2021c). Fundamental research funding nearly doubled, reaching 150 billion yuan, and the technology market contract turnover exceeded 2.8 trillion yuan in 2020. The Global Innovation Index released by the World Intellectual Property Organization shows that China's comprehensive ranking of innovation capabilities has elevated from 29th in 2015 to 14th in 2020, becoming the only middle-income economy among the top 30 (Wang, 2021).

China is one of the four ancient civilizations with a long history and culture. The continuous improvement of the cultural system has promoted the development of the cultural industry. The increasingly prosperous literature and art market provides good conditions for creation. By the end of 2020, there were 2027 art-performing groups, 3510 museums, 3.203 public libraries and 3327 cultural centers in the culture system throughout China. Radio broadcasting and television broadcasting coverage rates were 99.4% and 99.6% respectively. The new business model of "Internet Plus Culture" maintained rapid growth. The growth rates of the revenue of Internet advertising services, manufacturing of smart unmanned aerial vehicles for entertainment, and manufacturing of wearable smart cultural equipment all exceeded 20% (Xin, 2021). The "Building a Community of Shared Future for Mankind" and the "Belt and Road" initiative have strengthened China's international collaborations, and promoted the Chinese culture worldwide. Chinese language has been included in the mainstream education system in more than 60 countries, including the US, Canada and other developed countries. There are more than 40 million people studying and using Chinese overseas (Huang & Mao, 2019). Furthermore, national essences, such as Chinese opera, calligraphy and tea ceremony, are recognized by more countries and nations.

16.2   Overview of the Education Development

16.2.1   Education System and Policy

The Ministry of Education is a department of the State Council, which is in charge of education across the country. The Ministry of Education is responsible for the overall education planning, coordinating and managing educational institutes, as well as conducting education reforms. According to the Education Law, the Chinese school system includes pre-school education, primary education, secondary education and higher education. The length of compulsory education is nine years, including primary and junior high school. Governments at all levels should guarantee school-age children and adolescents have the access to compulsory education. There are also vocational education and adult education. The national education examination system is implemented. The types of national examination are determined by the administrative department of education under the State Council, which shall be undertaken by the institutions that conduct such education examination as approved by the State. Moreover, all kinds of measures shall be taken to carry out illiteracy education program. In addition, the State applies a system of education certificates and a system of academic degree, as well as an education supervision system and an education evaluation system for schools and other institutions of education (NPC, 2021).

Since the "Reform and Opening Up" initiative, China has focused on considering science and technology as a first stone for development. The Strategy of Invigorating China through Science and Education was proposed to enhance scientific and technological strength and to improve the scientific and cultural quality of the entire nation. The "Strategy of Invigorating China through Talented Persons" was put forward, by which the State strives to cultivate, attract and make good use of talents, and to build a large and high-quality talent team. In 1983, Deng, Xiaoping proposed that education should gear education to the needs of modernization, the world and the future. In 1986, the Compulsory Education Law of the People's Republic of China was enacted, and compulsory education was officially implemented. In 1993, some goals of education were stated in the Outline of China's Educational Reform and Development, including: the education level of the whole people would be significantly improved; the pre-employment and post-employment education of urban and rural laborers would develop significantly, and the creation of various talents would basically meet the needs of modernization.

With the rapid development of science and technology, at the end of the 20th century, the State Council proposed to deepen education reform and comprehensively promote quality education by integrating moral education, intellectual education, physical education, and aesthetic education in all educational activities (State Council, 1999). Since the beginning of the 21st century, China has continued to comprehensively promote the development of education. In 2019, China's Education Modernization 2035 put forward the overall goal of advancing the modernization of education; first, by 2020, the overall strength and international influence of education will be significantly enhanced, and the average number of education years of the working-age population will increase significantly. Second, by 2035, China will fully realize the modernization of education (State Council, 2019).

16.2.2 Statistics on the National Education[1]

In 2019, There were 530.1 thousand educational institutions in China, which has increased by 2.17% compared to 2018. The number of students enrolled in degree-granting institutions was 282 million, which has increased by 2.40% compared to the previous year.

In terms of compulsory education, in order to meet the needs education, governments at all levels made efforts to optimize the educational resources in urban areas and narrow the gap between education in urban and rural areas. In 2019, there were a total of 212.6 thousand compulsory education schools nationwide, with 154 million students and 10 million full-time teachers. The graduate rate at compulsory education level reached 94.8%. There were 52 thousand junior high schools in total, with 48.3 million students. The number of full-time teachers of junior high school was over 3.7 million, with a qualification rate of 99.88%.

The compulsory education continues to be thoroughly implemented. In 2019, the net enrollment rate in primary schools reached 99.94%. 99.5% primary school pupils and 94.5% junior high school graduates continued to pursue higher education. The number of schools and the enrollment numbers in the compulsory education stage have increased rapidly in cities.

---

[1] All data in this subsection are from the Ministry of Education, except government expenditure. The statistics herein is exclusive of data on Hong Kong and Macao Special Administrative Regions and Taiwan Province.

Table 16.1 Situation of schools offering upper secondary education (2019)

	Number of students in schools	Number of graduates	Number of full-time teachers
Regular Senior High Schools	24,143,100	7,892,500	1,859,242
Adult High Schools	41,200	34,200	1933
Secondary Vocational Schools	15,764,700	4,934,700	842,900

In 2019, there were 2.4 thousand schools offering upper secondary education in China, with 39.9 million students. The gross enrollment rate at the upper secondary level was 89.5%. Specifically, there were 14.0 thousand regular senior high schools, 333 adult high schools, and 10.1 thousand secondary vocational schools across the country. The situations of students, graduates, and full-time teachers of different kinds of schools are shown in Table 16.1 (Ministry of Education, 2020a).

For higher education, there were 2956 higher education institutions (HEIs) nationwide in 2019. 2688 were regular HEIs, including 1265 universities offering Bachelor's degrees and 1423 higher vocational education institutes. The number of adults HEIs was 268. There were 828 institutions offering postgraduate programs, including 593 regular HEIs and 235 scientific research institutions (Ministry of Education, 2020b). The total enrollment number of higher education across the country was 40.02 million, with the gross enrollment rate of 51.6%. The number of HEIs was 11,260, of which 15,179 for undergraduate universities and 7776 for higher vocational education institutes. The number of higher education students of higher is shown in Table 16.2 (Ministry of Education, 2020a).

The national educational spending has increased rapidly in China in recent years. In 2020, national spending on education totaled 5.30 trillion yuan, with an increase of 5.65% compared to the previous year. The distribution of funding and the public funding among different levels of education are respectively given in Table 16.3 and Table 16.4, respectively (Ministry of Education, 2021).

Table 16.2 Number of schools offering upper secondary education (2019)

	Number of students in schools	Number of enrollment	Number of graduates
Students for doctoral degrees	424,200	105,200	62,600
Students for master's degrees	2,439,500	811,300	577,100
Undergraduates in regular HEIs	30,315,300	9,149,000	7,585,300
Students in adult HEIs	6,685,600	3,022,100	2,131,400

Table 16.3 Distribution of funding of different levels of education (2020)

Educational level	Spending in total (billion yuan)	Rate of growth (%)
Preschool	420.3	2.39
Elementary and Junior High School	2,429.5	6.55
Senior High School	842.8	9.14
Higher education	1,399.9	3.99
Spending on other education	209	0.09

Table 16.4 Public funding per student at different levels of education (2020)

Educational level	Spending in total (yuan)	Rate of growth (%)
Preschool	12,954	9.14
Elementary School	14,103	4.43
Junior High School	20,342	3.94
Senior High School	23,489	6.10
Secondary Vocational School	22,568	6.51
Higher Education Institution	37,241	3.78

16.2.3   Educational Research and International Collaboration

The National Institute of Education Sciences (NIES) is a national research institution under the Ministry of Education. It was officially established in 1957 and changed to its current name in 2011. The mission of NIES is to identify the country's changing educational needs, and through innovative reforms, to make policies and carry out programs that will maximize the effectiveness of the nation's schools and educational resources. There are currently 237 in-service employees in NIES, including 213 professional and technical personnel, 69% of which have doctoral degrees. There are nearly 20 comprehensive educational reform experimental areas and several experimental schools set up by NIES in the east, middle and west of China. Moreover, NIES has established cooperative relations with educational research institutions of dozens of countries and international organizations (NIES, 2018). From 2018 to 2020, a total of 193 decision-making policies were made, and 1244 academic papers and 202 academic works were published. NIES is generally recognized as the top educational think tank in China.

There are many normal universities of higher education in China, which have continuously imported many talents for the development of education, among which Beijing Normal University (BNU) is one of the most prestigious and prominent bases. As a university directly under the Ministry of Education, BNU is a well-known university with the main areas of teacher education and science of education. For more than a hundred years, BNU has played an important role in the crucial historical moments of China. BNU has been included in the "211 Project" construction plan, the national "985 Project" construction plan and the national "world-class university" construction list. Moreover, 11 disciplines have entered the national "world-class discipline" construction list. There are currently 77 undergraduate majors, 38 first-level disciplines authorized for master's degree, 32 first-level disciplines authorized for doctoral degree, and 28 post-doctoral research stations. According to the results of the fourth discipline evaluation released by the Ministry of Education, 15 disciplines were rate A level, six of them were awarded for A+, including pedagogy, psychology, Chinese language and literature, Chinese history, drama and film and television, and geography. Equipped with excellent teachers, BNU is an important force in the humanities and social sciences scientific research as well as technological innovation, with dozens of relevant platforms at the national, provincial and ministerial levels (BNU, 2021). East China Normal University (ECNU) is also a strong normal university, with 12 A-class subjects such as pedagogy and geography, and has made outstanding contributions to many aspects, especially the development of education, for decades. In addition, Central China Normal University, Northeast Normal University, Shanxi Normal University, Southwest University and others have all made progress to education research in China.

The Department of International Cooperation and Exchanges is established under the Ministry of Education. Its main missions are to organize and guide international cooperation and exchanges in education, as well as to make policies for studying abroad, studying in China, and international cooperation in running schools. In 2003, the Ministry of Education passed the Regulations of the People's Republic of China on Chinese-Foreign Cooperation in Running Schools, and first-class institutions of higher education have already launched numerous Chinese-foreign cooperative education projects in recent years (Ministry of Education, 2003). In order to promote the long-term development of education and strengthen cooperation with foreign high-level universities, the Ministry of Education initiated the International Joint Research Laboratory Program in 2014, and subsequently issued a sequence of relevant management measures (Ministry of Education, 2014). To promote innovation among students and universities, the Belt and Road Initiative Action Plan for Science and Technology Innovation Services in Universities was released in 2018, so that international cooperation in education was further promoted (Ministry of Education, 2018).

The China Education Association for International Exchange (CEAIE), founded in 1981, is a nationwide non-governmental organization for international educational cooperation and exchanges in China. CEAIE actively develops exchanges and cooperation between the Chinese educational community and that of the other countries and regions, promotes the development of education, culture, science and technology, and enhances the understanding and friendship between people of all countries and regions of the world. CEAIE has established an extensive network within domestic associations and institutions, as well as long-term and stable working relationship with more than 170 authoritative educational organizations in over 50 countries and regions. CEAIE was granted the Special Consultative Status the Economic and Social Council by the United Nations in 2006, and became the NGO Partner of the United Nations Department of Public Information. Since its establishment, CEAIE has contributed to the development of international education of China (CEAIE, 2014).

Many normal universities are also actively participating in the "Opening-up Strategy" and extensively expanding international exchanges and cooperation. As an example, in 2020, Beijing Normal University issued the Global Development Strategic Plan, establishing a strategic vision to help build a global community of academic excellence, education innovation, youth development, and university social responsibility. Cooperative relations with nearly 500 universities and research institutions in more than 40 countries and regions have been established. BNU continues to strengthen cooperation with world-class universities such as Oxford University, Stanford University and University of Helsinki. In addition, BNU also actively promotes exchanges with the Belt and Road countries, and carries out a number of internationally influential projects and activities, which provides a broad platform for academic exchanges, further studies, and cultural exchanges (BNU, 2021).

16.3   Current Situation of Science Education

16.3.1   Policies and Standards

Since the end of 1990s, science and technology innovation has played an important role in promoting the development of the world. Therefore, many countries including China have attached great importance to science education which is a fundamental approach not only to improve citizens' scientific literacy but also to guarantee the construction of innovation society. In 1999, Decision on Deepening Education Reform to Promote Quality Education was issued by the CPC Central Committee and State Council. Since then, a series of policy documents have been released. More details are listed in Table 16.5.

China also focused on the creation of innovative environment and the cultivation of innovative talents. The government also realized the importance of the public's scientific literacy, since this ability represents the national overall understanding of scientific concepts and process, as well as the ability of applying them in solving real-life events and analyzing practical issues (OECD, 2021). Hence, recently, the State Council issued a new long-term national action plan for enhancing the public's scientific literacy over the next 15 years with goals of achieving 15% of the country's population as scientifically literate by 2025, and 25% by 2035 (Gov, 2021).

Table 16.5 Science education policies and standards

No	Year	Organization	Title
1	1999	The CPC Central Committee and State Council	Decision on Deepening Education Reform to Promote Quality Education
2	1999	Ministry of Science and Technology	2000–2005 Program for the popularization of science and technology
3	2000	Ministry of Science and Technology; Ministry of Education et al.	2001–2005 Guidelines for the Activities of Science and Technology Popularization among Chinese Youth
4	2001	Ministry of Education	Compulsory education science curriculum standards (Grade 3~Grade 6)
5	2001	Ministry of Education	Compulsory education science curriculum standards (Grade 7~Grade9)
6	2002	The National People's Congress (NPC)	Law of the People's Republic of China on Popularization of Science and Technology
7	2006	The State Council	National Action Plan for Scientific Literacy (2006–2010–2020)
8	2011	The State Council	National Action Plan for Scientific Literacy (2011–2015)
9	2016	The State Council	National Action Plan for Scientific Literacy (2016–2020)
10	2017	Ministry of Education	Compulsory primary school science education curriculum standards
11	2021	The State Council	National Action Plan for Scientific Literacy 2021–2035

16.3.2 Curriculums, Digital Resources and Teacher Training

Science courses in China are basically guaranteed, but not enough. The Ministry of Education requires that the fourth-grade students should have 2–3 science lessons per week, while eighth-grade students should have 2–3 physics and geography classes per week. Tian et al. (2021) showed that 69.3% students in the fourth grade only have two science lessons per week, and 12.8% students have less than two science lessons. 65.2% eighth grade students have biology classes twice a week, and 64.9% eighth grade students have geography classes twice a week. The data above shows that the science curriculum only meets the basic requirements. Judging from teachers' professional backgrounds, only 16.6% of fourth-grade science teachers have qualifications related to science education, while the remaining 83.4% of science teachers are qualified to teach Chinese, mathematics, foreign languages, music and other subjects. Moreover, 84.8% of physics teachers, 64.4% of biology teachers and 54.9% of geography teachers in eighth grade have qualifications related to science education. Therefore, training of science teachers is of urgency and necessity. Laboratory, experimental equipment and experimental materials are the important media to carry out experimental teaching and scientific education. According to the survey, 99.7% eighth grade physics teachers, 98.4% eighth grade biology teachers, and 88.1% fourth grade science teachers reported that their schools had a laboratory. However, the teachers' usage rate of laboratory is not high, with only 46.3% fourth grade science teachers, 60.3% eighth grade physics teachers and 36.3% eighth grade biology teachers reporting the usage. This similar situation also exists for experimental equipment and experimental materials usage.

16.3.3 Student Assessment and Achievement

Shanghai took the PISA test for the first time in 2009 and was ranked first in reading, math and science literacy. It was also ranked first in 2012. However, this result does not necessarily provide a factual and comprehensive picture of the problems in education of China. For example, although students have made outstanding achievements, the OECD assessment report shows that Shanghai students spent 13.8 h per week on average to do homework, which was also ranked first (OECD, 2015), and they spent 2–4 h to take extracurricular math tutoring every week. Therefore, Shanghai middle school students spent too much time and made a lot of effort on learning. Although students from Beijing, Shanghai, Hangzhou and Jiangsu provinces also came out on top in the 2015 financial literacy assessment, it does not represent the overall level of financial literacy among Chinese teenagers, as the financial education is not available in most of schools in China. China was also ranked in the top in the 2018 PISA test, but students were far behind high-performing countries in terms of social-emotional skills and measures of students' well-being. Therefore, China's achievements should be viewed objectively, and more students should be directed to shift their attention from "more study time" to "high-quality study time".

As for the results of TIMSS 2015, Hong Kong SAR (Special Administration Region), China and Chinese Taipei were ranked 5th and 6th respectively in fourth grade science. Chinese Taipei was ranked 3rd in eighth grade science, and Hong Kong SAR was ranked 6th. However, teacher satisfaction was lower than the international average.

16.3.4 Outreach Science Education

Informal learning environments also play an important role especially in science education. Meanwhile, science and technology museum learning have great values for learners. Gong et al. (2020) revealed that parents take their children regularly to visit children's museum has a positive effect on children's originality score, as well as creativity. Science and technology venues are important places for informal science education activities. They are highly attractive to the public, including students, because of their unique advantages such as rich exhibition and teaching resources,

Table 16.6 The number and scale of science and technology museums nationwide

Year	Number	Scale(m^2)	Total number of visitors/10,000 person-times
2010	101	146.0	2100.0
2019	302	373.8	8103.6
2020	345	399.9	3441.7

experiential science practice, advanced technology application, and free and open learning environment. It plays an irreplaceable role in cultivating scientific interest, popularizing scientific knowledge and enhancing scientific quality.

Over the recent years, the number of science and technology museum has been growing rapidly (see Table 16.6). Meanwhile, science and technology museums in different regions take measures in accordance with local conditions and gradually get out of the shackles of following others and lacking individuality. Professional science and technology museums are built which reflects the inheritance of regional history and culture. Figure 16.1 shows a famous exhibition from China Science and Technology Museum.

Fig. 16.1 "Flying Sky" exhibition from China Science and Technology Museum

16.3.5   Utilizing Emerging Technologies

In recent years, emerging technologies, such as AI, Augmented Reality (AR) and Virtual Reality (VR), has brought a new vision to solve the subject dilemma of science education. Especially, AR and VR technologies have been applied in education since the 1990's, particularly in mathematics (Kebritchi et al., 2010; Pasqualotti & Freitas, 2002), geometry (Hwang & Hu, 2013; Kaufmann & Schmalstieg, 2003), science (Kartiko et al., 2010), physics (Coller & Shernoff, 2009) and chemistry (Merchant et al., 2012). High-tech product such as Microsoft HoloLens has made these technologies more accessible. Cai et al. (2014) conducted further experiment to test the effectiveness of application of AR technology in Junior high school classroom. The results revealed that AR tool had positive effects on learning performance especially for low-achieving students and most students hold positive attitudes toward it.

China also focuses the construction of Virtual Simulation Experiment Platforms which could be important for information construction and experimental teaching demonstration. On August, 2013, the Ministry of Education issued a document about the Construction of National Virtual Simulation Experimental Teaching Center. The website for the national virtual simulation experiment teaching project sharing platform is www.ilab-x.com and individual users must register to use the service. In this way, virtual simulation environments are provided to college students from all over the country so that they can use all the teaching materials and conduct experiments while sitting at home.

16.4   Requirements for Future Development of Science Education

There are many challenges for science education development in China, including: (1) insufficient science education lessons per week, (2) inadequacy of professional science teachers; and (3) the low usage rate of experiment teaching resources. In order to fully cultivate students' innovative spirit and practical ability, it is urgent to promote the deep reform of science education especially in the stage of compulsory education.

Meanwhile, the number of science and technology educators is still insufficient, and the research on their professional development is relatively limited (Piqueras & Achiam, 2019). Existing studies show that the front-line educators of science and technology in foreign countries have diversified professional and academic backgrounds. Most of them come from liberal arts majors such as literature, history, economics and management, while fewer from science and technology majors (Tran & King, 2007).

16.5   Conclusion

This chapter introduces the current situation and the development needs of science education in China. At the policy level, China has issued a series of policies to promote the development of science education and improve the scientific literacy of its people. From the society level, China has also paid more attention to the positive role of informal education to enhance learners' scientific literacy. However, more attention should be paid to the professional development of teachers. At the school level, the number of science lessons per week might not be enough to teach the core concepts of science. In addition, the professional level of science teachers needs to be improved, which highlights the importance and necessity of science teacher training.

References

BNU. (2021). The Belt and Road School. Retrieved July 27, 2021, from https://brs.bnu.edu.cn/english/index.htm.

Cai, S., Wang, X., & Chiang, F. K. (2014). A case study of augmented reality simulation system application in a chemistry course. *Computers in Human Behavior. 37*, 31–40. ISSN 0747-5632. https://doi.org/10.1016/j.chb.2014.04.018.

Coller, B. D., & Shernoff, D. J. (2009). Video game-based education in mechanical engineering: A look at student engagement. *International Journal of Engineering Education, 25*(2), 308.

China Education Association for International Exchange (CEAIE). (2014). *General introduction of CEAIE*. Retrieved June 20, 2021, from http://www.ceaie.edu.cn/guanyuxiehui/14.html.

Gong, X., Zhang, X., & Tsang, M. C. (2020). Creativity development in preschoolers: The effects of children's museum visits and other education environment factors. *67*. https://doi.org/10.1016/j.stueduc.2020.100932.

Gov. (2021). *National Action Plan for Scientific Literacy 2021–2035*. Retrieved June, 28, 2021, from http://www.gov.cn/zhengce/content/2021-06/25/content_5620813.htm.

Huang, Q., & Mao, Z. (2019). China's construction of a cultural power of socialist modernization: Development, achievement and enlightenment. *Journal of Social Sciences of Shanxi Colleges and Universities, 31*(04), 35–38.

Hwang, W.-Y., & Hu, S.-S. (2013). Analysis of peer learning behaviors using multiple representations in virtual reality and their impacts on geometry problem solving. *Computers & Education, 62*, 308–319. https://doi.org/10.1016/j.compedu.2012.10.005

Kaufmann, H., & Schmalstieg, D. (2003). Mathematics and geometry education with collaborative augmented reality. *Computers & Graphics, 27*(3), 339–345.

Kartiko, I., Kavakli, M., & Cheng, K. (2010). Learning science in a virtual reality application: The impacts of animated-virtual actors' visual complexity. *Computers & Education, 55*(2), 881–891. https://doi.org/10.1016/j.compedu.2010.03.019

Kebritchi, M., Hirumi, A., & Bai, H. (2010). The effects of modern mathematics computer games on mathematics achievement and class motivation. *Computers & Education, 55*(2), 427–443. https://doi.org/10.1016/j.compedu.2010.02.007

Li, K. (2021). *Report on the Work Of The Government 2021*. Retrieved July 27, 2021, from http://www.xinhuanet.com/2021-03/12/c_1127205339.htm.

Merchant, Goetz, E. T., Keeney-Kennicutt, W., Kwok, O.-M., Cifuentes, L., & Davis, T.J. (2012). The learner characteristics, features of desktop 3D virtual reality environments, and college chemistry instruction: A structural equation modeling analysis. *Computers & Education, 59*(2), 551–568. https://doi.org/10.1016/j.compedu.2012.02.004.

Ministry of Education. (2018). The Belt and Road Initiative Action Plan for Science and Technology Innovation Services in Universities. Retrieved July 27, 2021, from http://www.moe.gov.cn/srcsite/A16/kjs_gjhz/201901/t20190102_365666.html.

Ministry of Education. (2020a). *Overview of Educational Achievements in China in 2019.* Retrieved July 27, 2021, from http://en.moe.gov.cn/documents/reports/202102/t20210209_513095.html.

Ministry of Education. (2020b). *Statistical Communique on National Education Development in 2019.* Retrieved July 27, 2021, from http://www.moe.gov.cn/jyb_sjzl/sjzl_fztjgb/202005/t20200520_456751.html.

Ministry of Education. (2021). *MOE releases 2020 statistical bulletin on educational spending.* Retrieved July 27, 2021, from http://en.moe.gov.cn/news/press_releases/202105/t20210512_531041.html.

National Bureau of Statistics of China. (2021a). *Communique of the seventh national population census.* Retrieved July 27, 2021, from http://www.stats.gov.cn/english/StatisticalCommuniqu/.

National Bureau of Statistics of China. (2021b). *National economy recovered steadily in 2020 with main goals accomplished better than expectation.* Retrieved July 27, 2021, from http://www.stats.gov.cn/english/PressRelease/202101/t20210118_1812432.html.

National Bureau of Statistics of China. (2021c). *Statistical communique of the People's Republic of China on the 2020 national economic and social development.* Retrieved July 27, 2021, from http://www.stats.gov.cn/english/PressRelease/202102/t20210228_1814177.html.

National Institute of Education Sciences (NIES). (2018). *Introduction to the basic situation of the National Institute of Education Sciences.* Retrieved July 27, 2021, from http://www.nies.edu.cn/gywm/lsyg/jbjs/201809/t20180910_334148.html.

National People's Congress. (2021). *Education Law of the People's Republic of China.* Retrieved July 27, 2021, from http://www.moe.gov.cn/jyb_xxgk/xxgk_jyfl/flfg_jyfl/202107/t20210730_547843.html.

OECD (2015). *PISA 2015 Result: Students' well-being[EB/OL].(2017-04-19) [2021-02-20].* http//www.oecd.org/pisa/Publications/pisa-2015-results-volume-iii-9789264273856-en.htm.

OECD (2021). *The future of education and skills education 2030 [EB/OL] (2018–04–05) [2021–3–22].* http://www.oecd.org/education/2030/oecd-education-2030-position-paper.pdf.

Pasqualotti, A., Freitas, & C. M. d. S. (2002). MAT3D: A virtual reality modeling language environment for the teaching and learning of mathematics. *Cyber Psychology and Behavior, 5*(5), 409–422.

Piqueras, J., & Achiam, M. (2019). Science museum educators' professional growth: dynamics of changes in research-practitioner collaboration. *Science Education, 103,* 389–417.

State Council of China. (1999). *The decision of the CPC central committee and the state council on deepening education reform and comprehensively promoting quality education.* Retrieved July 27, 2021, from http://www.moe.gov.cn/jyb_sjzl/moe_177/tnull_2478.html.

State Council of China. (2019). China's Education Modernization 2035.

State Council of China. (2021). Geography. Retrieved July 27, 2021, from http://www.gov.cn/guoqing/index.htm.

Tran, L. U., & King, H. (2007). The professionalization of museum educators: The case in science museums. *Museum Management and Curatorship, 22*(2), 131–149.

Wei, T., Tao, X., & Weiping, H. (2021). Science education in compulsory education: Key problems and countermeasures. *Journal of Beijing Normal University* (Social Sciences) (03), 82–91.

Wang, Z. (2021). *Commit to science and technology support and accelerate the development of science and technology.* Retrieved July 27, 2021, from http://www.qstheory.cn/dukan/qs/2021-03/16/c_1127209154.htm.

Xi, J. (2021). *Speech at a national conference to review the fight against poverty and commend individuals and groups involved.* Retrieved July 27, 2021, from http://www.cpad.gov.cn/art/2021/2/25/art_305_187439.html?isappinstalled=0.

Xin, J. (2021). *The growth rate of cultural enterprise operating income turned from negative to positive with the new business model maintaining rapid growth.* Retrieved July 27, 2021, from http://www.stats.gov.cn/tjsj/sjjd/202101/t20210129_1812935.html.

Xiangling Zhang is working as a lecturer of teacher professional development at Beijing Institute of Education. She finished her post-doctoral studies at Beijing Normal University in 2020. Her research focuses on educational technology, STEM education.

Yao Song is currently studying for a bachelor's degree at Beijing Normal University, majoring is Statistics and minoring in Computer Science.

Zhenyu Cai is currently a graduate student at Faculty of Education, Beijing Normal University. His research interest includes the field of technology application in education as well as STEM education.

Lixin Zhu is working as a senior engineer, Beijing Normal University. He is also the Co-director, Learning Environment Designing and Assessment & Evaluation lab, the National Engineering Lab for CyberLearning and Intelligent Technology. His research focuses on educational technology, STEM education.

Tianyue Sun is currently a graduate student majoring in Applied Behavior Analysis at the Department of Health and Behavior Studies, Teacher College, Columbia University. Her research interest includes the field of special education and developmental psychology.

Ronghuai Huang is a Professor in Faculty of Education and Dean of Smart Learning Institute in Beijing Normal University. He received "Chang Jiang Scholar" award in 2017, which is the highest academic award issued to an individual in higher education by the Ministry of Education in China. He now serves as the Director of the National Engineering Lab for Intelligent Cyber-Learning Technology, and Director of Beijing Key Laboratory for Educational Technology. He is also the president of International Association of Smart Learning Environments, Editor-in-Chief of Springer's Smart Learning Environments and Journal of Computers in Education, as well as Editor-in-Chief of Springer's series Lecture Notes in Educational Technology and Smart Computing and Intelligence.

Ahmed Tlili is the Co-Director of the OER Lab, Smart Learning Institute of Beijing Normal University. He is also the Associate Researcher of Beijing Normal University. His research areas include game-based learning, smart learning environments, technology enhanced learning, learner modeling, adaptive learning systems, learning analytics, and educational psychology.

Chapter 17
Science Education in Malaysia

Eng Tek Ong

Abstract The chapter begins with a snapshot of Malaysia in terms of its geographical location, population, political situation, economic and cultural development, and more importantly, its educational system. It then discusses various competitive research grants provided by the Malaysian Government in the quest to produce new theories or ideas, and to foster inter-varsity as well as international collaboration and networking. It then deliberates on the current science education scenario in terms of its policy, focusing the deliberation on the policies that have been drawn up to achieve the aspirations encapsulated in the National Education Blueprint (2013–2025). The standard-based science curriculum is then given adequate discussion and some downloadable examples of the digital resources are provided. The science assessments are discussed in the context of the Standardized National Examinations taken by the Malaysian students, and also in the context of TIMSS. The use of ICT in the Malaysian science classrooms is discussed in the context of the Malaysian Smart School Initiative, Virtual Online Learning System, and Google Classroom. It ends with recommendations in solving ubiquitous problems in students' interest in science, pedagogical approach and science teacher education programs.

Keywords Malaysia · Science education · Standard-based curriculum · National Education Blueprint

E. T. Ong (✉)
Department of Educational Studies, Faculty of Human Development, Sultan Idris Education University, Proton City, 35900 Tanjung Malim, Perak, Malaysia
e-mail: ong.engtek@fpm.upsi.edu.my

Lecture Notes in Educational Technology,
https://doi.org/10.1007/978-981-16-6955-2_17

17.1   Overview of the Country

17.1.1   The Geography, Population Situation and Political System

Malaysia is one of the countries in Southeast Asia. Comprising a total of 13 states and 3 federal territories, Malaysia is divided by the South China Sea into Peninsular Malaysia (or, West Malaysia) and East Malaysia. The former has 11 states (namely, Perlis, Kedah, Penang, Perak, Selangor, Negeri Sembilan, Melaka, Johor, Kelantan, Terengganu and Pahang) and 2 federal territories (namely, Kuala Lumpur and Putrajaya) while the latter consists of 2 states (namely, Sabah and Sarawak) and 1 federal territory (e.g., Labuan).

Out of the 13 states, 9 states are ruled by hereditary monarchies called *Sultan* (in Perlis, Kedah, Perak, Selangor, Johor, Kelantan, Terengganu and Pahang), *Raja* (in Perlis) or *Yang di-Pertuan Besar* (in Negeri Sembilan), while the remaining 4 states are ruled by *Yang di-Pertua Negeri* or Governors. These 9 monarchies and 4 governors form the Council of Rulers as officially established by Article 38 of the Constitution of Malaysia, with the responsibilities of, amongst others, electing the *Yang di-Pertuan Agong* (King) and his deputy, the *Timbalan Yang di-Pertuan Agong* (Deputy King) from among the 9 monarchies every five years or when the positions fall vacant due to death, resignation, or removal.

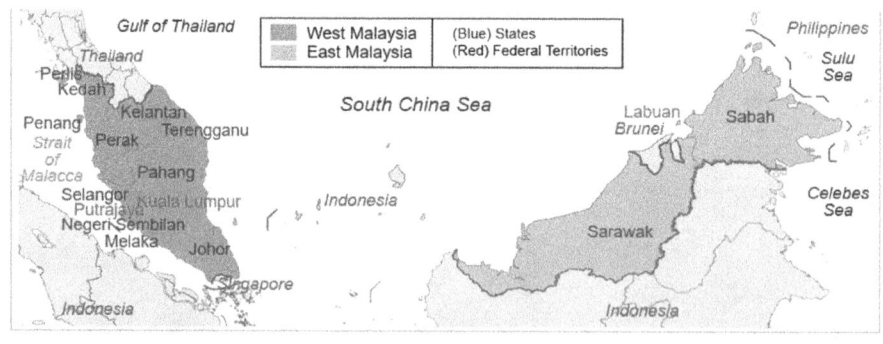

Source https://en.wikipedia.org/wiki/Malaysia

Malaysia's population in 2020 is estimated at 32.7 million as compared to 32.5 million in 2019 and 32.4 million in 2018 (Department of Statistics Malaysia, 2020; Educational Planning and Research Division, 2019). The country is multi-ethnic where 69.6% of the population is Bumiputera (which consists of Malay, Orang Asli, and the indigenous groups of Sabah and Sarawak), 22.6% Chinese, 6.8% Indians, and 1% others (Department of Statistics Malaysia, 2020). The official language and also the medium of instruction in Malaysia is *Bahasa Melayu* (the Malay Language) while the English Language is officially taught as the second language in schools.

The government in Malaysia is modelled on the Westminster parliamentary system and the legal system is based on common law. The Prime Minister is the head of the government. Currently, Dato' Sri Ismail Sabri bin Yaakob is the Prime Minister of Malaysia. Meanwhile, Dr *Mohd Radzi Md Jidin is the Minister of Education at the Ministry of Education and he is* responsible for the education system, compulsory education, pre-tertiary education, technical and vocational education and training (TVET), curriculum standard, textbooks, standardized national examinations, language policy, translation, selective schools, and comprehensive schools. Ministry of Higher Education, headed by the Minister of Higher Education, Dato' Dr. Noraini Ahmad, is responsible for higher education, polytechnic, community college, student study loan, accreditation, and student volunteer.

17.1.2   The Current Situation of Economic, Technologies and Cultural Development

In terms of economic development, the report from the Organization for Economic Co-operation and Development (OECD) Economic Surveys for Malaysia (OECD, 2019) indicated that Malaysia has performed very successfully in recent years as compared to other emerging market countries and that Malaysia is catching up rather rapidly towards the living standards prevailing in OECD countries. Malaysia's 2017 per capita GDP was close to two-thirds of the OECD average. This favourable economic development, in part, is due to diversification of export products and improved macroeconomic prudence.

Meanwhile, Malaysia is also moving forward towards a technology-driven and high-tech production-based pattern of economic development. This is unsurprising because Malaysia has been categorised as a group of countries that have the potential to create new technologies on their own (Mani, 2000). Kumar (2019) observes that technology may have been transforming Malaysia into a digital economy and that Malaysia has one of the most digitally connected societies in the world, with approximately 80% of the people having access to the internet, mainly through mobile networks. The devices used among Malaysians have in fact powered the rate of globalisation and according to Kumar (2019), one billion devices are expected to get connected by 2025. As such, the Malaysian Government has embarked on several initiates to spur the digital economy such as encouraging the digital entrepreneurship and promoting technology adoption among businesses.

Malaysia is a multi-cultural society, with Malays, Chinese, Indians, Orang Asli (the natives in West Malaysia), and Orang Asal (the natives in East Malaysia) living side by side. The Malays are the largest community and they are Muslims, speak the Malay Language called Bahasa Melayu, and are largely responsible for the political scenes of the country. Other races speak their respective mother tongues. Malay is the national and official language used in schools and government departments and Islam is the official religion of Malaysia, although English is widely spoken among

Malaysians. Other ethnicities besides the Malays in Malaysia are free to practise whatever religion that they deemed best and retain their customs and way of life.

17.2 Overview of the Education Development

17.2.1 The Education System and Policy

The Malaysian education system is structured on a K + 6 + 3 + 2 + 2 model, with one year of preschool education, six years of compulsory primary education beginning at age seven, followed by three years of lower secondary education, two years of upper secondary (that may take the form of academic secondary education, technical school education, vocational education, or religious secondary education), and one to two years of pre-university (which may take the form of Form 6, matriculation, foundation, or Malaysian Religious Higher Certificate (STAM)). The general levels of education in Malaysia are summarized in Table 17.1.

The education from the pre-school up to the upper secondary is provided free for all Malaysian children and that the Malaysian Government aspires full access to education with 100% enrolment across all levels of education up to Form 5 which is equivalent to Grade 11 (Ministry of Education, 2012). According to the professional circular (Ministry of Education, 2002a, 2002b) which is still in effect, the primary education for all children from age 7 till 12 (Years 1–6) is mandatory, and that this circular invokes Section 29A (2) of the Education Act 1996 which mandates the

Table 17.1 The levels of education in Malaysia

Level	Age (years)
Preschool and kindergarten	4–6
Primary (years 1–6)	6+ to 11+
Lower secondary (Forms 1–3)	12+ to 14+
Upper secondary	15+ to 16+
Academic secondary Education (Forms 4 & 5)	
Technical secondary education	
Vocational education	
Religious secondary education	
Post-secondary or Pre-University	17+ (for 1 or 2 years)
Form 6 (1½ years)	
Matriculation/Foundation (1 year)	
STAM or Malaysian religious Higher certificate (1 year)	
Higher education	

registration in a primary school when a child reaches 6 years of age and subsequently follow through the mandatory primary education beginning from age 7.

Several educational policies are in effect in Malaysia. However, given that this chapter discusses science education in Malaysia, therefore, it is pertinent to highlight the 60:40 Policy which was instituted in 1967 and implemented in 1970. This 60:40 Policy envisages that 60% of the Malaysian students would follow through with the science-based subjects while the remaining 40% studying arts-based subjects. To date, the policy has yet to attain the desired proportion between the science- and arts-based students. Given that the Malaysia Education Blueprint (2013–2025) (Ministry of Education, 2012) continues to emphasize this policy and that there is no specific year to indicate the policy's expiration, hence the inference that the 60:40 Policy will still be in force until further notice. This 60:40 Policy is a long-term plan towards achieving the developed nation status where human capital in terms of talents who are highly skilled, knowledgeable, competitive, and productive are greatly needed (Kamaludin, 2019). These science and technical-based talents are highly needed because they can meet the demand of a digitalization era (Alcacer & Cruz-Machado, 2019) and that the science-based graduates are better prepared to meet the challenges of the Industrial Revolution 4.0 (IR 4.0) (Vijaindren, 2018).

As for the percentage of actual total educational expenditure against total government expenditure, the data from the Educational Planning and Research Division or EPRD (2019) indicates that for the three consecutive years of 2017, 2018, and 2019, the corresponding Malaysian government expenditure were 16.92%, 21.25% and 19.03%. Meanwhile, the percentages of recurrent educational expenditure on emolument (e.g., the pay for the teachers and staff in the Ministry of Education) for years 2017, 2018, and 2019 were 81.51%, 67.94%, and 70.80% respectively (EPRD, 2019).

17.2.2 School, Students and Teachers' Profiles

The data provided by the Educational Planning and Research Development of the Malaysian Ministry of Education indicates that as of 1st January 2019, there was a total of 10, 208 schools, of which 7772 were the primary schools while the remaining 2436 were secondary schools (ERRD, 2019). Meanwhile, there was a total of 6152 pre-schools of which a majority of these pre-schools were attached to primary schools.

In terms of student enrolment, there were 1,398,201 male and 1,328,560 female students in the primary schools, signifying that there were more male students in the primary schools. However, in the secondary schools, there were 999,074 male and 1,008,422 female students, indicating that the number of females was more than the males. This indicates that the retention rate for the females to remain in secondary education after having completed primary education is higher than that of the males. In terms of the number of teachers, there were more female teachers than male teachers across the pre-school, primary, secondary education as shown in Table 17.2.

Table 17.2 Number of schools, student enrolment and teachers

	Schools	Student enrolment			Number of teachers		
		Male	Female	Total	Male	Female	Total
Pre-school	6152	103,628	101,572	205,200	1091	8220	9311
Primary	7772	1,398,201	1,328,560	2,726,761	70,025	167,292	237,317
Secondary	2436	999,074	1,008,422	2,007,496	53,717	128,870	182,587
Total	**10,208**	**2,397,275**	**236,982**	**4,734,257**	**123,742**	**296,162**	**419,904**

Note The data only covers schools by the Ministry of Education, and the Total excludes pre-school

17.2.3 Educational Research and International Collaboration

The Ministry of Higher Education does offer a wide range of competitive research grants to which lecturers from the Ministry of Education as well as the Ministry of Higher Education may apply. One of these research grants is the Fundamental Research Grant Scheme (FRGS) where its main purpose is to produce new theories, concepts or ideas for the sake of knowledge development, conducted within a two- or three-year duration. The applicants must ensure that the proposals meet one of the 9 National Priority Areas: Food Security; Energy Security; Plantation Crops; Cyber Security; Water Security; Biodiversity; Healthcare and Medicine; Environment and Climate Change; and Transportation and Mobility. Besides, there is a Trans-disciplinary Research Grant Scheme (TRGS) which may be fundamental or exploratory in nature and yet, it contributes towards the production of new theories, concepts or ideas within the 9 National Priority Areas.

Additionally, there is the Long Term Research Grant Scheme (LRGS) which is fundamental research, involving more extensive scope, a long period of time and requires a high commitment. LRGS must be able to produce theories, new ideas and innovative creations heading towards a strategic niche for the development of frontier knowledge; strengthen excellence and expand knowledge, especially in cutting edge fields, specifically the niche that has been identified; and contribute to Malaysia's strategic agenda. Besides, the LRGS promotes inter-varsity research because there will be at least 3 project leaders within a successful program, and that this research program must involve at least 3 universities in Malaysia and the projects within the program are interdisciplinary in nature.

Research funding is also given to the lecturers in the form of the Prototype Research Grant Scheme (PRGS). PRGS aims to encourage the development of prototypes that are demand-driven so as to meet the needs of the industries and society at large. The development of a prototype reduces the gap between scientific discovery and commercialisation. The PRGS may involve the collaboration between or among universities, and between the university and the industry (or the private agency).

In order to encourage international collaboration, there is a Matching Research Grant Scheme. For example, in Sultan Idris Education University, the Matching

Research Grant Scheme aims to boost research collaborations between the Education University and a Research University or a Public University or a Private University or an Industry or an Agency at national and/or international levels. Such kind of collaboration aims to catalyse the generation of new ideas and innovative inventions.

17.3 Current Situation Of Science Education

17.3.1 Policies and Standards

The Ministry of Education has developed a National Education Blueprint. (2013–2025) that encapsulates the 5 aspirations deriving from the outcomes of the Malaysian Education System: (i) Access (e.g., with 100% enrolment across all levels from pre-school to upper secondary); (ii) Quality (e.g., being placed at the top third countries in international assessments such as TIMSS and PISA); (iii) Equity (e.g., with at least 50% reduction in urban–rural, rich-poor, male–female achievement gaps); (iv) Unity (e.g., where children have shared values and experiences, embracing diversity); and (v) Efficiency (e.g., student outcomes are maximised within the current budget) (Ministry of Education, 2012).

These aspirations are rooted within The National Education Philosophy (NEP) for Malaysia that was written in 1988 and revised in 1996. The NEP enshrines the Ministry's and Government's vision of education as a means for the holistic development of all children. In essence, the NEP states that,

> Education in Malaysia is an ongoing effort towards further developing the potential of individuals in a holistic and integrated manner, so as to produce individuals who are intellectually, spiritually, emotionally, and physically balanced and harmonious, based on a firm belief in and devotion to God. Such an effort is designed to produce Malaysian citizens who are knowledgeable and competent, who possess high moral standards, and who are responsible and capable of achieving high levels of personal well-being as well as being able to contribute to the harmony and betterment of the family, the society, and the nation at large (Ministry of Education, 2012).

In the National Education Blueprint, it is envisaged that every Malaysian child, upon learning within the Malaysian Education System, should have eventually acquired the 6 attributes, namely, knowledge, thinking skills, leadership skills, bilingual proficiency, ethics and spirituality, and national identity.

Therefore, in order to achieve these aspirations of the Malaysian Education System, the Ministry of Education has conceptualised 11 Shifts of which the following two shifts are related to science education: Shift 1: Provide equal access to quality education of an international standard; and Shift 4: Transform teaching into the profession of choice.

Accordingly, one of the policies to achieve Shift 1 is to *benchmark the learning of Mathematics and Science to the international standards.* Hence, the revision of the science (and mathematics) curricula across grades and levels by benchmarking them to that of the international standards. While the primary science standard-based curricula have been in existence since 2011–2016, the Primary School Standard Curriculum (Revised) or KSSR (Revised) for Science Years 1–4 were first implemented in 2017, 2018, 2019 and 2020 respectively. In 2021, the KSSR (Revised) Science Year 5 will roll out nationwide.

One of the revisions made on the syllabuses is that the new version contains not only the standards for the curriculum, it also contains the evaluation standards which are divided into 6 levels, namely remember, understand, apply, analyze, evaluate, and create. The first two categories are considered as measuring lower-order thinking while the other four categories are reckoned as gauging higher-order thinking. In the old version, the assessment was that of school assessment that uses summative and formative assessments. By contrast, in the revised version, the assessment is that of school-based assessment which is wider in scope and entails the utilization of central assessment, assessment of physical, sports and co-curriculum, psychometric assessment, and school assessment. In terms of content coverage, the old version was limited while the revised version covers more topics across grades.

The framework employed in the development of a standard curriculum for science is based on six strands, namely, (i) Communication; (ii) Spiritual, Attitude, Value and Humanity; (iii) Personal Development; (iv) Physical Development and Aesthetic; and (vi) Science and Technology. The strands are interdependent and are also fused with critical, creative and innovative thinking. Such integration is in line with the National Philosophy of Education which aims at developing competent Malaysian children who are imbued with noble values and the ability to think critically, creatively and innovatively.

Generally, in a standard curriculum, there are three columns for each theme: content standard; learning standard; and performance standard. For example, in the Standard Curriculum for Science Year 1, 6 themes have been advocated: (i) Inquiry in Science; (ii) Life Science; (iii) Physical Science; (iv) Materials Science; (v) Earth and Space; and (vi) Technology and Sustainability of Life. As an illustration, let us look at the theme of "Physical Science". Under this theme, the content standards are magnets; electrical circuit; float and sink; sounds; volume; mass and time. For the content standard on magnets, there are 6 learning standards to achieve the expected conceptual understanding of magnets as shown in Fig. 17.1.

In terms of assessment, the performance standards are clearly dictated in the standard curriculum. There are 6 performance levels, which are Level 1 (Recall), Level 2 (Understand), Level 3 (Apply), Level 4 (Analyze); Level 5 (Evaluate) and Level 6 (Invent). These 6 levels are similar to the 6 levels of revised Bloom's taxonomy of educational objectives (Anderson & Krathwohl, 2001).

Another policy drawn up to help achieve Shift 1 (B*enchmark the learning of Mathematics and Science to the international standards*) is the strengthening of quality STEM (Science, Technology, Engineering and Mathematics) Education. A National STEM Centre has been set up at the Ministry of Education and its roles, among others, are to provide in-service training to key science and mathematics teachers in terms of pedagogy, and to promote the love and interest in science among Malaysian children. The main science pedagogical approach advocated by the National STEM Centre is

7. MAGNET				
CONTENT STANDARD	**LEARNING STANDARD**	**PERFORMANCE STANDARD**		**NOTES**
		PERFORMANCE LEVEL	**DESCRIPTOR**	
7.1 Magnet	Pupils are able to: 7.1.1 Give examples the usage of magnets in daily life. 7.1.2 Identify the shapes of magnets e.g. bar, cylinder, horseshoe, U-shaped, button and ring. 7.1.3 Make generalisation on reactions of magnets to various objects by carrying out activities. 7.1.4 Conclude that magnet attracts or repels between two poles through investigation. 7.1.5 Determine the strengths of magnet towards object through investigation.	1	Give examples of objects or tools that use magnet.	Pupils are encouraged to bring various tools that use magnets such as magnetic pencil boxes, fridge magnets and magnetic toys. Pupils conduct investigations by placing magnet near to the object and observe whether the objects are attracted or not. Pupils conduct a fair test to investigate the strength of magnets in terms of distance and the number of paper clips that are attracted, the shape and size of the magnets must be constant.
		2	Identify various types of magnets.	
		3	Make generalisation on reactions of magnets to various objects.	
		4	Make generalisation that magnet attracts or repels between two poles.	
		5	Conclude the strengths of magnets based on investigation done.	
	7.1.6 Explain observations about magnets using sketches, ICT, writing or verbally.	6	Design a game or a tool using magnets.	Pupils can design simple games such as nail racing using magnets without touch the nail.

Fig. 17.1 Excerpt of a content standard from the primary school standard curriculum (revised) for year 1 science. *Source* Curriculum Development Division (2017, pp. 42–43)

that of 5E Inquiry Learning that entails the phases of engage, explore, explain, elaborate, and evaluate (Ong et al., 2020a). This pedagogical approach is constructivist in nature because it elicits students' pre-instructional conceptions at the engagement phase and provides them with the sense-making experience through hands-on and minds-on activities at the exploration phase. On the basis of the learning experiences gained from the engagement and exploration phases, students then restructure their pre-instructional conceptions to the conceptions that parallel the school science view (Ong et al., 2020a, 2020b).

A policy that has been implemented and practised in achieving Shift 4 (*Transform teaching into the profession of choice*) is the recruitment of pre-service teachers. In other words, the entry bar for teachers is raised so that the best candidates are recruited to the pre-service teachers' training programme. These best candidates are amongst the top 30% of graduating classes. According to the Ministry of Education (2012), "the quality of an [educational] system cannot exceed the quality of its teachers" (p. 14). As such, one of the criteria expected is that the candidates should have obtained at least 5 distinctions in the Malaysian School Certificate Examination, which is a National Standardised Examination, and also at least a credit pass in the Malay Language, English Language, and History. This seems to echo the call by Carlo (2011) that we should be attracting the "best candidates into teaching.

17.3.2   Curriculums and Digital Resources

Science is taught as a compulsory subject right from Years 1–3 (Level 1), Years 4–6 (Level II), and Form 1–3 (Lower Secondary) in the Malaysian schools. However, in Forms 4 and 5 (Upper Secondary), while Science is still a compulsory subject for students who have opted out from up-taking the pure sciences, it is not a subject to be taken by students who are taking Biology, Chemistry and/or Physics.

There are three major types of schools for primary education in Malaysia. They are National Schools, National Type Schools (Chinese), and National Type Schools (Tamil) of which the mediums of instruction are Malay, Chinese, and Tamil correspondingly. However, the science curricular for Years 1–6 are the same for these three major types of schools, although the science is taught in different languages. Meanwhile, when the students converge in secondary education, the Malay language is used as the medium of instruction for the curriculum.

The currently used standard-based curricular for Science Years 1–6, and Science, Additional Science and Pure Sciences (Physics, Chemistry, and Biology) Forms 1–5 are developed by the Curriculum Development Division (CDC), Ministry of Education Malaysia. Based on the curricular prepared by the CDC, the Resource and Educational Technology Division (formerly, the Textbook Division), the Ministry of Education is tasked to prepare the science textbooks to be used across Malaysian schools. For the science textbooks that are used by the National Primary Schools, the Textbook Division liaises with the Institute of Language and Literature (or, *Dewan Bahasa dan Pustaka, DBP*) to publish the relevant textbooks. As for the science and pure science textbooks, the Textbook Division calls of a tender from the Malaysian publishers for submission of exemplar chapters for each science syllabus and calls of a committee comprising experts in the corresponding areas of specialization to do a selection, before the textbook writing tender is awarded.

The Malaysian Ministry of Education has produced many digital resources as guides for all teachers in performing their teaching and learning. With examples drawn from the science, mathematics, languages, and art lessons, these digital resources are uploaded to the Ministry's website where Malaysian teachers or any interested readers may download them for personal use. For example, on the notion of "Higher Order Thinking", the Ministry has produced many digital resources (in the Malay Language) such as:

(i)   HOTS: Higher Order Thinking Skills in Schools Initiative (Curriculum Development Division, 2013).

(ii)   Higher Order Thinking Skills: School Applications (Curriculum Development Division, 2014a).

(iii) HOTS Elements in the Curriculum (Curriculum Development Division, 2014b).

(iv)   HOTS Elements in Pedagogy (Curriculum Development Division, 2014c).

(v)   HOTS Elements in Assessment (Curriculum Development Division, 2014d).

(vi)   HOTS Elements in Co-Curriculum (Curriculum Development Division, 2014e).

(vii)   HOTS Elements in Effort Building (Curriculum Development Division, 2014f)

(viii) HOTS Elements in Resources (Curriculum Development Division, 2014g)

(ix)   The Support from the Community and Private to Boost HOTS (Curriculum Development Division, 2014h)

(x)   Exemplary Items in Assessing Higher Order Thinking in Primary Schools (Malaysian Examination Syndicate, 2015a)

(xi)   Exemplary Items in Assessing Higher Order Thinking in Secondary Schools: Science and Mathematics Areas (Malaysian Examination Syndicate, 2015b)

(xii)   Assessing Higher Order Thinking Skills (Malaysian Examination Syndicate, 2013)

(xiii) Application Guide for Higher Order Thinking Skills Based on Forms 4 and 5 Integrated Curriculum for Secondary Schools (Textbook Division, undated).

The Ministry of Education has even come up with QR Codes for those who would like to download the digital resources. For example, the following is the QR Code provided by the Malaysian Examination Syndicate, Ministry of Education Malaysia.

17.3.3 Student Assessment and Achievement

Malaysian students are assessed in science at the Standardized National Examinations (SNE) at Year 6, Form 3 and Form 5. In the SNE Science for Year 6, students sit for 2 papers, namely Paper 1 consisting of 40 multiple-choice items and Paper 2 consisting of 8 structured questions. In 2019, out of 431,635 candidates, 86.57% passed (Grades A-D), while 3.43% failed (Grade E), as opposed to 427,151 candidates in 2018 with 96.12% passed while 3.88% failed (Malaysian Examination Syndicate, 2020a).

In the Science paper at SNE which is also known as the Malaysian Certificate of Education (SPM) taken by the Form 5 students, they are assessed in 2 papers, namely Paper 1 consisting of 50 multiple-choice items and Paper 2 consisting of 12 structured questions. In the year 2019, 284,729 students sat for the science paper, with 89.9% passed (Grades A+, A, A−, B+, B, C+, C. D and E) while 10.1% failed (Grade G). There is a drop of 1.2% passes when it is compared to the achievement obtained in 2018 where there were 282,377 students who sat for the exam (Malaysian Examination Syndicate, 2020b).

As for the International Assessment, Malaysia inaugurally participated in the TIMSS 1999 and TIMSS 2003, achieving the average scores of 492 (22nd position)

Table 17.3 Trend in science achievement of Malaysia in TIMSS

	TIMSS (science achievement)				
	1999	2003	2007	2011	2015
Malaysia's average scale score	492^{+}	510^{*}	471^{**}	426^{**}	471^{**}
International standing	22nd	20th	21st	32nd	24th
Number of countries participated	38	45	48	63	57

$^{+}$No significant difference as compared to the international average
*Significantly higher as compared to the international average
**Significantly lower as compared to the international average

and 510 (20th position) correspondingly, as depicted in Table 17.3. However, in TIMSS 2007, the average score plummeted to 471 (21st position), and subsequently, in TIMSS 2011, the average score dipped further to that of 426 (32nd position) (Martin et al., 2012). In TIMSS 2015 (Martin et al., 2016), the average score did improve to that of 471 (24th position) although the average score achieved was still significantly below the International Average of 500.

Meanwhile, Malaysia is still waiting enthusiastically for the results of TIMSS 2019 which will be officially released in December 2020. Enthusiastically because the Malaysian Ministry of Education uses the achievement in TIMSS as the benchmark for measuring the quality education as officially stated in the Malaysia Education Blueprint 2013–2025 (Ministry of Education, 2012), According to the benchmark, an aspired quality of education is achieved if and only if Malaysia is placed at the "top third of the countries in international assessments such as … TIMSS in [the next] 15 years [from 2013]" (Ministry of Education, 2012, Executive Summary, p. 9).

17.3.4 Outreach Science Education and Infusion of Formal and Informal Science Education

Formal education or formal learning entails an education process that has features such as having physical school, prescribed curriculum, a teacher with students, assessment and certification. Informal education, by contrast, involves real-life examples of learning and it is not imparted by an institution or dictated by a fixed curriculum (Marsick & Watkins, 2001). Informal education may include, among others, activities such as (1) visits to museums, science centres, planetariums, or to science fairs and exhibits; (2) listening to radio broadcasting or watching TV programmes on educational or scientific themes; (3) reading texts on sciences, education, and technology in journals and magazines; and (d) participating in scientific contests or competitions.

In terms of formal learning of science, the directive from the Ministry of Education states that students in Level 1 (Years 1–3) and Level 2 (Years 4–6) in the primary schools should learn science for a total of 48 and 64 h respectively in a year (Ministry of Education, 2019a). Meanwhile, students in secondary schools should learn science for a total of 112 h in a year (Ministry of Education, 2019b).

In terms of informal learning, the Malaysian Ministry of Education has issued a professional circular (Ministry of Education, 2002a, 2002b) on "School Visits during Term Time" to all schools across Malaysia. This circular encourages science teachers to bring their students for educational visits to many places of interest such as the National Science Centre, The National Planetarium, and The Putrajaya Wetlands Park. These educational visits constitute informal learning for students.

There is a National Science Centre (http://www.psn.gov.my/) in Malaysia which provides many fun-learning science activities and exhibits. The main aim of the National Science Centre is to promote greater understanding and interest in science and technology. Among the themes featured in the exhibition, galleries are the environmental odyssey, pathways to discovery, future world and thinking machines. Additionally, there is a National Planetarium (https://www.planetariumnegara.gov.my/) which offers many attractions that include a space theatre that screens space shows and large format film, in addition to permanent exhibits related to space science. Among the exhibits is the Arianne IV space engine, which is also one of the engines used to launch MEASAT 1, Malaysia's first satellite into space. Besides, there is a 14-inch (360 mm) telescope located in the observatory.

In terms of environment and biodiversity, Malaysia Government has created a Wetlands Park in Putrajaya. This Wetlands Park was created to serve as a natural filtering system for the man-make networks of lakes so as to prevent the lakes from being polluted and stagnant. The city of Putrajaya was built around a network of lakes. The wetlands perform three vital functions, namely flood mitigation (e.g., by retaining and storing stormwater and spreading it out over a wide flat area), water filtration (e.g., the urban run-off is filtered through a series of catchment cells, where marsh plants trap sediment and pollutants while toxic substances such as pesticides, herbicides and metals are removed through complex bacterial and chemical processes. Hence, its acts are akin to the kidneys of Putrajaya), and habitat creation (e.g., the resulting filtered water becomes a functioning wetland, providing a home to various plants and organisms which form the base of the food chain, sustaining the biodiversity of invertebrates, frogs, reptiles, fish and water birds). To date, with the wetlands, the network of lakes in Putrajaya remains clean and it is full of wildlife, rendering the wetlands a popular eco-tourism destination.

The Ministry of Education also continues to promote interest in science among the students through the yearly held national science fairs, science practical competition, and robotic competition. For example, the science practical competitions are first held at a district level where various science teams from the primary schools within a district come together and compete. In the competition, there are many stations where each team needs to conduct the given practical in every assigned station within a stipulated time frame before moving to other stations. The winning team from each district will then compete with the winners from other districts at the science practical competition at the state level. The winners from the states will then compete at the national level of the practical science competition.

Besides, the Ministry of Education does organize the National Robotic Competition (NRC) on a yearly basis. The NRC has 3 categories of participation: Primary, Lower Secondary, and Upper Secondary. The Winners and the identified potential teams of the NRC will then participate in an international competition, such as the World Robot Olympiad. It is worthy to mention that the robotic teams from Malaysia won 1 Gold, 1 Bronze and 4 Excellence Awards when they participated in the 16th World Robot Olympiad 2019 held at Gyor, Hungary.

17.3.5 Utilizing Emerging Technologies

The use of ICT in education has been on the increase since the Malaysian Smart Schools Initiative which was conceptualised in the 90s and implemented in 1999 with the 3-year piloting of 90 Smart Schools, aiming to capitalise on the presence of leading-edge technologies. In essence, the Malaysian Smart School is defined as "a learning institution that has been systematically reinvented in terms of teaching–learning practices and school management in order to prepare [students] for the Information Age" (Smart School Project Team, 1997, p. 10). The preparation of students for the information entails the rampant and school-wide judicious use of ICT.

In 2014, the online learning system called Frog Virtual Learning Environment (VLE) was introduced and implemented in 10,000 schools across Malaysia through the 1BestariNet Project. The Frog VLE was adopted from the UK to develop the use of e-learning in schools. The virtual Frog VLE learning environment is a web-based learning system that mimics learning in the real world by integrating the conventional education system with the virtual approach. For example, a teacher teaches by capitalising on the instructional facilities such as graphics, videos, animations, sounds and hyperlinks, gives assignments and tests, and marks them virtually, while the students send in their assignments and check their marks through the Frog VLE.

Nevertheless, on 1 July 2019, all schools had to stop using the current internet provider and switch to the services of the major internet providers in Malaysia (e.g., Telekom, Celcom, and Maxis) as identified by the Ministry of Education. Besides, the 1BestariNet Project has come to an end, and as such, on 1 July 2019, the Ministry replaced the use of Frog VLE with that of the Google Classroom. Besides helping the students to communicate effectively, Google Classroom is easy to use and it is equally very effective. It acts as a digital organiser where teachers could store their materials and paperlessly share them with their students. It is also cloud-based which implies that there will be no more "losing" of assignments by students.

In terms of other emerging technologies, some individual Malaysian teachers use Augmented Reality (AR) technology to attract students' attention, arouse their interest in science and enhance their motivation (Izwan Nurli et al., 2018). AR technology is a computer-generated system that allows its users to see the objects in the real-world environment. Studies have consistently shown that verbal method of teaching such as chalk-and-talk and lecture is less effective and more likely to cause disinterest among students as compared to the verbal-visual method where students remember presented information (retention) and are able to use the information to solve problems (transfer) (Izwan Nurli et al., 2018; Mayer, 1997). As such, more science concepts should be taught via AR so that students' interest in science is boosted and that they have a better understanding of the science concepts at hand.

17.4 Requirements for Future Development of Science Education

The future, undeniably, is uncertain, just as we are currently facing a novel coronavirus pandemic that was not imaginable prior to stepping into the year 2020. Besides, many employers and bosses are complaining that the current workers in the workforce have not been adequately prepared in their undergraduate studies. Therefore, we must be mindful in re-thinking, re-conceptualising, and revamping (or, re-tweaking, if you like) the current science education programs for pre- and in-service teachers so that when these teachers teach in schools, colleges, or institutions, they are thoroughly equipped and are malleable, agile, and adaptable in the sense that they could be easily re-skilled in today and future's world of artificial intelligence, robotics, and the Fourth, Fifth, or even Sixth Industrial Revolution.

In addition to the IT skills, it is strongly proposed that students should be imbued with the 4Cs of the 21st Century Skills, namely critical thinking, creativity, collaboration and communication, instead of learning to memorize facts and figures, and to regurgitate when being tested in science (and other subjects as well). When a student is said to possess critical thinking, he or she manifests through his or her ability to discover, interpret and analyse, reason, construct arguments, problem solve and think systemically. If the student possesses creativity, he or she is able to generate ideas, design and refine, explore courageously, work creatively with others, and produce and innovate creatively. Meanwhile, if a student is said to have the skill of collaboration, it means the student is able to cooperate, has flexibility, is responsible and productive, and collaborate using digital media. Finally, the skill of communication is reflected in one's ability in listening effectively, delivering oral presentations, communicating using digital media, writing in to support, and supporting an argument with claims. Therefore, these important skills and abilities should be infused into the science teacher education program with a plethora of pertinent pedagogical approaches.

Equally, it is proposed that we should bring the working world into science education when re-looking and re-framing the current science curricula and science teacher education programs so that future science teachers are adequately prepared for a future world. Hence, the future pedagogy needs to be researched and conceptualised, and these findings could then be incorporated into the future science teacher education programs.

17.5   Discussion and Conclusion

Although the education system in Malaysia is on par with that of many other developed countries, there are still a number of challenges that need to be addressed. While Malaysia strives to achieve the 60:40 Policy, many students despite qualifying to uptake the science-based subjects shy away from taking science at higher levels of education. This shy-away-from-science phenomenon may be attributed to a number of reasons. Among these reasons is the lack of interest in science which the science teachers need to identify the factors that contribute to students' lack of interest and thereafter, help address the problem.

Lacking interest among students is not idiosyncratic to Malaysia, instead, it is a ubiquitous problem. For example, on the basis of students' responses on a 27-item questionnaire, Adu-Gyamfi (2013) identified the contributing factors to students' lack of interest in school science that include "less practical nature of science teaching and learning …[and] teaching of science is a transmission of knowledge from science teachers or textbooks to students" (p. 7).

Accordingly, science teachers should ensure that students are given every opportunity to carry out practical investigations that allow students to plan and decide on the procedures in which a hypothesis is tested: manipulating a variable of interest and measuring the responding variable while keeping other variables constant. By giving students the opportunity to plan the procedures for the practical investigation, we are providing them with the "hands-on" (psychomotor) and "minds-on" (cognitive) science learning experience which will lead to "hearts-on" (affective), where students will have a better interest in science. Such a science learning experience is in great contrast with the prevailing method of experimentation where a set of procedures has already been given to the students, detailing the step-by-step approach to the experimentation which is akin to a cookbook recipe. While the cookbook-recipe approach does promote "hands-on", it fails to promote "minds-on", let alone "hearts-on".

With regard to the problem of teachers using the transmission approach to science teaching, perhaps the pre- and in-service science teacher education programs need to be re-examined in terms of how the programs are conducted. Do the facilitators lecture to the pre- and in-service teachers as to how a certain science pedagogical approach is enacted in the classroom? It is strongly advocated that when a science pedagogical model, strategy or technique is to be introduced to the pre- and in-service teachers, it should be done via "modelling the model" where the facilitator teaches a particular teaching model through the use of the model itself. In "modelling the model", the participants take the role of the students while the facilitator or trainer models through each of the phases in the model or strategy using a suitable science concept as the context (Loughran & Berry, 2005; Ong et al., 2020a, 2020b).

Finally, there is yet another important point to note in the face of the current pandemic, declared by World Health Organisation (WHO) due to the global outbreak of novel coronavirus, which sees many countries declared lock-down with closures of, among others, universities and schools, teachers are mandated to teach via the Google Classroom (Kamenet, 2020). Therefore, on the basis of the current scenario of the on-going pandemic which requires online teaching and learning, and also the digitalization, as well as the Industrial Revolution 4.0 (IR 4.0) era in which we live in, the technological pedagogical content knowledge with the ability of harnessing the emerging technologies in one's teaching repertoire, is crucially needed. Hence, the current and future trends of pedagogical approaches need to be conceptualized and propagated.

Acknowledgements I would like to thank Ms Faridah Yang Razali, the Deputy Director, Mainstream Schools Management Division (*Bahagian Pengurusan Sekolah Harian*), Malaysian Ministry of Education for being so helpful when pertinent information is needed for this write-up. Equally, I would like to express my sincerest gratitude to the Research Management and Innovation Centre, UPSI for the project grant (Coded: 2019-0123-109-01) that has enabled the review of the pertinent literature for this book chapter.

References

Alcacer, V., & Cruz-Machado, V. (2019). Scanning the Industry 4.0: A literature review on technologies for manufacturing systems. *Engineering Science and Technology: An International Journal, 22*(3), 899–919.

Anderson, L. W., & Krathwohl, D. R. (2001). *A taxonomy for learning, teaching, and assessing: A revision of bloom's taxonomy of educational objectives.* Addison Wesley Longman Inc.

Carlo, M. D. (2011). Attracting the "best candidates" to teaching. Retrieved from https://www.shankerinstitute.org/blog/attracting-best-candidates-teaching.

Curriculum Development Centre. (2013). *HOTS: Higher order thinking skills in schools initiative.* Ministry of Education.

Curriculum Development Centre. (2014a). *Higher order thinking skills: School applications.* Ministry of Education Malaysia.

Curriculum Development Centre. (2014b). *HOTS elements in the curriculum.* Ministry of Education Malaysia.

Curriculum Development Centre. (2014c). *HOTS elements in pedagogy.* Ministry of Education Malaysia.

Curriculum Development Centre. (2014d). *HOTS elements in assessment.* Ministry of Education Malaysia.

Curriculum Development Centre. (2014e). *HOTS elements in co-curriculum.* Ministry of Education Malaysia.

Curriculum Development Centre. (2014f). *HOTS elements in effort building.* Ministry of Education Malaysia.

Curriculum Development Centre. (2014g). *HOTS elements in resources.* Ministry of Education Malaysia.

Curriculum Development Centre. (2014h). *The support from the community and private to boost HOTS.* Ministry of Education Malaysia.

Curriculum Development Centre. (2017). *Primary school standard document science year 1* (English ed.). Ministry of Education.

Educational Planning and Research Division (EPRD). (2019). *Quick facts 2019: Malaysia educational statistics.* Ministry of Education Malaysia.

Izwan Nurli, M. B., Syed Zulkarnian, S. I., & Salleh, A. B. (2018). The use of Augmented Reality technology in Perlis, Malaysia. *Journal of Physics: Conference Series, 1019*, 1–9.

Kamaludin, A. A. (2019). *New vision to lift Malaysia to developed nation status.* BERNAMA.COM, October 11, 2019. Retrieved from https://www.bernama.com/en/general/news_wkb2030.php? id=1774763.

Kamenetz. A. (2020). *'Panic-gogy': Teaching online classes during the Coronavirus Pandemic.* Retrieved from https://www.npr.org/2020/03/19/817885991/panic-gogy-teaching-onl ine-classes-during-the-coronavirus-pandemic.

Kumar, A. (2019). *How technology is shaping Malaysia's economy.* Retrieved from https://www. computerweekly.com/news/252462116/how-technology-is-shaping-malaysias-economy.

Loughran, J., & Berry, A. (2005). Modelling by teacher educators. *Teaching and Teacher Education, 21*(2), 193–203.

Mani, S. (2000). *Policy instruments for stimulating R&D in the enterprise sector: The contrasting experience of two MNC dominated economics from Southeast Asia.* United Nations University, Institute for New Technologies.

Malaysian Examination Syndicate (2013). *Assessing higher order thinking skills.* Ministry of Education Malaysia.

Malaysian Examination Syndicate. (2015a). *Exemplary items in assessing higher order thinking in primary schools.* Ministry of Education Malaysia.

Malaysian Examination Syndicate. (2015b). *Exemplary items in assessing higher order thinking in secondary schools: Science and mathematics areas.* Ministry of Education Malaysia.

Malaysian Examination Syndicate. (2020a). *Primary School Assessment Report.* Ministry of Education Malaysia.

Malaysian Examination Syndicate. (2020b). *Report of Results Analysis for Malaysian Certificate of Education (SPM).* Ministry of Education Malaysia.

Marsick, V. J., & Watkins, K. W. (2001). *Informal and incidental learning. New directions for adult and continuing education, No 89, Spring.* Jossey-Bass.

Martin, M. O., Mullis, I. V. S., Foy, P., & Stanco, G. M. (2012). The *TIMSS 2011 international results in science.* TIMSS & PIRLS International Study Center, Boston College.

Martin, M. O., Mullis, I. V. S., Goy, P., & Hooper, M. (2016). *TIMSS 2015 international results in science.* The International Association for the Evaluation of Educational Achievement (IEA).

Mayer, R. E. (1997). Multimedia learning: Are we asking the right questions? *Educational Psychologist, 32,* 1–19.

Ministry of Education. (2002a). *Professional circular No. 5/2002: School visits during term time.* Ministry of Education Malaysia.

Ministry of Education. (2002b). *Professional circular No. 14/2002: Implementation of compulsory education at lower level 2003.* Ministry of Education Malaysia.

Ministry of Education (2012). *Malaysia education blueprint 2013–2025 (Pre-school to Post-Secondary Education).* Ministry of Education Malaysia.

Ministry of Education. (2019a). *Professional circular 5/2019: Implementation of primary school standard curriculum (Revised 2017).* Ministry of Education.

Ministry of Education. (2019b). *Professional circular 5/2019: Implementation of secondary school standard curriculum effective from 2020.* Ministry of Education.

Ong, E. T., Luo, X. K., Yuan, J., & Yingprayoon, J. (2020a). Professional development programme on the use of STEM-based 5E inquiry learning model. *Science Education International, 31*(2), 179–184.

Ong, E. T., Keok, B. L., Yingprayoon, J., Swaran Singh, C. K., Borhan, M. T., & Tho, S. W. (2020b). The effect of 5E inquiry learning model on the science achievement in the learning of "magnet" among year 3 students. *Jurnal Pendidikan IPA Indonesia (indonesian Journal of Science Education), 9*(1), 1–10.

Organization for Economic Co-operation and Development (OECD). (2019). *OECD economic surveys: Malaysia.* OECD.

Smart School Project Team. (1997). *Smart School flagship application: The Malaysian Smart School – A conceptual blueprint.* Ministry of Education.

Vijaindren, A. (2018). *Malaysia needs 500,000 scientists and engineers by 2020*. News Straits Times. August 13, 2018. Retrieved from https://www.nst.com.my/news/politics/2018/08/400909/malaysia-needs-500000-scientists-and-engineers-2020.

Eng Tek Ong (王永德) is a Professor of Science Education at the Department of Educational Studies, Faculty of Human Development, Universiti Pendidikan Sultan Idris (UPSI). His qualifications include PhD (Cambridge, UK), M.Ed (Houston, USA), & B.Sc.Ed. (Hons) (USM). He was the recipient of the PhD Scholarship from the Commonwealth Scholarship Commission. He had also successfully followed through a two-year postdoctoral fellowship program at Cambridge University. He has published more than 100 journal articles and was the principal investigator for two UNESCO research grants, three Fundamental Research Grant Schemes (FRGS) from the Ministry of Higher Education, and a number of university as well as the NGO research grants. His research interests and publications include areas such as science process skills, inquiry learning, cooperative learning, problem-based learning, concept cartoon, flipped classroom, indigenous people (IP), Education for Sustainable Development (ESD), Science, Technology, Engineering and Mathematics (STEM) education, Higher Order Thinking Skills (HOTS), and HIV-AIDS preventive education.

Chapter 18
Science Education in Nepal: Problems and Prospects

Mohan Paudel and Rajani Rajbhandary

Abstract This paper attempted to present the current status of science education in Nepal in terms of the country-specific problems, challenges, and prospects. Despite government efforts in improving education, by implementing reforms programmes, developing schools infrastructure, increasing the education budget, training teachers, and others, Nepal's science education is lagging in many ways. Various challenges, obstacles, and problems stand in the way of science education development in Nepal. Nepal's education policies are highly focused on policy inputs but silent about the journey. Lack of collaborative efforts, poor management system, insufficient funding, less communication among the stakeholders, etc. are some reasons for the education policies and programmes just remaining as a mere paper declaration. Science education in Nepal is characterized by inadequate institutional setup for its promotion, having no national science competency standards, few science and technology research centers, limited funding in science and technology research, lack of professional science teachers and researchers, etc. The discourse of science education reform has received less attention while reframing government policy and programs. Proper implementation of approved education policies is essential to achieve the desired goal of education. Science education needs to improve significantly emphasizing the sustainable development of the nation. Nepal needs to develop and implementing science education reform policy and programmes, reframing science education curriculum, allocating dedicated funds in infrastructure development, and improving the quality of teachers.

Keywords Assessment · Curriculum · Policy · Science education · Teaching · Technology

M. Paudel (✉)
Department of Science and Environment Education, Central Department of Education, Tribhuvan University, Kirtipur, Nepal
e-mail: mohan@tucded.edu.np

R. Rajbhandary
Department of Science Education, Sanothimi Campus, Tribhuvan University, Bhaktapur, Nepal

18.1   Nepal's Geography, Population, and Political System

Nepal is a South Asian country that lies between 26° 22′ north to 30° 27′ north latitude and 80° 04′ east to 88° 12′ east longitudes. It occupies 0.03 percent land area of the world and borders with India in the east, west, and south, and with China in the north. Being a landlocked country, Nepal has to depend on India and China to get access to the sea. Kathmandu is the capital of the country which is famous for its historical heritage, monuments, and ethnic culture. Ecologically the country is divided into three regions viz., terai, hill, and mountain. The country has a diverse topography and climatic conditions, the huge potential of natural resources, attractive tourist destinations, and distinct cultural groups of people. According to National Population and Housing Census 2011, the country's population stands at 26,494,504 with a population growth rate of 1.35 per annum and a female to male ratio of 1.06 (Central Bureau of Statistics [CBS], 2011). Nepal has a population density of 181-person per square kilometer of area 147, 181 km^2. Nepal is ever independent, sovereign and secular country. The constitution of Nepal endorsed in 2015 has declared the country a federal democratic republican state with seven political and administrative provinces. President is the ceremonial head of the country and the prime minister is the head of the executive of the state's cabinet elected among the member of the House of Representatives.

18.2   The Current Situation of Economic and Cultural Development

Nepal is one of the least developed countries in South Asia. One-fourth of the population (25.16 percent) lives below the absolute poverty line (CBS, 2011). Agriculture remains Nepal's principal economic activity, employing about 65 percent of the population and providing 26.5 percent of GDP (Ministry of Finance, 2018).

Nepal is now officially a lower-middle-income country, an upgrade from its previous status as a low-income nation, according to the world's bank's latest country classifications by income level (World Bank, 2020). As per UNDP's Human development report, Nepal's HDI value increased from 0.380 to 0.579 between 1990 and 2018, an increase of 52.6 percent, basically driven by the increased life expectancy of people and years of schooling, still a country with low human development status (National Planning Commission [NPS], 2015; United Nations Development Programme, 2019). Nepal's Gross National Income Per Capita increased by about 130.5 percent between the same periods, as per the report. Nepalese economic development is facing so many challenges, such as absolute poverty, income inequalities; slow-growing economy, low industrial development, etc. (NPS, 2015). The country's economy is becoming more remittance dependent due to the continuous growth of remittance inflow. Due to low economic return from agriculture, the rural population is now switching over to other occupations.

Nepal is a culmination of different ethnicities, religions, values, and beliefs. Nepal is a multilingual country. CBS (2011) reported 125 diverse ethnic groups with diverse cultural practices and 123 mother tongue languages. Right to education in mother tongue is guaranteed by law in basic level education while the language of instruction in public schools is predominantly Nepali and private schools often use English.

18.3 Overview of the Education Development

The Citizens' right to education is included in Article 31 of the constitution of Nepal 2015. The constitution has guaranteed the right to free and compulsory basic education and free secondary education for all citizens from the state. Special provisions had been made by law to educate and uplift marginalized and vulnerable communities, peoples living in poverty, and disable peoples (Constitutional Assembly, 2015). The Government considers education as a center for human resource development and a priority sector for the Federal Government of Nepal. In addition, education has also been defined as the concurrent competence of the three levels of the government—federal, provincial and local—indicating that the respective layers of the government have to share the responsibilities and functions in the administration and management of education (See The Constitution of Nepal 2015). Despite power-sharing between state government, provincial government, and local level government, the education system is over-centralized and the central administration has absolute authority over almost every aspect of education activity in the country.

The Ministry of Education, Science, and Technology (MoEST) is an apex body responsible for formulating policies, plans and implementing them for the overall development of education, and science and technology in the country through the institutions under it. There are separate central units under the ministry for curriculum development, teacher training, examination, non-formal education, education review, teachers' records management and science and technology development. Universities are governed and managed by the respective university acts.

The formal education structure of Nepal consists of two tiers of school education, eight years of basic education (grades 1–8) and four years of secondary education (grades 9–12). In addition, there is a provision of separate technical education and vocational training of one and half years to three years TSLC and diploma. Eleven universities and four medical academies are providing higher education across the country. Ten universities are of general type offering general science and applied science courses along with other disciplines. Agriculture and Forestry University is a single university that offers applied science courses only, particularly agriculture, animal husbandry, fisheries, forestry, and others. The country needs a dedicated Science and Technology University that can concentrate on producing S & T competent human resources. Recently the government has allocated some budget to establish the first Science and Technology University in the country (see Budget speeches of Ministry of Finance, 2020 & 2021).

18.4   School, Teacher and Students' Profile

Education data published by the Department of Education (2017) of the year 2017/18 showed there are in total 35,601 General School Units, 283 Technical & Vocational stream schools (9–12), 724 Technical Schools/Institutions for TSLC Level Programs, 796 Technical Schools/Institutions in Diploma Level Programs. Qualified and professional teachers are the foundation for quality science education. In Nepal, 3, 25,519 teachers are in school jobs from primary to higher secondary (grade 11 and 12) in all types of schools, nearly half the number of female teachers than their male counterparts. Below 25 percent of total school teachers are working in a permanent position of community schools nationwide. Nearly 75 percent of school teachers of community schools had received teacher's professional development (TDP) training, but minimal effects of training seen on classroom teaching (Center for Education and Human Resource Development [CEHRD], 2018; Khaniya et al., 2015). In the school year 2018/19, community schools share nearly eighty percent of grade 1–12 students while the remaining twenty percent study in institutional schools across the country. The percentage shares in enrolment of community schools decreased by 2 percent than in the previous year. The Gross Enrollment rate (GER) in the year 2018/19 at a lower basic level (1–5) is 118.8 percent and remained at 45 percent at the secondary level (11–12)- indicating that a significant proportion of enrolled students drop out after Grade 8 and especially after Grade 10. It was found that gender parity has been maintained overall at all levels of community schools. On the flip side, there are more boys than girls in all levels of institutions schools. The gender parity index (GPI) in Gross Enrollment Rate (GER) remained at 1 or above in all levels of school education. Net Enrollment Rate (NER) at basic education (grades 1–8) was 92.7 percent, while it remained at 46.4 percent for secondary level (9–12) in the year 2018/19. The NER values suggesting more attention needs to be paid to increase the enrolment of students within specified age for the grade levels. Student–teacher ratio (STR) for all types of schools has been within the national standard level with values 19:1 for lower basic level, below 35:1 in upper basic and secondary levels, however, school wise disparities have been observed in community and institutional schools. The overall survival rate to grade five is 89.6 percent, which sharply decreased to 58.4 percent to grade ten and 22.2 percent to grade twelve. Data indicates the tenacity of problems and issues in school education (CEHRD, 2018; Ministry of Education, Science and Technology [MoEST], 2017). Tribhuvan University (TU), the oldest and biggest national university, alone shares 75.95 percent of students in higher education (University Grants Commission [UGC], 2020). It is reported that only 23.03 percent of students enrolled in technical programs in the year 2018/19 in higher education programs of all universities. Gender parity of HE was found to be 1.09 in 2018/19. In terms of science specialized faculty, medicine has a GPI of 1.6, S & T has 0.56 and engineering has 0.2 (UGC, 2020). The data suggest a significantly low number of students enrolled in the science stream in higher education, particularly a minority of female students in engineering.

18.5 Policies and Standards

The Nepalese education system is governed by the education act, 1971 and education rules, 2002. The Ministry of Education, Science and Technology is a responsible apex body for looking after the science, technology, and education sector in the country.

The constitution of Nepal 2015 Schedule-5.15 gives power to the state government regarding the central universities, universities' standards, and regulation of central libraries. Schedule-6.8 provides power to the federal government regarding state universities, higher education, libraries, and museums. Similarly, Schedule-6.18 grants power to the State as protection and use of languages, scripts, cultures, fine arts, and religions. Part-4.51.h.1 of the constitution prescribes policies of the state to prepare human resources that are competitive, ethical, and devoted to national interests while making education scientific, technical, empirical, vocational, employment, and people-oriented. The constitution authorizes the local government to govern the education sector up to grade 12 where school education is the responsibility of the local level government.

Compared to other countries, Nepal does not have a long history of science education. Trichandra College for the first time in the history of Nepal had started intermediate studies in Science in 1919, which was limited to the members of the ruling Rana and some influential families. Science and technology education came to educational discourses in Nepal since the 1950s, particularly with the start of the democracy movements. The National Education System Plan (NESP) 1971 adequately emphasized the development of science, technology, and vocational education in the country. It also introduced science as a compulsory subject from grade four. The plan had focused on providing special teacher training in science and mathematics and the establishment of a science laboratory in every school (Ministry of Education, 1971). In 1977–1990, Government tried to improve technical education and vocational training, and science education by taking three loans support from Asian Development Bank (Asian Development Bank, 2004). National Council for Science and Technology was established during the sixth five-year plan (1980–1985). The Ninth Plan (1998–2003) had incorporated science from the beginning of school education (primary level) throughout the secondary level (grade 10) as a compulsory subject. National Science and Technology Policy (2005), National Nuclear Policy (2007), Information and Communication (ICT) in Education Master Plan (2013–2017), and National Science, Technology and Innovation Policy- 2019 emphasized developing science education in Nepal by modernizing and integrating ICT in education (MoEST, 2019).

School Sector Development Plan (2016–2023) is now guiding most of the activities of Nepal's education system. Ministry of Education, Science, and Technology (MoEST) has come up with the "National Education Policy-2019". Nepal does not have a concrete science education policy as such. It is accommodated within the broader umbrella of the National Education Policy 2019. The policy gives significant importance to STEM education from school to university degrees. The policy emphasized on development and expansion of STEM education as an integral part of the overall education system by adopting an appropriate policy, raising invest-ment, focus on human resource development, curriculum reform, and modernization of teaching–learning methods. Some important future arrangements of the policy involved: inclusion of STEM education from the earliest levels, fifty percent of the school curriculum from STEM-related fields, development of a need-based STEM curriculum at the bachelor's and master's. National Science, Technology and Inno-vation policy - 2019 put special focus on developing scientists and technical human resources encouraging research-oriented science education. Nepal lacks a national teaching competence standard which is being practiced in the USA since 1996 (Alake-Tuenter et al. (2012), such standard is needed from school to university science education.

National Examination Board has started standardization of grade 8, grade 10, and grade 12 examinations in three major areas science, mathematics, and english as per provisions and program set under School Sector Development Plan (SSDP). There is a Teacher Service Commission (TSC), an autonomous government body, to select and recommend teachers for permanent positions in schools. It has also given a role to recommend teachers for promotion based on the set criteria. It has its act, rules, curriculum, and policy provisions for selecting teachers for the government funding community schools in Nepal. Persons willing to teach science and other subjects at schools, particularly in community schools must have a teaching license issued by TSC. Only the license owner candidates with required qualifications can take part in primary, lower secondary and secondary levels teacher examinations held separately. There is a shortage of license-holder science teachers in the country as many community schools have vacant science teacher positions.

Universities have their autonomous service commissions to appoint university professionals. In higher education, UGC has given a significant role in supporting and regulating the universities in Nepal. The UGC has launched the Quality Assur-ance and Accreditation (QAA) program, as an important aspect of reform in higher education in Nepal. Nepalese higher education sector needs improvement; mainly in policy and its implementation. Although the country is declared as a federal demo-cratic republic country, however, the policy has envisioned centralized institutional arrangements. At the same time, to improve the current status of higher education, the government has proposed to draft University Umbrella Act. The government has also decided to restructure UGC as Higher Education Commission in nearest future.

18.6   Educational Research and International Collaboration

Science and technology have never been on government's top priority. For too long, Nepal has lagged behind the rest of the developing world in the field of research and development (R & D). The lack of investment is hurting our ability to drive productivity, stimulate economic growth. The total investment in R & D has been less than 1 percent (0.11–0.48 percent) of the total budget (Bajracharya et al., 2006). To improve the status quo, educational institutions need to collaborate with other institutions inside and outside of the country. To inculcate a culture of research, UGC has been providing various kinds of research support to scholars, professional researchers in universities, and faculty research grants. Nepal Academy of Science and Technology (NAST) is an autonomous apex body to promote science and technology in the country. NAST provides regular grants to support university professionals researching in the field of science and technology. It honors Ph. D fellowship as well as support M. Sc. students providing research assistantship reward or dissertation grants. Nepal Agricultural Research Council (NARC) is another autonomous organization to conduct agriculture research in the country to uplift the economic development of the country. Tribhuvan University (TU) and other universities collaborate in their capacities with foreign universities and research institutes. There are four research centers of TU. Research Centre for Applied Science and Technology (RECAST) helps to contribute to the rapid and sustainable development of the country through R & D in science and technology. Research Centre for Education Innovation and Development (CERID) researches to upgrade the quality of education.

The Government has sought financial and technical help from various international organizations to ensure access to quality education. Some important executed projects that contributed to timely reform and expansion of education in Nepal were Primary Education Project (PEP), Secondary Education Support Programme (SESP), Community School Support Programme (CSSP), Teacher Education Project (TEP), Secondary Education Development Project, Basic and Primary Education Project Phase II (BPEP II, 1999–2004), Education for All (EFA), School Sector Reform Programme (SSRP), School Sector Development Programme (SSDP)., Higher Education Reform Project (HERP), Higher Education Reform Project, etc.

Various international aid agencies like WHO, Asian Development Bank, UNESCO, DFID, USAID, and others have made a significant contribution in national education policy-making, infrastructure development, increasing access and equity, and ensuring quality education.

Science Education Project (SEP, 1982) was the first project in science education to strengthen science in school education through upgrading school facilities and science teaching skills. Under this project, one central Science Education Development Center (SEDEC) and 25 Science Education Development Units (SEDU) were established in selected districts in the country. Later in 1992/93, SEDEC converted into Secondary Education Development Project (SEDP). Secondary Education Development Project (1993–2001) contributed more significantly in improving the quality of secondary education. Some achievements relating to science education involved the development of and implementation of teacher training programs, development of curriculum and textbooks in core subject including science for grades 6–10, renovation of science laboratory of Faculty of Education, construction of science laboratories in 25 higher secondary schools. There were 1025 schools supported with science equipment, furniture, library books, and other learning materials and 981 schools were trained in the use/maintenance of science equipment (ADB, 2004). It was reported that some higher secondary schools were unable to well use the provided equipment, laboratories, and library books due to a lack of resources for maintenance (ADB, 2004). Later SEDUs merged under Education Training Centers

(ETCs), Now ETCs under CEHRD are responsible to conduct school teacher training programs.

18.7 Current Situation of Science Education

As a human enterprise, science aims at acquiring objective knowledge of the world. Science is generally considered as the study of facts related to the natural and material world. The improvement in science is an accelerator for the economic and social development of a nation. This improvement in science can be achieved through a carefully planned science education that caters to the needs of the population.

Science and technology have not been the first choice of students in Nepal as evidenced from lower students' enrollment in the science stream in grade 11 as compare to enrollment in the management stream (36.72 percent) and education stream (41.72 percent), only 13.92 percent of total grade 11 students enrolled in science stream in the year 2017. Out of which, 9.68 percent of students drop out and did not get enrollment in grade 12 in the year 2018 (National Examination Board, 2019). The percentage enrollment in the science stream in 2018/19 slightly increased and maintained at nearly 15% (CEHRD, 2018). The school education lacking behind to arouse interest, prepare mentally and attract academically towards science education in higher-level studies. Many of the students, parents, even teachers portrayed science as a difficult discipline so they rarely prefer science for further study.

Students' Proportion in higher education science courses of all universities and academies in the year 2018/19 were found to be 8.38 percent in science and technology, 6.57 percent in engineering, 6.39 percent in medical sciences, 1.00 percent in agriculture, 0.43 percent in forestry and 0.25 percent in animal science, veterinary sciences and fisheries (UGC, 2020). Data revealed students' number in all faculties of science remained at low but increased slightly in comparison to the previous year. Limited subject choice, limited quotas, high cost, poor quality, and limited job opportunities are some compelling reasons that pushing youths to go abroad for higher education, especially for S & T, medicine, engineering, technology, and other applied science education. Every year increasingly large numbers of students are leaving the country to pursue higher education abroad as per government data relating to the 'No Objection Letter' granted for abroad studies (CEHRD, 2018).

The development of a country like Nepal cannot be imagined without the advancement of science and technology. The yearly average of domestic patents registration was less than one (0.8), which is a very poor indicator of science education effectiveness (Ministry of Finance, 2019). The discourse of science education reform has received less attention in government policy and programs. A small amount of budget has been allocated by the state in the development of required infrastructure and programs for science education (see Ministry of Finance budgets). Improvement of science education at the school level and university level has not received comprehensive and sustained discussions among different level stakeholders (Government authorities, university and school executives, teachers, and the general public).

18.8   Curriculums

Curriculum Development Centre (CDC), an academic center under the MoEST is responsible for developing curricula, teacher guides, textbooks, and various other instructions materials for school education. The school curriculum is as per the National Curriculum Framework of School Education 2019 endorsed by MoEST.

Science is included in the school curricula from class one. In the primary levels (grades 1–3), science is integrated with other subjects such as health and physical education, social studies, moral science, and creative arts activities. In grades 4–10, science is integrated with technology (5 credit hours). In addition, students can choose one more science (4 credits) subject as an optional subject in grades 9 and 10. In grades 11 and 12 subject-specific science courses like physics, chemistry, biology, environmental science, computer science etc. are offered as optional subjects. Universities are offering general science courses and applied science courses at bachelors', masters', and Ph. D. levels.

The science education curriculums in school to university education are highly loaded with factual information; delivered through didactic teacher-centered lectures-which emphasize on memorization and rote learning. Assessment encourage reproduction of knowledge. Despite a provision in curriculum, laboratory work in school is neglected. Textbooks are often the only resource materials students can utilize in learning science. Textbooks are loaded with factual information with a focus on declarative knowledge rather than procedural knowledge of science.

Only four universities have science education programs (teacher education) under the education stream. Faculty of Education, TU offering four-year B. Ed. in science education, one-year B. Ed. in science education (for B. Sc. graduates), two years M. Ed. in science education with three separate major areas chemistry education, physics education, and biology education, aimed at producing a competent human resource for school and university teaching. Recently two semesters of M. Ed. in science education have been launched by the Faculty of Education for M. Sc. graduates who are willing to or pursuing a profession in teaching and education sectors. Tribhuvan University significantly contributes to producing skilled human resources required for education sectors. Kathmandu University, School of Education has endorsed 3 semesters M. Phil. in STEAM (Science, Technology, Engineering, Arts and Mathematics) Education. Institute of Open Learning, Purbanchal University and Faculty of Social Science and Education, Nepal Open University runs similar one-year B. Ed. in science education program for B. Sc. graduates through distance mode.

University graduates who become a teacher after completing master degree still lack conceptual understanding and analytical capability in science and research. Poorly prepared graduates often become poor science teachers, who in turn teach inadequately in a cycle that reinforces a limited view of science. A degree is not enough (Kind, 2013) but they need to develop professional skills required by 21st-century science teachers.

18.9   Assessment

Assessment is regarded as a component of a curriculum. CDC is responsible for designing the assessment framework of school education. The letter grading system has been adapted from the beginning of the basic level to the secondary level grade 12, shifted from percentage evaluation system. A subject-specific specification grid is being used at all levels that guide on designing assessment tools. Formative and summative assessments are suggested in the education policies, curriculum, and teacher training courses in the Nepalese education system. According to Education Act, the district-level examination is administered at the end of basic education level (grade 8), a regional examination at the end of grade 10, and a national examination at the end of secondary (grade 12). Despite provision, the province-level examination is not implemented yet. Formative assessment tools such as classwork, project work, community work, unit tests, observation, and formative and innovative work are included in students' assessment but not counting in district level and national level final evaluation. Schools, municipalities/rural municipal, the provincial examination office, and the Office of Controller of Examinations (OCE) are the responsible authorities to administer the respective levels of examination. Universities have their own examination controller office to carry students' evaluations. Universities adopted annual systems or semester systems or both systems for different levels with their evaluation framework and practices. TU and other Universities had tried to pave the way for more progressive assessment practices.

Nepal is not participating in international high-stakes testing systems such as Programme for International Student Assessment (PISA), Trends in International Mathematics and Science Study (TIMSS), PIRL, etc. however, it has its own national assessment system framed by Education Review Office (ERO). A study about the feasibility of Nepal's participation in international assessment suggests that Nepal in the current context is not ready to participate in international testing systems due to some factors related to policy and legal requirements, cost, poor institutional and technical preparedness, inadequate human resources and technical set-up at ERO, etc. (Center for Educational Research & Social Development, 2016). ERO has developed its assessment as well as audit framework, tools, and standards for national-level evaluation of school education. ERO conducts different types of studies including national assessment of students' assessment (NASA) study, early learning, and development assessment, classroom-based early grade assessment (CBEGRA), performance audit (PA) of schools and institutions, and educational research. Altogether ERO has completed seven NASA studies so far beginning from 2011 to 2019, a high-stake assessment of students learning in subject areas Nepali, English, Mathematics, and Science at grade levels three, five, eight, and ten with discrimination to examine the quality, equity and efficiency of the educational system in Nepal. Science was included for the first time in the second cycle of grade eight NASA study in 2013. NASA Study- 2013 revealed the poor achievement of grade eight students' in science with a national average achievement score of 41 out of 100. NASA-2017 study found a significant downfall of grade 8 students' achievement in science as compared to the previous 2013 study. Both reports depict disparity in learning science as per ethnicity, geography, gender, availability of textbooks, school type, parents' education, and location. In both cases girls perform lower than boys; average achievement (36 percent in 2013) of Madheshi students was lower than the national average (41 percent in 2013); urban students had performed far better than students from rural areas, with the proportion 57 percent and 37 percent in the NASA 2013. Students reflect significantly weak performance in higher ability skills, one-third of the students (35 percent) being unable to answer any of the items under HA level (ERO, 2015; ERO, 2018). Study results suggest that school science needs special attention with science education reform programs to enhance quality and achieve equity in learning science. A study on Student Performance in SLC (Mathema & Bista, 2006) showed that mathematics, science, and english are the three main subjects failed by a significant number of students in SLC, and average performance scores in these three subjects were lower than the overall average score in the SLC examinations. This study also found a disparity in learning achievement in SLC in terms of gender, ethnicity, ecological regions, location, etc.

18.10 Outreach Science Education and Infusion of Formal and Informal Science Education

Science fairs, exhibitions, olympiads, search for scientific talents shall be conducted at provincial and national levels as per National Science, Technology and Innovation policy (2019). Robotics club Pulchowk Campus and Centre for Applied Research and Development have organized Robocon, a robotic contest. Nepal Physical Society, a member association of Asia Pacific Physical Societies, conducts International Physics Olympiad every year for secondary level (11 & 12) students. Nepal participates in the international chemistry olympiad every year. Nepal had its first school-level STEAM challenge organized by MoEST in collaboration with other organizations, where students had engaged to design prototypes focused on the UN SDGs. Natural history museum conducts various education/awareness activities for disseminating knowledge among students and the community. It includes school programs, programs for the community; mobile science exhibitions, etc. Information technology (IT) park was established in 2003 to promote information technology but the service has failed to kick start. IT park has not been able to run its activities may be due to political instability, insufficient funding, and poor management.

18.11   Utilizing Emerging Technologies

National Curriculum Framework of School Education 2019 has emphasized using ICT for the education transformation in school education. The key focus areas are developing content for digital literacy; digitalization of curriculum and text-book, development of interactive digital learning materials, technology-based learning facilitation, etc. Lack of high-quality internet, electricity and supporting infrastructure are major challenges in Nepal's rural schools.

Nepal needs expansion of internet infrastructures throughout Nepal and should increase the possibility of e-learning and enable school teachers to access the latest online information and so facilitate rural school students' education. Internet penetration in Nepal stood at 36.7 percent in January 2021 and there were 38.61 million mobile connections and 10.78 million internet users. Nepal Wireless Networking Project (NWNP), an initiative of Mahabir Pun, bring the internet service to the isolated remote mountain region of Nepal. NWNP uses ICT in various sectors such as telemedicine, distance education, e-commerce, communication, etc. has potentially improved many aspects of rural communities. Digital technology and internet service is too costly for the majority of the population in the country. The poor economic status of many individuals is a barrier to afford expensive digital technology. Therefore, even when there is access to internet service and mobile data internet, the high expenses associated with using the them will make it diffi-cult for children from low-income families, which eventually is creating a digital divide. Now days there are increasing number of students who opt for IT educa-tion in the country and carrier in IT sector is very luring. CEHRD has recently launched an internet-based Integrated Education Management Information System (IEMIS) that facilitates for timely gathering, recording, and analyzing school-level data and bringing together people, practice, and technology for planning, monitoring, and supporting decision making. Now schools are using various new technologies for instruction and administrative work. Global pandemic due to COVID -19 has adversely affected all sectors of life including education. As the schools continued to close since mid—April 2020 majority of school students remained out of school, while few children from urban locations got an opportunity for digital learning. This might seriously impact the performance gap between the rural and urban students that already exist. One positive impact of COVID-19 in country case is that many teachers became aware of using ICT in the teaching–learning process, several teachers learned and practiced to use ICT in their teaching, Many schools in urban and rural loca-tions started online classes using LMS, Zoom, Google meets, Microsoft teams, etc. Nepal's government has emphasized the role of technology in curriculum and also on teaching/learning methods. In the past, most of the teachers depended on *chalk and talk* as their teaching method. With the modernization and digitalization of educa-tion teachers has access to different teaching learning materials and digital contents in the form of audio, video, animated and simulated objects by the use of software tools, internet/web tools, computer, laptop, projectors, smart boards, etc.

18.12   Requirement for Future Development of Science Education

The purpose of teaching science in schools is that it can inculcate in children certain values and attitudes–scientific temper, rationality, reasoning, critical thinking, creative thinking, problem-solving, methods of science, and so on. The state of science teaching in schools and colleges in Nepal is far from satisfactory. Science is losing its appeal in an alarming shift of choice. There are fewer science-based jobs in the country. The teaching profession, especially in the primary and secondary levels at school, is often viewed as a last choice. If we want to bring change to the education system, we need to make teaching a more lucrative career so that top talents are willing to make a career in teaching. We have to make teaching rewarding and challenging. Increasing the number of qualified teachers is a natural need for expanding the system of education. The government should take definitive steps to upgrade the pedagogical capacity of science teachers and ensure that every school has a science laboratory equipped with fundamental facilities.

Our education system is not substantially preparing the youth for the 21st century. Nepal needs progressive educational institutes that promote creativity, innovation, and scientific problem-solving. Instead of a traditional classroom that rewards memorization and standard testing, progressive education helps develop skills that foster critical thinking, problem-solving, project management, collaborative thinking and inquiry skills. Science education shall be made more practical and applied and it shall be linked with evidence-based research. The government needs dedicated wings that work particularly on designing science education policy, programs and also monitor overall progress and achievement. There has to be a wider collaboration between industry and colleges, colleges and schools, and other government bodies to help students learn from diverse people and foster collaborative values. The government has to invest more in science education research, science teacher professional development, setting and improving science laboratories, managing teaching materials, school and university curriculum reform, and science popularization activities at the community level.

18.13   Achievements

Nepal has made significant progress in achieving universal enrollment in primary education, ensuring gender parity in education, and widening adult literacy. The current achievement in education development is due to the decade-long efforts through policy reform and programs implementation. The biggest achievement has been an education for all goals. Nepal has significantly expanded and improved education for vulnerable and disadvantaged children. Increased enrolment in pre-primary education, increase in net enrolment rate at all levels, increase in the number of trained teachers, and schools' infrastructure development are notable progress.

School merger scheme is being implemented to solve the problem of teachers' quota and reduce the student-to-teacher ratio. An integrated curriculum for school education has been made.

Although the importance of S & T was realized from the first development plan, however, the sixth development plan (1980–85) gave due importance in S & T sectors. National Council for Science and Technology was established in 1976, various S & T related departments, institutions, and organizations were established in Nepal to promote S & T. Number of acts, rules, and regulations were formulated in different sectors of S & T. Science and Technology Policy -2005 have been already endorsed, National Science and Innovation policy 2019 has been devised and implemented. Now 100 more professional societies are working in the field of S & T.

18.14   Discussion

Science education is very important in developing competent human resources for the development of the country. Science education is instrumental to promote citizens' awareness towards environmental protection, sustainable use of natural resources, developing a sense of social responsibility, solve daily life problems scientifically, etc.

Nepal is rich in indigenous knowledge and technology that being practiced from history. Such kinds of knowledge and technologies need to be explored, studied and researched and integrated with the school science curriculum. Science has to be made "relevant" to the students from the point of view of their lives and their futures.

The biggest challenge of science education is the lower level of students' interest and preference to study science at higher levels. Many secondary levels (grade 12) graduates with strong potential in physics, chemistry, biology, and mathematics are opting for business studies. There are ample job opportunities for people pursuing business courses, however, the graduate from natural science streams hardly find jobs inside corporate offices. There is a poor image of science-related careers. The main job available to them is teaching. Nepal labor market cannot employ science graduates at salaries worth their investment-resulted due to lack of required voca- tional and technical skills among the youths (Ministry of Youth & Sports, 2015). The government and the colleges offering science need to devise strategies to make students job-ready. The other challenge is the lack of good teachers and attrition as many new teachers quit working as a teacher in few years. There is no provision for formal training for private school and college teachers. Private schools and colleges teachers should go through rigorous teachers training programmes.

Most of the decisions regarding administrative and pedagogical aspects are taken by the central authority. The national science curriculum failed to address the need of diverse communities. Science teaching methodologies are to be designed in such a way that helps students appreciate and understand the importance of science educa- tion. At the same time, teachers need to understand science as a consensus-building discipline that brings people from different backgrounds and cultures to make science a team and collaborative effort. Science should be taught as a way of knowing. One should learn how scientific knowledge is generated.

Nepal's education system lacking impactful research activities. The government falls behind in allocating sufficient funds for research in the education sector. Government has to invest more on research activities and science popularization to uplift the current state of science education in Nepal.

18.15   Conclusion

Science is more than a particular subject embedding a diverse applied body of knowledge, way of living that is not only important to an individual but the country as a whole. Science education in the country cannot be considered as its strength. However, it looksquite promising and better at present despite so many interruptions and problems surrounding it. The ministry of education has numerous policies for the betterment of education in the country. But when it comes to execution and implementation, it has a very poor track record. The constitution of Nepal has considered science and technology as a means to achieve sustainable development goal and overall development of the country, which requires boosting investment in research, science education promotion, and technological innovation. To promote science education in Nepal, there has to be significant funding and government participation in science literacy and education. Nepal needs a science education policy that guides, promotes, and will give a formal and much-needed push to develop science education in the country. A dedicated government body that promotes invests, and helps develop science education needs to be established. Therefore, a systematic approach is a prerequisite to promote science education in Nepal and is possible only through policy initiatives, need-based programs, sufficient investment, collaboration, cooperation, and shared practices at the national level and international level.

References

Alake-Tuenter, E., Biemans, H. J. A., Tobi, H., Wals, A. E. J., Oosterheert, I., & Mulder, M. (2012). Inquiry-based science education competencies of primary school teachers: A literature study and critical review of the American National Science Education Standards. *International Journal of Science Education, 34*(17), 2609–2640. https://doi.org/10.1080/09500693.2012.669076

Asian Development Bank. (2004). *Secondary education development project: Project performance audit report*. Retrieved from http://www.oecd.org/countries/nepal/35183393.pdf.

Bajracharya, D. M., Bhuju, D. R., & Pokhrel, J. R. (2006). *Science, research and technology in Nepal*. UNESCO.

Center for Education and Human Resource Development. (2018). *Flash I report 2018/19*. Author.

Center for Educational Research and Social Development (2016). *Feasibility study on Nepal's participation in international assessment*. Education Review Office.

Central Bureau of Statistics (2011). *National population and housing census 2011*: National report. Author.

Constitutional Assembly (2015). *The constitution of Nepal*. Nepal Gazette, Department of Printing, Ministry of Information and Communication.

Department of Education. (2017). *Flash Report I (2017–2018)*. Author.

Education Review Office. (2015). *National assessment of student assessment 2013*. Author.

Education Review Office. (2018). *National assessment of student achievement-2017: Public report*. Author.

Khaniya, T. R., Parajuli, T. R. & Nakarmi, S. S. (2015). *Background study: Curriculum, textbooks, and student assessment and evaluation*. Ministry of Education.

Kind, V. (2013). A degree is not enough: A quantitative study of aspects of pre-service science teachers' chemistry content knowledge. *International Journal of Science Education, 36*(8), 1313–1345. https://doi.org/10.1080/09500693.2013.860497

Mahtema, K. B., & Bista, M. B. (2006). *Study on student performance in SLC: Main report*. Ministry of Education and Sports.

Ministry of Education. (1971). *The national education system plan for 1971–76*. Author.

Ministry of Education Science & Technology (2017). *Education in figure 2017 at a glance*. Author.

Ministry of Education Science and Technology. (2019). *National science, technology, and innovation policy 2019*. Author. Retrieved from https://moe.gov.np/assets/uploads/files/NSTI_Policy_2019_English.pdf.

Ministry of Finance. (2018). *Economic survey 2018/19*. Author. Retrieved from https://mof.gov.np/uploads/document/file/compiled%20economic%20Survey%20english%207-25_20191111101758.pdf.

Ministry of Youth and Sports. (2015). *Youth vision 2025 and tenth year plan*. Nepal: Author. Retrieved from http://moys.gov.np/sites/default/files/nitiheru/Youth%20Vision-2025_2.pdf.

National Examination Board. (2019). *NEB result at a glance-2074–2075*. Author.

National Planning Commission. (2015). *National development goals (2016-2030): National (preliminary) reports*. NPS, Government of Nepal.

The World Bank. (2020). *Climbing higher: Toward a middle-income Nepal*. Retrieved from https://www.worldbank.org/en/region/sar/publication/climbing-higher-toward-a-middle-income-country.

United Nations Development Programme. (2019). *Human development report 2019*. Author. Retrieved from: http://hdr.undp.org/sites/default/files/hdr2019.pdf.

University Grants Commission. (2020). *Education management information system: Report on higher education 2018/19*. Author. Retrieved from: https://www.ugcnepal.edu.np_uploads_publicationsAndReports_HSFpPB.pdf.

Chapter 19
Science Education in Pakistan: Existing Situation and Perspectives for Planner

Muhammad Yasir Mustafa, Ali Gohar Qazi, and Afaq Ahmed

Abstract This chapter provides a picture of science education, from past to present, in Pakistan, and examines how science education has developed so far in the wider context of educational research and development. Technology's impact on developing national capacity, in emerging science and technology fields, is highlighted through various scientific and technological programs, international collaborations, government, and research organizations that are operating to develop human capital in sciences. Lastly, few ways in which existing issues in science education in Pakistan might be resolved are proposed.

Keyword Science education · Policy · Reforms · Research and development · Emerging technologies

19.1 Overview of the Country

19.1.1 Geographical Location, Population, and Political System

Pakistan, officially the Islamic Republic of Pakistan, appeared on the world map on 14th August 1947. It covers 796,096 (km^2) land, situated in the South Asia region, sharing borders with China, India, Iran, and Afghanistan. It is the 6th-most populous country globally, with a population exceeding 212.2 million (Pakistan Bureau of

---

The original version of the book was revised. In this chapter, order of the authors "Muhammad Yasir Mustafa, Afaq Ahmed, Ali Gohar Qazi" are changed to "Muhammad Yasir Mustafa, Ali Gohar Qazi, Afaq Ahmed". The erratum to this chapter can be available at https://doi.org/10.1007/978-981-16-6955-2_31

M. Y. Mustafa (✉)
School of Educational Technology, Beijing Normal University, Beijing, China

A. G. Qazi · A. Ahmed
Institute for Educational Development, The Aga Khan University, 1-5/B-VII, F. B. Area
Karimabad, P.O. Box No.13688, Karachi 75950, Pakistan

R. Huang et al. (eds.), *Science Education in Countries Along the Belt & Road*,
Lecture Notes in Educational Technology,
https://doi.org/10.1007/978-981-16-6955-2_19

Statistics, 2018). It has the second-largest Muslim population in the world. Pakistan is a democratic state administered by federal and provincial governments where provincial governments also exercise greater autonomy and residual powers. Executive power is delegated to the federal cabinet led by the prime minister, who deals with both the bicameral parliament and the judicature in a coherent way (Article 25A-25. Right to Education: Government of Pakistan, 2012). Pakistan is a federation of four provinces, along with Islamabad capital city, and two autonomous regions of Azad Jammu and Kashmir and Gilgit Baltistan. The four provinces are Baluchistan, Punjab, Sindh, and Khyber Pakhtunkhwa (KP). Punjab is the most populous province in Pakistan, accounting for 50% of the population (Pakistan Bureau of Statistics, 2018). Sindh is the second-most populated province; it includes the sprawling Karachi metropolis, which is one of the largest cities in the world with some 15 million residents. By comparison, Baluchistan is a big, lightly populated, and mountainous province of about 12 million people.

19.1.2   Current Situation of Economic, Technologies, and Cultural Development

Since the inception of the index in 1995, the economy of Pakistan has been mostly unfree. However, GDP growth has been robust during the last five years, driven by textile industry exports. Digital development in Pakistan is undergoing rapid evolution. The information technology (IT) industry is one of the fastest-growing sectors in Pakistan, contributing around 1% of Pakistan's GDP at about USD 3.5 billion (Ministry of Finance, 2020). Over the past four years, it has increased, and analysts expect it to grow to $7 billion in the next four to five years by another 100%.

From the cultural development perspectives, Pakistani society is culturally very complicated and mixed. It comprises of numerous ethnic groups including the Punjabis, Pathans, Sindhis, Sarikis, Balochis, Baltis, and Kashmiris, etc. There are differences among the ethnic groups in various aspects such as language, sect, and caste, culture, religion, etc. This means that Pakistani culture is not monolithic, it is diverse. Currently, the country can best be described as cultural mosaic, where conservatism and traditionalism reside side by side with a mix of extremism, secularism and liberalism. Therefore, the government of Pakistan envisions to promote a united culture driven by progressive notions of Islam, the modern sciences and technology.

19.2   The Educational Development

19.2.1   The Education System

The 2005/06 National Education Census (NEC) was the first census of education conducted in Pakistan's history, specifically designed to gather information on all types of education. It thus provided a full and detailed image of the existing education system in the country and provided a comprehensive baseline of knowledge for assessing future development. The education system in Pakistan is generally divided into six levels: pre-school (from 3 to 5 years of age); primary (from 1 to 5 years of age); middle (from 6 to 8 years of age); high (from 9 to 10 years of age leading to a Secondary School Certificate or SSC); intermediate (from 11 to 12 years of age leading to a Higher Secondary (Graduate) Certificate or HSC); and university programs leading to university education (Rashid, 2012).

19.2.2 Statistics on National Education

According to the latest statistics, the education sector of Pakistan has 180,846 public and 80,057 private institutions (see Fig. 19.1). There are 69% public and 31% private schools where 1,535,461 teachers teach 41,018,384 students in 260,903 schools (see Fig. 19.2) (ASER, 2019).

The education department's annual statistical report for 2017–18 depicts that the dropout rate at the primary and secondary levels is 44 and 40%, respectively. As noted by the report, 685,000 students admitted to the preparatory classes in 2012–13, but after six years in 2017–18, the schools had 369,163 students as 315,837 dropped out (ASER, 2019) (Fig. 19.3).

With 298,000 students enrolled in sixth grade in 2013–14, the dropout rate at high schools was 40%. Conversely, after five years in 2017–18, students completing 10th grade reached 169, 782 after the 128, 218 dropout (ASER, 2019).

On the other hand, Pakistan's university system is progressing since the first university—the Punjab University in Lahore—was founded in 1882 in the territory

Fig. 19.1 Total no of Schools in Pakistan

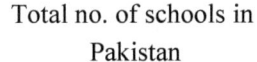

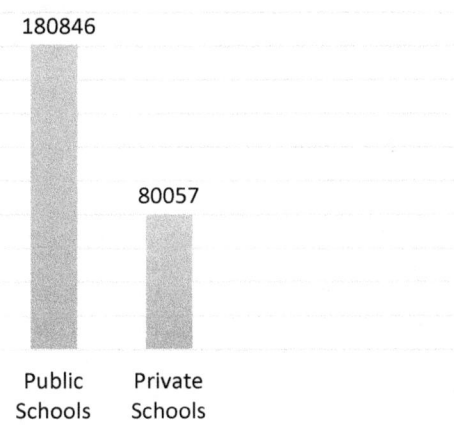

No. of schools, teachers, and
students

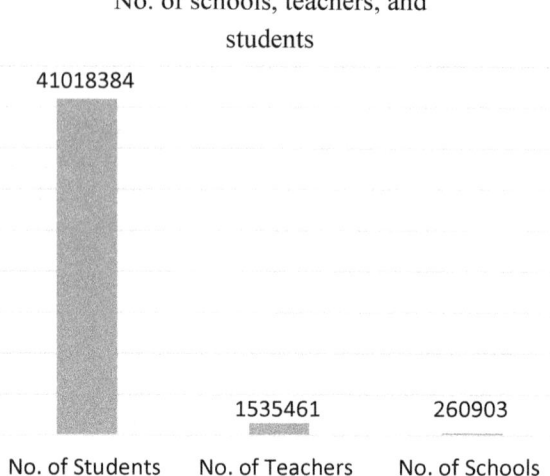

Fig. 19.2 No. of schools, teachers, and students

No of enrolled and drop out students from 2012-13 to 2017-18

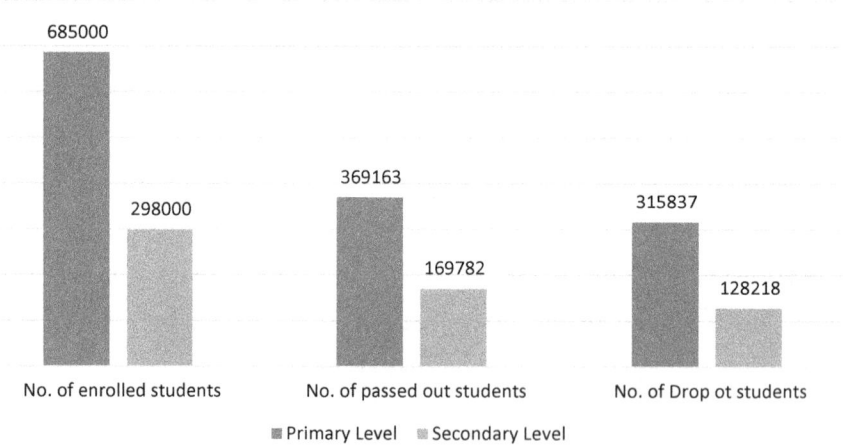

Fig. 19.3 Enrollment and dropout rate

that now represents Pakistan. In 1947, the university grant system (UGS) was established to deal with the university's affairs regarding funding and supervising. In 2002, the Higher Education Commission (HEC) replaced the UGS. Now, the HEC is responsible for overusing, funding, and accrediting the higher education institutes (HEIs)/universities in Pakistan. The number of universities in Pakistan has grown significantly since the establishment of the HEC in 2002, by accrediting both pre-existing and newly formed academics. A total of 174 private and public universities

Fig. 19.4 Total no. of universities in Pakistan

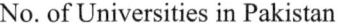

No. of Universities in Pakistan

■ Public Sector:137   ■ Private sector:37   ■ Total:174

are operating to cater to the needs of 1,355,649 enrolled students (National Education Management Information System [NEMIS] & Academy of Educational Planning and Management [AEPAM], 2017) (Fig. 19.4).

Additionally, the public universities have around 3,000 affiliated colleges, which are provincially governed and act as feeders to their affiliated universities, in which 45,358 faculty members are working (31,719 in public and 13,639 in private universities) (Higher Education Commission, 2020) (Fig. 19.5).

Presently, Pakistan's university system produces a meager amount of PhDs per university. Public universities graduated 12 Ph.Ds in 2014 on average, and private-sector universities graduated 1.4 Ph.Ds (a net average of 7.6 Ph.Ds per university) (Higher Education Commission, 2020). The HEC also offers scholarship support for overseas research to complement and improve the academic standard. The HEC does not publicly provide details on its expenditure on scholarships. Still, this expenditure is significantly greater than its budget for research funding, and the HEC is a net transmitter of revenue to international universities. Since September 2018, the HEC (Higher Education Commission) fully funded 1,614 Ph.D. candidates enrolled in

Fig. 19.5 No. of Students, teachers, and universities

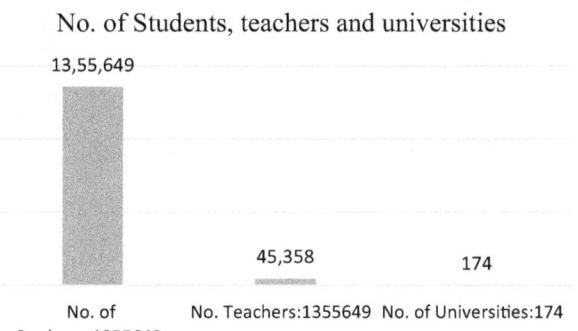

No. of Students, teachers and universities

13,55,649

45,358   174

No. of Students:1355649   No. Teachers:1355649   No. of Universities:174

universities abroad (Higher Education Commission, 2020). Given that the HEC only publishes limited data on the return on investment of its international scholarship program, its effectiveness and influence are very difficult to estimate.

19.2.3 Educational Research and International Collaboration

Pakistan has dozens of educational research institutes whereby the research tasks are carried out regularly. Recently, around 15 different research programs have been started to support educational research under the HEC's National Research Program for Universities (NRPU)—a major research funding program lead by HEC (Knowledge, 2018). NRPU supports and provides funds to the various programs of educational research, including faculty start-up research grants (FSRG), textbook and monograph writing scheme (TAMWC), scientific instrumentation (SI), social integration outreach program (SIOP), Pak-France PERIDOT research program, knowledge economy partnership (KEP), patent filings, Pakistan program for collaborative research (PPFCR), faculty outstanding research awards (FORA), ORIC support, business incubation centers, Pak-U.S. science and technology collaboration (PUSTC), and HEC digital library. These programs cover a wide range of tasks and activities including, to support new PhDs research scholars to set up new research projects, to organize seminars and conferences to promote research culture, to provide support for text-books design and developments as well as monographs, to support service providers for shared scientific instruments, to develop a relationship between universities and communities, to foster science and technology cooperation between Pakistan and other countries and international organizations, to support outstanding research, innovation, and publication among faculty, and to enable universities to build ORICS and business incubators, etc.

Furthermore, in 2015, different research centers (namely US-Pakistan Centers for Advanced Studies [USPCAS]) were established, with the help of USAID, to promote research culture and reinforce a community of applied Science (Knowledge, 2018). The primary objective of the centers is to meet advanced studies in water, energy, agriculture, and food security. These centers are established in various universities in Pakistan. For instance, (1) USPCAS in Water at Mehran University of Engineering and Technology, Jamshoro, in partnership with the University of Utah, (2) USPCAS in Energy at National University of Science and Technology, Islamabad, in partnership with Arizona State University, (3) USPCAS in Agriculture and Food Security at the University of Agriculture Faisalabad in partnership with University of California Davis, and (4) The US-Pakistan Center for Advanced Studies in Energy at the University of Engineering and Technology, Peshawar, in partnership with Arizona State University. These centers provide offerings to update curricula and other changes to make university education and research more important to fulfill the needs of industry and government, promoting effective policy dialogue and improvements for each technical sector headed by policy think tanks at every center, establishing robust scholarship and exchange programs and building strong ties with the private sector to upgrade world-class research facilities.

19.3 Current Situation of Science Education

19.3.1 Policies and Standards

The policies and standards regarding Science can be traced back from the early years of the inception of Pakistan, whereby the due consideration to Science was given at a higher level (e.g., intermediate or bachelor). In the 1950s, the focus at primary or secondary level remained dominantly on liberal art subjects, and very little was taught at the primary and secondary level (Iqbal & Mahmood, 2000). At this stage, no proper policies existed or comprehended Science education. The first educational conference 1947 and 1952 developmental plan majorly talked about providing free and compulsory education at the primary level (Ahsan, 2003). There was no clear policy and curriculum of Science as a subject in the education system of Pakistan focused on traditional reading, writing, and arithmetic (Government of Pakistan [GoP], 1975). However, it was believed that science and technological education is essential for the economic development of the country. Realizing the importance, the commission in 1959 put its first effort to recognize the teaching of Science a strong base in schools and recommended Science and mathematics compulsory for grades 6–10. Thus, the equal emphasis on Science, mathematics, and liberal arts were given (GoP, 1959). Iqbal and Mahmood (2000) mention that Science education was a compulsory part of the curriculum in the late sixties at the grade level from 1 to 8. Still, the poor quality of instruction is being imparted due to the lack of science teachers.

The National Education Policy (1979) advocated the discipline of Science as a separate component at the secondary level, emphasizing the quality of Science at the lower level. The National Education Policy (2009) and National Education Policy Draft (2017) emphasize promoting Science and technology for economic development. Nevertheless, no explicit framework is mentioned in educational policies (2009) and policy draft (2017) regarding science education at the primary and secondary levels. Currently, Science is a multi-disciplinary subject taught as a compulsory subject at the primary and middle level, which has embedded subjects such as physics, biology, and chemistry.

The National Commission of Science and Technology (NCST), established in 1984 at the apex of policy-making, is responsible for coordinating with inter-ministerial and inter-provincial stakeholders to put efforts into the production and development plans of Science and technology. Despite national policies, the 18th amendment in 2010 gave provinces the autonomy to make their policies. As a result, the provincial policies in education vary, which also directly affects science education (discussed below). For instance, the recruitment policies of teachers are different in four provinces of Pakistan (e.g., Khyber Pakhtunkhwa eliminated professional qualification from teaching posts). Likewise, books are developed at the provincial level differently.

19.3.2 Curriculums, Digital Resources, and Teacher Training

In Pakistan, the curriculum framework has undergone several revisions of its first national curriculum in 1975–1976, which was revised in 1984–1985, 1994–1995, respectively, and again reviewed in 2002. A comprehensive curriculum, known as National Curriculum (2006) for 25 core subjects (grades 1–12), is currently being followed in Pakistan. According to National Curriculum for General Science (2006), learning science, teachers play a significant role in enabling students to achieve scientific literacy with support from the education system in terms of training, resources (materials, labs), and promoting a conducive environment. The General Science Curriculum (2006) aimed to develop an understanding of scientific literacy through inquiry-based, student-centered approaches, and an outcome-based curriculum. However, the challenge always remained at the implementation level in the Pakistani context.

A wide body of literature indicated the poor quality of science education imparted to the students at the primary and secondary levels in Pakistan (Ali, 2012; Mohammad & Kumari, 2007; Najmonnisa & Saad, 2017; Saeed et al., 2005). The contributing factors to the unfortunate situation of science education are manifold. The most prominent in under-developing countries, especially Pakistan, are (1) shortage of trained science teachers and inadequate preparedness of teachers, (2) lack of resources (laboratories, teaching materials), and (3) low quality of textbooks and examinations. According to Halai (2017), the subject-content knowledge gap remained at all levels before implementing pedagogies. Therefore this lack of confidence in the understanding of Science does not allow teachers to apply innovative methods.

As a consequence, the teaching of Science proceeds through rote memorization. Usually, teachers from any academic background are allocated at primary and middle levels. This means a lack of science specialists in schools at the lower level.

On the other side, an overview of the teachers' development progress in Pakistan states a considerable quantitative expansion in training institutes established by Pakistan and international donor agencies. Currently, the country has about 300 teacher training institutes with 3477 teacher educators catering to the needs of 0.724 million enrollment (Gopang, 2016; NEMIS & AEPAM, 2018). However, the training delivered has little or no impact on teachers' pedagogical development (Aslam, 2013; Rizvi, 2015). Loose teacher education programs are provided to pre-service and in-service teachers unable to change teachers' pedagogical beliefs previously constructed. Further, it becomes more problematic at the policy level as provinces being autonomous have their recruitment policies. For instance, Khyber

Pakhtunkhwa (KP)—a province of Pakistan—has introduced an induction policy that eliminated professional qualification. Therefore everyone with 14 years of education is eligible for teaching at the primary level. Since the 18th amendment, the textbooks are developed through private publishers register with provincial governments. A separate wing within the education department looks for the approval of textbooks, i.e., the Directorate of Curriculum and Teacher Education (DCTE) and the Khyber Pakhtunkhwa Textbook Board (KPTBB) in KP Province and Bureau of Curriculum and Extension Wing (BCEW) and Sindh Textbook Board (STBB) in Sindh. Society for the Advancement of Education [SAHE] (2014), reviewing the quality of textbooks and curriculum, mentioned that with a lack of staff and no rigorous process of recruitment, the textbook writers lack the expertise and appropriate subject specialization, hence undermined the quality of textbooks. These writers are not trained to develop textbooks nor have pedagogical expertise to associate curriculum with students learning outcomes.

Besides, the unavailability of science equipment and resources in most public schools is negatively clobbering. According to NEMIS and AEPAM (2018), the non-availability of electricity affects the performance of students, having computer and science labs, with 62% availability of electricity at the primary level and 79% at the middle level. These statistics revealed unstable and low maintenance of schools and 73% at primary to 85% at the middle level (some provinces below 50%) have access to clean water in schools.

19.3.3 Student Assessment and Achievement

The consequences of the above-stated factors are evident in students' achievement in Science subjects. For instance, Javed (2017) examined the effect of the classroom environment found a significant positive effect of the classroom environment, teacher feedback, and motivation on students' academic achievement. According to NEAS (2016), Science remained a difficult subject for students across all provinces with no optimal improvement than other subjects such as mathematics and languages. Simultaneously, teaching science kits remained below 50% for fourth grade and 55% for eighth-grade classes. Likewise, Saeed et al. (2005), assessing the achievement of primary graders in Punjab province, found the overall low performance of students with an average of 29 for grade 3 (9 mean for Science) and 31 for grade 5 (13 mean for Science). Chang and Jilani (2015), looking at the students' achievement test (SAT) III funded by World Bank in Sindh, found overall average score significantly lower for Mathematics and Science of 8th and 5th grades. Overall performance in Mathematics and Science was found below 20% in class VIII. It was stated that Science scores do not have a strong variation in Science content strands based performance of students. However, life science and earth and space science were a little above, and physical Science fell slightly below the subject average.

There is a general perception of teachers in Pakistan that they are largely using rote memorization methods to help students score higher in examinations. This assumption might slightly vary to the type of schools (public or private). However, empirical evidence has well-documented traditional methods of assessment dominantly used in classrooms (Ahmad & Malik, 2011; Gouleta, 2015; Hussain et al., 2019). According to the Education Policy of Pakistan 2009, various assessment tools should be used in traditional examinations to incorporate formative techniques with summative assessment and assessing students' skills in a specialized area. Unfortunately, schools in Pakistan assess the memorized knowledge from the textbook of the students (Shiekh et al., 2013). Those students who best produce knowledge on paper is declared as successful. Assessment in the schools of Pakistan does not examine the skills/abilities of the learners with no ongoing assessment of the students at the primary level; therefore, the exams conducted produce very little feedback for students' improvement. Also, self-assessment activities and questions that aim at higher-order skills are consistently missing in Pakistani schools (Sarwar et al., 2011). These summative assessment tools include multiple-choice questions, open-ended questions, true/false, and so on. At a higher level, universities, especially teacher education programs, use formative techniques such as portfolios, quizzes, assignments, reflections, and observation techniques.

19.3.4 Outreach Science Education

Recognizing the importance of science and technology, the Ministry of Science and Technology (MoST), formed in 1972, is responsible for planning, coordinating, and directing efforts to initiate scientific and technological programs as per national agenda (e.g., laboratories, universities, research institutes). Recently MoST is undertaking initiatives through bilateral and multilateral cooperation that include young scientist programs, training, workshops, and so on to enhance the scientific industry (see MoST, 2020). According to the recent available Pakistan Council for Science and Technology databook (2009), 134 higher education institutes and 85 organizations are operating to develop human capital in sciences. Moreover, the following science and technology venues and centers promote formal and informal science education in the country.

The Pakistan Council of Scientific and Industrial Research (PCSIR): PCSIR is owned by the government and research organization for promoting Science and industrialization that focuses mainly on industrial research and development. It carries out research publications, administration of research laboratories, human resources development (HRD) centers, research funding, research cooperation, and accreditation criteria for material testing.

Ignite is a company founded by the Information Technology and Telecommunication Ministry, which controls an innovation fund. The objective of Ignite is to promote research and innovation. Ignite now is engaging in three key fields of activity: (1) supporting university, corporate and non-profit R&D and innovation projects; (2) developing IT skills with the help of scholarships, competitions, and a large online Digi skills program; and (3) financing incubation centers to support Pakistan's development and exciting start-up community.

Pakistan Science Foundation (PSF): The Pakistan Science Foundation is the leading agency of the federal government committed to supporting study and sharing knowledge. PSF is responsible for supporting scientific research in universities, encouraging science popularization, setting up science centers, organizing scientific conferences, organizing scientific exchanges, helping scientific societies, and granting science awards and fellowships. Although PSF does not have data on its quantity of funding for the study, it is limited compared to the HEC. To date, PSF has sponsored 1,031 research projects, which have resulted in 612 research papers being published in international and national journals and five patent filings. One hundred sixty-nine students also received MPhil degrees (with a further 115 Ph.Ds) due to PSF-supported projects.

Moreover, PSF promotes and popularizes Science in the country at the grassroots level by introducing various outreach science projects in formal and informal settings. Details of these projects are mentioned below:

- **Inquiry-Based Science Education**: PSF initiated an inquiry-based science education program with the collaboration of Academic Des Sciences, France, intending to improve the teaching and learning of sciences at the primary and secondary levels. This program focuses on developing the questioning attitude of children while conducting experiments.
- **Mobile Science Exhibition**: To popularize Science in the country, PSF initiated science caravans or mobile science exhibitions. These are specially designed vehicles that display items like panel exhibits, equipment, inflatable planetarium, documentaries, microscopes, computers, laser holograms, and working models reflecting various phenomena of physics, chemistry, mathematics, and biology.
- **Science and Technology Fairs and Expos**: PSF regularly organizes science and technology fairs and expos to stimulate public interest and provide opportunities to scientific institutions in the country to showcase the progress and accomplishments of scientific discoveries.
- **World Science Day**: PSF celebrates world science day for peace and development (WSDPD) annually on 10th November. Various activities such as the convention of scientists, speech competition among students, and science exhibitions are carried out to demonstrate the value of Science.
- **Science Essay and Poster Competition**: PSF collaborates with all Boards of Intermediate and Secondary Education (BISE) organizes the science essay and poster competitions among high school students each year. Moreover, to encourage the students, PSF awards certificates, and cash prizes to the winners.
- **Distribution of Scientific Literature**: PSF distributes popular science magazines such as "Monthly Global Science" and quarterly "Urdu Science Magazine" and scientific books to more than 1500 schools all over the country popularize science education.

- **Science lectures**: PSF organizes popular science lectures on different scientific themes and issues. For this, they invite eminent scientists, technologists, and educationists to deliver Science related lectures in non-technical language to a wider body of audiences.

  National Centers of Innovation and Respective fields: In 2017–2018, the HEC (with the Planning Commission) launched four national centers (see Table 19.1) to develop national capacity in emerging science and technology fields. Investment and emphasis promoted by these National Centers could provide Pakistan with a competitive edge in cloud computing, artificial intelligence, cybersecurity, data analytics, robotics, and automation. These newly formed organizations have a national center for each discipline and series of 'associated laboratories' at various universities to facilitate inter-university collaboration.

19.3.5 Utilizing Emerging Technologies

Pakistan ranked 105 out of 129 in the Global Innovation Index for 2019, representing very little development for the region. The tech industry is making strides all around the world only because of creativity and innovation. To some extent, in recent years, the Pakistani government has been very successful in convincing leading technology masters, such as Facebook, Innovation Lab, SheMeansBusiness, and WeThink Digital, to contribute and cooperate at various technical levels to enhance Pakistan's hi-tech growth.

19.4 Requirements for Future Development of Science Education

Evident from statistics, policies, and previous studies show that science education is far behind in terms of qualitative development in Pakistan, especially at the primary and secondary levels. We foremost argue that uniform policies are needed much more than ever before in Pakistan. The first step towards improvement in science education is to explicitly include science education in educational policies at primary and secondary levels, so far mostly missing and unclear in policies. Ali (2006) highlighted unclear goals, political commitment, governance, centralization, resources, and foreign aids at the center of policy failures. We also urge provincial and federal governments to implement uniform policies regarding science education. The continuous changes in policies are unstable and unsustainable failing policies. Because policy implementation is a process to "actualize, apply, and utilize it in the world of practice" (Bhola, 2004, p. 296) that requires time in output.

Table 19.1 Pakistan's national centers of innovation

Research areas	National centers	Name and location of research labs
Artificial intelligence (A.I)	NUST, Islamabad	Medical Imaging and diagnostics(CIIT, Islamabad)
		Deep Learning (NUST, Islamabad)
		Intelligent Field Robotics (NUST Islamabad)
		Smart Cities (NED UET, Karachi)
		Neuro-computing(NED UET, Karachi)
		Intelligent Information Processing (UET Peshawar)
		Intelligent System Design (UET Peshawar)
		Intelligent Criminology (UET Lahore)
		Agent-Based Modeling (Punjab University Lahore)
Robotics and automation	NUST, Islamabad	Robot Maker (NUST, Islambad)
		Robot Design and Development(NUST, Islamabad)
		Human-centered Robotics, (UET Lahore and UCP Lahore)
		Industrial Monitoring and Automation(ITU Lahore)
		MEMs Sensor Design and Testing (AIR University, Islambad)
		Agricultural Robotics(LUMS, Lahore)
		UAV Dependability(FAST, Islamabad)
		Swarm Robotics(UET Taxila)
		Haptics Human Robotics (NED Karachi and MUET)
		Control Automotive and Robotics (BUITEM, Quetta and MUST Mirpur)
		Advanced Robotics and Automation(UET Peshwar)
Cyber security	AIR University, Islamabad	Cyber Security (AIR University Islamabad)
		Cyber Security (Bahria University, Islamabad)

(continued)

Table 19.1 (continued)

Research areas	National centers	Name and location of research labs
		Cyber Security (UET Texla)
		Cyber Security (UET Lahore)
		Cyber Security (ITU Lahore)
		Cyber Security (NUST, Islamabad)
		Cyber Security (PIEAS, Islamabad)
		Cyber Security (UETP, Peshwar, and Nowshera)
		Cyber Security (LUMS, Islamabad)
Data analytics and cloud computing	LUMS, Lahore	Digital Pakistan (NUST Rawalpindi)
		Distributed Computing (LUMS, Lahore)
		Language Engineering (UET Lahore)
		Exascale Open Data (NED Karachi)
		Precision Medicine (FAST Karachi)
		Video Surveillance (NUST, Islamabad)
		Sustainable Energy (LUMS, Lahore)
		Predictive Analytics (SZABIST, Karachi)
		Precision Agriculture and Analytics (UAF, Faisalabad)
		Crime Investigation Prevention (ITU, Lahore)
		Data Acquisition (Ziauddin University, Karachi)
		Data Analytics (UETP. Peshawar)

Consequently, a strong science teaching force is required to teach at the lower levels. The need is the development of teachers both in terms of content and pedagogical expertise. It has been documented that teachers' training programs in Pakistan are insufficient in preparing prospective teachers (Ali, 2018; Ullah et al., 2008; Yadav, 2011). Therefore teacher education programs need to be robust and coherent to prepare effective science teachers. These programs should be further comprehended with on-job mentoring programs of science teachers. In the context of Pakistan, empirical evidence suggests mentoring as effective for science teaching (Halai,

2006, 2018; Manzar-Abbass et al., 2017). Finally, material and technical resources will contribute to the implementation of policies, imparting inquiry-based teaching methods, and provide a conducive learning environment for science education. Therefore, we suggest providing proper infrastructure, resources like laboratories, scientific centers, and the promotion of Science through competitions, exhibitions, scientific programs, etc.

It is important to know that numerous donor-funded programs are implemented, but the long-term impact remains questionable. Much of these programs focus on quantitative expansion rather than qualitative and sustainable change (Ali, 2011). According to Coll and Taylor (2012), disconnection from contextual realities (e.g., culture, religious issues, social issues) in developing countries context in the mere implementation of imported western science curricula fails. In this regard, the Road and Belt initiative by the Chinese government to create regional economic cooperation, spanning around some 68 countries, can strengthen the education sector, particularly science education in Pakistan. This initiative could be influential in providing collaborative and partnership platforms to work together and understand the complex contextual realities that are often neglected in implementing these programs. In doing so, working and understanding in developing and developed context will give hybrid models through cooperation, knowledge networks by creating and sharing scientific knowledge at the national and international levels. We argue that it is necessary for the capacity development in terms of quantity (e.g., expansion of labs, scientific centers) and quality (e.g., human capital, STEM integration), and a more need-based development. Consequently, fostering the collaborative partnerships will bring closer the scientific and business communities for bringing innovations in technology and economic development among Belt and Road countries.

19.5   Conclusion and Suggestion

The science education in the context of Pakistan faces critical challenges in its development and implementation; therefore, to strengthen science and technology, we foremost suggest taking concrete steps to improve science education, particularly at the primary and secondary levels. Because these are the foundational stages of education that are highly neglected for science education. Additionally, the broader context for science education can be provided through legislation in policies. We suggest uniform policies for science education across Pakistan contemplating theoretical and practical instruction in Science. Ali (2005), in this regard, directed our attention that short-term experiences implemented at the national level do not meet long-term goals. This implies that the implementation process could be achieved by providing sufficient resources, techniques, and equipment at the grassroots level, i.e., at the educational institutes' settings. Further, Tseng and Seidman (2007) noted that social processes and settings are affected by change and resources. Therefore, efficient classroom practices of teachers are crucial in improving science education in Pakistan. Nevertheless, classroom practices of teachers in Pakistan had remained at the lower end. The resources essential for instruction rely on teachers' competencies to efficiently utilize in their instructional strategies. To get out of conventional teaching approaches, we need to build a strong science teaching force accompanied by developing curricula and new ideas. This possibility arises from teacher development that can change the traditional beliefs of teachers and move towards more constructivist pedagogies, strong subject-mastery, and assessment. However, such teacher development occurs with rigorous and coherent teacher education programs.

References

Ahmad, S. I., & Malik, S. (2011). Examination scheme at secondary school level in Pakistan: Composite vs Split. *Canadian Social Science, 7*(1), 130–139.
Ahsan, M. (2003). An analytical review of Pakistan's educational policies and plans. *Research Papers in Education, 18*(3), 259–280. https://doi.org/10.1080/0267152032000107329
Ali, S. (2005). The influence of globalisation on the national education policies of developing countries. *Journal of Educational Research, 8*(1), 14–21.
Ali, S. (2006). Why does policy fail? Understanding the problems of policy implementation in Pakistan-a neuro-cognitiveperspective. *International Studies in Educational Administration, 34*(1), 2–20.
Ali, T. (2011). Understanding how practices of teacher education in Pakistan compare with the popular theories and theories and narrative of reform of teacher education in international context. *International Journal of Humanities and Social Sciences, 1*(8), 208–222.
Ali, T. (2012). A case study of the common difficulties experienced by high school students in chemistry classroom in Gilgit-Baltistan (Pakistan). *SAGE Open, 2*(2), 1–13. https://doi.org/10.1177/2158244012447299
Ali, T. (2018). Raising teachers' voices: An in-depth qualitative inquiry into teachers' working conditions and professional development needs in Khyber Pakhtunkhwa, a province of Pakistan. *Teacher Development, 22*(1), 78–104. https://doi.org/10.1080/13664530.2017.1308432
ASER. (2019). *Annual-status-of-education-report-ASER-PAKISTAN-2019*. Retrieved from https://palnetwork.org/wp-content/uploads/2020/02/Annual-Status-of-Education-Report-ASER-PAKISTAN-2019.pdf.
Aslam, H. D. (2013). Analysis of professional development practices for school teachers in Pakistan: A comparative case study of public and private schools of Pakistan (Punjab). *International Journal of Human Resource Studies, 3*(4), 311–326. https://doi.org/10.5296/ijhrs.v3i4.6251
Bhola, H. S. (2004). Policy implementation: Planning and actualization. *Journal of Educational Planning and Administration, 18*(3), 295–312.
Chang, H. F., & Jilani, B. D. (2015). Standardized Achievement Test (SAT) III: Sindh government schools achievement class V & VIII—Subjects mathematics, science, and languages. *Education and Literacy Department, Government of Sindh.* https://doi.org/10.13140/RG.2.2.24119.01440
Coll, R. K., & Taylor, N. (2012). An international perspective on Science curriculum development and implementation. In B. J. Fraser, K. G. Tobin, & C. J. McRobbie, *Second international handbook of science education* (Vol. 2, pp. 771–782). New York: Springer.
Gopang, I. B. (2016). Teacher education and professional development programs in Pakistan. *The International Journal of Research in Teacher Education, 7*(1), 1–14.
Gouleta, E. (2015). Educational assessment in Khyber Pakhtunkhwa Pakistan's North-West Frontier province: Practices, issues, and challenges for educating culturally linguistically diverse and exceptional children. *Global Education Review, 2*(4), 19–39.
Government of Pakistan. (1959). *Report of the national commission on education.* Karachi: Ministry of Education.

Government of Pakistan. (1975). *Elementary science curricula for classes VI–VIII.* Islamabad: Ministry of Education, Curriculum Wing.

Government of Pakistan. (1979). *National education policy and implementation program.* Islamabad: Ministry of Education.

Government of Pakistan. (2006). *National curriculum for general science grades IV–VIII 2006.* Ministry of Education, Islamabad, Pakistan. Retrieved from https://bisep.com.pk/downloads/curriculum/Grades-IVVIII/pk_prsc_sc_2006_eng.pdf.

Government of Pakistan. (2009). *National education policy 2009.* Ministry of Education, Islamabad: Pakistan.

Government of Pakistan. (2012). Article 25A-25. Right to education: The constitution of Islamic Republic of Pakistan. *National Assembly of Pakistan, 211.* Retrieved from http://na.gov.pk/uploads/documents/1333523681_951.pdf.

Government of Pakistan (2017). *National education policy 2017.* Islamabad: Ministry of Federal Education and Professional Training.

Halai, A. (2006). Mentoring in-service teachers: Issues of role diversity. *Teaching and Teacher Education, 22*(6), 700–710.

Halai, N. (2017). *Science education for the future: Five recommendations a policy brief for Pakistan.* Aga Khan University, Institute for Educational Development.

Halai, N. (2018). *Enhancing primary science teaching through school-based mentors: A study from Pakistan.* Oxford University Press.

Higher Education Commission, P. (2020). *HEC recognized universities statistics.* Retrieved from https://hec.gov.pk/english/universities/Pages/AJK/UniversitiesStatistics.aspx.

Hussain, S., Shaheen, N., Ahmad, N., & Islam, S. U. (2019). Teachers' classroom assessment practices: Challenges and opportunities to classroom teachers in Pakistan. *Dialogue, 14*(1), 87–97.

Iqbal, H. M., & Mahmood, N. (2000). Science teacher education in Pakistan. In S. K. Abell (Ed.), *Science teacher education: An international perspective* (pp. 75–92). Kluwer.

Javed, T. (2017). Effect of classroom environment, motivation and teacher feedback on academic achievement of secondary school students in Pakistan. Mohi-Ud-Din Islamic University Nerian Sharif AJ&K, Pakistan.

Knowledge, P. (2018). *The University research system.*

Manzar-Abbass, S., Ahmed, S., & Malik, N. A. (2017). Primary school teachers' perceptions about mentors' role in their professional development. *New Horizons, 11*(2), 47–59,109. Retrieved from https://search.proquest.com/docview/1985549784?accountid=6724.

Ministry of Finance, G. of P. (2020). *Pakistan economic survey 2019–20.* Retrieved from http://www.finance.gov.pk/survey_1920.html.

Mohammad, R., & Kumari, R. (2007). Effective use of textbooks: A neglected aspect of education in Pakistan. *Journal of Education for International Development, 3*(1), 1–12. Retrieved from http://citeseerx.ist.psu.edu/viewdoc/download?doi=10.1.1.533.1185&rep=rep1&type=pdf.

MoST. (2020). *Year book 2018–19.* Ministry of Science & Technology, Government of Pakistan Islamabad. Retrieved from https://most.comsatshosting.com/Publications/YearBook2018_19.pdf.

Najmonnisa, K., & Saad, I. (2017). The role of cooperative learning method in teaching of science subject at elementary school level: An experimental study. *Bulletin of Education & Research, 39*(2).1–17.

NEAS. (2016). *National assessment report 2016.* Ministry of Federal Education & Professional Training Islamabad (Pakistan). Retrieved from http://www.neas.gov.pk/SiteImage/Misc/files/National_Stakeholder_2016.pdf.

NEMIS & AEPAM. (2017). *Pakistan education statistics 2015–2016.* Premier Printers, Islamabad. Ministry of Federal Education and Professional Training Government of Pakistan.

NEMIS & AEPAM. (2018). *Pakistan education statistics 2016–2017: 25th annual publication since 1992–93.* Premier Printers, Islamabad. Ministry of Federal Education and Professional Training Government of Pakistan. Retrieved from http://library.aepam.edu.pk/Books/Pakistan%20Education%20Statistics%202016-17.pdf.

Pakistan Bureau of Statistics. (2018). Pakistan statistical year book. In *Pakistan Beureau of statistics*.

Pakistan Council for Science and Technology. (2009). *Pakistan science and technology data book*. Pakistan Council for Science and Technology Islamabad. https://www.pcst.org.pk/docs/S&T%20Data%20Book%202009.pdf.

Rashid, K. (2012). Education in Pakistan : Problems and their Solutions. *International Journal of Academic Research in Business and Social Sciences*.

Rizvi, M. (2015). Teacher education pedagogies related to preparing preservice teachers as leaders in Pakistan. In L. Orland-Barak, & C. J. Craig, *International teacher education: Promising pedagogies (Part B)* (pp. (Vol. 22B, pp. 7–30)). Bingley: Emerald Group Publishing Limited.

Saeed, M., Gondal, M. B., & Bushra. (2005). Assessing achievement of primary grader students and factors affecting achievement in Pakistan. *International Journal of Educational Management, 19*(6), 486–499. https://doi.org/10.1108/09513540510617436.

Sarwar, M., Yousuf, M. I., & Ranjha, A. N. (2011). Usefulness and level of interest in Pakistan National curriculum subjects: Secondary school students' perceptions. *International Journal of Academic Research, 3*(1), 964–969.

Shiekh, M. A., Chohan, B. I., & Jawad, A. (2013). A comparative study of the assessment practices and proposed curriculum objectives in revised teacher education programs. *Bulletin of Education and Research, 35*(2), 75–89.

Tseng, V., & Seidman, E. (2007). A systems framework for understanding social settings. *American journal of community psychology, 39*(3–4), 217–228.

The Society for the Advancement of Education [SAHE]. (2014). Education monitor: Reviewing quality of key education inputs in Pakistan. Society for the Advancement of Education, Lahore, Pakistan. Retrieved from http://www.sahe.org.pk/wp-content/uploads/2016/01/4.-EM-I.compressed.pdf.

Ullah, S. Z., Farooq, M. S., & Memon, R. A. (2008). Effectiveness of teacher education programs in developing teaching skills for secondary level. *Online Submission, 4*(1), 33–38.

Yadav, S. K. (2011). Comparative study of pre-service teacher education program at secondary stage in Bangladesh, India, Pakistan and Sri Lanka. *Educational Research and Reviews, 6*(22), 1046–1050. https://doi.org/10.5897/ERR10.066

Muhammad Yasir Mustafa is a Ph.D. student in Educational Technology from school of Educational technology, Beijing Normal University, China. He is also affiliated with the Smart learning institute of Beijing Normal University. Before that, he did his MS in Innovative technologies in education from National University of Science and Technology NUST, Islamabad. He secured his BS Honors in Information Technology from department of computing at Bahuddin Zakariya University, Multan. His research areas include Learning Management Systems, Online Learning, Online merge offline (OMO) learning, Artificial Intelligence, Open Educational Resources, and Gamification in Education.

Ali Gohar Qazi is an experienced individual with expertise in teaching, research, management, and evaluation. His professional work has been in the social sectors with several donor (USAID, DFID) and non-profit organizations. Currently, he is pursuing his Ph.D. in Education from the AKU-IED, Karachi. Previously, he received his MS and BS degrees in Innovative Technologies in Education and Information Technology, respectively, from the NUST, Islamabad. His areas of interest include ICT in Education, curriculum design and development, instructional technologies, mobile-supported teacher professional development, blended/e/m-learning.

Afaq Ahmed is Research Associate at The Aga Khan University—Institute for Educational Development (AKU-IED), Karachi. His research area includes teacher education, quality of teaching, and policies.

Chapter 20
Science Education in the Philippines

Robert John D. De La Cruz

Abstract This report gives an overview of the current situation of science education in the Philippines. In addition, this features the requirements and initiatives done by the government in terms of keeping up its science education program with the changes brought by the twenty-first century. The science curriculum in the Philippines was implemented to produce scientifically literate individuals who are responsible decision makers and can apply scientific knowledge to look for solutions to problems of the community. However, in the latest results of Programme for International Student Assessment (PISA) 2018, the Philippines ranked last among participating countries in which Science was one was one of the subjects tested. The results of the first participation of the country in PISA has paved the way for the Department of Education to propose more programs to address the deficient academic performance and advance the quality of education in the Philippines. Furthermore, the rapid change brought by Industry 4.0 also brings challenge to science education implementers for the country needs to ensure that it can adapt with the emerging technologies like A.I. and Robotics. Several initiatives were done by the Department of Science and Technology and Science Education Institute through their responsive and tailor-fit programs and projects.

Keywords Science curriculum · Emerging technologies · Fourth industrial revolution · Philippines

20.1 Overview of the Country

20.1.1 The Geography, Population Situation, and Political System

The Philippines is a Southeast Asian archipelago with 7,641 islands. It has a total land area of more than 300,000 km^2. Luzon, Visayas, and Mindanao are the three

R. J. D. De La Cruz (✉)
Rizal National Science High School, J.P. Rizal, 1940 Batingan, Binangonan, Rizal, Philippines

© The Author(s), under exclusive license to Springer Nature Singapore Pte Ltd. 2022   331
R. Huang et al. (eds.), *Science Education in Countries Along the Belt & Road*,
Lecture Notes in Educational Technology,
https://doi.org/10.1007/978-981-16-6955-2_20

main islands that make up the Philippines. Luzon is the largest island group situated in the northern part of the country. It is spanning 47% of the total land area. Mindanao is the second-largest island group accounting for 34% of total land area. Visayas is a smaller group of islands found between Luzon and Mindanao with 19% of land area.

The Philippines is divided into 17 regions: Luzon has eight (Regions I, II, III, IV-A, IV-B, V, CAR and NCR), Visayas has three (VI, VII, and VIII) and Mindanao has six (IX, X, XI, XII, XIII and BARMM). Sixteen urbanized cities and one municipality which contribute to 36.5% of the country's GDP are located in Metro Manila area. The Philippines' proximity to the regions' largest economies, China and Japan, provides it with more opportunities for trade and commerce (Oxford Business Group, 2017a, 2017b).

As of 2020, the Philippines has an estimated population of 109 million based on the population projections from the 2015 Census of Population conducted by the Philippine Statistics Authority (Commission on Population and Development, 2020). With an annual increase of 2 million people, it is expected that the population will double in 40 years (WorldPopulationReview, n.d.). The Philippine population is the 13th largest in the world according to worldometer.com. In July 2020, a labor force participation rate of 61.9% was registered in the country as revealed by the Philippine Statistics Authority. According to Commission on Filipinos Overseas, there are a total of 10.2 million overseas Filipinos as of 2020. These groups of Filipinos are considered one of the largest diasporas in the world and contribute to the economy by way of their remittances.

The Philippines has a democratic government in the form of a constitutional republic where the President is acting as the head of the state and head of the government. The President is elected by popular vote for a single six-year term, while senators and congressmen are elected for a three-year term. The Supreme Court holds the judicial power presided by the Chief Justice and composed of fourteen associate justices appointed by the President of the Philippines. The incumbent President is Rodrigo Roa Duterte who hails from Davao City.

20.1.2   The Current Situation of Economic, Technologies, and Cultural Development

The economy of the Philippines is considered as one of the most dynamic in the East Asia Pacific region (World Bank.org, 2020). The Philippines has transitioned to a nation that merely relies on agriculture to an economy that is based on services and manufacturing. Its primary exports include coconut oil, fruits, semiconductors, electronic products, transport equipment, garments, and petroleum products. The Philippines' economic dynamism has its roots in strong consumer demand coupled with an increase in labor market and remittances.

The Philippine economy has shown progress towards delivering inclusive growth as evidenced by a decline in poverty rates and its Gini coefficient. In 2015, the country has a poverty rating of 23.3% which decreased to 16.6% in 2018 while the Gini coefficient also showed a decline from 44.9 to 42.7 over the same period. The impact of the current health crisis brought by the COVID-19 pandemic may also hinder poverty reduction in the Philippines (World Bank.org, 2020).

According to the Philippine Institute of Development Studies (PIDS) the Philippine economy would improve by an estimated value of 10% if emerging technologies and knowledge of the Fourth Industrial Revolution can be used to speed up productivity (Gatpolintan, 2018).

In addition, it is now accepted that science, technology, and innovation (STI) are important elements in the development of a country. This poses a challenge for the Philippine government on how science, technology and innovation will be utilized in overcoming global and local challenges. These include (a) intensified competition from globalization and regional integration; (b) climate change, environmental degradation and natural disasters; and persistent poverty and increasing inequality.

Moreover, in the latest Technological Readiness Ranking published by the Economist Intelligence Unit, the Philippines has moved to the 55th spot out of 82 economies, showing that the country has improved in terms of preparedness for technological change (Cigaral, 2018). To promote the culture of creativity and inventiveness, the government encouraged young students to take courses related to STI and creative arts. In the academic year 2016–2017, the number of students who enrolled in Science, Technology, Engineering and Mathematics (STEM) related courses in state colleges and universities reached 1.27 million which is higher than the target of one million enrollees.

In like manner, the National Economic and Development Authority (NEDA) revealed that the Filipino culture and common values are essential in bringing positive changes in the country. It is mentioned that the culture and values of the Filipino people are vital in making well-targeted plans and effective policies. NEDA conducted a National Values Survey that aims to understand the cultural values that characterize the Philippine society, which can be used to examine the country's progress in values formation and its contribution to national development. The initial results of this survey show that the Philippines is a relatively conformist and interdependent society. In addition, the survey also indicates that Filipinos are generally proud of being Filipino. They are accustomed to tradition, religion, and trust for others regardless of sexuality (NEDA, 2019).

20.2   Overview of the Education Development

20.2.1   Education System and Policy

The Philippine educational system is a culmination of its history from the influences of foreign invaders and curricular reforms made as the years progress. The current system focuses on the K-12 program which was signed into law by former President Benigno Aquino III in 2013. The law adds additional two years of senior high school and a mandatory kindergarten to all students of the basic education program. This change makes the country's educational system to at par with other countries in terms of duration.

In the Philippines, the government is assisted by the private sector to deliver education from the early years up to college and university. The Department of Education manages basic education in the country. Some kindergarten, international schools and religious institutions belong to the private sector.

The Philippine educational system comprises the Basic Education (Kindergarten, Elementary, Junior High School and Senior High School), Alternative Learning System (learners with special needs and out-of-school learners), Technical Vocational Education and Training (TVET) and Higher Education (Bachelor level, post-baccalaureate, master's level, and doctorate level education). The Technical Education and Skills Development Authority oversees TVET in the country while the Commission on Higher Education manages the state colleges and universities.

The education department receives the largest portion of the national budget by order. In 2020, the government allotted P551.7 billion of appropriations for the Department of Education. This comprises 82% share of the budget for the education sector. The budget was spent for basic education facilities such as school buildings and laboratories, school furniture, and electrification. Other priorities for funding are major programs for access to education like the Alternative Learning System and programs for quality such as computerization, human resource development, learning tools and equipment, hiring of new teachers, and purchase of instructional materials.

20.2.2   School, Students and Teachers' Profiles

In 2019–2020, about 27.7 million students enrolled from Kindergarten to Senior High School across public and private schools, state colleges and universities that offer basic education, Philippine Schools Overseas and Alternative Learning System. Table 20.1 shows the total number of enrolled students in the school year 2019–2020.

Recent data show that the Department of Education also prioritized the construction of additional school buildings. At the elementary level, the total number of schools (49,593) in the school year 2015–2016 is 14% larger than the school

Table 20.1 Key statistics on Philippine education

Indicator	Baseline	Latest	Data source agency
Proportion of children at the end of primary achieving at least a minimum proficiency level in reading and mathematics, both sexes	Reading 40.4 (2016) Mathematics 34.8 (2016)	Reading 40.4 (2016) Mathematics 34.8 (2016)	NAT, DepEd
Proportion of children at the end of lower secondary achieving at least a minimum proficiency level in reading and mathematics, both sexes	Reading 46.0 (2016) Mathematics 37.3 (2016)	Reading 46.0 (2016) Mathematics 37.3 (2016)	NAT, DepEd
Elementary completion rate	Elementary 93.1 (2016) Female 95.5 (2016) Male 90.8 (2016)	Elementary 97.2 (2018) Female 99.1(2018) Male 95.3 (2018)	EBEIS DepED
Secondary (Junior High School) Completion Rate	Secondary 80.9 (2016) Female 85.6 (2016) Male 80.9 (2016	Secondary 88.4 (2018) Female 93.0 (2018) Male 84.7 (2018)	EBEIS DepED
Cohort survival rate elementary	Elementary 93.8 (2016) Female 96.0 (2016) Male 91.8 (2016)	Elementary 97.4 (2018) Female 99.1(2018) Male 95.7 (2018)	EBEIS DepED
Cohort survival rate secondary (Junior High School)	Secondary 83.1 (2016) Female 87.4 (2016) Male 78.7 (2016)	Secondary 89.5 (2018) Female 93.6 (2018) Male 85.4 (2018)	EBEIS DepED
Dropout rate or school leavers rate elementary	Elementary 1.5 (2016) Female 1.0 (2016) Male 2.0 (2016)	Elementary 0.5 (2018) Female 0.1(2018) Male 0.8 (2018)	EBEIS DepED
Dropout rate or school leavers rate secondary (Junior High School)	Secondary 6.2 (2016) Female 4.4 (2016) Male 7.9 (2016)	Secondary 3.4 (2018) Female 2.2 (2018) Male 5.3 (2018)	EBEIS DepED
Net enrolment rate in elementary education	Elementary 96.2 (2016) Female 96.1 (2016) Male 96.2 (2016)	Elementary 94.1 (2018) Female 93.9 (2018) Male 94.3 (2018)	EBEIS DepED
Net enrolment rate in secondary education (Junior High School)	Junior High School 74.2 (2016) Female 79.9 (2016) Male 68.8 (2016)	Junior High School 81.4 (2018) Female 85.8 (2018) Male 77.2 (2018)	EBEIS DepED

(continued)

Table 20.1 (continued)

Indicator	Baseline	Latest	Data source agency
Net enrolment rate in secondary education (Senior High School)	Senior High School 37.4 (2016) Female 44.1 (2016) Male 31.0 (2016)	Senior High School 51.2 (2018) Female 58.7 (2018) Male 44.2 (2018)	EBEIS DepED

Source Philippine Statistics Authority SDG Watch as of March 2020 *Classified under Open Data with Creative Commons Attribution License (cc-by)*

year 2006–2007. At the secondary level, there are 13,574 schools throughout the archipelago.

The government also noted the need for hiring additional teachers as the number of schools in the country rises. As of 2020, over 900,000 regular personnel from the Department of Education where 800,000 of them are teachers.

Table 20.1 shows basic education data from Sustainable Development Goals (SDG) Watch as reported by the Philippine Statistics Authority in March 2020.

20.2.3 Educational Research and International Collaboration

One of the notable educational research institutions in the country is the Educational Research Program of the University of the Philippines Center for Integrative and Development Studies. This program aims to make educational research relevant to communities. One of the strands in their research agenda is on functional literacy and non-formal education. This area of inquiry is determined to establish community-based development programs through educational interventions for adults. The other strand was on indigenizing the curriculum to suit knowledge, culture, skills, values, and practices of indigenous or cultural groups. The center also specializes in family studies and education inquiries as research projects. The program aims to address the changes brought by the institutionalization of the K-12 program in the country, which directly impacts higher education and technical education.

Considering research on teacher quality as a priority for development, the Philippines and Australian government inked a partnership after bilateral discussions. This led the Philippine Normal University and the Science, ICT and Mathematics Education for Rural and Regional Australia (SiMERR) National Research Centre based at the University of New England (UNE) Australia to establish a joint project, the Philippine National Research Center for Teacher Quality (RCTQ) in October 2012. The RCTQ now supports PNU in its mandate to provide technical support to the Department of Education and Commission on Higher Education in their programs and projects that tackles teacher training, teacher education, continuing professional development of teachers, school heads and supervisors and teacher education curricula.

20.3 Current Situation of Science Education

20.3.1 Policies and Standards

In recent times, our world is now filled with inquiry products that require scientific literacy for an individual to make choices and decisions. Scientific literacy is what school science education must prioritize to help develop students that are engaged in scientific inquiry. Science will also allow them to develop the skills, values and attitudes such as habits of mind, curiosity, objectivity, honesty and critical thinking. These things will be useful in the development of students and their future goals or careers in life. Most importantly, the inquiry skills can make innovations that can help the community and the nation to prosper.

In addition, science learning can also be linked with the country's cultural development and its identity. The usefulness of Science can be observed when it is applied to solve the nation's challenges and problems. This will preserve the nation's cultural uniqueness and peculiarities. As a result, several countries have adopted this norm to associate culture with science teaching and learning.

In 2011, the Science Framework for Philippines Basic Education was released by the Department of Science and Technology (DOST) through the Science Education Institute (SEI). This framework is developed to serve as a resource for curriculum developers, faculty of teacher education institutions, teachers, policy makers and school administrators to design, implement and evaluate the science curricula. The framework has the following guiding principles: (1) science is for everyone; (2) science is both content and process; (3) school science should emphasize depth rather breadth, coherence rather than fragmentation, and use of evidence in constructing explanation; (4) school science should be relevant and useful; (5) school science should nurture interest in learning; (6) school science should demonstrate a commitment to the development of a culture of science; (7) school science should promote the strong link between science and technology, including indigenous technology; and (8) school science should recognize that science and technology reflect, influence, and shape our culture.

Moreover, the Department of Education issued Policy Guidelines on the K to 12 Basic Education Program. This explains the framework and components of the Philippines Basic Education. This policy also includes an explanation of its programs for Science Education. Special curricular programs were offered to elementary and secondary school students to provide varied learning experiences that will cater to their needs and interests. These are the Special Science Elementary School (SSES) and Science, Technology and Engineering (STE) programs. SSES is a nationwide initiative that aims to provide a learning environment through special science curriculum for primary learners. On the other hand, the STE offers learners with enriched science and technology oriented curriculum that will prepare them for higher education or work in areas of science, technology and engineering.

20.3.2 Curriculums and Digital Resources

The aims of science education in the Philippines are reflected in the K to 12 Curriculum Science 2013. In its conceptual framework, science education envisions to develop scientific literacy among Filipino learners. This will prepare them to be critical thinkers who can make wise decisions and judgment regarding the applications of scientific knowledge in different fields.

The K-12 science curriculum provides learners with competencies that are relevant in a knowledge-based society and the world of work. It is also a goal of the Philippine science curriculum to develop these skills among learners: (1) critical problem solver; (2) responsible stewards of nature; (3) innovator; (4) informed decision maker; and (5) effective communicator.

The current science curriculum presents learner-centered and inquiry-based lessons. The content is in spiral progression which means that concepts and skills in Biology, Chemistry, Physics and Earth Science are presented with increasing complexity in succeeding grade levels. This integration of science and other disciplines is important for learners to realize the real life applications of science concepts.

In addition, the students can easily learn science concepts if they are actively engaged in the lessons. It is now imperative for science teachers to utilize technology such as digital resources in teaching. The Department of Science and Technology has launched DOST Starbooks, the first digital science library in the country. DOST STARBOOKS also created the mobile version of this app in June 2020. The mobile app contains a collection of international scientific research, video lessons, and other resources anchored in the K-12 Science curriculum. Meanwhile, the Department of Education created the Learning Resources Management and Development System (LRMDS) which is intended for providing access to teaching and learning resources. DepEd also created an LR Portal where teaching and professional development resources can be accessed and downloaded. DEPED Commons was also created in 2020 to make education to continue amid pandemic. This serves as a direct solution to give access to digital resources and other open educational resources during class suspensions and related circumstances.

20.3.3 Student Assessment and Achievement

The learners' progress and achievement in Science can be gauged using several types of assessments. According to National Science Education Standards, the assessment process can be a useful tool to inform those concerned of science education outcomes on how students are learning, how teachers are doing and what resources are to be allocated. The National Achievement Test (NAT) is administered to pupils/students in the Philippines by the Department of Education through the National Education Testing and Research Center (NETRC) to assess learning outcomes and support quality education in the country. In the international assessments, the Philippines have participated in Trends in International Mathematics and Science (TIMSS) in 1999 and 2003 and for the first time, in the Programme for International Student Assessment (PISA) in 2018. The NAT results for the year 2010–2011 revealed that Grade 6 pupils got a mean percentage score (MPS) of 68.2 below the minimum competency level of 75% and requires remediation. On the other hand, high school students obtained an MPS of 44.3 that is far below the minimum competency level. In international tests like TIMSS 2003, the country ranked only 34th out of 38 countries in secondary Math, and 43rd out of the 46 participating countries in secondary Science. For elementary, fourth grade participants ranked 23rd out of the 25 countries in both math and science. The Philippines did not participate in TIMSS since 2007.

In addition, the 2018 PISA results also showed that the Philippines got a score of 357 in Science which is significantly lower than the OECD average of 489. This is the first time that the Philippines participated in PISA which opened doors for evaluating the deficiencies of the educational system. Fifteen year old students in the Philippines scored lower in reading, mathematics and science than those and most of the countries participated in PISA 2018. Key findings revealed that over 80% of students in the Philippines who participated in PISA failed to achieve the required level of reading competence. The expenditure per student in the Philippines is found lowest among PISA participating countries (OECD, 2019).

20.3.4   Science Technology Venues and Centers

One way to enhance students' knowledge and engagement in Science is through visiting centers or museums. In the Philippines, science museums can either be public or privately owned. The following table shows the list of science museums in the Philippines This also includes other science centers found in the country.

In support of enhancing the quality of science education in the Philippines, the Department of Science and Technology initiated several programs and activities to promote science and technology awareness to the students. In 2019, President Rodrigo Roa Duterte signed Proclamation 780 which declares the conduct of National Science and Technology Week every fourth week of November. The seven-day event features various awards for basic and applied research, scientific meetings, technical and investment forums, technology demonstrations, technical tours, and science exhibits. On the other hand, the Department of Education also conducts the National Science and Technology Fair (NSTF) each year. This activity aims to promote science and technology consciousness to secondary school students (Grade 7–Grade 12) by doing scientific researches that will benefit their community. The winners of the NSTF will be the country's representatives to Intel International Science and Engineering Fair (ISEF), the largest international research competition for pre-university students. The research competition will start at the school level advancing to division, regional, national then international level. The official participants in NSTF are the first place winners of regional science and technology fairs in each of the categories.

Table 20.2 List of science
and technology Museums in
the Philippines

Science Museum	Exhibits	Location
Philippine Science Centrum (PSC)	Basic science	Marikina City (Metro Manila)
Manila Ocean Park	Marine ecosystem	Manila
Philippine Air Force Museum (PAF)	Military science	Parañaque City
National Museum of the Philippines	Herbarium, zoology	Manila
Museum of Natural History	Botanical and zoological collections	Los Baños (Laguna)
IRRI Rice World	Rice-related exhibits	Los Baños (Laguna)
Coca-Cola Pavilion	Coca-Cola related exhibits	Sta Rosa (Laguna)
The Mind Museum	Life science and Physical science	Taguig City
Ninoy Aquino Parks and Wildlife Center	Zoology	Quezon City
Mayon Planetarium and Science Park	Volcanology and Geography	Albay
National Planetarium	Astronomy	Manila

These categories are life science, physical science, robotics and intelligent machines and science innovation expo. The National Scientific Review Committee (NSRC) must approve the research papers of the regional winners before being considered a qualified participant to the NSTF. The qualified papers will be submitted to the identified board of judges for evaluation. Only those shortlisted research papers will enter the actual national fair. Aside from the research competitions, the participants in the national level will undergo training at the NSTF Science Academy. This will give the qualifiers an opportunity to learn from expert speakers on innovation, creativity and excellence in science and research. The winners of the national fair will be the Philippine delegates to ISEF in the United States. The DOST Science Education Institute also spearheads contests like "Indie-Siyensya" for science documentary making and National Olympiads for Informatics, Mathematics and Sciences that select the Philippine representatives to International Olympiads. There are also initiatives conducted by SEI in reaching various individuals in promoting science and technology literacy. Project Science Teacher Academy for the Regions (STAR) is a capacity-building activity for teachers which promotes professional development in STEM all over the country. Science camps were also held for senior citizens to ensure the improvement of their well-being and their full participation in the society. The activities included lectures, demonstrations and hands-on activities on health care, arts and crafts on livelihood, effective response during emergencies and other topics that presented the connection of science to everyday living. The Science Explorer, the Philippines' first and only mobile learning facility visited provinces in the island of Luzon to bring informal science learning elementary and high school students. It even cross waters to reach other provinces in the Visayas. The mobile learning facility offers modules different fields of Science like Geology, Robotics, Biology, Physics and Chemistry. As of 2018, it had served a total 4,647 students and 136 schools. DOST SEI also gives scholarship opportunities to undergraduate students through R.A. 7687 also known as Science and Technology Scholarship Act of 1994 which provides scholarships to talented and deserving students whose families' socio-economic background do not exceed the set cut-off values of certain indicators. All qualified scholars shall pursue a degree in science and technology, engineering and mathematics in several institutions in the country accredited by the Commission on Higher Education. In 2020, there are a total of 9,788 passed the DOST-SEI Undergraduate Scholarship Examination which grants them slots as qualified scholars. In the private sector, the Unilab Foundation established STEM + PH, a flagship program that envisions a future-ready Philippines through producing innovators that can create solutions using the principles of integrated STEM. The organization initiated STEAM Innovation program, a collaboration with Center for Integrated STEM Education in the Philippines (CISTEM) which aims to establish a culture of innovation and collaboration through partnership with local community, sustained professional development for teachers and community expo which involves schools that offer STEM program.

20.3.5   Utilizing Emerging Technologies

The DOST in cooperation with private and government research and development organizations, industry and other concerned agencies formulate the Harmonized National Research and Development Agenda (HNRDA) 2017–2022 that aims to ensure that the results of studies related to science and technology will benefit the people, economically and socially. One of the priority research areas for the year 2017–2022 is in emerging technologies which comprise these four research areas: creative industries, space technology applications, human security and defense and artificial intelligence and data science.

In response to the educational requirements of the twenty-first century, science teachers are deemed to revolutionize their teaching methods for the Philippines to keep up with the global standards and realize the benefit of emerging technologies.

In this regard the Science Education Institute (SEI) allocated support to Space Science scholars to produce a workforce trained in doing research and development activities in technologies related to space science. Teacher scholarship grants were also given to deserving faculty to study MS and Ph.D. programs in the country and abroad. The courses taken by the scholars are not yet offered to any university in the country. Several teacher training programs were also conducted for Science teachers in engineering design process and assessment that will enhance learning in STEM. The SEI also entered into partnerships with the private sector to produce an application utilizing advanced technologies. DOST CMApp is a collection of mobile applications that includes DOST developed coursewares in Science. The Strategic Intervention Material for Teaching Augmented Reality (SIMaTAR) was also pilot tested in some schools in the Philippines. SIMaTAR is a compilation of science-based teaching and learning materials that utilizes augmented reality technology that presents the learners with 3D and 4D experience of the lesson content. Likewise, Project ARISE was also launched by SEI to foster STEM learning in schools. ARISE is described as a culmination of twenty-first century ecosystem of learning environments and ICT based resources to support the K-12 curriculum and STEM. Another initiative was inaugurated from ARISE which is the 21st Century Learning Environment Model (21st CLEM) that showcases a new generation classroom equipped with ICT resources and creative technologies. Several training programs were also created to capacitate the 21st CLEM teachers in robotics, ICT, and Media and Information Literacy.

20.4   Requirements for Future Development of Science Education

This report presented the current features of science education in the Philippines. According to DepEd Science curriculum guide, the aim of science education is to produce scientifically literate citizens that can contribute to the country's human capital. These individuals are needed to innovate solutions to pressing problems of the community. This can be achieved if the science education system will be successful in its delivery of its programs to the target clientele. The dismal results of the National Achievement Test and international assessments like TIMSS and PISA is a reflection of the state of education in the country. This predicament should not stop the Philippines to look for ways to improve its science education program. The government should use the results as basis in developing policies that will address the deficiencies in school buildings, quality instructional materials, science laboratories, modern facilities and teaching personnel.

As the Philippines gears towards the shift to Fourth Industrial Revolution (Industry 4.0), modern science education equipped with the emerging technologies is needed to produce competitive individuals having the twenty-first century skills. Science teaching must be revitalized to keep up with the emergence of digital technologies. The government must prioritize further development of science education through continuously supporting innovations in Science teaching and learning. Science education must prepare our learners with the skills and competencies that they will need to face the future. For this to happen, the schools must also transform. For this reason, the industry, academe and the scientific community must convene to improve the adaptability of schools and its students to this rapid change in our society. The road towards Industry 4.0 will not be easy, but even small steps can make a change to prepare the learners.

20.5 Discussion and Conclusion

Science education in the Philippines has undergone a series of changes since the beginning. The current science curriculum aims to produce scientifically literate citizens who are informed and active participants of the society, critical thinkers, and responsible decision makers. The government also allotted a big chunk of the national budget for the improvement of the education sector. This is to be utilized in purchasing tools and equipment, ICT infrastructure, hiring teachers and other personnel, and constructing additional school buildings. Despite these initiatives, the Philippines still performs low in national and international tests like TIMSS and PISA. The current results of the 2018 PISA was welcomed by the Department of Education and led to the launching of "Sulong EduKALIDAD" that leads the national effort to quality education. The Sulong Edukalidad will implement reforms in these areas: (1) K-12 Curriculum analysis and upgrading, (2) enhancement of learning facilities, (3) skills development of teachers and school leaders through a reformed professional development program, and (4) involvement of all stakeholders for collaboration and support. In this program the Department of Education calls the nation to actively participate in this cause to advance the quality of education in the country. Likewise, DOST-SEI the primary organization for the improvement of science education in the country, spearheaded numerous programs and projects to foster the culture of science to schools and learners. Scholarship programs, national and international competitions, research and development initiatives, development of resources based on emerging technologies and teacher training programs were accomplished to keep up science education in modern times.

Furthermore, the success of all these initiatives lies in the continuous support of the government to science education. The greater emphasis being applied in the field of science and technology would mean that more opportunities for innovations. A nation that invests more in Science and Technology Education would likely produce more experts who can help solve its pressing problems at present and in the future.

Acknowledgements I want to thank our God Almighty for giving me wisdom and perseverance to write this report. I also want to extend my gratitude to Prof. Lee Shok Mee of Penang, Malaysia for this opportunity. To my wife Cara, thank you for the love and support that you have given me in the writing process. To my friend, Mr. Romnick P. Nicolas, thank you for proofreading this paper.

References

Albert, J. R., & Raymundo, M. J. (2016). Trends in out-of-school children and other basic education statistics. Retrieved from https://dirp3.pids.gov.ph/websitecms/CDN/PUBLICATIONS/pidsdps1639.pdf.

Bautista, D. (2011). Promoting science culture through science Museums. Retrieved from https://www2.gsid.nagoya-u.ac.jp/blog/anda/files/2011/08/58-dexter-bautista.pdf.

Cigaral, I. (2018). Philippines climbs in technological readiness ranking. Retrieved from https://www.philstar.com/headlines/2018/06/06/1822109/philippines-climbs-technological-readiness-ranking.

Commission on Population and Development. (2020). Philippine population clock. Retrieved from https://popcom.gov.ph/.

Department of Education. (2019a). National science and technology fair for school year 2019–2020.Retrieved from https://www.deped.gov.ph/wp-content/uploads/2019/09/DM_s2019_113.pdf.

Department of Education. (2019b). Policy guidelines on the K-12 basic education program. Retrieved from https://www.deped.gov.ph/wp-content/uploads/2019/08/DO_s2019_021.pdf.

Department of Science and Technology. (2017). Harmonized National Research and Development Agenda 2017–2022. Retrieved from https://www.dost.gov.ph/phocadownload/Downloads/Journals/HNRDA_booklet_FINAL3_2018-10-23.pdf.

DOST Science Education Institute. (2011). Science framework for Philippine basic education. Retrieved from http://www.sei.dost.gov.ph/images/downloads/publ/sei_scibasic.pdf.

Gatpolintan, L. (2018). Technology can grow PH economy 10% faster: PIDS. Retrived from https://www.pna.gov.ph/articles/1046790.

Hailaya, W. (2014). Teacher assessment literacy and student outcomes in the province of Tawi-Tawi, Philippines. Retrieved from https://digital.library.adelaide.edu.au/dspace/bitstream/2440/99098/2/02whole.pdf.

Llego, M. (2020). DepEd basic education statistics for school year 2019–2020. Retrieved from https://www.teacherph.com/deped-basic-education-statistics-school-year-2019-2020/.

Magno, C. (2010). A brief history of educational assessment in the Philippines. Retrieved from https://files.eric.ed.gov/fulltext/ED511798.pdf.

National Economic and Development Authority. (2019). Understanding PH culture is key to create positive changes. Retrieved from https://www.neda.gov.ph/understanding-ph-culture-is-key-to-create-positive-changes-neda/.

Organisation for Economic Cooperation and Development. (n.d.). Programme for international student assessment results from PISA 2018. Retrieved from https://www.oecd.org/pisa/publications/PISA2018_CN_PHL.pdf.

Oxford Business Group. (2017a). Education reform in the Philippines aims for better quality and more access. Retrieved from https://oxfordbusinessgroup.com/overview/thorough-examination-substantial-reform-has-brought-it-variety-challenges#:~:text=K%2D12%20extends%20compulsory%20schooling,system%20of%20just%2010%20years.

Oxford Business Group. (2017b). The report the Philippines country profile. Retrieved from https://oxfordbusinessgroup.com/node/921607/reader.

Pacquing, G. (2018). Goals of science education in the Philippines. Retrieved from http://www.depedbataan.com/resources/4/goals_of_science_education_in_the_philippines.pdf.

Philippine National Research Center for Teacher Quality. (n.d.). History of RCTQ. Retrieved from http://www.rctq.ph/?page_id=42.

Philippine Statistics Authority. (2019). Population projection statistics. Retrieved from https://psa.gov.ph/statistics/census/projected-population.

Philippine Statistics Authority. (2020). Sustainable Development goals watch as of March 2020. Retrieved from https://psa.gov.ph/sites/default/files/phdsd/PH_SDGWatch_Goal04.pdf.

San Buenaventura, P. (2019). Education Equality in the Philippines. Retrieved from https://unstats.un.org/sdgs/files/meetings/sdg-inter-workshop-jan-2019/Session%2011.b.3_Philippines___Education%20Equality%20AssessmentFINAL4.pdf.

Science Education Institute (2020). Programs and projects. Retrieved from http://www.sei.dost.gov.ph/index.php/programs-and-projects?limitstart=0.

Science Education Institute. (2018). Annual report 2018. Retrieved from http://www.sei.dost.gov.ph/images/ts/seiAR2018.pdf.

The Free Library. (2014). Gokongwei brothers foundation sponsors DepEd science fair winners to US intel science fair. Retrieved from https://www.thefreelibrary.com/Gokongwei+Brothers+Fou ndation+sponsors+DepEd+Science+Fair+winners+to...-a0583979185.

The National Academies Press. (n.d.). Assessment in science education. Retrieved from https:// www.nap.edu/read/4962/chapter/7.

The Philippines. (2020). About the Philippines. Retrieved from https://www.gov.ph/about-the-phi lippines.

The World Bank. (2020). The World Bank in the Philippines. Retrieved from https://www.worldb ank.org/en/country/philippines/overview#1.

UP Center for Integrative and Development Studies. (2018). Education research program. Retrieved from https://cids.up.edu.ph/programs/education-research-program/.

Robert John D. De La Cruz The author works as a Senior High School teacher in the Philippines. He handles subjects such as Biology and Research. He was sent at SEAMEO RECSAM Penang, Malaysia as a scholar on April 1 to 26, 2019 for a Regular Course. At present, he is taking up Doctor of Philosophy in Biology Education.

Part IV
European Countries

Chapter 21
Science Education in Bulgaria

Malinka Ivanova

Abstract The chapter discusses the current state of science education in Bulgaria, taking into account some factors like: geographical location, population, cultural characteristics, technological development, economics, educational system, and international collaboration. The performed exploration shows that the science education in Bulgaria, following the European educational policy, national priorities and the world best practices, is characterized with great progress and achievements, challenging issues and well defined future directions, documented in strategies, policies and standards.

Keywords Science education · Policies and standards · Teachers training · ICT adoption

21.1 Overview of the Country

21.1.1 Geographical Location, Population and Political System

Bulgaria is located on the eastern Balkan Peninsula and is placed on territory of 110 994 square kilometers. Bordering neighbors are the countries: Greece and Turkey to the south, North Macedonia and Serbia to the west and Romania to the north, as the east border is Black Sea. The border with Romania is realized through the river Danube. Bulgaria is the sixteenth-largest country in Europe, occupying its southeast part.

The Bulgarian population consists of 6 951 482 persons that is 1.4% of the European population according to statistics for 2019 year (National Statistical Institute, 2019a). The number of female persons in Bulgaria is 51.5% and the male population

M. Ivanova (✉)
Faculty of Applied Mathematics and Informatics, Department of Informatics, Technical University of Sofia, Kl. Ohridski blvd. 8, 1000 Sofia, Bulgaria
e-mail: m_ivanova@tu-sofia.bg

© The Author(s), under exclusive license to Springer Nature Singapore Pte Ltd. 2022   349
R. Huang et al. (eds.), *Science Education in Countries Along the Belt & Road*,
Lecture Notes in Educational Technology,
https://doi.org/10.1007/978-981-16-6955-2_21

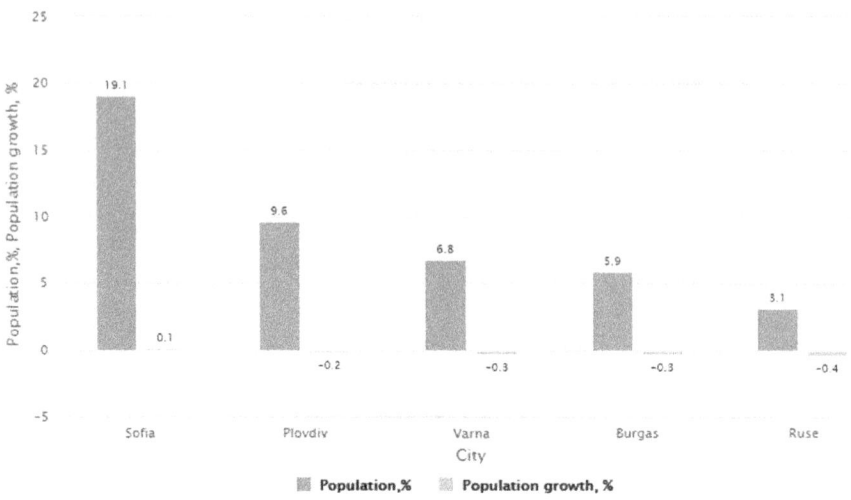

Fig. 21.1 Population and population growth in the largest five cities in Bulgaria

is 48.5, 21.6% are persons aged 65 and over, 14.4% are children up to 15 years, 59.8% are persons at their working ages. The most populated urban center is the capital Sofia where 19.1% of Bulgarian citizens live. The next largest cities are Plovdiv, Varna, Burgas and Ruse (Fig. 21.1). The population growth in almost all Bulgarian cities is negative and it ranges from −0.1% (Shumen) to −2.4% (Vidin). The positive population growth for 2019 year is calculated only for cities Sofia (0.1%) and Kurdzhali (3.5%). The gender gap index for Bulgaria in 2020 year is 0.727 and it presents the disadvantages of women in comparison to men taking into account their participation in four important areas: education, health, economy and politics (World Economic Forum, 2020). According to this criterion, Bulgaria occupies 49th position in a rank list of 153 countries.

Bulgaria is a parliamentary democratic republic based on the principle of powers separation. The executive power is performed by the government with the head of a Prime minister. The legislative power is exercised by National Assembly and the judiciary power is responsible for interpretation and application of law at solving legal cases. The head of the state of Bulgaria and the commander-in-chief of the Bulgarian Army is the president of the Republic of Bulgaria.

21.1.2   Current Situation of Economic, Technologies and Cultural Development

The Republic of Bulgaria is a member of European Union since 2007 and possesses industrialized economy based on the principle of free market, where the private sector

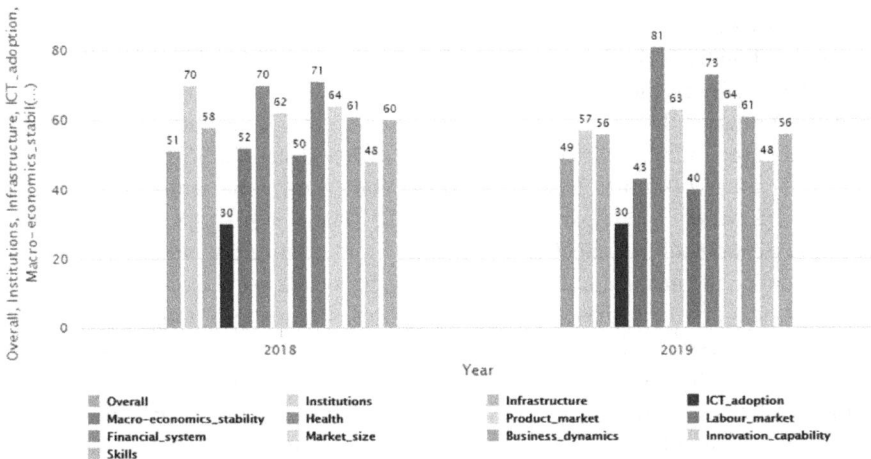

Fig. 21.2 Bulgarian competitiveness performance

is medium developed. The country is characterized with currency board regime and the national currency is bind to the euro.

According to the innovation strategy for smart specialization, Bulgaria has to resolve serious economic challenges through implementation of an appropriate economic policy and should increase the labor productivity, decrease the dependency of industrial production from energy, should increase the technology transfer by foreign direct investment, should increase the internationalization of national enterprises and should improve the Bulgarian export that currently includes low-tech products (Ministry of Economy, 2015).

The Global Competitiveness Report for 2019 year of the World Economic Forum reveals that Bulgaria takes the 49th place (among the 141 economies) considering its economic profile (Fig. 21.2).

It means that the country competitiveness is improved in comparison to the 51th place from the previous report from 2018 year (World Economic Forum, 2018, 2019). The main indicators in this analysis are: (1) Institutions (57th)—budget transparency, judicial independence, legal framework, press freedom, etc.; Infrastructure (56th)—transport, and utility; (3) ICT adoption (30th)—Internet and mobile technologies; (4) Macroeconomics stability (43rd)—inflation, debt dynamics; (5) Health (81st)—healthy life expectancy; (6) Skills (56th)—years of schooling, skills at current and future workplace; (7) Product market—(63rd)—domestic competition and trade openness; (8) Labor market (40th)—flexibility and meritocracy; (9) Financial system (73rd)—depth and stability; (10) Market size (64th)—gross domestic product and imports of goods and services; (11) Business dynamics (61st)—administrative requirements and entrepreneurial culture; (12) Innovation capability (48th)—interaction and diversity, research and development and commercialization.

It can be seen that the indicators Institutions, Infrastructure, Macro-economic stability, Skills and Labor market are improved, while Health, Product market and

Financial system are worsen when a comparison between 2018 and 2019 year is done. Regarding the ICT adoption, Bulgaria takes the 30th place that is excellent and it is related to bigger number of users (76th position) with access to high-speed and broad-band Internet, bigger number of subscribers to mobile-cellular services (69th place) and bigger number of users to mobile broadband services (28th place).

Overall values of export and import from and to Bulgaria are decreased with around 1.1% when a comparison between 2018 and 2019 year is performed. Bulgarian export includes electrical machinery and equipment, mineral fuels and oils, machinery and mechanical appliances, copper, cereals, vehicles and accessories, pharmaceutical products, plastics and articles, ores, slag and ash, articles of apparel and clothing accessories and mainly is directed to Germany, Romania, Italy, Turkey, Greece, France, Belgium, China, Spain and Netherlands (Trend Economy, 2020). Import is related to petroleum oil and gas, medicaments, copper ores, motor cars, transmission apparatus, chemical products, tractors and mainly come from Germany, Russia, Italy, Romania, Turkey, Greece, Chine, Spain, Netherlands, and Hungary.

Bulgaria possesses rich culture treasure that is transferred from generation to generation, saving national traditions, and habits. The UNESCO World Heritage Sites in Bulgaria are ten: seven of them are cultural heritage sites: ancient city of Nessebar, Boyana Church, Madara Rider, Rila Monastery, Rock-Hewn Churches of Ivanovo, Thracian Tomb of Kazanlak, Thracian Tomb of Sveshtari and three are natural heritage sites: Srebarna Nature Reserve, Pirin National Park and Ancient and Primeval Beech Forests of the Carpathians and Other Regions of Europe (UNESCO, 2020).

Among the 7,360,000 artefacts in the Bulgarian museums and galleries the world-famous is Panagyurishte gold treasure.

The Bulgarian theatrical and cinema production and organized national and international festivals attract people from different age groups every year.

The Bulgarian folk music is extremely popular all over the world, combining a wide range of sounds typical for national instruments like: gaida, kaval, gadulka and tupan that is bases for creation of synchronized traditional dances and songs.

Bulgarian cuisine is characterized with some specific and delicious foods which are prepared for special occasions or consumed during the whole year.

The Ministry of Culture follows the policy for development and protection of Bulgarian culture, collaborating with European and international institutions to facilitate authors, different arts and cultural events (Bulgarian Ministry of Cultural, 2009).

21.2   Overview of the Education Development

21.2.1   Education System and Policy

Education in Bulgaria is national priority and is based on an integrated state educational policy for equal right and access to high-quality education with a wide range of opportunities for lifelong competences development.

Educational system in Bulgaria covers all degrees and levels for education and vocational training and they can be described through the following stages (Fig. 21.3):

- *Early childhood education and care*, comprising pre-kindergarten for children with ages between 0 and 3 and kindergarten for 4–6 aged children. There is a possibility the children from age of 2 also to enter in the kindergarten where the education is performed according to educational standards for pre-school and school. A big part of the kindergartens are public and small ones are private. The type of the kindergarten can be: all-day, half-day, weekly, seasonal, remedial, special kindergartens for children with special educational needs and/or chronic

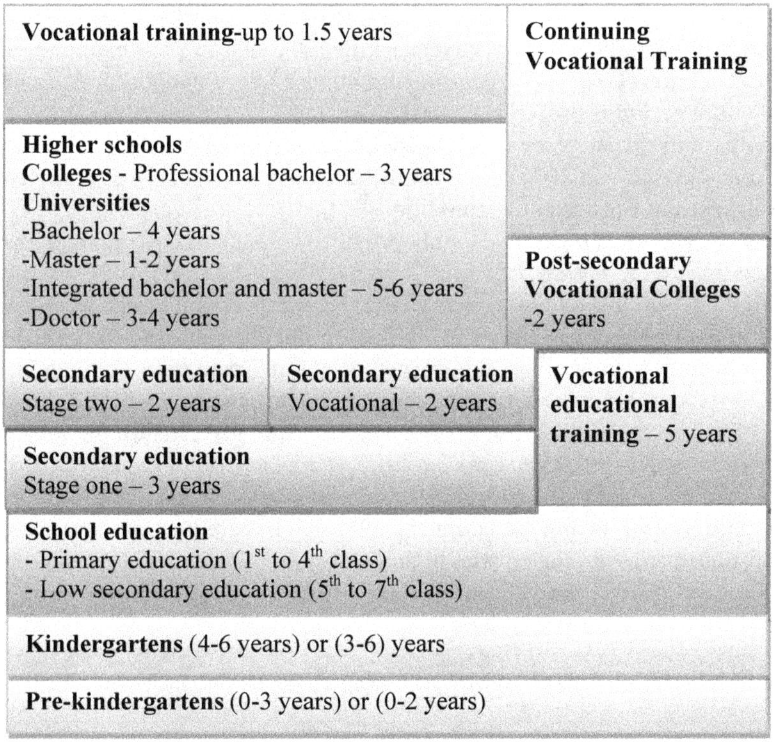

Vocational training-up to 1.5 years	**Continuing Vocational Training**
Higher schools **Colleges** - Professional bachelor – 3 years **Universities** -Bachelor – 4 years -Master – 1-2 years -Integrated bachelor and master – 5-6 years -Doctor – 3-4 years	**Post-secondary Vocational Colleges** -2 years
Secondary education Stage two – 2 years / **Secondary education** Vocational – 2 years	**Vocational educational training** – 5 years
Secondary education Stage one – 3 years	
School education - Primary education (1st to 4th class) - Low secondary education (5th to 7th class)	
Kindergartens (4-6 years) or (3-6) years	
Pre-kindergartens (0-3 years) or (0-2 years)	

Fig. 21.3 Bulgarian educational system

diseases. Education in kindergartens is conducted taking into account the pre-school educational standards, principles and objectives that are described in the Pre-school and School Education Act (Ministry of Education & Science, 2016).

- *School education* is mandatory for children from 7 ages, but 6 aged children also can go to school after the parents' permission. School education is provided at three stages: primary education with pupils from 1st to 4th class, low secondary education with pupils in 5th–7th class and secondary education—from 8 to 12th class. The following types of school education are available: primary (1st–4th grade), basic (1st–7th grade), integrated (1st–10th grade), gymnasiums (8th–12th grade), secondary (1st–12th grade), vocational schools that provide training for obtaining third and second level of professional qualification (art schools, sport schools, vocational gymnasiums, religious schools).
- *Education in vocational colleges* that provided education with students after their successful accomplishment of secondary education.
- *Higher schools* that includes universities, colleges and special higher schools and educate students after their secondary education in professional bachelor, bachelor, master and doctorate degrees. The institutions from the system of Higher education have the rights of an academic autonomy that is provided according to the national law and strategies developed by the Ministry of Education and Science (2020a). Further modernization and internationalization of Higher education in Bulgaria is formulated as priority and a detailed action plan explained in the *Strategy for development of Higher education in the Republic of Bulgaria for 2014–2020 period* is proposed (2014a). The objectives, principles, activities and measures of the Strategy are synchronized with the project "Science and Education for Smart Growth—2020" and with Partnership Agreement between the Republic of Bulgaria and European Commission.
- *Adult education* in Bulgaria is mainly provided by Licensed Vocational Training Centers in a wide variety of forms—from class-based training to self-directed learning. According to report of CEDEFOP, the first Bulgarian presidency of the Council of European Union in 2018 is outlined the priority for ensuring the adequate sets of skills to prepare learners for their future work through involvement in vocational education and training (VET) programmes (CEDEFOP, 2018). The VET priority is with compliance with the initiative *New skills agenda for Europe, European framework for quality and effective apprenticeships, Erasmus* + programme for mobility of VET learners and professionals. VET in Bulgaria is an opportunity for facing the national challenges like: reduction the early school leaving, ensuring a provision with high quality and relevance VET, and providing support to teachers and trainers.

21.2.2 Statistics on the National Education

In Bulgaria, during the 2019/2020 school year in kindergarten are enrolled 217.9 thousand children that is 0.4% less than the previous school year (National Statistical

Institute, 2019b). The average number of children entered in one kindergarten is 118, as this number is significantly greater for the cities—153 in comparison to this number for villages where it is 61. One kindergarten group is formed on average by 23 children, respectively 25 in the cities and 19 in the villages. The pedagogical staff employed in the kindergartens is 20.1 thousand and in comparison with the previous school year it is increased by 251 teachers, or 1.3%. During the 2019/2020 school year, there are 106 licensed private kindergartens, or 5 more than the previous year. They are attended by 4 982 children, or 2.3% of enrolled ones in the kindergartens.

In 2019, the enrolled school students are 572.5 thousand in 1 963 schools as 133 of them are primary schools, 1 162 are low secondary schools, 70 are integrated schools, 114 are gymnasiums, and 484 are secondary schools. The average number of students in one class is 21. The private schools are 86 (1.9%) with enrolled 10 907 students.

For the 2018/2019 school year, the number of the drop-out students is 15.8 thousand, as the: 7 thousand (44.4%) are students leaved the country, 6.4 thousand students (40.7%) are lived the school, because of the family reasons, and 1.6 thousand students (10.2%) do not wish to continue their education.

Teachers (including principals and deputy principals) in the general education schools in the 2019/2020 school year are 53.1 thousand. 94.8% of them are with completed educational qualification degree bachelor or master, 4.6% of them possess professional bachelor degree and 0.7% of them have secondary education. The teaching profession is dominated by women, who make up 85.6% of the total number of teachers.

The two-year professional education after the secondary education is provided by 21 art schools, 25 sport schools, 358 professional gymnasiums, and 23 professional colleges as the involved students are 135.1 thousand. The private professional gymnasiums and professional colleges are 31 with entered 1040 students. The teachers in professional education are 12.3 thousand as 95% of them possess bachelor or master degree.

During the academic 2019/2020 year, 226.6 thousand students are entered in colleges and universities in professional bachelor (8 233 students), bachelor and master degree programmes. The doctorate students are 6 440 as their number is decreased in comparison to the previous year. The number of educators in colleges is 676, in universities and special higher schools is 20.2 thousand.

21.2.3 *Educational Research and International Collaboration*

The biggest research center with 151 years of successful functioning is Bulgarian Academy of Sciences. It provides high quality research and education that contributes to further development of our science and society in the scope of national priorities and according to European Union directions. Bulgarian Academy of Sciences is structured in three main branches: Natural, mathematical and engineering sciences; Biological, medical and agrarian sciences and Social sciences, humanities and art and each of them include scientific institutes and laboratories. The Academy collaborates with Academies of sciences, Universities, well-established research centers and laboratories from courtiers in European Union and from countries all over the world to promote excellence of science as a base for achieving smart and sustainable economy.

The responsible body for monitoring, evaluation and accreditation the teaching and research quality in universities, colleges and research centers is *National Evaluation and Accreditation Agency*. Its mission is to encourage all institution in the area of Higher education to follow high academic standards and excellence in research. In Bulgaria 48 educational and research institutions are accredited by National Evaluation and Accreditation Agency as in the ranking list the first five are: Sofia University St Kliment Ohridski, New Bulgarian University, Technical University of Sofia, Angel Kanchev University of Ruse and Medical University–Varna.

Sofia University St. Kliment Ohridski is the first Bulgarian higher academic institution (founded in 1888 year) with well-established academic, scientific and cultural standards, recognized at national, European and international scene. It is the biggest educational and research organization in Bulgaria with 16 faculties and 119 teaching special subjects. The students could choose their education among 96 bachelor degree, 308 master degree, 183 doctorate degree programmes. The postgraduate qualification is organized for students who wish to specialize in their current topic or specialize new one without obtaining a degree. Lifelong learning is managed by the following educational structures: Center for Educational Services, Department of Language Learning, Department for Information and In-Service Training of Teachers, Centre for Slavo-Byzantine Studies "Prof. Ivan Dujcev", CEDES-Germanikum Centre and Networking Cisco Academy. Research fields in Sofia University St. Kliment Ohridski are adhered to the national and European priorities as well as to the university traditions. The research strategy is focused on projects development in collaboration with national, European and international scientific research organizations.

New Bulgarian University is the youngest university in Bulgaria, founded in 1991 year and since 2004 it is an accredited partner of the Open University, UK. Its mission is related to delivering liberal art education through proposing affordable opportunities for high quality interdisciplinary and specialized education and research. The main aim is the students to be prepared for the challenges of our modern society, training them in critical thinking, creativity and problem solving in the subject areas like: human sciences and humanities, natural sciences and modern technologies and art. New Bulgarian University is organized in four schools: School of general studies, School of undergraduate studies, School of graduate studies, School of distance, online and continuing education. The students could be involved in 22 bachelor degree, 91 master and 39 doctorate degree programmes. The scientific research in the University is facilitated through bilateral national and international collaborations, development of joint research projects participation in European and world reputable organizations, mobility of faculty members and students.

The Technical University of Sofia is the largest and the first university in Bulgaria with specialization in technical sciences (established in 1945 year). Its mission is to contribute to the socio-economic and cultural prosperity of society through preparing high qualified engineers and performing valuable research. Its structure is organized in 18 Faculties, 3 Departments, 3 Colleges and 2 Schools proposing to students' education in 7 professional bachelor degree programmes, 68 bachelor, 83 master and 93 doctorate degree programmes. The Technical University of Sofia was the first one which established foreign language teaching Faculties, offering double degrees with Universities from Germany, UK and France. The University has huge experience in research partnerships and relationships with European and world recognized scientific institutions, training organizations and industrial companies. The results of research and educational projects lead to development of unique innovative solutions not only at national level, but also covering the standards of Central and Easter Europe, to enhancement of the research facilities and practical laboratories. Many innovative products and patents are presented at the most famous international exhibitions all over the world.

"Angel Kanchev" University of Ruse is established in 1945 year in the form of higher technical school. Nowadays, its strategic priorities are related to excellent students' preparation for the contemporary labor market, creation of internal university system for control of quality in education, continuous staff development for competences enhancement in teaching and research, realization of activities concerning European integration and international collaboration, development of modern university facilities and flexible university system. The University is organized in eight Faculties, two branches and a Bulgarian-Romanian inter-university Europa Centre. The students are admitted in 48 bachelor, about 80 master and 43 doctorate degree programmes. The research teams are participated of many multilateral and bi-lateral scientific and educational projects as the partnership network is built among 40 countries from Europe, America and Asia.

Medical University–Varna is a modern and high-tech oriented university, implementing a wide variety of teaching and training methods, including explorative and practical simulations. It is structured in four Faculties of Medicine, Dental medicine, Pharmacy and Public health, one Medical college, one Department for linguistic education, and four affiliates and the enrolled students come from approximately 50 countries. There is 1 professional bachelor degree program, 2 bachelor degrees programs, and 2 master degree programs. Also, 3 master programs are offered in English. The Research Institute that was founded in 2016 is directed to perform foundational, clinical and translational research. It consists of 6 departments that conduct scientific explorations according to the national and University priorities.

21.3   Current Situation of Science Education

21.3.1   Policies and Standards

Many national and European documents draw the current state of science education in Bulgaria and describe directions for future development.

The National Roadmap for Research Infrastructure 2017–2023 is a strategic document that proposes key instruments for implementation a modern research infrastructure that significantly could increase the scientific potential of Bulgarian scientists and students (Ministry of Education and Science, 2017). The diagnostic review concerning the research infrastructures in Bulgaria reveals the following findings: the available research facilities with potential for pan-European projects are concentrated in the capital; the highest percentage of modernized laboratories are in the areas of physics, material science and engineering; with the highest regional significance are the infrastructures in medical and agro-bio sciences; the infrastructures in social sciences and humanities are gained the smallest funding resources. In Bulgaria are identified 161 research infrastructures—7% with European significance, 52% with national importance, 40% with regional impact. The implementation of this Strategy is planned to be realized in three stages: synchronization (2017–2018), building (2019–2020) and development (2021–2023) as the balance between national and European sources is expected.

The National Strategy for development of scientific research in Republic Bulgaria 2017–2030 (2017b) gives an analysis of the current state of the scientific and research activities in Bulgaria, outlining the main challenges and bottlenecks. It is seen than the volume of the scientific production, the number of citations, the number of collaborative publications—joint papers with foreign scientists, participation in European projects, are characterized with decreased indexes. This could lead to serious risk for slowing the economic and social processes in Bulgaria. The strategic vision is outlined as in the center are placed the development of human resources; the infrastructures modernization; the balance in scientific research—fundamental and applied, balance in regions and scientific directions; better integration in European research and international scientific society; better connections between the science and other social spheres. For implementation of this strategy several policies, activities and measures are proposed—policy for adequate and effective financing, policy for legal regulatory changes, policy for reformation the science and research system, policy for development of human resources, policy for encouraging the fundamental and advanced scientific achievements, policy for integration with European research and development the international collaboration, other.

Other strategic documents that describe the current landscape of science education in Bulgaria are: Strategy for educational integration of children and students from minorities, Strategy for effective implementation the information and communication technologies in education and science in Republic Bulgaria, National strategy for encouraging and increasing the literacy, Strategy for development of professional

Table 21.1 Science education disciplines in school and secondary education

1st and 2nd class	World around us
3rd and 4th class	Man and nature
5th and 6th class	Man and nature, geography and economics, history and civilization
7th, 8th, 9th, 10th, 11th and 12th class	Physics and astronomy, biology and health education, chemistry and environmental protection, history and civilization, geography and economics

education and learning in republic Bulgaria, National strategy for development of pedagogical staff, Strategy for reducing the share of early school leavers, others.

The educational standards for pre-school and school education are described in the national law that regulates all aspects: from educational curriculums and educational results assessment to institutions and their quality control (National Law for pre-school and school education, 2017). Science education in these early childhood years is focused on the topics in the area of Environment as the aims are related to development of social and cognitive skills through world exploration, mainly in the form of games.

All curriculums as well as these in science education follow the national educational standards and they are published on the web site of Ministry of Education and Science (2020b). The teaching disciplines in school and secondary education are summarized in Table 21.1. In classes from 1st to 4th the science is presented as one integrated subject as the title of the disciplines in the 1st and 2nd class and 3rd and 4th class are respectively World around us and Man and nature. In 5th and 6th class in addition to the discipline Man and nature, two other disciplines are studied by students: Geography and economics and History and civilization. During the 7th–12th class the students receive more advanced knowledge in: Physics and astronomy, Biology and health education, Chemistry and environmental protection, History and civilization, Geography and economics.

Science education subjects are also included in the curriculums of some Higher education institutions. More commonly the studied disciplines are Physics and Chemistry.

21.3.2 Curriculums, Digital Resources and Teacher Training

The curriculum reform in Bulgaria was successfully performed at all educational levels in the period after 1999 (Psifidou, 2008) and currently the syllabuses are continuously updated to include new knowledge and skills regarding the state of ICT, economics and modern society requirements; educational content is actualized with emerging concepts and critical issues; the objectives are pre-defined taking into account the regional, national, European and international dimensions; the courses are re-organized in order to be more flexible and interdisciplinary.

Contemporary curriculums, including these in the area of science education for pre-school and school education are approved by Ministry of Education and Science and each science programme concerns: science content standards, expected learning outcomes, topics, activities and interdisciplinary links, teaching methods and guidelines, assessment forms and methods. The science education is more general from 1st to 4th class, revealing to students physical, chemical and biological facts and phenomena. The students from 5 to 6th class receive more specific knowledge in the fields of: man and nature, geography and economics, history and civilization. The science education from 7 to 12th class presents and explores problems in physics and astronomy, biology and health education, chemistry and environmental protection, history and civilization, geography and economics. The science education for children in kindergarten is related to learning about environment around them through explorative methods and games. Higher education also concerns sciences as in the curriculums of some specialty are included mainly physics and chemistry subjects.

An exploration regarding the usage of digital technologies in Bulgarian schools reveals that only 40% from our schools utilize learning management systems that are considered as a factor for facilitation the access to digital resources (Терзиева et al., 2016). More often, the teachers rely on existing digital educational resources (82%) as 73% of them adapt resources to the specific teaching needs. The teachers use Internet sources, extend and adapt them to prepare presentations. More experienced teachers create educational games. The most utilized sources for preparation the digital educational resources are: existing digital books of Bulgarian publishers, Bulgarian and foreign educational web sites, groups of interests in social media, educational games, shared educational resources in VBOX7, YouTube, SlideShare, special multimedia educational software, authoring software for creating digital educational resources and teachers' portals and blogs. The digital resources are used by teachers to teach new knowledge, to deliver practical exercises, to perform summary, in assessment, in home works, in activities performed out classes, in self-learning (Паунова et al., 2018).

The teachers training in Bulgaria like in the most European countries is performed according to a model, combining theoretical and practical professional training with basic training on a set of courses (Ministry of Education and Science, 2014). The professional training comprises around 25–30% of the educational curriculum. In Bulgaria, the universities train future teachers in bachelor degree programmes with possibilities for further specialization in master and doctor degree programmes. There is not a general framework or standard for needed teachers' knowledge, skills and competences, but Bulgaria adopts obligatory continuous professional development. The strategic aim is to be built a unified and effective educational system for training and continuous professional qualification of teachers through national regulation and following the European educational policy.

21.3.3 Student Assessment and Achievement

In the year of 2018, 600,000 Bulgarian students are assessed by PISA.

The assessment measures the acquired knowledge and skills by 15-years-old students in mathematics, science and proficiency in reading. The global competence in innovative domain is assessed too and the students' well-being is also explored. The assessment procedure includes computer-based tests which are performed for two hours. The test consists of multiple-choice questions and questions with free answer. Reading assessment is adaptive and multi-staged as the assigned block with items is consistent with the previous students' achievements. The students also share information about their attitudes, beliefs, their school and home activities, about their learning performance.

The students present lower scores in reading, mathematics and science in comparison to the OECD average scores (PISA, 2018). Level 2 or higher in reading is achieved by 53% of the students (77% is OECD average), 56% of the students in mathematics (76% OECF average) and 53% of them in science (78% is OECD average). Only 2% of students are characterized with top performance (Level 5 or 6) in reading (9% is OECD average), 4% of students in mathematics (11% is OECD average) and 2% of them in science (7% is OECD average).

The results show that the socio-economic conditions influence on the students learning performance. In Bulgaria, the socio-economic advantaged students have better outcomes in reading (106 score points) in comparison to disadvantaged students which number is bigger than OECD average (89 score points). In all countries, girls are significantly better in reading that boys. In Bulgaria, the difference in reading between girls and boys is 40 score points (OECD average 30 score points). In mathematics, girls and boys have similar results. In science, girls outperformed boys—the difference is 15 score points. In Bulgaria, 65% of the students are satisfied in their lives (67% is OECD average); 87% of the students are sometimes or always happy and 84% of the students agreed or strongly agreed that they usually solve difficult situation (84% is the OECD average).

21.3.4 Outreach Science Education

Bulgarian students with expressed interests in science and technology have opportunity to participate in national, European and international contests and research projects. Bulgaria was a host of the *EU Contest for Yong Scientists* (EUCYS) in 2019, where students from more than 40 countries presented their scientific ideas. Another famous contest in science is *FameLab* that every year brings many young people and their projects at one place. The contest *Science on Stage Bulgaria* is organized every two years with aim to drive and stimulate the students' interests in STEM topics. The students together with their teachers present innovative projects and create connections with other scientific teams. Also, Bulgaria is a leader in many

Olympiads as the students won 694 medals, taking part in international Olympiads in Astronomy, Biology, IT, Mathematics, Physics, Chemistry, Mathematical Linguistics. Bulgaria has a well-developed model and rules for contests organization in science and technology every year at national level, where many students from schools and universities are involved in several rounds, showing their deep knowledge and future ambitions. The main organizers of the annually conducted contests and Olympiads in Bulgaria are Ministry of education and science, Union of Bulgarian Mathematicians, Bulgarian Academy of Sciences.

Also, science education in Bulgaria is promoted among students from 1st to 12th class through organization of outclass activities and informal training in different sciences: chemistry, physics and astronomy, history. Recently, innovative teaching methods are applied in support of learning different science topics and in this way to facilitate formal education and to extend the obtained by students' knowledge and skills. Every student with interests in different science area have possibilities to participate in a wide variety of outclass educational activities in his/her school, in private educational centers, in university centers or other training organizations. Popular form for informal training is specialized summer schools, which from several days to 1–2 weeks introduce school and university students, teachers, researchers to well-accepted science theories and concepts, novel science achievements and challenging problems.

21.3.5 Utilizing Emerging Technologies

The emerging technologies like virtual reality (VR) and augmented reality (AR) are well accepted by Bulgarian educators from schools and universities and the evidence is scientific publications, projects and created laboratories. The advantages of AR and VR in geography and economics education are discussed by Латева (2018) and interactive technologies are seen as a factor for increasing the students' interest and motivation to learn. Another exploration for usage of AR/VR technologies in mathematics is proposed by Лебамовски and Петков (2017) who define their main benefits for improving the educational quality: intuitive perception of the explored objects, development of spatial thinking, interactive contacts with objects, retention the students' attention, possibility for stimulating the students' creativity. The AR is used as a tool for better understanding the studied topics and learning performance improvement of Bulgarian school students as it is discussed in (Petrov & Atanasova, 2020). A project "Virtual and augmented reality in educational environment" of the Center for creative education (2019) attracted 21 Bulgarian schools and more than 40 teachers who apply AR/VR technologies in classroom through usage of cloud-based platform G Suite, tablets Chromebook Tab 10 and application Google Expeditions. A laboratory for research in augmented and virtual reality is created with applications in science, education, industry and social life (2020).

21.4 Requirements for Future Development of Science Education

A report of European Union *"Education and Training Monitor 2019 Bulgaria Report"* (2019) that evaluates the state of Bulgarian education regarding a set with indicators concludes that: the modernization of Bulgarian education and training system is a process in progress as much is done, but some issues should be improved and enhanced; more attention should be placed on improvement of educational quality; Bulgaria needs to invest more resources in continuous development of competences and skills of all current and future participants in educational process; the models regarding the lifelong learning should be improved for skilling and re-skilling the adult population; the profession of teachers should be promoted among young people and among our society at whole, explaining its important role for current and future well-being; more steps need to synchronize the vocational education and training with the requirements of the labor market.

Science education is planned to be strengthen through international co-operation and students' exchange, teachers training and mobility as well as scientific networks building according to different international initiatives, one of which is the "One Belt, One Road".

At European level, it is important the Education and Training Framework 2020 to be mentioned, which main aim is to distribute knowledge about best practices in educational policies, facilitating the educational reform at the regional and national levels. Among the created working groups are these focused on Early Childhood education and care with priorities in teachers continuing development and decreasing the gap in skills of children; Schools educational system development through promoting quality assurance and teachers profession presenting; Higher education with measures for better students preparation to the societal challenges and world dynamics.

21.5 Discussion and Conclusion

The performed analysis about the current state of science education in Bulgaria reveals big achievements, huge efforts, but also serious gaps. The aim of the identified challenging issues is the competitiveness of education and science in the world scale to be increased. Bulgarian Ministry of Education and Science creates and actualizes policies, action plans, strategies to facilitate the processes for achievement of high quality, technology-supported and competency driven educational programmes at all educational levels. Modernization of Bulgarian schools and science will continue with improvement of the educational and research infrastructures, with increased collaboration among training organizations, scientific centers, government bodies, industries and society, with creation of contemporary educational models that will reflect on the demands of current digitally-oriented learners. Bulgarian institutions in support of European and international best practices should continue to perform specific activities for preparing the conditions for appropriate changes in society and economy, considering the local characteristics and requirements.

References

Bulgarian Ministry of Culture. *Culture Heritage Act*. (2009). http://mc.government.bg/files/3696_C ulturalHeritageAct-Bulgaria.pdf

CEDEFOP. (2018). *Vocational education and training in Bulgaria: short description*. Luxembourg: Publications Office. https://www.cedefop.europa.eu/files/4161_en.pdf.

Centre for creative education. Virtual and Augmented Educational Environment Project. (2019). https://vr.cct.bg/.

European Union. (2019). *Education and training monitor 2019 Bulgaria report*. https://ec.europa. eu/education/sites/education/files/document-library-docs/et-monitor-report-2019-bulgaria_en. pdf.

Ministry of Education and Science. (2014). *Национална стратегия за развитие на педагогическите кадри*. https://www.mon.bg/upload/6550/strategy_razvitie_pedagogic heski_kadri.pdf.

Ministry of Economy. (2015). *Innovation strategy for smart specialisation the Republic of Bulgaria 2014–2020*. https://www.mi.government.bg/files/useruploads/files/innovations/ ris3_26.10.2015_en.pdf.

Ministry of Education and Science. (2016). *Pre-school and school education act*. http://lll.mon.bg/ uploaded_files/ZAKON_za_preducilisnoto_i_ucilisnoto_obrazovanie_EN.pdf.

Ministry of Education and Science. Republic of Bulgaria. (2017). Bulgaria National roadmap for research infrastructure 2017–2023. https://www.mon.bg/upload/4013/Roadmap_2017_ENG. pdf.

Ministry of Education and Science. (2020a). *Закон за висшето образование*. https://lex.bg/laws/ ldoc/2133647361.

Ministry of Education and Science, Educational Curriculums. (2020b). https://www.mon.bg/bg/28.

National Law for Pre-school and School Education. (2017a). http://odz-prolet.com/wp-content/upl oads/2019/11/zkn_PUObr_180717.pdf.

National Strategy for development of scientific research in Republic Bulgaria 2017–2030. (2017b). https://www.mon.bg/upload/6527/SStrategy_2030_BG.pdf.

National Statistical Institute. Republic of Bulgaria. (2019a). *Population and demographic processes in 2019*. https://www.nsi.bg/sites/default/files/files/pressreleases/Population2019_en_ XE8MEZL.pdf.

National Statistical Institute. Republic of Bulgaria. (2019b). *Образованието в Република България през учебната 2019/2020 година*. https://www.nsi.bg/sites/default/files/files/pressr eleases/Education2019_G60QZG9.pdf.

Psifidou, I. (2008). *School curriculum reform and mentalities in transition: looking into the Bulgarian case*. https://www.cedefop.europa.eu/events/curriculum-innovation-2011/images/sto ries/files/I.Psifidou_Bulgarian%20Society_July%202008.pdf.

Petrov, P., & Atanasova, T. (2020). The effect of augmented reality on students' learning performance in STEM education. *Information, 11*, 209. https://doi.org/10.3390/info11040209

Strategy for development of Higher Education in the Republic of Bulgaria for the 2014–2020 period. (2014a). https://www.mon.bg/upload/6537/STRATEGY_MS_29_09_2014.pdf.

Терзиева, В.,Тодорова, К. &Кадемова-Кацарова, П. (2016). *Преподаване чрез технологии – споделеният опит на българските учители*. IX Национална конференция „Образованието и изследванията в информационното общество 2016, 185–194.

Лебамовски, П. & Петков, Е. (2017). Анализ на технологии, подходи и методики за 3D обучение на ученици в стереометрия. *Eastern Academic Journal*, (4), 76–86. ISSN: 2367-7384.

Латева, Р. (2018). Добавена реалност и живо съдържание в класната стая. *Образователен форум*, бр.2. https://doi.org/10.15547/PF.2018.011.

Паунова, Е., Терзиева, В. & Тодорова, К. (2018). *Проучване на използването на общодостъпни образователни ресурси в българските училища*. https://npict.bg/sites/default/files/documents/Poster_%D0%A0%D0%9F%202.1_20.09-3.pdf.

Trend Economy. (2020). *annual international trade statistics by country*. https://trendeconomy.com/data/h2/Bulgaria/TOTAL.

The Programme for International Student Assessment (PISA). Bulgaria. (2018). https://www.oecd.org/pisa/publications/PISA2018_CN_BGR.pdf.

UNESCO. (2020). *World herritage*. https://whc.unesco.org/en/statesparties/bg/.

Virtual and Augmented Reality Lab. (2020). https://sofiatech.bg/en/laboratory-complex/laboratories/virtual-and-augmented-reality-lab-2/.

World Economic Forum. (2018). *The global competitiveness report 2018*. http://www3.weforum.org/docs/GCR2018/05FullReport/TheGlobalCompetitivenessReport2018.pdf.

World Economic Forum. (2019). *The global competitiveness report 2019*. http://www3.weforum.org/docs/WEF_TheGlobalCompetitivenessReport2019.pdf.

World Economic Forum. (2020). *Global gender gap report 2020*. http://www3.weforum.org/docs/WEF_GGGR_2020.pdf.

Malinka Ivanova is Associate Professor in Communications and Computer Science at Technical University of Sofia, Faculty of Applied Mathematics and Informatics, Department of Informatics. She posses two Ph.D. degrees: in Automation of Engineering Labour and Systems for Automated Design and in Informatics, both earned from Technical University of Sofia. She has M.Sc degree in Electronics and Automation from Technical University of Sofia; M.Sc in Engineering Physics, Microelectronics and Information Technologies from Sofia University "St. Kl. Ohridski" and LLM, Law from South-West University "N. Rilski". She is reviewer of British Journal of Educational Technology, Human-Centric Computing and Information Science, International Journal of Distance Education Technologies (IJDET), Computer Science and Information Systems, Journal of Big Data, others. She is committee member of International Conference on Computer Supported Education CSEDU 2020, International Conference on Educational Technologies IceDUTech 2020, International Conference on Big data BIC2020, International Conference Information Visualisation IV2020, eLearning eL2020, eLearning and Software for education ELSE2020, others.

Chapter 22
Science Education in Cyprus

The Future of Education and Generic Skills: Policies for Science Education

Zehra Altinay, Fahriye Altinay, Gokmen Dagli, and Ayten Yıltanlılar

Abstract There is intensified need to act policies for the future of education. While the world becomes interconnected and collaborative on policies and actions with the impact of new technologies, every region has its own policies and strategies which could be the benchmarking strategies for others. For the future of education, there is a demand for interdisciplinary and inclusive approaches to work and collaborate for policies on specific subject field to propose equality and accessibility to educational activities. Change and transformation in education has butterfly effect which all countries are affected from each other which outcome of the education is the learners who have mobility and work power all around the world. Science education is the core discipline of education which proposes language and artistic literacy, digital literacy, personal accomplishment, responsible citizenship, social and economic development, innovation, and entrepreneurship skills with its interdisciplinary perspective in all levels of education. Furthermore, science education sheds a light to meet with decision making, critical thinking, problem solving and other social skills as well where new technologies enhances science learning and provide a ground for real life experiences. The quality of teaching and professional

Z. Altinay (✉)
Faculty of Education, Societal Research and Development Center, Near East University, Nicosia, Cyprus
e-mail: zehra.altinaygazi@neu.edu.tr
URL: http://www.neu.edu.tr/

F. Altinay
Institute of Graduate Studies, Faculty of Education, Societal Research and Development Center, Near East University, Nicosia, Cyprus
e-mail: fahriye.altinay@neu.edu.tr
URL: http://www.neu.edu.tr/

G. Dagli
Faculty of Education, University of Kyrenia, Kyrenia, Cyprus
e-mail: gokmen.dagli@kyrenia.edu.tr
URL: http://www.neu.edu.tr/

G. Dagli · A. Yıltanlılar
Institute of Graduate Studies, Near East University, Nicosia, Cyprus
URL: http://www.neu.edu.tr/

© The Author(s), under exclusive license to Springer Nature Singapore Pte Ltd. 2022   367
R. Huang et al. (eds.), *Science Education in Countries Along the Belt & Road*,
Lecture Notes in Educational Technology,
https://doi.org/10.1007/978-981-16-6955-2_22

development, training and collaboration are critical success factors for the policies and success on science education. This chapter covers national practices and policies of Cyprus and highlights policies, standards, curriculum, digital resources, teacher training, and emerging technologies.

Keywords Digital literacy · Science education · Professional development · Teacher training · Transformation

22.1   Overview of the Country

22.1.1   *Geographical Location, Population and Political System*

Cyprus is an island that is in the Mediterranean Sea, which is at the crossroads of Asia, Europe and Africa. Cyprus, the island, is geographically already part of Asia. Cyprus is also located approximately 80 km south of the coast of Turkey, west of Lebanon, and Syria, north of Egypt, northwest of Israel, and east of Greece. Cyprus is the third most populous and largest island after Sicily and Sardinia in the Mediterranean Sea.

Official internationally recognized name of Cyprus is Republic of Cyprus. Cyprus was a former British colony then it became independent and became a unitary state country of both Greek and Turkish Cypriots in 1960. There were strong tensions between the Greek Cypriot majority and Turkish Cypriot minority communities. All foreign governments and also the United Nations, which are peacekeepers between Greek and Turkish communities on the island. Today, Cyprus Island has a presidential republic government structure. The president is the representative of both head of state and government.

In short of political background, Turkey, Greece, and England are the guarantors of Cyprus Island and United Nations are the peacemakers, which are ensuring the Green line, which was established as ceasefire-line, has influentially partitioned the two communities on the same Island since 1974. The official languages spoken on the Island are Greek and Turkish. However English is also commonly spoken although it is not official status. In addition to official languages, there are also minority languages such as Armenian and Cypriot Maronite Arabic.

According to the latest United Nations data projections, the current population of Cyprus is 1,209,406. The UN estimates this population rate as 1,207,359 since July 1, 2020. Population as whole compromises the North (pop 61,000) and the South (pop 55,000), which is the official division of Nicosia, the capital in Cyprus.

22.1.2 Current Situation of Economic, Technologies and Cultural Development

Cyprus became a member of the EU in 2004. Cyprus adopted the euro as its national currency in 2008 as a result of this membership. After this money adoption, Cyprus was listed as one of the 31 advanced economies among other countries by the IMF in 2016. Cyprus economy has had effects of external shocks throughout its history that have been followed by surprising regrowth. Economy in Cyprus has gone through many transformations such as minerals and agricultural products export in 1961–73, manufactured goods starting from late 1970s to the early 1980s, economy transformation into business, international tourist and services center since 1980–1990s. Today the economy in Cyprus is basically based on the services sector like tourism, higher education, real estate and financial services, which accounts most of over 80% of employment rate and total GDP.

The culture in Cyprus is divided between the two communities as northern Turkish and the southern Greek sections of the Cyprus Island. Turkish has been promoted by Turkish community in northern Cyprus as its own official language and also representative of Islamic culture. Northern Cyprus supported its own newspapers and periodicals and has changed many place-names to Turkish since 1974. Traditional Muslim holidays; the anniversary of the proclamation of the TRNC is celebrated in the north every November 15 since 1983. On the other hand, Greek Cypriots have been preserving their traditional culture and also celebrating holidays such as pre-Easter Carnival and as Easter, and also a spring flower festival which is called Anthestiria. Greek and Turkish Cypriots as two communities living on Island separately both enjoy various tradition of handicrafts and folk arts. Cypriot lacework stands among the internationally best-known expressions of this art.

In terms of STEM subjects, Cyprus has a low proportion of graduates. Much more broadly, digital skill levels in Cyprus are low when considering its population. When comparing Cyprus to the benchmark countries around the world, Cyprus has the lowest basic digital skills. On the other hand, 26% of firms in Cyprus which are comparable to most benchmark countries provide ICT training, which is counted slightly above the EU average (European Commission, 2018).

According to the data shared by European Union in 2018, innovation performance in Cyprus ranks below when comparing to the EU average. The European Innovation Scoreboard indicated that Cyprus is obviously behind Northern European countries, but it is still comparable to other small countries and Mediterranean countries too. When considering the index value for Cyprus, it fell between 2013 and 2014 years, which lead widening the existing gap between Cyprus and the Innovation Scoreboard leaders around the world. However, Cyprus has much improved its innovation performance and has started to catch-up the innovation performance leaders in recent years.

Although current need for STEM or digital skills is moderate, the increasing significance of digital technologies across the world and in all sectors make it inevitable that Cyprus may not catch up with other benchmark counties in terms of technological integration in all work areas. If the workforce and the educational system in Cyprus are not fully and effectively prepared to adapt to new and emerging trends, Cyprus may be under the risk of falling behind comparing to benchmark countries.

To boost digital skills in Cyprus, the National Coalition has introduced cost-free digital certification programs for students and school competitions for Digital Jobs such as coding and robotics. Furthermore, primary schools and secondary schools have course like computer science in their curriculum, which is taught in all-day' schools.

As a result of advances in technology cultural developments in Cyprus are becoming more and more visible day by day. There is no doubt that one of the most astonishing applications of IT with its futuristic traits all of us to make sense of the past and its unknown dimensions. In order to preserve cultural heritage and learn more about its mysteries undiscovered parts the Cyprus Institute's Science and Technology in Archaeology and Culture Research Center (STARC) puts together many conducted researcher from a wide range of subject such as history, art, chemistry, computer science archaeology and physics. Archaeological discoveries throughout the years led to new applications such 3D documentation technique for heritage assets and research libraries. Traditional documentation techniques have become much more technology based and this transformed libraries into digital libraries. In today's world, archaeological researches has accelerated with the help of effective photographic techniques, mapping methods with valuable insights and many more virtual investigation opportunities. As a result, this can be enable countries to extend to be more recognized globally in terms of cultural landmarks.

22.2   Overview of the Education Development

22.2.1   Education System and Policy

22.2.1.1   Primary Education

Education is compulsory between ages 5 to 15 and it is free in Cyprus. Usually, following possibly 1 year of optional pre-primary school, children at the age of 6 continue to primary school. Then, children who attended to primary school for 6 years of, receive a school-leaving certificate at the end. School subjects taught compatible with the British tradition and all are academic. The purpose of primary education is to establish secure learning opportunities for children, regardless of their social background, gender, family, age and academic ability. Skills such as languages and mathematics, environmental education, health, creative and artistic expression have a strong emphasis. In order to increase use of modern technology, to expand the all-day schools effectively and step forwards in terms of integrating pupils into the mainstream along with their special educational needs, current innovations have also been included (Table 22.1).

Table 22.1 Education system in Cyprus

Education	School/level	Grades	Age	Years
Primary	Primary education	1–6	6–12	6
Notes: Primary school leaving certificate—Greek-Cypriot system				
Primary	Primary education	1–5	6–11	5
Notes: Turkish-Cypriot system				
Middle	Gymnasio	7–9	12–15	3
Lower				
Notes: Secondary education—Greek-Cypriot system				
Secondary	Lower secondary education	6–8	12–14	3
Notes: Turkish-Cypriot system				
Secondary	Eniaio Lykeio	10–12	16–18	3
Notes: Eniaio Lykeio—Ενιαιο Λυκειο certificate/diploma granted: Apolytirion—Greek-Cypriot system				
Secondary	Lise	9–12	14–18	4
Notes: Turkish-Cypriot system				
Vocational	Technical school	10–12	15–18	3
Notes: Certificate/diploma awarded: Apolytirion				
Tertiary	Pre-primary and primary/basic school teachers	–	–	–
Notes: All teachers in the public sector must have a University of Cyprus degree (Ptychio) or equivalent degree in pedagogical studies				
Tertiary	Tertiary	–	–	–
Tertiary	Bachelor	–	–	4
Notes: Leads to the award of a Ptychio (240 ECTS)				

(continued)

Table 22.1 (continued)

Education	School/level	Grades	Age	Years
Tertiary	Master; post-graduate degree	–	–	2
Notes: University of Cyprus confers a master's degree eighteen months after the bachelor's degree in most fields				

22.2.1.2 Middle Education

Middle schools alternately known as gymnasiums are school types in which children take three years of compulsory education. There is still an ongoing single stream of academic subjects learning and teaching.

22.2.1.3 Secondary Education

There are 3 forms of Secondary education, although all successful students have qualification to enter university. Three types of secondary education are as secondary general, secondary technical and vocational education in Cyprus. In other words, two three-year cycles of education is offered in Secondary general education. It is also known as Gymnasium which appoints learners to the lower secondary education process. Moreover, the Lyceum designate the students to the upper secondary general education. Learners may tend to choose to attend Secondary Technical and Vocational Education (STVE) instead of the Lyceum. STVE is a three-year education cycle of upper secondary education, which is offered in the technical and vocational education. These two are private and public schools of secondary education. The curriculum of these mentioned schools differes each other. For example, the curriculum at lykeion schools, remains academic. Scientific subjects at Technical schools are emphasized, while workshop practices and industrial trainings at vocational schools are more focused on. Like in above mentioned schools, secondary school is also free during six years of duration and it is a compulsory education up to the age of 15. Learners are awarded variety of certificates when they fulfill theirs subjects in these schools.

22.2.1.4 Vocational Education

Ongoing vocational training is under The Human Resource Development Authority responsiblity in Cyprus. This happens both through setting standards and presenting courses.

22.2.1.5 Tertiary Education

Both private and state universities including 3 public ones are available in Cyprus. They include the Open University of Cyprus, the University of Applied Sciences and Arts, and the University of Cyprus. The University of Cyprus is the oldest of these three. It was established in 1989. The number of the students are approximately 6,000, and teaching happens basically in Greek. Regulation of admission happens by residence, cultural/religious group and also nationality.

Other than above mentioned and detailed explained schooling types, there are other kinds of schools as below.

22.2.1.6 Pre-school Level

Children under the age of three can go to day nurseries; likewise, children who are at the age of 6 in the case of they are integrated with a kindergarten can also attend day nurseries schools. The purpose of day nurseries is to provide child care rather than education. In day nurseries, children's parents pay fees.

22.2.1.7 Pre-primary Education

Children between 3 to 6 years old are allowed to attend Pre-primary education. Children between 5 to 6 years can get this education as chargeless and compulsorily, while children in the rage of 3–5 years old are accepted but their parents pay fees. Pre-primary education has its education purpose as satisfying the children's needs for the development of a wholesome personality by establishing an experiential environment.

22.2.1.8 Adult Education

There are public, semi-government and private institutions for adult education offered according to the type of training they provide. To name adult education institutions types: formal adult education; non-formal adult education and vocational training. In these institutions some courses are offered chargesless, but most of the courses must be paid.

22.2.1.9 Special Education

At school level, children integration with special needs into mainstream schooling is emphasized. These children follow the normal curriculum; however, their curriculum be adjusted to match their particular needs. On the condition that full-time participation in a mainstream class is not possible for the child's needs, special tuition for specified periods per week may be recommended in a resource room or, alternatively, participation in a mainstream school at a special unit is also possible. Such special units provide more intensive special educational support pupils who need especially small number of students (usually a maximum of six), whereas sustaining contact and some level of integration with a mainstream class. If none of the options above is considered suitable, the children are oriented to attend a public special education school.

22.2.1.10   Home Education

Learners of pre-primary, primary and secondary education, who are either handicapped or with special needs, are allowed to have Home Education according to Law 113 (I) in 1999. Moreover, students who have very serious health issues, great difficulties to continue to attend their classes and also students due to some reasons who have serious problems for short term which really hinder them to attend classes also have opportunities for Home education in Cyprus according to education policy. Home Education is possible when only approved by the Ministry of Education and Culture. Teachers, who are approved by only the Ministry of Education, is allowed to teach and provide simpler versions of exams to students in home education in order to support students be successful for the next grade. Students who have special needs are the ones that have serious learning functional or adaptive difficulties due to their physical, mental, sensory, or other cognitive deficiencies. Therefore, there is definitely a need to have practices for special treatment and education.

22.2.2   Statistics on the National Education

According to the comprehensive survey of Statistical Service in Cyprus, which covers recent years 2017–2018, statistical data on the national education system in Cyprus which includes teachers and students numbers and dropouts are summarized as below:

1.   There were 1.338 educational institutions at all levels of education. The number of teachers was 14.982 and pupils/students were 191.787, which indicates a pupil/student teacher ratio of 12,5(1).
2.   66,8% was enrolled in public/communal educational institutions. 32,2% of the total percent were enrolled in private. For all level of education the respective percentages are as follows: 51, 7% of the total pupils/students were enrolled in public/communal institutions whereas 48, 3% out of 51,7% were enrolled in private when considering Pre-School and Pre-Primary education. Furthermore, when taking Primary education into account 90,4% of the total pupils/students were enrolled in public institutions whilst 9,6% out of 90,4% were enrolled in private. When it comes Secondary education, the total pupils/students in 81,4%

were enrolled in public institutions and 18,6% out of the total percent were enrolled in private. As final notes, Tertiary education had pupils/students at about 31,5% enrolled in public institutions whereas 68,5% of the total percent were enrolled in private.

3. The enrolments summary of pupils and students by level of education respectively were as follows: 32.522 for Pre-School and Pre-Primary, 56.699 for Primary, 54.966 for Secondary, 47.169 for Tertiary and 431 for Special education.

22.2.3 Educational Research and International Collaboration

"CREF" namely The Cyprus Research and Educational Foundation was established in February 2004, which is driven by the aim of supporting the improvement of knowledge. The vision of The Cyprus Research and Educational Foundation is to highlight the act of Cyprus as a portal of innovation and knowledge between the region and EU by adopting humanist and tender applications over the course of Eastern Mediterranean. By this way, it is aimed to lead a transformation in the impression of Cyprus from economy in a knowledge-based kind economy.

To actualize its vision and aims, The Cyprus Research and Educational Foundation, from the initial, grounded a plan of setting up of an advanced research and educational institution, which will support research and education in Cyprus and its zone. CREF created its new institution called The Cyprus Institute (CyI) with the aim at getting benefit from the public involvement at large. CyI was established to be globally known in terms of its membership and collaborated with settled most advantageous institutions. The Cyprus Institute (CyI) is independent in its partnership but public in its duty and service. In addition to its independent commitment, CyI represents itself as the devotion in terms of the wellbeing of the regions and the international community profile of Cyprus.

22.3 Current Situation of Science Education

A more influenceable science education can provide extensive cooperation in terms of knowledge-based innovation which reaches the ethical standards in the highest level and also provide opportunities for sustainable communities into the future.

Studies conducted on the integration of coding education into education curriculum aim to raise awareness on increased mobility, social network, cloud computing, increase in open-source commercial values and coded software, which are the most promising concerns in Cyprus and they open a gateway of increase the urge for information sector in the country. Therefore, "Informatics Island" view is being run as a result of these advancements in 21st Education field. Universities on

Cyprus Island have an active role to achieve their missions in terms of developing science education like in other developed countries abroad.

One of the fundamental parts of learning should be science education, which would provide continuum for all individuals, starting from early ages like pre-school to engaged and active citizenship. In today's world with no doubt that either local or global communities are going under an intense shift in terms of their cultural, economic, political and most importantly educational structures. Therefore, science education became this much urgent for all countries. It should put emphasis on gaining competences on learning through science and transforming from STEM to STEAM by engaging science with other fields and disciplines.

22.3.1 Policies and Standards

Cyprus entrance the European Union in 2004 brought development as a result of globalization on both cultural and economic aspects, which led to new demands to update science education in Cyprus (Zembylas, 2002). Global Education Network Europe has a big role to supports peer learning, policy research, networking, national strategy development and quality enhancement in education field of all European countries. Thus EU has certain influence on education system in Cyprus with its advancements. Cyprus has had its own national curriculum in science after gaining its independence from the British in 1960. The latest curriculum, which was the science curriculum, was completed in 1994 (Ministry of Education in Cyprus, 1996). Today, there is a National Curriculum in Cyprus education system that was last revised in 1996 (Ministry of Education in Cyprus, 1996).

Science education policy and system suggest that science should be compulsory education for all students. Moreover, science education should provide support for schools, teacher educators, teachers, and also students of all ages. Most importantly, it should provide everyone equal opportunities to gain excellence in learning process and learning outcomes. One of the priorities of science education is to give value to all disciplines and indicate that to what extend inter-disciplinarity (STEAM instead STEM) can bring contributions to individuals' perception and scientific principles knowledge by enabling them to solve societal challenges in their own communities as active citizens.

22.3.2 Curriculums, Digital Resources and Teacher Training

Since then the latest curriculum created, piagetian hierarchical- developmental philosophy of learning has been adopted in Cyprus education system. Its basic under-lying philosophical view was 'guided discovery'. This approach puts emphasis on students' active engagement in discovering concepts related science on their own. The aims of the curriculum are based on promoting the learners with the scientific skills, which will enable them to make investigation and create their own under-standing of science (Zembylas, 2002). In other words, the new curricula contain the extensive objectives of education in Cyprus; fundamental principles of peda-gogy; subjects, aims, objectives and the content for every curriculum subject for all levels and grades. The curricula also includes teaching and learning method-ology elements; achievement indicators; methods of evaluation; and, representative teaching activities.

Based on the descriptions above, the implementation of STEM, which is a pilot program (Science, Technology, Engineering, and Mathematics) has been planned by the Ministry of Education, Culture, Sport, and Youth. STEM program has been also planned to be operated in nine Primary Schools and Secondary Education. The purpose of this program states the educational policy and pathway of promoting an integrated interdisciplinary approaches into the learning process. Children are expected to involve in activities which enable them with opportunities to activate their knowledge, utilize concepts, and engage actively into all learning processes in Science, Technology, Engineering and Mathematic fields, which address real life situations and issues connected to everyday life and today's the world. Moreover, the students fulfill their projects that they wanted, they design, elaborate and also bring out innovative solutions to different situations as a result of the 1-day experience. This occurs in a context of synergy and coexistence of the four applied disciplines.

Moreover, the teachers' mentorship as the main tool for teachers lesson manage-ment, detailed lesson analysis also are help teachers to prepare their teaching the material too (Ministry of Education & Culture, 2003). Curriculum developers expect teachers to follow the shared suggestions but at the same time each teacher can differ-entiate some of the aspects depending on their grade or level they teach. Therefore, guidelines for teaching methodologies are not prescribed by the Ministry of Educa-tion and Culture for teachers to use in public primary schools. Since all teachers are university graduates, they are supposed to adopt modern teaching methodologies and techniques. In addition, teachers are also expected to choose the appropriate method according to the needs of the students and the curriculum requirements. The inspections system ensures that teachers are utilizing appropriate and up-to-date methodology in their teaching. In today's education conditions, based on the current science education policies to increase teaching quality, continual improve-ment, greater emphasis on teacher competences and teachers disciplinary knowledge has become much more prerequisite in Cyprus education too. Continuous Profes-sional Development (CPD) is highly recommended to become an essential compo-nent for all educators during their teaching career. Thus, Pedagogical Institute of Cyprus offers either compulsory or optional courses for teachers at all school levels to engage them into continuing education. Differently, academic staff at the higher education level are not offered regulated continuing education like in state schools. Teachers are expected to improve their self-efficacy, from their induction process through pre-service preparation and in-service professional development stages in order to be the agent of increase in quality of teaching and learning outcomes of education they are involved in. For example, the Department of Teachers' In-service

Training provides compulsory courses, which are held in morning hours, as in the following:

- Compulsory recurrent course for newly promoted headteachers in primary education;
- Compulsory recurrent course for newly promoted headteachers in secondary education;
- Compulsory course for newly promoted deputy headteachers in secondary education; and,
- Compulsory course for newly appointed teachers and their mentors (induction course).

Apart from above mentioned, the Pedagogical Institute offers optional seminars, which are usually held in afternoon or evening hours, and they are public and free for all teachers like in previously mentioned courses. A wide range of topics including Pedagogy, Psychology, Sociology, and Information Technology are covered.

As latest implementation, The Digitally Supported Learning Environment (Vocational and Learning Center) was started in October 2019 in order to develop the digital ability of teachers.

22.3.3 Student Assessment and Achievement

According to OECD statistics which were published by the recently, pupils managed slightly better in mathematics and science but slightly worse in reading in Cyprus when comparing the current situation to 2015. However, Cyprus still ranked below average in the Programme for International Student Assessment, which is known as PISA. The assessment every three years is carried out and the most recent report, which is covering almost 78 countries, was published in 2018. The summarized findings are as below:

Reading: 487 was The OECD average. Pupils' score of Cyprus was 424 points, which led Cyprus to rank 50th among 77 countries. Cyprus ranked 47th in 2015 when comparing is performance to the latest OECD findings.

Maths: 489 was The OECD average. Cyprus got 451 points and shared its own place, which stands 43rd, with Greece among 78 countries. Cyprus stood 49th in 2015.

Science: The average point of OECD was 489. Cyprus had 439 points and ranked 47th out of 78 countries, which remained two places leveled up from 49th when considering 2015. There have been several increase and decline of education performance of Cyprus throughout the years. Hence, science education is expected to support the inefficacies in education field so that Cyprus catches up with other benchmark countries in terms of education quality for its sustainable community for the future. The Ministry of Education concerned about PISA international competition results, although Cyprus has much improved its rank slightly when compared to 2015 results, yet performance of student remains below the European average. Although

Cyprus scored below average when compared to another international student assessment in other 77 countries took place in the test, still this helps education policy makers to measure school quality standards in Cyprus, but yet stakeholders have disagreements on the idea that tests are not a clear enough to reflect education system of the country (OECD, 2018).

Apart from PISA, The MoEC Inspectorates in Cyprus, as part of Primary and Secondary Education, provide quality control, improvements and also system wide changes in the education system. Three inspectorate teams actively operate at primary education level, secondary education level, technical and vocational education levels.

22.3.4 Science and Technology Venues and Centers

The need for technology integration is becoming a prerequisite day by day in all areas of life. Education is among these areas, which is most affected by advancements in technology. There are various science technology venues and centres. For example, RISE is the first Research centre in Cyprus, which was designed to perform as mediator of academic research and industrial innovation by enhancing technological, scientific, and economic growth of Cyprus and Europe. Furthermore, "Cyprus Science Research Center (CSRC). Centre for STEAM Education Research, Science Communication and Innovation- TEAMING HORIZON 2020" was established as the first time to meet the needs of 21st Century requirements in all sectors of information.

Cyprus Science Research Centre allows universities on the island to join together to design an area for Innovation and Center of Excellence in Cyprus, in accompany with Cypriot companies. The primary focus of CSRC will be on promoting of technology, science communication, engineering, mathematics, arts and teaching and learning in science by means of interactive tools. In order to improve innovation, research, and entrepreneurial culture in technology field, most importantly the innovative communication of scientific accomplishments and the technology for teaching and learning in science.

Apart from above mentioned collaborations, The British Council has been collaborations with Cyprus since 1940. Due to 10th consecutive year, cultural relations and exchange between Cyprus and the UK will be celebrated through, FameLab, which has been organized in accompany with Scico Cyprus and the Research and Innovation Foundation (IDEK). The competition has promoted occasions to many young and talented scientists since 2011. Individuals over 18 years old who are studying, working, or graduating in technology, physics, chemistry, engineering, biology, medicine, mathematics, etc. fields, are welcome to participate. Over all, FameLab offers opportunities for networking and collaboration with other scientists in Europe and Cyprus, as well as opportunities to take part in British Council events in Cyprus and also abroad.

22.3.5 Utilizing Emerging Technologies

In today's education atmosphere, virtual reality (VR) and augmented reality (AR) can change the classroom experience into a desired 21st leaning environment. Cyprus is on its way to integrate emerging technologies in to its all sector especially education in order to have sufficient performance to compete with other highly ranked countries. There is no doubt that in this digitalized world VR provides much practical, easiest and fastest way to reach a lot of concepts, which can boost students' curiosity, increase their engagement much into learning, and discovering new experiences.

Some examples that all of us are now closely familiar with after pandemic case are doing lessons through virtual classrooms platforms, coding was integrated into education curriculum as part of educational activities is given one hour a week starting from 1st year of primary schools in Cyprus. Especially technological tools and implications in all sectors most importantly in education can ease the handicapped and disabled peoples' lives in terms of reaching information whenever and wherever they can. Therefore, online lessons brought advantages into learning and teaching environment. Educational materials have started to change form from pen and paper to digitalized forms. Learning activities have become to be much more fun, engaging and attention grabbing when considering especially introverted students.

22.4 Requirements for Future Development of Science Education

Although there are still some handicaps of technology based education, which almost all of us experienced during pandemic time, the idea itself is promising for better the future education atmosphere for the whole world. Educational system in Cyprus is going under educational reformation, therefore, weaknesses identified in the current education system may show some next steps which need to be followed for the future.

In order to establish well-functioning science education starting from the government to all parties who are included into education filed should gather into a common ground in terms of the current needs, requirements to meet the needs as expected, participate actively in decision-making process to clarify the pathways of future implications that are into consideration. In other words, shared responsibilities and strategic problem-solving should be put into effect during the development of science education. Most importantly, a lot of future research and needs analysis must be done in order to determine what further steps need to be taken during this long term transformation.

22.5   Discussion and Conclusion

As a member of Belt and Road countries, it is necessary to strengthen the cooperation among all countries not only by improving science education but also building bridges between continents, nations and citizens by establishing cooperations through trading, establishing infrastructure networks, strengthening the multilateral cooperations. For example, the Erasmus programs has crucial contributions to international cooperation and mutual exchange of students and educators among higher education institutions. It enhances academic recognition and transparency for the programs. Therefore, international connectivity and multilateral cooperation in different sectors including education, economy, cultural and social implications can increase the possibility of overall developments in all areas within the island.

No doubt that various challenges which schools have faced with determined the future pathways of establishing effective actions towards the long term improvement in education. One of the biggest challenges that was come across during the pandemic time is related with how to integrate eLearning into all educational platforms in order to develop the traditional learning processes quality, but at the same time to find out and adopt new learning approaches which are compatible with latest ICT advancements, which support education quality in general. All educational parties such as researchers, policy makers, stakeholders, educators, technology developers put so much great efforts on how to prepare learners in all ages to involve in innovation and knowledge formation, enable them to bring contributions into knowledge-based society concept.

Since there is a lack of self-awareness in Cyprus in terms of the advantage that the mass usage of ICT in education can ease the existing learning methods, it took a while to meet the expectations of society about eLearning during integration into education process. Additionally, there was another crucial challenge experienced, which was related to evaluation methods and also educators and learners getting accustomed to operate the new approaches of ICT-based learning in the education process but also in the lifelong learning process.

Finally, the third and the most significant component of setting up e-learning environment was the sustaining a compelling monitoring system. When considering an education system, it is needed to include administration, analysis, synthesis, design, implementation and evaluation, regarding budget delivery, time, quality and but also effectiveness and efficiency of project.

The future of education and policies in science education are the significant paradigms to develop life standards in society and understand 'science' as functional subject field to fulfill generic skills. Although current reports and case studies shed a light for the current situation, there is intensified need to integrate technology based science education even online education practices for the success. For that reason, there is more intensified need for partnership, collaboration between scholars, researchers and stakeholders to set policies and act on those policies with strategic planning. There is butterfly effect between countries where best practices could be shared and benchmarking strategy could be optional attempt to put forward action plan. Under the framework of policies and strategies, grounded knowledge need to be transferred to the values within society and learners need to gain generic skills within science education. Therefore, there is potential need for new understanding of science education and technology integration into science and life skills to reach out entrepreneurship, creativity, innovation, social skills for the future of education. Science is not simple single discipline, as it has interdisciplinary perspective to affect all part of lives. Being in active engagement during learning, being knowledgeable person with social skills, there is need for science education in order to be prepared life for the future. The quality of education, quality of teacher, professional development programmes are critical success factors in national science education policies. Delegating mentoring, sharing and self-evaluation, transferring knowledge into practice are crucial. Therefore, multi professional teams should work for program course design, technical infrastructure, intellectual capacity, learning communities for the future of science education.

Acknowledgements We would like to thanks all scholars who actively work for science education in the field for the policies and strategies.

References

Cyprus Institute of Technology. http://www.technology.org.cy/.

Cyprus Ministry of Education and Culture. http://www.moec.gov.cy/.

European Commission. (2005). National summary sheets on education systems in Europe on ongoing reforms. Eurydice, Directorate-General for Education and Culture.

European Commission. (2019). 2nd survey of schools. ICT in education: Cyprus country report. https://publications.europa.eu/en/publication-detail/-/publication/092c1496-46d6-11e9-a8ed-01aa75ed71a1/languageen/format-PDF/source-99674504.

European Commission/EACEA/Eurydice. (2018). Teaching careers in Europe access, progression and support. https://publications.europa.eu/en/publication-detail/-/publication/435e941e-1c3b-11e8-ac73-01aa75ed71a1/languageen/format-PDF/source-97869117.

Ministry of Education and Culture. (2003). The development of education: National report of Cyprus. Nicosia, International Bureau for Education.

Ministry of Education and Culture. (2017). The referencing of the Cyprus qualifications framework to the European qualifications framework for life long learning. http://www.cyqf.gov.cy/index.php/el/dimosiefseis.

OECD. (2018). Effective teacher policies: Insights from PISA. https://doi.org/10.1787/978926430 1603-en.

Zembylas, M. (2002). The global, the local, and the science curriculum: A struggle for balance in Cyprus. *International Journal Science in Education, 24*(5), 499–519.

Zehra Altınay is Director of Educational Sciences Department and Chair of Societal Research and Development Center at Near East University. She is Vice Director of Graduate School of Educational Sciences. She completed a master degree in the field of Educational Technology, Distance Education in Educational Sciences. Further to this, she completed PhD degree in Educational Administration and Management at Near East University. Then, she completed another doctorate programme in the field of Educational Technology at Middlesex University. She has started teaching and research activities while she was research assistant. She became a full time lecturer at Near East University and she is working as Prof Dr. at Education Faculty. She is teaching doctorate, graduate and undergraduate degree courses. Zehra Altınay is the member of academic journals and she has national and international books, publications and research projects. Zehra Altınay is chair of ethical committee in the university and chair of monitoring rights in disability in the field of education.

Fahriye Altinay is Director of Graduate School of Educational Sciences and she is Chair of Societal Research and Development Center as a board member in the university and the faculty. Fahriye Altınay is the member of academic journals and she has national and international books, publications and research projects. Prof. Dr. Fahriye Altınay is member unobstructed information technology platform and actively works in her studies about disability in smart society. Current researches consider the importance of diversity management, sensitivity training, disability, global citizenship for smart society.

Gokmen Dagli Vice Rector of University of Kyrenia. Dean of Faculty of Education at University of Kyrenia. Head of Department in Educational Management at Near East University.

Ayten Yıltanlılar Graduated from English Language Teaching. She is PhD student in educational management. She is working technology integration to educational programmes and management.

Chapter 23
Science Education in Estonia

Külli Kori

Abstract This chapter presents the current situation of science education in Estonia. Estonian students have shown good results in international assessments of science achievement (e.g., PISA, TIMMS) which has raised international interest towards the Estonian educational system. Estonian educational system is based on the Lifelong Learning Strategy and the science curriculum puts emphasis on using technology to support learning and developing students' inquiry skills. The future of science education in Estonia seems to be moving towards integrating science, technology, engineering, arts and mathematics (STEAM), developing students' complex problem solving skills and general competences (e.g., digital competence, cultural and value competence, social and civic competence), connecting science with the real life and developing active citizens who use scientific thinking and creativity in their everyday lives. This can be done through solving socio-scientific issues and participating in citizen science projects. However, Estonian science and mathematics teachers are rather old and several challenges should be solved on the way.

Keywords STEAM · Inquiry-based learning · Socio-scientific issues · Citizen science · General competences

23.1 Overview of the Country

23.1.1 Geographical Location, Population and Political System

Estonia is located in Northern Europe. It borders the Gulf of Finland, Latvia and Russia. Estonia is the most northern county of the Baltic states and the country is known for its medieval cities, lush forest, sandy beaches, and developments in

K. Kori (✉)
School of Digital Technologies, Tallinn University, Narva mnt 25, 10120 Tallinn, Estonia
e-mail: kulli.kori@tlu.ee

technology. The territory of Estonia consists of the mainland and 2,222 islands in the Baltic Sea, covering a total area of 45,227 km².

Estonian population is about 1.3 million people (47.4% male and 52.6% female) (Statistics Estonia, n.d.). The national language is Estonian which is spoken by 68.5% of the people. The second most spoken language in Russian (29,6%). The capital of Estonia is Tallinn which has a population of 437 619 (Statistics Estonia, n.d.).

Estonia is a unitary parliamentary republic. The unicameral parliament serves as the legislative and the government as the executive. The president is elected by the parliament and the 101 members of parliament are elected by Estonian citizens for a four-year term.

Estonia joined the North Atlantic Treaty Organization (NATO) in 2004 and the European Union (EU) in 2005. Since 2011 the currency in Estonia has been the euro.

23.1.2   Current Situation of Economic, Technologies and Cultural Development

The economy of Estonia is an advanced economy and the country is a member of the European Union and the eurozone. Estonia has a modern market-based economy. It has one of the higher per capita income levels in Central Europe and the Baltic region, but its economy is highly dependent on trade. Estonian government has pursued a free market, business-friendly economic agenda, and sound fiscal policies that have resulted in balanced budgets and the lowest debt-to-GDP ratio in the EU (Central Intelligence Agency, n.d.).

The economy of Estonia benefits from strong electronics and telecommunications sectors and strong trade ties with Finland, Sweden, Germany, and Russia. The government is making efforts to boost productivity growth with a focus on innovations that emphasize technology start-ups and e-commerce (Central Intelligence Agency, n.d.).

The preservation and development of Estonian culture is one of the main goals in Estonia. This goal is often interpreted rather strictly as the state supports exclusively professional and folk culture. Recently, digital culture has become more and more important in the development of education, participatory culture and cultural heritage (Tamm, n.d.).

23.1.3   Education System and Policy

Estonian educational system is based on the Lifelong Learning Strategy 2020. The general goal of the Lifelong Learning Strategy is "to provide all people in Estonia with learning opportunities that are tailored to their needs and capabilities throughout their whole lifespan, in order for them to maximize opportunities for dignified self-actualization within society, in their work as well as in their family life" (Ministry of education and research, n.d.). Lifelong learning begins with general education where the common curriculum is taught at all levels.

The educational system consists of four levels: pre-school education, basic education, upper-secondary education, and higher education. Pre-school education is delivered to 3–7-year-old children and the especially dedicated educational institutions follow the state curriculum. Pre-school education is not mandatory. The focus of pre-school education is supporting the child's family through fostering the child's growth and development by taking into account their individuality (Ministry of education and research, n.d.).

Studying at basic education level is mandatory until graduating basic school or until the student will be 17 years old. Basic education is divided into three stages: I stage: grades 1–3, II stages: grades 4–6, III stage: grades 7–9. Schools that provide basic education follow the national curriculum. The length of the study period consists of at least 175 teaching days (35 weeks) and four intervals of school breaks. The schools must provide places for compulsory school children living in their service area and parents can influence the school's development through the school board. To graduate basic school, students have to learn the curriculum on at least a satisfactory level and pass three exams: (1) Estonian language or Estonian as a second language, (2) mathematics, (3) a subject of the student's choice. In addition, students have to complete a creative assignment to graduate basic school (Ministry of education and research, n.d.).

After basic school it is possible to acquire general secondary education at upper-secondary school, vocational secondary education at vocational education institutions or acquire an occupation.

The study program at upper-secondary school is divided into mandatory and voluntary courses. To graduate upper-secondary school students have to complete the curriculum (consisting of at least 96 individual courses) on at least a satisfactory level, pass the state exams of the Estonian language or Estonian as a second language, mathematics and a foreign language; pass the upper-secondary school exam and complete a student research paper or practical work. After graduation it is possible to continue studies at a higher educational institution or to obtain vocational education (Ministry of education and research, n.d.).

At basic and upper-secondary education level the school's running costs will be covered by the school manager, which in most cases is the local government. The local governments are authorized to establish, rearrange and close general education schools. Also, they ensure school attendance control, make arrangements for school transport and provide school meals. The amount of state subsidies for schools are calculated based on the number of students. The state subsidy is used for covering expenses on teachers' salaries, social taxes, training and textbooks (Ministry of education and research, n.d.).

23.1.4   Statistics on the National Education

This chapter gives an overview of the statistics on Estonian national education based on the data of HaridusSilm (n.d.).

In the school year 2019/2020 pre-school education was provided in 614 kindergartens to 66 330 children. The number of teachers working in these kindergartens was 6654 and only 4 of the teachers were male. At the same time, general education was provided in 516 schools. These included 53 primary schools, 306 basic schools, and 157 upper-secondary schools. During the last 15 years the number of general education students has varied between 173 822 and 134 975 while being the highest in school year 2005/2006 and lowest in school year 2012/2013. The number of students decreased from 2005/2006 to 2012/2013, but has been increasing again after that. After graduating basic school, about 69% of the students continue studying in upper-secondary school, 27% in vocational school, and 4% do not continue their studies.

The number of teachers who were teaching at primary, basic and upper-secondary schools in the school year 2019/2020 was 15 822. 6222 of the teachers were teaching grades 1–3, 10 367 were teaching grades 4–6, 9574 were teaching grades 7–9, and 4421 were teaching grades 10–12 (one teacher can teach in more than one grade group). In general, female teachers are the majority in Estonia (about 85%).

Estonian teachers are rather old—8.4% of all the teachers are younger than 30 years old, 16.5% are 30–39 years old, 24.2% are 40–49 years old, 29.6% are 50–59 years old and 21.3% are 60 years old or older. This issue is critical especially in science and mathematics subjects as every fifth mathematics, chemistry, geography and biology teacher is at least 60 years old and every fourth physics teacher is at least 60 years old (Kutsekoda, 2018).

After graduating upper-secondary school 62% of the students continue studying in higher education institutions and 10% continue in vocational education schools. Higher education is provided in 36 institutions (including public and private universities, public and private vocational schools, public and private institutions of professional higher education). The number of students starting higher education first level studies has decreased during the past 15 years, while being the highest in 2007/2008 when 15 297 students started their studies and being the lowest in 2019/2020 when 8694 students started their studies. However, the number of students starting masters level studies has slightly increased (from 3499 in 2006/2007 to 4346 in 2019/2020). Dropout is a serious issue in higher education institutions, as about 42% of the students drop out. Dropout is especially high in the fields of computer science, technology and natural sciences.

In the school year 2019/2020 vocational education was provided in 37 institutions. The number of vocational education students has also slightly decreased during the past 15 years, while being the highest in 2009/2010 when the total number of students was 28 363 and being the lowest in 2018/2019 when the total number of students was 23 387.

23.1.5 Educational Research and International Collaboration

Estonian Ministry of Education and Research aims to support science-based policy making, and therefore, the ministry is constantly analyzing, researching and evaluating the situation in education. Also, the ministry orders studies from experts and researchers from universities.

The Ministry of Education and Research develops an annual plan for research projects, including projects within its own jurisdiction, wider national projects and international projects. E.g., the plan for 2020 includes participation in several international studies like PISA, TALIS, PIAAC, ICCS, research about the use of digital learning resources, about learners' individual learning paths, about the effect of implementing anti-bullying programs in schools, and a survey about satisfaction with youth work (Ministry of education and research, n.d.).

In addition, Estonian Research Council aims to support research and innovation in Estonia. The Estonian Research Council provides research and mobility grants in order to facilitate high-level research projects to strengthen international competitiveness of Estonian research and development; to promote cooperation between the government, businesses and research institutions; and to contribute to the internationalization of research and support the next generation of researchers (Estonian Research Council, n.d.). Also, researchers from universities are constantly looking for international project funding for educational research (e.g., Erasmus + and Horizon funding).

23.2 Current Situation of Science Education

23.2.1 Policies and Standards

Estonian educational system is based on the Lifelong Learning Strategy 2020 (Ministry of education and research, n.d.). Lifelong learning begins with general education where common a curriculum is taught in all levels of education. The Lifelong Learning Strategy has five strategic goals: change in the approach to learning, competent and motivated teachers and school leadership, concordance of lifelong learning opportunities with the needs of the labour market, a digital focus in lifelong learning (including using modern digital technology for learning and teaching, improving digital skills of the population), and equal opportunities and increased participation in lifelong learning (Ministry of education and research, n.d.).

Science education in Estonian basic and upper-secondary schools follow the national curriculum. The responsibility for curriculum development is at the Ministry of Education and Research who chooses a certain group of people for curriculum development. Educational Forum, educational researchers at universities and practitioners at schools have usually strong influence on curriculum development.

While developing Estonian national curriculum, good examples from other countries have been used. For example, in 2005 Finnish curriculum was taken as guidance in curriculum development, because Finland had been successful in international ratings of education.

23.2.2 Curriculums, Digital Resources and Teacher Training

The last curriculum reform in Estonia took place in 2011 and the current national curriculum for basic school and upper-secondary school were last revised in 2017. The curriculum consists of the general part and appendixes. The general part provides an overview of the basic values of education, learning and educational goals, the concept of learning and the learning environment, overview of competencies required at each school level, learning organization, assessment and completion of class and school etc. Learning and educational goals include 8 general competences that are needed in all subject areas. These competences are developed in all subjects and through extracurricular activities. The general competences are: (1) cultural and value competences, (2) social and civic competence, (3) self-determination competence, (4) learning competence, (5) communication competence, (6) mathematics, science and technology competences, (7) entrepreneurial competence, (8) digital competence (Basic school national curriculum, 2011; Upper-secondary school national curriculum, 2011).

Appendix 4 of the national curriculum focuses on science education. It presents the subject area plans, subject curricula and descriptions of cross-cutting topics. The science subjects taught in basic school are: general science (is taught in grades 1–7), biology (taught in grades 7–9), geography (taught in grades 7–9), physics (taught in grades 8–9), and chemistry (taught in grades 8–9). The science subjects taught in upper-secondary school are: biology (divided into 4 courses), chemistry (divided into 3 courses), geography (divided into 3 courses), and physics (divided into 5 courses). Several cross-cutting topics between the science subjects are also presented in the curriculum. These are for basic school: (1) environment and sustainable development, (2) citizens' initiative and entrepreneurship, (3) cultural identity, (4) information environment, (5) technology and innovation, (6) health and safety, (7) values and morals. One more cross-cutting topic is added to upper-secondary science curriculum, this is (8) lifelong learning and career planning.

Science curriculum in both basic school and upper-secondary school emphasizes the use of technology and inquiry-based learning. Inquiry-based learning can be defined as a process of discovering new relations between different variables through formulation hypothesis and testing the hypothesis in experiments or by collecting data through observations (Mäeots, Pedaste & Sarapuu, 2011). It means that students are following methods that are similar to those that scientists use in order to construct knowledge (Keselman, 2003). Inquiry-based learning is a student-centered, active learning approach that is focusing on questioning, critical thinking and problem solving (Savery, 2015).

Estonian science teachers can use a variety of digital resources to teach science in the classroom. The biggest repository of digital resources is called e-schoolbag (www.e-koolikott.ee). The portal included different learning materials for kindergarten, basic school, upper-secondary school and vocational schools. Every teacher can freely use this portal to search learning materials, used materials that someone else has created and share their own materials (including tasks, work sheets, games, videos etc.). All these learning resources are linked with the national curriculum. In addition, teachers and students can access digital textbooks and digital workbooks (https://www.opiq.ee/).

Science teachers are trained at two universities in Estonia. Science teachers, among other subject teachers, are required to have a Master's degree and teacher profession to teach at school.

23.2.3   Student Assessment and Achievement

On the national level, students' science achievements are assessed through standard-determining tests. The general science tests are created to map students' knowledge and skills at grade 4 and grade 7 to provide help and tools for the teachers for further organizing the teaching. The test is carried out electronically by the Examination and Qualification Centre. In 2016, the concept of the general science test was revised and the new concept focuses on evaluating students' inquiry skills and decision making skills (Pedaste et al., 2018).

Estonian 15 years old students are participating in the PISA assessment and they are showing good results. In 2018, Estonian students were in first place among European and OECS countries in science achievement. In general, the Estonian students' science achievement has been on a high level since 2016 when Estonia participated in PISA for the first time (Tire et al., 2019). In addition, Estonian students have participated in TIMSS and have shown good results there as well.

23.2.4   Outreach Science Education

Science and technology venues and centers are located mostly in the bigger Estonian cities. Here are some examples of the activities that science and technology centers in Estonia offer. The Science Centre AHHAA (www.ahhaa.ee) is located in Tartu and the center aims to encourage studying through the joy of discovery. More than 3 million people of different ages have visited their exhibitions and other science events. The center offers different workshops, planetarium, science theatre shows and special programs for schools. The Energy Discovery Center (www.energiakesku s.ee) is located in Tallinn and it has over 100 exhibits, which help to study physical phenomena (e.g., electricity). They also offer different programs and workshops for schools. Both of these science centers are involved in international cooperation (e.g., are members of the European Network of Science Centers and Museums) and collaborate with universities.

In addition, several other organizations offer school visits and science theaters for students, and science competitions are organized. For example, a student research festival takes place once a year. At this festival, students make poster presentations about their research, participate in workshops and science shows (Estonian Research Council, n.d.).

Estonian Research Council has initiated TeaMe + program which aims to increase the ratio of informal science education activities (e.g., after school activities) and train people who are able to supervise these activities (Estonian Research Council, n.d.). In addition, Estonia has an active STEM Education Union whose activities include: sharing knowledge, organizing workshops and events, developing training programs for STEM educators, participating in policy making, raising awareness about STEM education and STEM after school activities, giving awards to STEM educators etc. (Eesti TEADUSHUVIhariduse Liit, n.d.).

23.2.5   Utilizing Emerging Technologies

Estonian schools are using more and more emerging technologies in teaching science. This is often happening thanks to the financial support of governmental organizations.

Based on a study where data was collected with Digital Mirror tool (https://dig ipeegel.ee/) we can see that, in general, schools have computer classrooms and/or mobile computer classrooms (i.e. laptops), computer and data projector in all classrooms, smartboards in some classrooms, tablets that students and teachers can use, different robotics tools (especially in pre-school and primary school level), sensor-based technology (e.g., Vernier sensors and Globisens Labdiscs), video equipment, and in some cases even drones, augmented reality (AR), virtual reality (VR) or artificial intelligence (AI) tools. A study showed that 64,1% of Estonian teachers have the opportunity to use smartboard in their classes (Tomson, 2018). However, AR and VR tools are regularly used in only a few schools, and even these schools are just testing the opportunities (Dremljuga-Telk, 2018). In addition, 97% of basic school students have their own smartphones which enables teachers to use the bring your own device (BYOD) approach (Hiiesalu, 2016). Most of the schools have also hired an educational technologist who provides support to the teachers in using technology.

During the past decades several technology-enhanced learning environments have been developed that Estonian science teachers can use in their teaching practices. For example, SCY (de Jong et al., 2010), Young Researcher (Kori et al., 2014; Mäeots & Pedaste, 2009), Go-Lab platform (https://www.golabz.eu/). In addition, some teachers find new ways to teach outside the classroom and use mobile applications, e.g., to learn in a zoo (Mettis & Väljataga, 2019). However, when outdoor learning scenarios created by Estonian K12 teachers were analyzed, it was found that teachers tend to design learning scenarios, which hardly embrace learning contexts and enable them to support higher order knowledge building (Väljataga & Mettis, 2019).

Some Estonian schools also have sensor-based technology available for science teachers, for example, Globisens Labdiscs are used for inquiry-based learning and citizen science activities (Kori, 2020). However, it is challenging for the teachers to start using this new technology (Kori & Pata, 2020). Also, robotics and mathematics are integrated in Estonian primary schools. Mathematics and computer science teachers sometimes teach these lessons together (Stepanova et al., 2020).

In addition, AR shows good potential for supporting science learning in Estonian schools. Some schools already have the AR technologies and systematic review conducted by Estonian researchers shows that in the context of inquiry-based learning, AR can be implemented successfully to achieve cognitive, motivational and emotional learning goals (Pedaste et al., 2020). VR and AI also show good potential for supporting science education. Using these emerging technologies in science education can support the development of students' computational thinking skills and digital competences.

23.3 Requirements for Future Development of Science Education

The future developments of Estonian science education seems to be moving towards integrating different science related subjects together (e.g., carrying out a project that requires knowledge of biology, chemistry, physics and geography), integrating science classes with technology, engineering and mathematics (STEM), and even integrating art with science, technology, engineering and mathematics (STEAM).

Integrating STEAM subjects may help to develop two of the general competences emphasized in the Estonian national curriculum. One of them is mathematics, science and technology competence which is defined as following "ability to use mathematical language, symbols, methods in school and everyday life; the ability to describe the world around us using scientific models and measurement tools and to make evidence-based decisions; understand the importance and limitations of science and technology; use new technologies in a targeted way". This competence is related to complex problem solving which includes three types of skills: inquiry skills, computational thinking skills, and mathematical problem-solving skills (Pedaste et al., 2019). The other one is digital competence which is defined as "the ability to use renewable digital technologies to cope in a rapidly changing society, both through learning, citizenship and community interaction; to find and store information by digital means and to assess its relevance and reliability; participate in digital content creation, including the creation and use of texts, images, multimedia; use digital tools and techniques suitable for problem solving, communicate and collaborate in different digital environments; be aware of the dangers of the digital environment and be able to protect their privacy, personal data and digital identity; follow the same moral and value principles in the digital environment as in everyday life".

In addition to mathematics, science and technology competence and digital competence, science education has an important role in developing other general competence emphasized in the Estonian national curriculum as well. These are cultural and value competence, social and civic competence, self-determination competence, learning competence, communication competence, and entrepreneurial competence.

The future of science education also seems to move towards guiding students to become active citizens who use scientific thinking and creativity in their everyday lives. This can be done in school through learning scenarios which focus on context-based learning (Kang et al., 2019) and solving socio-scientific issues (Sadler, 2011). This means that learning should be relevant to some aspects of students' lives and they can use scientific knowledge and procedures to solve societal problems. Research has shown that including socio-scientific issues in science education has an important role in promoting global citizenship (Lee et al., 2012, 2013). According to Chowdhury et al. (2020a) science education has an important role in developing active informed citizens who are scientifically literate, aware and able to conceptualize from a scientific perspective, who are willing to take action in scientific activities, and who collectively contribute to science embedded social issues (not only in national level, but also in global level). Science education should be more than promoting cognitive development within a science frame; in the future science and society interactions should provide a meaningful learning context for students (Chowdhury et al., 2020b).

The shifting focus on solving socio-scientific issues brings science education outside the classroom. Citizen science is one new approach that is reaching into Estonian schools right now. Citizen science means that regular people are carrying out scientific research together with professional scientists (Silvertown, 2009) and using this approach in school has found to have a positive effect on learners (e.g., Kermish-Allen et al., 2018; Eastman et al., 2014; Hilelr & Kintsantas, 2014). However, often students are only involved in the data collection and analysis phase of the citizen science projects. To involve the students more, researchers have merged citizen science and inquiry-based learning into one term—citizen inquiry (Sharples et al., 2017). Therefore, in the future students will probably carry out more citizen inquiry projects where they come up with a question or topic of interest and they start an investigation around it. They ask other people and scientists to join the investigation, collect data which are shared and available for everyone, discuss the topics online and try to reach consensus about the findings (Sharples et al., 2017).

23.4 Discussion and Conclusion

In conclusion, the future of science education seems to be moving towards integrating science classes with technology, engineering, art and mathematics, developing students' general competences and guiding students to become active global citizens who use scientific knowledge and procedures to solve societal problems. However, there are several challenges that we are facing in this process.

Firstly, changes in the natural curriculum are needed to support the integration of biology, chemistry, physics, geography, and mathematics. Teachers claim to be busy following the topics in the curriculum and often do not have time in the lessons for using new approaches, for example, to solve tasks that develop learners' general competences.

Another challenge is that Estonian teachers are rather old. Every fifth mathematics, chemistry, geography and biology teacher is at least 60 years old and every fourth physics teacher is at least 60 years old (Kutsekoda, 2018). The teachers have a lot of experience in teaching, but it may be challenging for them to keep up with innovations, new technologies and new teaching methods. Often schools have different technologies that can be used for teaching but teachers will not use them. Teachers' attitudes have found to be a key aspect of technology use in school. A study among Estonian STEM teachers found that when teachers see that a mobile device is useful then it is strongly related to how easy it is to use it and how enjoyable it is to use it. Performance expectancy, self-efficacy and anxiety predicted 52% of the variance in teachers' willingness to use mobile devices in teaching and social aspects served as facilitators for teachers' attitudes and behavioral intention (Adov et al., 2020). Therefore, more effort should be put on changing teachers' attitudes. Of course using technology should not be the aim by itself, but the technology should be used in a way that it supports the learning process effectively.

A variety of learning scenarios that teachers can use freely have been developed in the context of socio-economic issues, outdoor learning, inquiry-based learning, citizen science, context-based learning etc. and the number of the scenarios will probably increase in the future. However, the challenge is that we do not have evidence of which learning scenarios are effective and which are not. Therefore, we need to find a way to measure effectively what makes a good learning scenario (Kang et al., 2019).

References

Adov, L., Pedaste, M., Leijen, Ä., & Rannikmäe, M. (2020). Does it have to be easy, useful, or do we need something else? STEM teachers' attitudes towards mobile device use in teaching. *Technology, Pedagogy and Education, 29*(4), 511–526. https://doi.org/10.1080/1475939X.2020.1785928

Basic school national curriculum. (2011). https://www.riigiteataja.ee/akt/114022018008.

Cental Intelligence Agency. (n.d.). The World Factbook. Retrieved 14. September, 2020, from https://www.cia.gov/library/publications/resources/the-world-factbook/geos/en.html.

Chowdhury, T. B. M., Holbrook, J., & Rannikmäe, M. (2020a). Addressing sustainable development: promoting active informed citizenry through trans-contextual science education. *Sustainability, 12*(8). https://doi.org/10.3390/su12083259.

Chowdhury, T. B. M., Holbrook, J., & Rannikmäe, M. (2020). Socioscientific issues within science education and their role in promoting the desired citizenry. *Science Education International, 31*(2), 203–208.

Dremljuga-Telk, M. (2018). Hariduse tehnoloogiakompass. Haridus- ja Noorteamet. Retrieved 29 Sept 2020, from https://kompass.hitsa.ee/

Eastman, L., Hidalgo-Ruz, V., Macaya-Caquilpán, V., Nuñez, P., & Thiel, M. (2014). The potential for young citizen scientist projects: A case study of Chilean schoolchildren collecting data on marine litter. *Journal of Integrated Coastal Zone Management, 14*, 569–579. https://doi.org/10.5894/rgci507

Eesti TEADUSHUVIhariduse Liit (n.d.). Retrieved 29 Sept 2020, from https://teadushuvi.ee/

Estonian Research Council (n.d.). Retrieved 29 Sept 2020, from https://www.etag.ee/en/estonian-research-council/

Estonian Research Council (n.d.). Retrieved 29 Sept 2020, from https://www.etag.ee/tegevused/tea dpop/teamepluss/teadushuviharidus/

HaridusSilm (n.d.). Retrieved 14 Sept 2020, from https://www.haridussilm.ee/.

Hiiesalu, T. (2016). Nutiseadmed muudavad iga klassi arvutiklassiks. Retrieved 14 Sept 2020, from https://opleht.ee/2016/12/nutiseadmed-muudavad-iga-klassi-arvutiklassiks/.

Hiller, S. E., & Kitsantas, A. (2014). The effect of a horseshoe crab citizen science program on middle school student science performance and STEM career motivation. *School Science and Mathematics, 114*, 302–311. https://doi.org/10.1111/ssm.12081

de Jong, T., van Joolingen, W., Giemza, A., Girault, I., Hoppe, U., Kindermann, J., Kluge, A., Lazonder, A., Vold, V., Weinberger, A., Weinbrenner, S., Wichmann, A., Anjewierden, A., Bodin, M., Bollen, L., d'Ham, C., Dolonen, J., Engler, J., Geraedts, C., Grosskreutz, H., et al. (2010). Learning by creating and exchanging objects: The SCY experience. *British Journal of Educational Technology, 41*(6). https://doi.org/10.1111/j.1467-8535.2010.01121.x.

Kang, J., Keinonen, T., Simon, S., Rannikmäe, M., Soobard, R. & Direito, I. (2019). Scenario Evaluation with Relevance and Interest (SERI): Development and Validation of a Scenario Measurement Tool for Context-Based Learning. *International Journal of Science and Mathematics Education*, 1−22. https://doi.org/10.1007/s10763-018-9930-y

Kermish-Allen, R., Peterman, K., & Bevc, C. (2018). The utility of citizen science projects in K-5 schools: Measures of community engagement and student impacts. *Cultural Studies of Science Education, 14*, 1–15. https://doi.org/10.1007/s11422-017-9830-4

Keselman, A. (2003). Supporting inquiry learning by promoting normative understanding of multivariable causality. *Journal of Research in Science Teaching, 40*, 898–921.

Kori, K. (2020). Four steps towards implementing citizen science approach in school. In *2020 IEEE 20th International Conference on Advanced Learning Technologies (ICALT)* (Vol. 1). IEEE, 191−193.https://doi.org/10.1109/ICALT49669.2020.00062.

Kori, K., Mäeots, M. & Pedaste, M. (2014). Guided reflection to support quality of reflection and inquiry in web-based learning. In: *The European procedia social and behavioral sciences* (pp. 242−251). Elsevier. doi: https://doi.org/10.1016/j.sbspro.2014.01.1161.

Kori, K. & Pata, K. (2020). Training teachers to use globisens labdiscs for citizen science projects in school. *Journal of Strategic Innovation and Sustainability, 15*(3). https://doi.org/10.33423/jsis.v15i3.2951.

Kutsekoda (2018). Tulevikuvaade tööjõu- ja oskuste vajadusele: Haridus ja teadus. Uuringu Lühiaruanne. Retrieved 14 Sept 2020, from https://oska.kutsekoda.ee/wp-content/uploads/2016/12/oska_HT_veeb.pdf.

Lee, H., Chang, H., Choi, K., Kim, S. W., & Zeidler, D. L. (2012). Developing character and values for global citizens: analysis of pre-service science teachers' moral reasoning on socioscientific issues. *International Journal of Science Education, 34*, 925–953. https://doi.org/10.1080/09500693.2011.625505

Lee, H., Yoo, J., Choi, K., Kim, S. W., Krajcik, J., Herman, B. C., & Zeidler, D. L. (2013). Socio-scientific issues as a vehicle for promoting character and values for global citizens. *International Journal of Science Education, 35*, 2079–2113. https://doi.org/10.1080/09500693.2012.749546

Mettis, K., & Väljataga, T. (2019). Mapping the challenges of outdoor learning for both students and teachers. In: T. Väljataga, M. Laanpere (Ed.). *Digital Turn in Schools—Research, Policy, Practice—Proceedings of ICEM 2018 Conference* (pp. 51−65). Springer.

Ministry of education and research (n.d.). Pre-school, basic and secondary education. Retrieved 14 Sept 2020, from https://www.hm.ee/en/activities/pre-school-basic-and-secondary-education.

Ministry of education and research (n.d.). Estonian Lifelong Learning Strategy 2020. https://www.hm.ee/en/estonian-lifelong-learning-strategy-2020.

Ministry of education and research (n.d.). Uuringud ja statistika. Retrieved 14 Sept 2020, from https://www.hm.ee/et/tegevused/uuringud-ja-statistika-0.

Mäeots, M., & Pedaste, M. (2009). Uurimuslike oskuste arendamine õpikeskkonnas "Noor teadlane". LoTe. *Ajakiri loodusteaduste õpetajatele*, 4–6.

Mäeots, M., Pedaste, M., & Sarapuu, T. (2011). Interactions between inquiry processes in a web-based learning environment. In *Proceedings of the 2011 11th IEEE international conference on advanced learning technologies: 11th IEEE international conference on advanced learning technologies* (pp. 331–335). https://doi.org/10.1109/ICALT.2011.103.

Pedaste, M. Brikker, M., Rannikmäe, M., Soobard, R., Mäeots, M., & Reiska, P. (2018). Loodusvaldkonna õpitulemuste hindamine, 2018. Retrieved 28 Sept 2020, from https://www.innove.ee/wp-content/uploads/2018/09/Loodusvaldkonna_e_hindamise_kontseptsioon_august_2018.pdf

Pedaste, M., Mitt, G., & Jürivete, T. (2020). What Is the effect of using mobile augmented reality in K12 inquiry-based learning? *Education Sciences, 10*(4). https://doi.org/10.3390/educsci10040094.

Pedaste, M., Palts, T., Kori, K., Sormus, M., & Leijen, Ä. (2019). Complex problem solving as a construct of inquiry, computational thinking and mathematical problem solving. In *2019 IEEE 19th International Conference on Advanced Learning Technologies (ICALT)* (pp. 227–231). IEEE.https://doi.org/10.1109/ICALT.2019.00071.

Sadler, T. D. (2011). Situating socio-scientific issues in classrooms as a means of achieving goals of science education. In *Socio-Scientific Issues in the Classroom: Teaching, Learning and Research*; Sadler, T.D., Ed.; Springer Netherlands: Dordrecht, The Netherlands, 1–9.

Savery, J. R. (2015). Overview of problem-based learning: Definitions and distinctions. *Essential readings in problem-based learning: Exploring and extending the legacy of Howard S. Barrows, 9*, 5–15.

Sharples, M., Aristeidou, M., Villasclaras-Fernández, E., Herodotou, C., & Scanlon, E. (2017). The Sense-it App: A Smartphone Sensor Toolkit for Citizen Inquiry Learning. *International Journal of Mobile and Blended Learning, 9*, 16–38. https://doi.org/10.4018/IJMBL.2017040102

Silvertown, J. (2009). A new dawn for citizen science. *Trends in Ecology & Evolution, 24*, 467–471. https://doi.org/10.1016/j.tree.2009.03.017

Statistics Estonia (n.d.). Retrieved 14. September, 2020, from https://www.stat.ee/en.

Stepanova, J., Leoste, J. & Heidmets, M. (2020). Co-teaching robot-supported math lessons in the third grade. In: *International Conference on the Advancement of STEAM 2020. Borderless Connectivity* (37–42). International Conference on the Advancement of STEAM 2020. Borderless Connectivity. Korea National University of Education, Korea. 26 June 2020. the International Society for the Advancement of STEAM.

Tamm, M. (n.d.). Estonia's Cultural Changes in an Open World. Retrieved 14. September, 2020, from https://2017.inimareng.ee/en/estonias-cultural-changes-in-an-open-world/.

Tire, G., Puksand, H., Leppman, T., Henno, I., Lindemann, K., Täht, K., Lorenz, B. & Silm, G. (2019). PISA 2018 Eesti tulemused. Eesti 15-aastaste õpilaste teadmised ja oskused funktsionaalses lugemises, matemaatikas ja loodusteadustes.

Tomson, T. (2018). *Interactive Whiteboards in Estonian Schools – Current Situation and Challenges*. Master thesis. Tallinn university.

Upper-secondary school national curriculum (2011). Retrieved 14. September, 2020, from https://www.riigiteataja.ee/akt/129082014021.

Väljataga, T. & Mettis, K. (2019). Analyzing Integrated Learning Scenarios for Outdoor Settings. In: Michael A. HerzogZuzana KubincováPeng HanMarco Temperini (Ed.). *Advances in Web-based learning - ICWL2019* (287–294). Springer. doi: https://doi.org/10.1007/978-3-030-35758-0_27.

Külli Kori is working as a lecturer of research methods in IT at the Tallinn University, School of Digital Technologies. She received her Ph.D. in educational sciences at the University of Tartu in 2017 and finished post-doctoral studies at Tallinn University in 2020. She also has a master's degree in biology didactics and a bachelor's degree in genetic engineering from the University of Tartu. Her research focuses on citizen science in education, educational technology, STEM education, inquiry-based learning, and student retention.

Chapter 24
Science Education in Greece

**Charalampos Karagiannidis, Angeliki Karamatsouki,
and George Chorozidis**

Abstract Science education is vital to promote scientific thinking, knowledge, and skills with the aim to prepare students participate actively in the today's complex scientific and technological society. This study aims to examine the current situation of science education in Greece. Greece support many separate initiatives for the development of science education, but there is a lack of an overall strategy to improve science education. The performance of the Greek students in all PISA cycles was below the OECD average. To this extent, further changes are required especially in science education curricula, in teaching approaches and in teachers' professional development to fully equip and engage Greek students to scientific thinking and learning.

Keywords Science education · Digital Resources · Greek Educational System

C. Karagiannidis (✉)
Department of Special Education, ICT Applications in Learning and Special Education,
University of Thessaly, Volos, Greece
e-mail: karagian@uth.gr

A. Karamatsouki
Department of Education, University of Nicosia, Nicosia, Cyprus
e-mail: karamatsouki.a@unic.ac.cy

G. Chorozidis
Department of Special Education, University of Thessaly, Volos, Greece
e-mail: gchorozidis@uth.gr

© The Author(s), under exclusive license to Springer Nature Singapore Pte Ltd. 2022   399
R. Huang et al. (eds.), *Science Education in Countries Along the Belt & Road*,
Lecture Notes in Educational Technology,
https://doi.org/10.1007/978-981-16-6955-2_24

24.1   Overview of the Country

24.1.1   Geographical Location, Population and Political System

Greece is located in south-eastern Europe, on the southern end of the Balkan Peninsula; it lies at the meeting point of three continents—Europe, Asia and Africa. Greece is bordered to the east by the Aegean Sea, to the south by the Mediterranean Sea, and the west by the Ionian Sea. There are land borders to the north and northeast with, from west to east, Albania, the Republic of North Macedonia, Bulgaria, and Turkey. The total area of Greece is 131,957 sq. km and consists of three main geographic areas, a peninsular mainland being the biggest geographic feature of the country, the Peloponnese peninsula that is separated from the mainland by the canal of the Corinth Isthmus, and around 6.000 islands and islets, scattered in the Aegean and Ionian Sea, most of them grouped in clusters, that constitute the unique Greek archipelago. The country has the longest coastline in Europe and is the southernmost country in Europe.

The population of Greece is 10.8 million permanent residents, and of these 5,303,223 are men and 5,513,063 are women (census 2011). The average age is 41.9 years. The table shows some of the population characteristics (Hellenic Statistical Authority, Census 2011).

Of the total of 10.8 million Greek citizens, 91.6% people have Greek citizenship, 1.8% are citizens of other E.U countries, 6.5% people are citizens of other countries and 0.04% people are without citizenship or have no specified citizenship. Based on population forecasts of ELSTAT, the population structure per age groups will be rather different in the upcoming decades due to adverse demographic changes and the trends of low birth rate and population aging (Table 24.1).

Greece is a Parliamentary Republic. The President, elected by Parliament every five years, is Head of State. The Prime Minister is Head of Government. The legislative body is the Greek Parliament, which has mainly regulatory competences. The judiciary branch is independent of the legislative branch.

24.1.2   Current Situation of Economic, Technologies and Cultural Development

24.1.2.1   Economic and Technologies Development

The current political status has been influenced by the economic crisis. From late 2009—early 2010, due to both international and domestic factors, Greece confronted serious economic hardships. In 2010 a Memorandum of Understanding signed between Greece and the International Monetary Fund (IMF), the European Union

Table 24.1 Population
situation

	Total	%
Population	10.816.28	
Gender		
Male	5.303.223	49
Female	5.513.063	51
Age		
0–14	1.514.280	14
15–64	7.214.278	67
65 +	2.044.278	19
Average age	41.9	
Citizenship		
Greek	9.904.286	91.6
EU countries	199.121	1.8
Other countries	708.054	6.5
Without citizenship or no specified	4.825	0.04

(EU) and the European Central Bank (ECB) to receive financial assistance for reducing its debt.

The economic and political situation in Greece has had a marked impact on the education system as well. The public education expenditure as a share of GDP was 3.8% in 2017, below the EU average (4.6%) (OECD, 2018).

The COVID-19 pandemic and containment measures are projected to reduce GDP by 8% in 2020 if there are no further virus outbreaks. If there is a second virus outbreak, the fall in GDP will amount to 9.8%. The negative impact on tourism, investment and public finances is a setback to Greece's longer term recovery (OECD, 2020).

Innovation is a sector of science and technologies in Europe. Greece is ranked 20th in Europe (9th from the end) and is ranked among the countries of the third category—it is a country of "moderate innovation". Greece's relatively strong points are the innovative sector (37.6% increase from 2012) and the linkages sector (an increase of 43.4% from 2012). Innovators and Linkages are the strongest dimensions and Greece is above the EU average. Small and medium-sized enterprises in Greece seem to be innovative and offer innovative solutions to the market, while at the same time, there is an improvement in the collaborations between the public and the private sector in projects aimed at innovation and research (European Commission, 2020).

It should be noted that in recent years, when Greece is experiencing a major economic crisis, there has been a mass escape of highly skilled people. This is due to rising unemployment and austerity measures that have affected both education and the labor market, while government funding for research has been severely curtailed, which, combined with reduced wages, has discouraged scientists working abroad from doing so. return to their country (Ifanti et al., 2013).

24.1.2.2 Cultural Development

In Greece, the Ministry of Culture and Sports has started the implementation of a program focuses on mapping the whole cultural and creative industries of the country, aiming to a comprehensive understanding of the cultural creation and the development of specific supporting policies. Additionally, it operates a database where private cultural institutions can be registered and a platform, which includes information on Intercultural Dialogue issues as well as, all the information about festivals and other cultural events nationwide. The Ministry of Culture and Sports and its agencies have set up or participate in a number of inter-ministerial committees or joint programmes (The Compendium of Cultural Policies & Trends, 2019):

- support for modern Greek studies abroad involving also the General Secretariat of Greeks Abroad and the Hellenic Foundation for Culture
- the cultural heritage digitisation programmes
- supervision of arms-length organisations, such as the Unification of the Archaeological Sites of Athens SA
- there is co-operation with the Department of Planning on architectural and urban conservation and cultural landscape projects (e.g., Rhamnous, Patmos).
- issues related to international conventions, such as the Council of Europe's European Convention for the protection of audiovisual heritage are managed in co-operation with the General Secretariat for Communication.

Funding for cultural activities in Greece is mainly from public sources, either from the Ministry of Culture and Sports in the central government or from the local budgets of authorities at local and regional levels (Mergos & Patsavos, 2017).

24.2 Overview of the Education Development

24.2.1 Education System and Policy

The central administrative body for the Greek education system is the Ministry of Education and Religious Affairs, which is responsible for taking the key decisions related to long-term objectives (Eurydice, 2018/2019). The Greek Education System is mainly divided into three levels, primary education, secondary education and tertiary education. Compulsory education in Greece is mandatory for all children between the ages of 4 to 15 and lasts 11 years.

Primary education includes pre-primary and primary schools. Primary school concerns children in the age range of 6–12 years. School year 2020/21 foresees the integration of the two-year compulsory pre-primary school. *Secondary* education includes the lower secondary school (compulsory) and the general or vocational upper secondary schools. In lower secondary school attendance starts at the age of 12 and lasts 3 years, while pupils enrol at the age of 15 to the upper secondary school, which lasts another 3 years. *Higher* education is the last level. Most undergraduate degree programs take four academic years of full-time study. Postgraduate courses last from one to two years and doctorates at least three years (Eurydice, 2018/2019) (Fig. 24.1).

The new educational policy agenda foresees changes for all levels of education, such as new curricula, pilot skills workshops in the compulsory school programme, professional training and development of teaching staff, digital technology and exploitation and many others (Eurydice, 2018/2019).

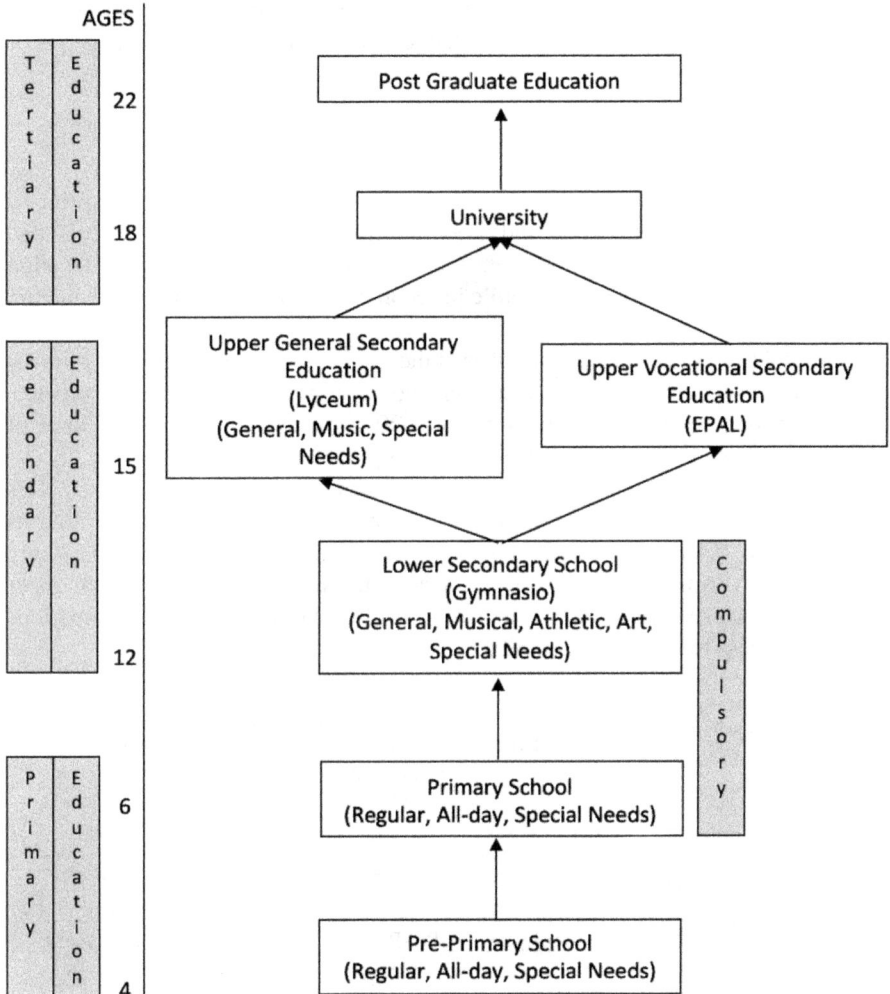

Fig. 24.1 Structure of the Greek Education System

24.2.2 Statistics on the National Education

Table 24.2 presents the teacher' level distribution in Greece in 2018, both in public and private institutions, by level of education. In 2018, the number of primary teachers was 69.766, the number of lower and upper secondary teachers was 76.625 and academic staff in tertiary education for Greece was 17.288.

Teachers of all educational levels in Greece are graduates of higher education. For all other levels except Tertiary, teachers must complete a first cycle degree (UNESCO, 2015). Academic staff in higher education must have at least a doctoral degree and their doctoral dissertation and the overall research and scientific work must be consistent with the field of knowledge of the announced position.

24.2.2.1 Enrollment Rate

Table 24.3 presents the overall number of the enrolled students in Greek pre-primary primary and secondary education and its evolution during the years 2018–2019 and 2019–2020. The number of students enrolled in pre-primary and primary education showed an increase of 9% and 8%, while in secondary education showed a reduction of 10%.

Table 24.4 presents the overall number of the graduated students in Greek primary and secondary education (compulsory and non-compulsory). The collected data showed a reduction in 2019 compared to 2020 regarding the overall number of graduated students both in primary and secondary education levels.

In tertiary education level, Greece has the highest enrolment rates in bachelor's programmes among 19–24 years olds, in comparison to other OECD countries (OECD, 2019a, b, c). Admissions to bachelor programmes are not open, since only a fixed number of student places is set at the central level. Students of the final upper secondary school year must perform at a minimum level in compulsory examinations in order to access tertiary institutions.

Table 24.2 Number of teachers in different levels

School year	Primary education teachers	Secondary Education teachers	Tertiary education teachers
2018	69.766	76.625	17.288

Table 24.3 Enrolled students per education level

School year	Pre-Primary education	Primary education	Secondary Education
2018–19	138.035	593.721	618.125
2019–20	151.804	645.250	557.516

Source Hellenic Statistical Authority

Table 24.4 Graduated
students per education level

School year	Primary education	Secondary education
2018–19	96.242	185.267
2019–20	101.771	175.815

Source Hellenic Statistical Authority

In 2017, 735,027 students were enrolled in Greek universities, undergraduate and postgraduate, active and "eternal". This is a very high number, as it corresponds to 6.83% of the total population of the country. Specifically, in the last place in Europe is our country and in the index of the percentage of graduates on the total of students: the index in Greece was 9.41% in the year 2017, while the European average of "28" was 24.15%. Of course, as pointed out by ADIP (Hellenic Authority for Higher Education), the percentages are "burdened" by the fact that the "eternal" students are included, the existence of which is not understood in the universities of Europe. Of the 735,027 students in Greek universities, 86.26% are in the undergraduate cycle, 9.84% in postgraduate studies and 3.9% in doctoral studies.

24.2.3 Educational Research and International Collaboration

The educational research centers in Greece, which are under the supervision of the General Secretariat for Research and Technology (GSRT) of the Ministry of Education and Religious Affairs, are the following:

- "Athena" Research and Innovation Center in ICT and Knowledge Technologies, (https://www.athena-innovation.gr/)
- Centre for Research and Technology Hellas (CERTH) (https://www.certh.gr/root. en.aspx)
- National Research Centre for Scientific Research "Demokritos" (NSCR) (http:// www.demokritos.gr/)
- National Center for Social Research (https://www.ekke.gr/)
- Foundation for Research and Technology—Hellas (FORTH) (https://www.forth. gr/index.php?l=e)

Each of the above research centers exploit the outcomes of its research activities by developing mock-up technologies, ready to use services or other products.

There are also two institutes controlled by the Ministry of Education, the Pedagogical Institute (PI) and the Institute for Technological Education (ITE). The PI is responsible among others for research relating to primary and secondary education.

24.3   Current Situation of Science Education

24.3.1   Policies and Standards

The improvement of science education has been high on the political agenda of many European countries during the last decades and many programs and projects have been set up in order to identify and share best practices for promoting science education and ensure sustainability. In Greece, there is an absence of an overall national strategy for science education. On the other hand, Greece has developed and participated in activities and projects for the development of science education. These initiatives promote science education through school partnerships with government institutions, higher education institutions or stakeholders outside the school setting, such as museums or science centres (Eurydice, 2011).

24.3.2   Curriculums, Digital Resources and Teacher Training

The curricula and timetables of Primary and Secondary education in Greece are drawn up by the Institute of Educational Policy (IEP), which is responsible for issues related to study programmes, school textbooks and other teaching material. In Greece, the science curriculum area consists of separate science subjects, in both primary and secondary education. More specifically, in Primary Education there are separate science subjects such as: mathematics, study of the environment, geography, natural sciences, information and communication technologies (Eurydice, 2019a). In Lower-Secondary Education, the school timetable includes science-related courses, such as physics, chemistry, biology, geography, mathematics, technology, and information technology subjects (Eurydice, 2019b). In General Upper Secondary Education, there are general education science-related courses (mathematics, natural Sciences, introduction to the principles of computer science) and science studies specialization courses (physics, mathematics, and chemistry) (Eurydice, 2019c). Similarly, in Vocational-Upper Secondary Education there are general education science-related subjects (mathematics, physics, chemistry, biology) and elective subjects (introduction to electrical & electronic engineering, introduction to mechanical engineering). The curriculum also includes technological-vocational science-related sectors, such as Electrical & Electronic Engineering and Automation sector, Mechanical Engineering sector and the sector of Informatics (Eurydice, 2019d).

In Greece, the highly centralized education system is leaving little autonomy to the education staff. Recently, the significant reduction of the amount of syllabus enable teachers to participate in research programs and develop strategies to integrate and implement STEM approaches and activities into schools (Patrinopoulos & Iatrou, 2019). Nevertheless, all these actions are taking place individually and non-systematically. In this context, the New Greek Science Curriculum (NGSC) for Primary Education includes topics from physics, biology, chemistry, geography, technology, and environment aiming to provide high quality science education and to develop students' scientific skills, understanding and competences in both formal and informal settings (Plakitsi et al., 2013). Greece has currently been engaged in curriculum reform to meet the needs of the latest developments of each scientific field. This reform arises from the necessity for quality improvement of the education system in order to be in line with the European Union key competences approach (Eurydice, 2018).

Digital resources play a substantial role for the promotion of science education both in Greece and in other counties. Photodentro is the Hellenic National Educational Content Aggregator for Primary and Secondary education and it represents the core part of the Ministry's digital infrastructure for integrated research and distribution of digital educational content for schools. Photodentro provides a central catalog of the digital educational material for school education and contains a series digital repositories of Open Educational Resources (OERs):

- Photodentro Learning Object Repository (http://photodentro.edu.gr/lor/): it contains approximately 9000 Learning Objects (LO), which are autonomous and reusable units that can be used for teaching and learning, such as experiments, interactive simulations, explorations, educational games, 3D maps etc. These autonomous and reusable Learning Objects cover a wide range of subjects (mathematics, Geography, Chemistry, Biology, Literature, History etc.).
- Photodentro Educational Video (http://photodentro.edu.gr/video/): it contains approximately 1000 educational videos that can be integrated into educational activities to support teaching and learning.
- Photodentro Educational Software (http://photodentro.edu.gr/edusoft/): it contains educational software packages with educational activities, free of charge.
- Photodentro User Generated Content (http://photodentro.edu.gr/ugc/) it acts as a repository where teachers can post their own Learning Objects or search for other users' Learning Objects.
- Photodentro Open Educational Practices (http://photodentro.edu.gr/oep/): it acts as a repository where teachers and other members of the wider educational and scientific community can post and share open educational practices for the use of digital educational content.

Inspiring Science education (https://inspiring-science-education.net/) is a project aiming to provide digital resources and opportunities for science teachers to help them making their teaching more attractive and relevant to students' lives. This project has received funding from the European Union's ICT Policy Support Programme as part of the Competitiveness and Innovation Framework programme. Through this platform students will have the opportunity to use interactive tools and digital resources to learn STEM related subjects in a practical and imaginative way. On the other hand, European science teachers will have the opportunity to attend workshops and participate in communities of practice to make their teaching of science more inspirational.

The Go-Lab initiative (https://www.golabz.eu/) took its name from the Go-Lab project which took place between 2012–2016. The aim of this initiative is to facilitate the use of innovative learning technologies in science education. It offers a unique and wide range of Labs (virtual labs, remote labs and datasets) and inquiry learning applications. The Go-Lab platform is free and can be used by any teacher from any country. Science education teachers can use the Go-Lab ecosystem in their classroom to create highly interactive and personalized exploratory learning experiences for their students.

24.3.3   Student Assessment and Achievement

According to the programme for International Student Assessment (PISA) scientific literacy is *"the capacity to use scientific knowledge, to identify questions and to draw evidence-based conclusions in order to understand and help make decisions about the natural world and the changes made to it through human activity"* (OECD, 2003, p. 133). The above definition emphasizes on the importance of being able to apply scientific knowledge in real-life situation. PISA aims to measure the essential skills and the knowledge acquisition of 15-year-old students to fully participate effectively and productively in a knowledge-based society. The assessment focuses on subject areas such as reading, mathematics and science. The PISA science literacy scale includes different levels of proficiency, from the low-achieving students (Level 1) to the more developed scientifically literate students (Level 6) to measure what students can do in the aforementioned subjects at different levels of proficiency (OECD, 2019a, b, c).

According to OECD (2019b), the PISA assessment of mathematics focuses on measuring students' capacity to formulate, use, and interpret mathematics in real-life contexts. Figure 24.2 shows the percentage of students at different levels of Mathematics proficiency between Greece and the average across OECD counties. 36% of students in Greece were below Level 2 in mathematics, while the OECD average was 24%. These students can answer clearly defined questions according to direct instructions in explicit situations. Similarly, in Greece 64% of students scored at level 2 or above in mathematics, while the OECD average was 76%. These students can interpret and recognize how a situation can be depicted mathematically. On average across OECD countries, 11% of students scored at level 5 or higher in mathematics, while in Greece only 4% of students attained this level. These students can model complex situations mathematically and can conceptualize and develop problem-solving strategies to handle them. Figure 24.3 shows the performance trends of the Greek students in Mathematics by presenting the mean score points of 15-year-old students over the 2006–2018 period. The performance in mathematics shows upward trends over the period 2006–2009, while over the period 2012–2018 the performance remained stable.

Similarly, *"the PISA assessment of science focuses on measuring students' ability to engage with science-related issues and with the ideas of science, as reflective*

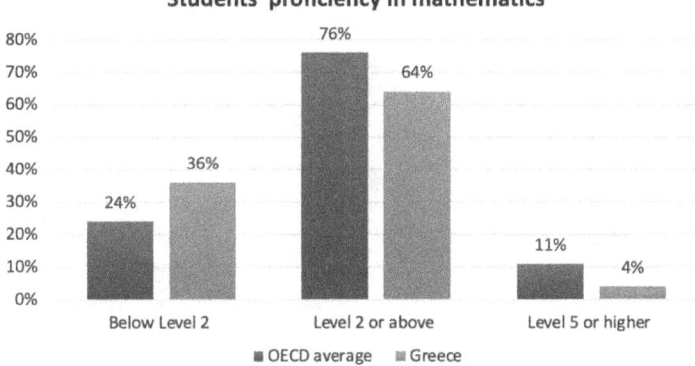

Fig. 24.2 Students' proficiency in mathematics. *Source* OECD, PISA 2018 Database, https://doi.org/10.1787/888934028634

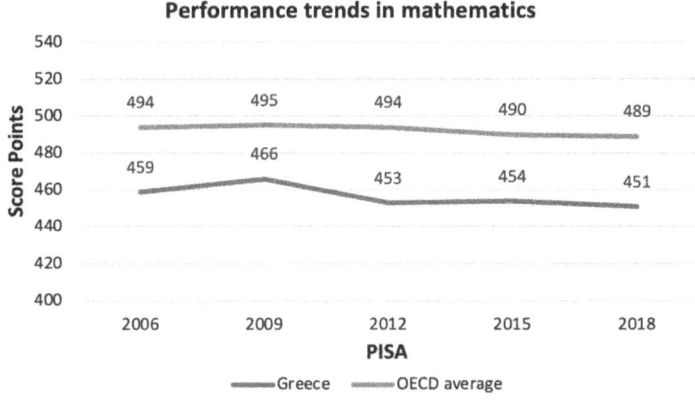

Fig. 24.3 Performance trends in mathematics. *Source* OECD (2020), Mathematics performance (PISA) (indicator), https://doi.org/10.1787/04711c74-en

citizens" (OECD, 2019b, p.112). Figure 24.4 shows the percentage of students at different levels of science proficiency between Greece and the average across OECD counties. 22% of students in Greece were below Level 2 in science, while the OECD average was 32%. These students can recognize aspects of familiar or simple phenomena. Similarly, in Greece 68% of students scored at level 2 or above in science, while the OECD average was 78%. At minimum, these students "*demonstrate the science competences that will enable them to engage in reasoned discourse about science and technology*" (OECD, 2018a, b, c, p.72). On average across OECD countries, 7% of students scored at level 5 or higher in science, while in Greece only 1% of students attained this level. These students can apply scientific ideas in order to explain more complex phenomena, including unfamiliar ones. Figure 24.5 shows the performance trends of the Greek students in science by presenting the mean

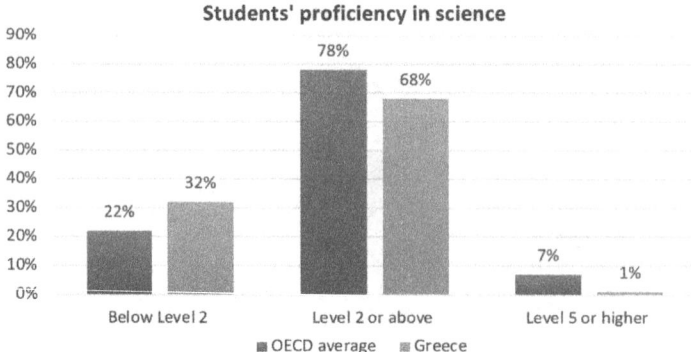

Fig. 24.4 Students' proficiency in science. *Source* OECD, PISA 2018 Database, https://doi.org/ 10.1787/888934028653

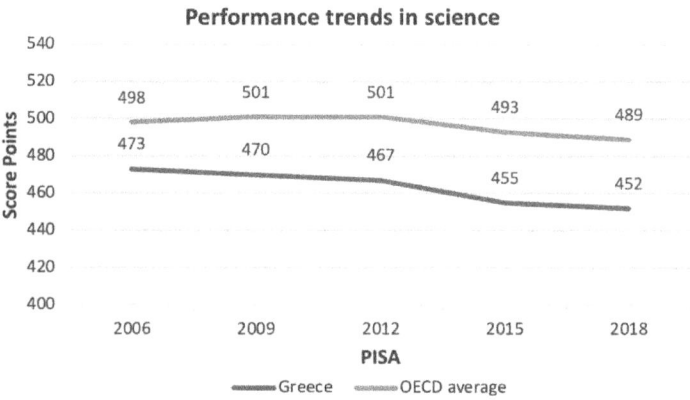

Fig. 24.5 Performance trends in science. *Source* OECD (2020), Science performance (PISA) (indicator), https://doi.org/10.1787/91952204-en

score points of 15-year-old students over the 2006–2018 period. The performance in science shows downward trends over the period 2006–2018, by an average of 5. It is concluded that, the performance of the Greek students was lower than the OECD average in science and mathematics in every year of its participation in PISA.

Students' assessment methods in science education related subjects in compulsory education do not differ from the assessment methods that apply to the other subjects of the school curriculum. Generally, student's assessment appears in a number of forms. Initial or diagnostic assessment (performed at the beginning of a course), formative or gradual assessment (performed during the learning process) and final or overall assessment (performed after the completion of a course). It is worth mentioning that the Greek Ministry of Education and Religious Affairs plans for the introduction of the descriptive assessment to compulsory education (Ministerial Decision 21072α/Γ2/28–02-2003).

24.3.4   *Outreach Science Education*

Science museums and science centres can promote science education outside the school classroom by offering a wide range of activities (Eurydice, 2011). The most known science centre and technology museum in Greece called NOESIS. The Noesis—Thessaloniki Science Center and Technology Museum (https://www.noe sis.edu.gr/) is a cultural and educational non-profit organization aiming to the popularization of modern knowledge and its dissemination to the public through exhibitions, films, seminars, and lectures. The activities of the center are directed towards school groups, educators, young people, and any other members of the public who are interested in Natural Sciences and Technology. The activities provided include guided tours and organized demonstrations of the exhibits, presentations of experiments in the fields of Physics and Chemistry, Film screenings, slide shows and other audio-visual presentations etc. In 2018, the total staff members of NOESIS were 42 employees (21 men and 21 women) and the total number of visitors was approximately 167,500 (https://bit.ly/32EMdqe).

Apart from science museum and centres, many countries develop partnerships and education activities with a view to promoting science education (Eurydice, 2011). The Computer Technology Institute and Press "Diophantus" (CTI) in Greece in cooperation with other technological institutions and academic organizations in Europe launched a Horizon 2020 project called UMI-Sci-Ed "Exploiting Ubiquitous Computing, Mobile Computing and the Internet of Things to promote Science Education" (http://umi-sci-ed.eu/). This project aims at enhancing the attractiveness of science education and careers for young people (14–16 year old) via the use of latest technologies, such as Ubiquitous and Mobile Computing and the Internet of Things (UMI). The ultimate goal of this project is to empower youngsters to think creatively, apply new knowledge in an effective way.

The Institute of Educational Policy in Greece in cooperation with other agencies, which varies from school authorities to science centres and museums, policy making organizations, universities and European associations, from both European and non-European countries launched a Horizon 2020 project called OSOS "Open Schools for Open Societies" (https://www.openschools.eu/). The main goal of this project is to become aware of and to promote the open schooling model, in which schools are going to be teaching students more about science education in collaboration with educational providers, enterprises, parents and local communities. The OSOS project aims to provide innovative ways to explore the world: not simply to automate processes, but to inspire, to engage, and to connect. It supports the development of innovative and creative projects and other educational activities. It transforms schools to innovative ecosystems, acting as shared sites of science learning in which leaders, teachers, students, and the local authority cooperate. The main goal of this initiative is to adopt a genuine collaborative approach to promote the collaboration and expand opportunities for science learning, in formal, non-formal and informal settings in order to ensure relevant and meaningful engagement of all societal actors with science and increase the uptake of science studies and science-based careers, employability and competitiveness (Sotiriou et al., 2017).

In order to continue promoting the open schooling approach, National and Kapodistrian University of Athens in Greece with partners from ten nations, launched an EU funded project called PULCHRA (https://pulchra-schools.eu/). This project aims to explore the open schooling concept through the theme "Cities as urban ecosystems" and creates new partnerships in local communities to foster science education for all citizens. The main objectives of this initiative are to build good scientific knowledge, promote expert and community participation, and encourage active engagement in shared living environments and futures.

Science contests and competitions are another type of activities for raising students' interest and enthusiasm for science (Eurydice, 2011). In 2020, the "STEM STARS GREECE" educational competition took place in Athens. The competition was held under the auspices of the Ministry of Education and Religious Affairs of Greece with the aim to supporting, highlighting, and rewarding 14 to 18-year-old students with a special inclination towards science, technology, engineering, and mathematics (STEM). The participation in the competition was free and the winners received special prizes, including the opportunity to represent Greece at the International Science and Engineering Fair (ISEF).

24.3.5 Utilizing Emerging Technologies

Scientific education refers, among others, to (a) the ability to use science processes to solve problems and make decisions; and (b) to understand the relationship between science and technology and to society. Taking both into account, we observe that both characteristics are reflected in the Greek curriculum for teaching and learning. More specifically: "science teaching should support students in developing skills related to scientific thinking and scientific methodology (observation, data collection, hypothesis testing, experimental testing, data analysis and interpretation, drawing conclusions, generalization ability and emerging patterns) "and also" science teaching must support students in developing the ability to appreciate the unity and continuity of scientific knowledge as well as the ability to recognize the interdependence of the natural sciences "(Greek Government Gazette, 2003, pp. 4170). In other words, science teaching aims to help students (a) build their understanding of science as a set of practices (experimental research, modeling, critique) and to attract students to them, and (b) to contribute to the connection of knowledge from different scientific fields (National Research Council, 2012).

In recent years, a number of studies have been conducted on the effectiveness of robotics, virtual reality, augmented reality and other emerging technologies in teaching and learning. These experiments cover all levels of education, while their results are very positive. In several cases, virtual laboratories have been created for students to understand phenomena related to physics, chemistry, biology, etc.

To teach Astronomy, a 3D virtual environment was created with models of rockets, spacecraft, vehicles used to explore the Moon and Mars, as well as a full launch platform of Appollo 11 (Fokides & Atsikpasi, 2016), while in another case, 3D models of the planets were given to the students (Zerva, 2019). Students could freely navigate the virtual world and study the information included in it. In Biology field, a 3D virtual reality biology lab was designed to offer a high level of realism regarding microscopy. Students could perform virtually a microscopy experiment, instead of passively watching a demonstration or passively listening to the instructor. The research results provided an initial proof that virtual laboratory simulations are very promising tools for biology education (Paxinou et.al., 2019). To enable students to enhance their understanding of the notions and phenomena of Quantum Physics, virtual workshops and mental experiments were designed (Margiolas, 2018). The findings of the studies in general, showed that virtual laboratories, especially when combined with practical laboratories, support students and help them gain a better understanding of each scientific field (Sypsas et.al., 2019; Paxinou et.al., 2020; Grivas et.al., 2016).

Through educational robotics, students built a model of a solar car, a model of a wind turbine and a traffic light, in order to experiment and become familiar with the concepts of solar and wind energy as well as light and colors (Gripeou, 2018), while in other experiments, students built robots in order to study Galileo's experiments on linear smooth motion, free fall, and lateral shooting (Gialavouzidou, 2020) or concepts such as speed, force, mass, weight and friction (Kosta, 2019; Papadopoulou, 2020). All studies confirmed the active participation of students, who stated that conducting experiments and creating constructions helped to better understand the concepts of teaching.

Other researchers focused on the integration of emerging technologies in science education in Greece, referring mostly to table devices, smart mobile devices, hand-held touch devices, robots, i.e. at the device/platform level (Kalogiannakis et.al., 2018; Psycharis & Kotzampasaki, 2019; Psycharis, 2019; Dorouka et al. 2020). The research of Fokidis and Foniadakis (2017) in the teaching of Geography with the use of augmented reality and the research of Nakka (2019) and Tsiavos and Sofos (2019) in the field of Physics, have highlighted the positive effects. Each textbook was augmented with multimedia elements by presenting additional information in the form of videos and pictures and 3D objects. In other cases, tablets were used to teach Biology. The students were given tablet devices and with the augmented reality technology, they were able to project certain elements in space, to observe the internal structures of the three-dimensional organs, visualizing even better the subject to be taught. The students expressed very positive impressions from their involvement with the application, while they stated that they would like to use the tablets in other lessons as well (Mastrokoukou & Fokides, 2017).

All findings came up to very positive results, such as promoting and improving the educational process and improving the performance of students in each different field. As a result, teachers and students involved, view positively the prospect of integrating these methodologies into teaching.

24.4   Requirements for Future Development of Science Education

Promoting science education and raising student's interest in science remains a key objective of many countries. The declining interest in science studies and related professions in conjunction with the rising demand for qualified researchers in science has led to discussions on developing more holistic strategies to improve public knowledge of science and promote a positive image on science. Implementing curriculum reforms, creating partnerships between schools, companies and science centres, cooperating with universities to improve initial teacher education, and providing particular guidance measures to encourage young people following a scientific career are key requirements for future development of science education (Eurydice, 2011).

According to the Framework for Science Education for Responsible Citizenship six key objectives are needed for the development of science education (European Commission, 2015, pp 8–13).

1. Science education should be an essential component of a learning continuum for all, from pre-school to active engaged citizenship.
2. Science education should focus on competences with an emphasis on learning through science and shifting from STEM to STEAM by linking science with other subjects and disciplines.
3. The quality of teaching, from induction through pre-service preparation and in-service professional development, should be enhanced to improve the depth and quality of learning outcomes.
4. Collaboration between formal, non-formal and informal educational providers, enterprise and civil society should be enhanced to ensure relevant and meaningful engagement of all societal actors with science and increase uptake of science studies and science-based careers to improve employability and competitiveness.
5. Greater attention should be given to promoting Responsible Research and Innovation (RRI) and enhancing public understanding of scientific findings and the capabilities to discuss their benefits and consequences.
6. Emphasis should be placed on connecting innovation and science education strategies, at local, regional, national, European and international levels, taking into account societal needs and global developments.

In order to put the aforementioned objectives into practice, the following indicative actions could be implemented (European Commission, 2015, pp 29–34). Education policies should ensure that science education in an essential component of school curricula in compulsory education. The co-creation of innovative curricula, with defined learning outcomes involving teachers, teacher educators, researchers and representatives from enterprise and civil society could enhance the collaboration between formal, non-formal and informal science education. Initiatives, such as curricular and/or extracurricular programmes and outreach activities can provide pathways to higher, vocational and professional science education and training to enrich science education, for all ages, in and beyond the school setting. Connecting science education to society is essential and actions such as projects and educational programmes that promote creativity and innovation can enhance the relevance of science education to people's lives. Emphasis should be placed on expanding international cooperation to introduce initiatives to support learning continuum for all citizens through sustainable "communities of learning" to collaborate internationally, improve practice and enhance science education opportunities for wider publics. On the other hand, actions should be taken to continually improve teaching quality with greater focus on teacher competences, through teachers' continuing professional development in order to improve the development of competences for science, innovation and employability for all students.

24.5   Discussion and Conclusion

This study attempts to examine the current situation of science education in Greece. The existing national policies support many separate initiatives and lacks an overall strategic framework for promoting science education. School science partnerships between Greece and other European countries embrace various aspects of science education and promote scientific culture and knowledge. Science centres in Greece also play a crucial role to the promotion of science education by providing students with activities that go beyond the traditional school setting through the combination of formal, non-formal and informal science education. However, student achievement in science-related subjects according to PISA results was lower than the OECD average, thus, several changes need to be done to strengthen student's scientific thinking (European Commission, 2015; Eurydice, 2011). To this extent, there is a necessity to redesign the science education curricula in both primary and secondary education to put an emphasis on students' real life experiences. Moreover, there is a need to improve the quality of science education teaching in Greece through teachers' continuous professional development in order to shift the focus on raising students' interest in science.

References

Dorouka, P., Papadakis, S., & Kalogiannakis, M. (2020). Tablets and apps for promoting robotics, mathematics, STEM education and literacy in early childhood education. *International Journal of Mobile Learning and Organisation, 14*(2), 255.

European Commission. (2015). Science education for responsible citizenship. (Publication No. EUR 26893 EN). Brussels: Belgium. Directorate-General for Research and Innovation Science with and for Society. Retrieved from http://ec.europa.eu/research/swafs/pdf/pub_science_education/KI-NA-26-893-EN-N.pdf.

European Commission. (2020). European Innovation Scoreboard 2020. Retrieved from: https://ec.europa.eu/commission/presscorner/detail/en/QANDA_20_1150.

Eurydice (2011). *Science education in Europe. National policies, practices and research.* Brussels: Education, Audiovisual and Culture Executive Agency.

Eurydice (2018)."Greece overview", National Education Systems, Eurydice website, https://eacea. ec.europa.eu/nationalpolicies/eurydice/content/greece_en.

Eurydice (2018/2019). The Structure of the European Education Systems 2018/19: Schematic Diagrams. Eurydice Facts and Figures. Luxembourg: Publications Office of the European Union. European Innovation Scoreboard Edition (2019). Main report: Greece

Eurydice. (2018). "Council Recommendation on Key Competences for Lifelong Learning", About education in EU. Eurydice website https://ec.europa.eu/education/education-in-the-eu/council-recommendation-on-key-competences-for-lifelong-learning_en

Eurydice. (2019a). "Teaching and learning in primary education", National Education Systems. Eurydice website https://eacea.ec.europa.eu/national-policies/eurydice/content/teaching-and-lea rning-primary-education-20_en

Eurydice. (2019b). "Teaching and learning in general lower secondary education", National Education Systems. Eurydice website https://eacea.ec.europa.eu/national-policies/eurydice/content/tea ching-and-learning-general-lower-secondary-education-16_en

Eurydice. (2019c). "Teaching and learning in general upper secondary education", National Education Systems. Eurydice website https://eacea.ec.europa.eu/national-policies/eurydice/content/tea ching-and-learning-general-upper-secondary-education-23_en

Eurydice. (2019d). "Teaching and learning in vocational upper secondary education", National Education Systems. Eurydice website https://eacea.ec.europa.eu/national-policies/eurydice/con tent/teaching-and-learning-vocational-upper-secondary-education-22_en

Fokides, E., & Foniadaki, I. (2017). Tablets, augmented reality and geography in elementary schools. *e-Journal of Science & Technology, 12*(3), 7–23.

Fokides, E., & Atsikpasi, P. (2016). Virtual Reality in Education. Results from the Pilot Implementation of an Application for Teaching Space Exploration Concepts and Facts to High School Students. *Theory and Research in Education, V9/2016*, 77–92.

Gialavouzidou, P. (2020). Galileo's historical experiments using educational robotics in the context of change laboratory method. Master Thesis, Aristotle University of Thessaloniki, Department of Physics.

Greek Government Gazette. (2003). FEK (B-303), Retrieved from: http://www.pischools.gr/dow nload/programs/depps/fek303.pdf.

Gripeou, L. (2018). Educational Robotics in the Science Education field. Master Thesis. University of Piraeus, Department of Digital Systems.

Grivas, G., Kounavis, V., Andreopoulou, A. & Voudoukis F. N. (2016). Augmented reality and science education. In *Conference: ICT in Education, November 2016, Athens, Greece.*

Ifanti, A., Argyriou, A., Kalofonou, F., & Kalofonos, H. (2013). Financial crisis and austerity measures in Greece: Their impact on health promotion policies and public health care. *Health Policy, 113*(1–2), 8–12.

Kalogiannakis, M., Ampartzaki, M., Papadakis, S., & Skaraki, E. (2018). Teaching natural science concepts to young children with mobile devices and hands-on activities. A case study. *International Journal of Teaching and Case Studies, 9(2)*, 171. https://doi.org/10.1504/ijtcs.2018. 090965.

Kosta, M. (2019). Experimental activities with educational robotics and use of virtual experiments with ALGODOO software: Comparative study of learning outcomes for the concept of friction in primary education, Post Graduate Thesis, Aristotle University of Thessaloniki, Department of Primary Education.

Margiolas, G. (2018). Introduction to Quantum Physics, an approach for Upper Secondary Students. Master Thesis, Hellenic Open University.

Mergos, G., & Patsavos, N. (Eds.). (2017). *Cultural heritage and sustainable development economic benefits, social opportunities and policy challenges.* Chania: Technical University of Crete. Retrieved from: http://ayla.culture.gr/wpcontent/uploads/2017/10/INHERIT-BOOK-PRINT-2. pdf

Mastrokoukou, A. & Fokides E. (2017). Tablets in education. Results from a pilot program for teaching the human body elements in elementary school students. *Research in Education, 6.*

Nakka, E. (2019). Development use and evaluation of a multimedia application for tablets for teaching physics unit to 6th grade primary school students. Master Thesis, Aegean University, Department of Primary Education.

National Research Council (2012). *A framework for K-12 science education: Practices, crosscutting concepts, and core ideas.* National Academies Press.

OECD (2003). The PISA assessment framework. mathematics, reading, science and problem solving knowledge and skills. Paris, OECD.

OECD. (2018a). "PISA for Development Science Framework", in PISA for Development Assessment and Analytical Framework: Reading. *Mathematics and Science, OECD Publishing, Paris.* https://doi.org/10.1787/9789264305274-6-en.

OECD (2018b), Online Education Database. Retrieved July 20, 2019, from https://stats.oecd.org/Index.aspx. See Digest of Education Statistics 2018, table 605.20.

OECD. (2018c). Education at a Glance 2018: OECD Indicators. *OECD Publishing, Paris.* https://doi.org/10.1787/eag-2018-en.

OECD. (2019a). PISA 2018 Assessment and Analytical Framework. *PISA, OECD Publishing, Paris.* https://doi.org/10.1787/b25efab8-en.

OECD. (2019b). PISA 2018 Results (Volume I): What Students Know and Can do. *OECD Publishing, Paris,*. https://doi.org/10.1787/5f07c754-en.

OECD (2019c). Education at a Glance 2019: OECD Indicators, OECD Publishing. https://www.oecd.org/education/education-at-a-glance/EAG2019_CN_GRC.pdf.

OECD (2020). OECD Economic Outlook Volume Issue 1: Preliminary Version. Retrieved September 15, 2020, from https://www.oecd.org/economy/greece-economic-snapshot/.

Patrinopoulos, M., & Iatrou, P. (2019). Implementation of STEM tinkering approaches in primary school education in Greece. *Sino-US English Teaching, 16*(12), 510–516.

Papadopoulou, A. (2020). Educational robotics in physics. Master Thesis, University of Piraeus, Department of Digital Systems.

Paxinou, E., Panagiotakopoulos, T. C., Karatrantou, A., Kalles, D., & Sgourou, A. (2019). Implementation and evaluation of a three-dimensional virtual reality biology lab versus conventional didactic practices in lab experimenting with the photonic microscope. *Biochemistry and Molecular Biology Education, 48*(1), 21–27.

Paxinou, E., Georgiou, M., Kakkos, V., Kalles, D., & Galani, L. (2020). Achieving educational goals in microscopy education by adopting virtual reality labs on top of face-to-face tutorials. *Research in Science & Technological Education.* https://doi.org/10.1080/02635143.2020.1790513

Plakitsi, K., Spyrtou, A., Klonari, K., Kalogiannakis, M., Malandrakis, G., Papadopoulou, P., & Kolios1, N. (2013). New Greek Science Curriculum (NGSC) for primary education: Promoting educational innovation under hard conditions. In *E-Book Proceedings of the ESERA 2013 Conference: Science Education Research For Evidence-based Teaching and Coherence in Learning. Nicosia, Cyprus: European Science Education Research Association (Vol. 31, p. 2017).*

Psycharis, S., & Kotzampasaki, E. (2019). The impact of a STEM inquiry game learning scenario on computational thinking and computer self-confidence. *EURASIA Journal of Mathematics, Science and Technology Education, 15*(4), 1–18.

Psycharis, S. (2019). Computational thinking, engineering epistemology and STEM epistemology: A primary approach to computational pedagogy. *International Series in Operations Research & Management Science*, 689–698. https://doi.org/10.1007/978-3-030-11935-5_65.

Sotiriou, S., Cherouvis, S., Zygouritsas, N., Giannakopoulou, A., Milopoulos, G., Mauer, M., et al. (2017). *Open schooling roadmap, a guide for school leaders and innovative teachers.* Pallini: Ellinogermaniki Agogi

Sypsas, A., Paxinou, E. & Kalles, D. (2019). Reviewing inquiry-based learning approaches in virtual laboratory environment for science education. In: *10th International Conference in Open & Distance Learning - November 2019, Athens, Greece.*

The compendium of cultural policies & trends (2019). Cultural policy system greece. Retrieved from: https://www.culturalpolicies.net/database/search-by-country/country-profile/cat egory/?id=16&g1=1.

Tsiavos P., & Sofos A. (2019). The user of augmented reality in education: development and use of application for the course "Physics – Explore and discover" in the 5th class of primary school. *The Journal for Open and Distance Education and Educational Technology, 15*(2).

UNESCO (2015). Education for All 2015 National Review Report: Greece, UNESCO, Paris.

Zerva, E. (2019). Augmented reality in teaching: Case of solar system. Master Thesis, Aristotle University of Thessaloniki, Department of Primary Education.

Chapter 25
Science Education in Italy

Maria Giulia Ballatore⦿ and **Anita Tabacco**⦿

Abstract This chapter provides an overview of Science Education in Italy. Firstly, the nation is introduced describing the geographical, political and economic context. Then, the whole education system is presented together with some statistics. These figures allow to analyze the current Science Education situation in greater depth. The study includes students' performance on Science indicators of TIMSS and PISA results as well as general information about the environment and approaches used in Science Education. Figures highlight the importance of science teachers training that supports students' interest in Science, Technology, Engineering, Mathematics and Medicine (STEMM) careers. Further details about future development and the key role of this needed effort are provided.

Keywords STEMM · Science teacher training · Science career · School system

25.1 Overview of the Country

25.1.1 Geographical Location, Population and Political System

Italy is a South-European country that overlooks the Mediterranean Sea. According to Unesco it holds the largest number of world heritage sites, exactly 55 on a par with China. From a geopolitical point of view, to the north the Alps separate Italy

---

The original version of this chapter was revised. The authors' "Martino Bernardi and Andrea Gavosto" have been removed from this Chapter. The erratum to this chapter can be available at https://doi.org/10.1007/978-981-16-6955-2_31

M. G. Ballatore (✉) · A. Tabacco
Department of Mathematical Science "G. Lagrange", Politecnico Di Torino, Corso Duca degli Abruzzi, Torino, Italy
e-mail: maria.ballatore@polito.it

A. Tabacco
e-mail: anita.tabacco@polito.it

© The Author(s), under exclusive license to Springer Nature Singapore Pte Ltd. 2022, corrected publication 2022
R. Huang et al. (eds.), *Science Education in Countries Along the Belt & Road*,
Lecture Notes in Educational Technology,
https://doi.org/10.1007/978-981-16-6955-2_25

from France, Switzerland, Austria, and Slovenia. Elsewhere, as it is a peninsula, it is surrounded by the sea. The country includes also a variety of islands of different dimensions, of which two major ones. It is divided into 20 regions, 5 of which are autonomous, that is with enhanced powers. Regions can make independent decisions on specific subjects such as health and local transports. As far as Education is concerned, regions have the power to open and close schools institutions; in principle, Constitution allows extra powers on Education to be assigned to regions by the State following specific legislation, but this has never occurred so far.

Italy is a democratic Republic since 1946 with Rome as a national capital. It is a charter member of NATO and one of the founder states of the European Economic Community and its subsequent successor the European Union (EU). It joined the Economic and Monetary Union in 1999 and it adopted the Euro as a currency from the beginning of its use.

The current population is 60,4 millions based on Eurostat 2019, thus, it counts as 0.78% of the total world population. Looking at the geographical distribution, 69.5% lives in the urban context and the life expectancy is 84 years with a median age of 47.3 years, among the highest in the world (EUROSTAT, 2020b). The country is recording a population decline and in 2019 it has reached the lower number of births since the 1946.

25.1.2 Current Situation of Economic, Technologies and Cultural Development

With a Gross Domestic Product (GDP) of 1.789.747 million € (2019), Italy is the third largest economy in the EU and the twelfth largest in the world (ISTAT, 2020a). See Fig. 25.1 for details.

In general, the secondary sector is driven by the manufacture of high-quality consumer goods produced by small and medium-sized enterprises, many of them family-owned. However, this economic development is not equally spread geographically, apart from the tourism sector (tertiary) that involves the entire nation. The majority of the industrial sector, dominated by private companies, is located in the

Fig. 25.1 Share of GDP for each sector (Agriculture, Industry and Constructions, Services)

North, while the South lags behind. Indeed, in 2018 GDP per capita in the North-West was equal to 36,200 €, twice as much as in the South (13,700). This has a direct impact on lower employment and higher youth unemployment rates (EUROSTAT, 2020c). On average the national unemployment rate in the first fourth months of 2020 is 9.4%. Considering the age distribution, the 28.1% between 20 and 24 years old are unemployed (ISTAT, 2020c). The regional unemployment rate is shown in Fig. 25.2.

On the technological side, Italian Research and development (R&D) intensity (expenditure on R&D as a percentage of GDP), sees a slightly increase in the last years passing from 1.37 in 2017 to 1.43 in 2018. The figure is below the EU average, mainly driven by Germany (3.14), Sweden (3.32), and Denmark (3.13) (OECD, 2019). To foster technological development, the government has set dedicated investments called "Industry 4.0". These involve different sectors from Education to Industry. The main goal is to support the new challenge that the factories need to face in term of skills, instruments, technologies and services.

Referring to the last available statistical analysis on cultural behaviors by ISTAT (*Istituto Nazionale di Statistica*), in 2018 cultural participation recorded a slight

Average Unemployment Rate

2,92   18,53

Fig. 25.2 Average unemployment rate by regions of 15–74 years old

increase compared to the previous year, going from 64.1% to 64.9% (ISTAT, 2020d; OECD, 2020a).

The observed increase is mainly driven by visitors to monuments and archaeological sites (+2% compared to 2017) and by those who attended concerts of music other than classical (+1.4%).

Considering the geographical distribution, residents in the Center-North are the most active in terms of cultural participation and are distinguished by the lowest overall abstention rates.

State museums institutes registered over 55 million visitors, an increase of about 10% compared to 2017. More than 60% of the visits concerned facilities in the Center of Italy.

The editorial production in 2017 remains concentrated in large publishing houses (80%). In the same year, the publication production amounted to over 70 thousand books, an increase compared to 2016 in the number of titles (+9.3%) and in print run (+14.5%).

The expenditure allocated by Italian families to culture and leisure remains virtually unchanged as a percentage of total consumption expenditure (just under 7%).

25.2 Overview of the Education Development

25.2.1 Education System and Policy

Education is compulsory between 6 (corresponding to the first year of the primary school) and 16 years old (typically, the second year of the upper secondary school). However, young people have the right to receive—and employers have the obligation to provide—formal training until they are 18 years old, when they reach the legal age.

Before compulsory schools, children can attend *asilo nido* (nursery or kindergarden), normally provided by municipalities or private institutions, until they are 3. From 3 to 6 years old they can attend *scuola dell'infanzia* (pre-school). At 6 years pupils start primary school which lasts for 5 years; thence, they join lower secondary school, which lasts for 3 years. Both levels are common to all students and share a nationwide curriculum. At the beginning of the upper secondary school, students choose between three tracks, all of which last for five years and lead to a final exam: *liceo* (Lycée), more academically oriented; technical school; professional school. Alternatively, some pupils can follow 3 or 4 years of vocational courses delivered by regional authorities. After having completed their final exam, students can join universities that follow the Bologna process structure: the Bachelor degree is obtained after three years, whereas the Master degree after further two. Some courses, like Law or Medicine, have a full track of five or six years.

It should be noted that 86% of students complete the full track of secondary school (in most cases up to 19 years of age) (EUROSTAT, 2020a; MIUR, 2020a). Little above 50% of each cohort attend tertiary (university) education and just 34% reach a degree (ALMALAUREA, 2020).

Preliminary figures for 2020–21 suggest that in the new school year—which follows a lockdown of 18 weeks due to Covid-19–8.3 million students from 3 to 19 years will attend classes: 7,507,484 in state schools and 860,000 in private ones (MIUR, 2020c). Official university enrollments are not yet known, as they can take place later in the year. However, preliminary information from university suggest that the number of overall students should be close to that of last year (1,750,000 total enrolled and 300,000 new entrants) despite the impact of the Covid-19.

Government policies in the recent years have focused on:

- expanding childcare opportunities between 0 and 2, from the current 24.7% of the relevant population to the European target of 33% (ISTAT, 2020b);
- fighting against school dropout, which amount to 13.7% of the young population, with spikes above 20% in Southern regions (EUROSTAT, 2020a);
- raising the percentage of young people who obtain a university degree, currently at 27% (OECD, 2020b).

Other areas of concern are post-secondary vocational education and training, which is largely absent in Italy, and the low average level of learning outcomes vis-à-vis other advanced countries.

25.2.2   Statistics on the National Education

In Table 25.1, we recap the whole gamut of school and university levels, the corresponding ISCED level and the number of attending students in the 2017–18 school year, the most recent to include both public and private pupils (MIUR, 2020a).

Table 25.1 Students and teachers distribution referring to 2017–2018 school year (OECD, 2020a)

Age	Level	Isced 2011	#students	#teachers
0–2	Kindergarten	0	354,641	
3–5	Pre-school	0	1,491,000	101,136
6–10	Primary	1	2,754,000	278,640
11–13	Lower secondary	2	1,731,000	196,770
14–18	Upper secondary*	3	2,688,000	295,722
19–21	Bachelor	6	1,045,893	
22–23	Master	7	644,941	

* This figure does not include students in vocational education and training who attend courses provided by regional agencies up to 16 and 17 years old

Schools were 8,636 in 2019 over the entire country: most include more than one building and encompass from pre-school to lower secondary. Overall, there are 40,000 buildings, most of which built from the Sixties to the Eighties.

School teachers in 2018 were 872,268, of which 135,025 on a temporary basis. The latter figure is likely to overcome 200,000 in the current year. The ratio students/teachers is around 10, one of the lowest among OECD countries. About 150,000 teachers are fully allocated to support 270,000 special needs students. Teaching or researching university personnel includes: 33,969 full or associate professors who are obliged to teach; 27,759 adjoint professors; 12,601 full time researchers; 6,216 researchers on a temporary basis (MIUR, 2020b). The number of students per university lecturer is around 30, one of the highest among advanced countries (ANVUR, 2018).

25.2.3 Educational Research and International Collaboration

The Italian university field relating to the education and training of teachers is located mainly in the Departments of Education. Each Department has 5-year courses for pre-school and primary school teachers as well as educational science research centers for school and extracurricular issues (including theory and history of educational and training processes, teacher training, interculturality, interventions on diversity, education and socialization processes, multimedia learning, adult education) and specifically methodological, didactic and experimental research. At the same time, in all the other departments related to secondary education (i.e., Mathematics, Literature, Physics, Biology, etc.) there exist some thematic research areas concerning teaching.

The Italian National Research Council has one dedicated institute, the *Istituto per le Tecnologie Didattiche*, devoted to the study of educational innovation achieved through the use of Information and Communication Technologies. The Ministry of Education has its own research agency, INDIRE, focused on education innovation. Moreover, at the national level, there exist private foundations that have educational research as a main objective.

All these research centers can receive competitive national or European funding related to the innovative and challenging projects (i.e., PRIN, SIR, H2020, Erasmus + , etc.). Moreover, they can count on strong networks either national (i.e., *Avanguardie Educative*) or international (i.e., E-Twinning), both centered around INDIRE. These relations focus both on the training of teachers and research for implementation and teaching improvements as well as for sharing good practices.

25.3   Current Situation of Science Education

25.3.1   Policies and Standards

Currently, there are not professional standards for science educators and teachers in general. Pre-school and primary teachers are enabled to teach with the degree. Secondary teachers, instead, are divided into *Classi di Concorso* (CdC) based on subject taught. To have the right to teach, they need to pass a national exam for each CdC.

Starting from 2004 the Ministry of Education, in collaboration with the Conference of Deans of Science and Technology and Confindustria, established a policy in favour of science education call *Piano Lauree Scientifiche,* Plan for Science Degrees (MIUR, 2004).

This plan, currently fully implemented, aims to:

- promote enrollment in scientific degree courses, also aiming at achieving a gender balance, by strengthening the offer of guidance;
- reduce university dropouts and improve students' careers through the introduction of innovative teaching tools and methodologies;
- implement training, support and monitoring of the activities of the first years of university;
- carry out self-assessment activities for students of upper secondary schools to verify the preparation for entry into universities concerning the required requirements and increasing the awareness of their knowledge to choose the training path;
- provide science teachers opportunities for professional growth through active participation in the planning of activities carried out jointly with the University.

25.3.2   Curricula, Digital Resources and Teacher Training

At primary and lower secondary level the curriculum is defined by the National Guidelines (MIUR, 2018a), which provide for Science and any other subject goals for the development of skills, that should be reached by the end of the grade 3 (corresponding to the third year of primary school), by the end of the grade 5 (end of primary school) and by the end of grade 8 (end of lower secondary). Learning objectives cover the following content areas:

Primary level:

- Exploring and Describing Objects, Materials and Transformations
- Observing and Experimenting in the Field
- Man, Living Things, and the Environment

Lower secondary level:

- Physics and Chemistry
- Astronomy and Earth Science
- Biology

At upper secondary level curricula are defined by National Guidelines for Lycée, technical, professional and vocational tracks (MIUR, 2018b).

In order to become a teacher at primary school, candidates are required to obtain a degree in Primary School Education. The academic component of the degree is the same for all graduates, usually with some exams that cover element of physics, chemistry and biology; only after graduation teachers specialize in a particular disciplinary field during their school internship.

From lower secondary education, teachers are subject specialist and are required to get a degree related to the subject taught. For example, to teach mathematics and science, teachers must hold a mathematics, physics, biology, life sciences or geology degree.

25.3.3 Student Assessment and Achievement

Italian student achievement in science are measured by TIMSS (TIMSS & PIRLS International Study Center, 2015) and PISA (OECD, 2018a), two international surveys that base their assessment respectively on "what students know" and "what students are able to do with their knowledge". Moreover, there exists a national survey managed by INVALSI, the testing agency of the Ministry of Education, which involves only Mathematics, Italian and English language (Istituto Nazionale per la VALutazione del Sistema educativo di Istruzione e di formazione, 2020). For this reason, this assessment tool will not be taken into consideration in this chapter.

Latest and available results come from 2015 round of TIMSS and 2018 round of PISA, the former on grade 4 and 8 performances and the latter on 15 years students, thus "covering" primary, lower and upper secondary school.

In TIMSS, at grade 4 the Italian average score is 516 points (overall average score is 500, with standard deviation of 100), in line with other European countries like Serbia, Netherlands, and Spain but significantly lower than worldwide top performers like Singapore (590 points), South Korea (589), Japan (569) and European ones like Finland (554 points) and Poland (547).

At grade 8 the Italian average score is 499 points, behind all European countries with the exception of Malta and far away from Slovenia (551), England (537) or Sweden (522).

From a time perspective, the trend of grade 4 achievements is negative, with average score steadily decreasing from the 537 points in 2007 and the 524 in 2011 round, while it remains quite stable for grade 8, 495 in 2007 and 501 points in 2011 round.

TIMSS also provides results according to content domains. Thus, while grade 4 students perform better in Life Science (519 points) with respect to Physics (513 points) and Earth Science (510 points), grade 8 students suffer the most in Chemistry (487 points), the less in Earth Science (514 points) and are in line with overall results in Biology (496 points) and Physics (496 points).

Grade 4 achievements are characterized by a relevant gender gap: the 9 points difference in favor of boys (521) with respect to girls (512) places Italy at the top of an undesirable ranking, only exceeded by the 10 and 11 points difference observed in Hong Kong and Korea. Grade 8 results confirm a gender gap issue in lower secondary school, with a similar magnitude of 10 points difference between boys (504) and girls (494) achievement levels.

In PISA, the Italian students' score in science assessment is 468, lower than the average OECD one (489) and lower than those of vast majority of participating European countries. Further national comparisons are shown in Fig. 25.3 (OECD, 2018b).

Moreover, 2018 round confirms and exacerbates a path of decline observed between 2012 (494) and 2015 (481), thus offsetting the positive trend started in 2009 (489) (see Fig. 25.4).

Figure 25.5 shows the negative gradient in science achievements once we move from northern to southern regions of Italy.

Figure 25.6 describes the differences among tracks: while academically oriented track performs better than OECD average score, technical, vocational and professional tracks lag behind.

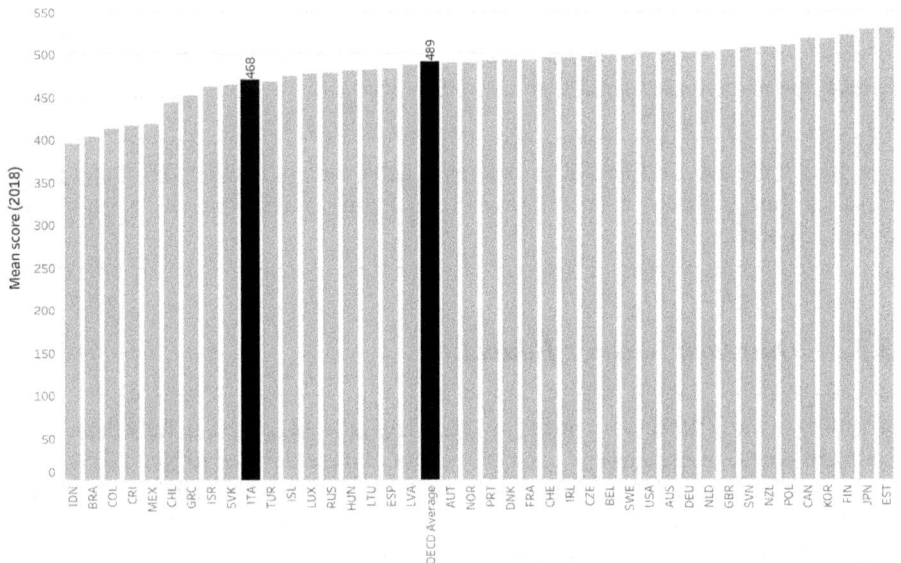

Fig. 25.3 Average PISA science score by nations

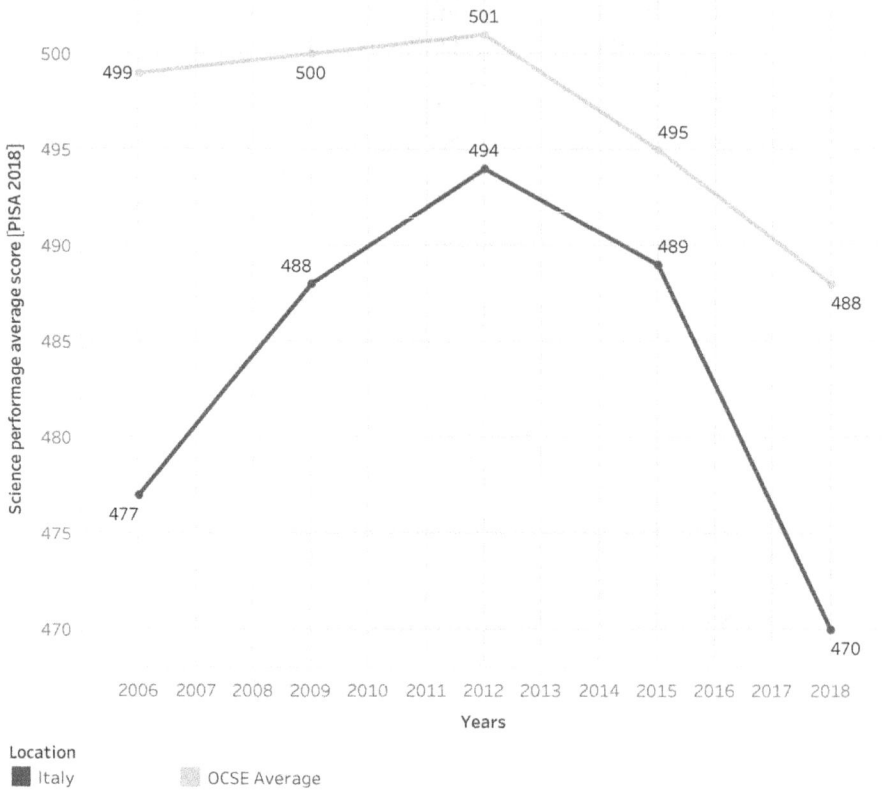

Fig. 25.4 Italian science performance score on PISA compared to the OCSE average

In Italy there are no differences in science achievement between boys and girls, a result in line with overall picture, where girls outperform boys by a negligible amount, 2 points.

25.3.4 Outreach Science Education

According to the 2019 survey carried by the ISTAT (ISTAT, 2018), in Italy there exist about 450 science museums and centers, two out of three devoted to Natural Science and the rest to Science and Technology. They are no homogeneously widespread over the country, with regions like Emilia Romagna, Lombardia and Trentino-Alto Adige that stand out for their museum and exhibition centers.

Unfortunately, ISTAT does not provide any aggregate number on annual visitors; in spite of this lack of information, for some of the most important science museum in Italy data are available and collected in Fig. 25.7.

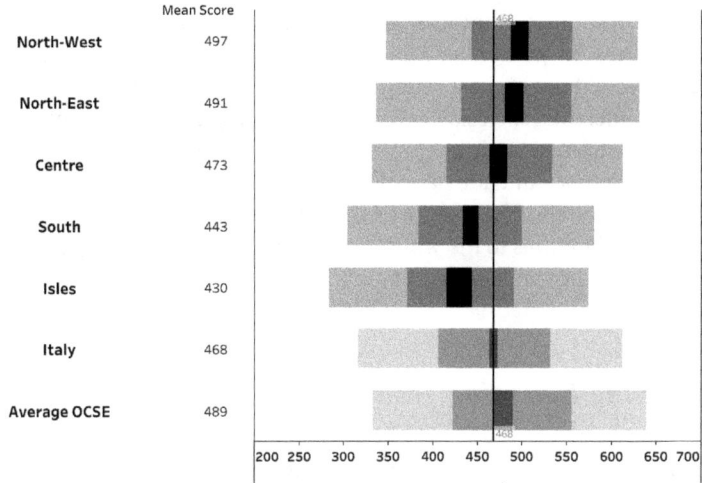

Fig. 25.5 Science score distribution among Italian geographic areas compared to the national and the OCSE values

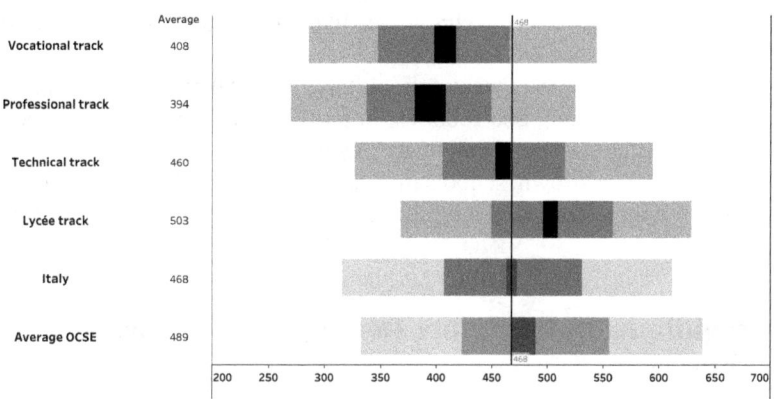

Fig. 25.6 Science score distribution among upper secondary schools' tracks compared to the national and the OCSE values

Fig. 25.7 Number of annual visitors in science museum (2018)

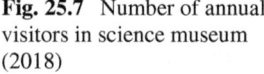
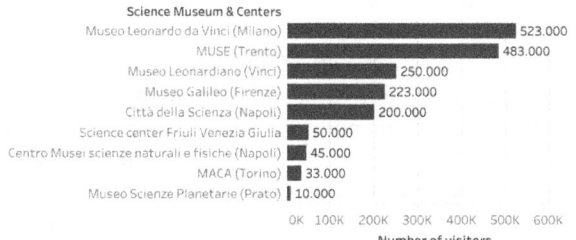

Other important sources of informal science education are TV shows, podcast and scientific magazines. In particular, a famous scientific popularizer is Piero Angela that last year recorded the highest ratings peak with his night show.

Considering other outreach science education programs, each city organizes a variety of periodic events, such as *Giovedì Scienza* in Torino with weekly appointments or *Galileo Festival* in Padova on yearly basis. One important event, organized at a European level, is the *Notte dei Ricercatori* that aims at explaining the role of research and its applications to the whole populations with streets experiments, open laboratories, talks and exhibitions.

25.3.5   Utilizing Emerging Technologies

In general, Education recently has seen a variety of incentives (i.e., *Piano per la Scuola Digitale, Industria 4.0, Programma Operativo Nazionale,* etc.) to foster technological development. This mainly involved structural equipment and infrastructure, regardless of teaching methodologies. Science Education does not benefit from any systematic strategy to introduce technologies like augmented reality, virtual reality or artificial intelligence.

Nevertheless, in recent years many schools have undertaken an increasing number of projects related to coding and robotics, in particular at the primary school level where coding is introduced in a funny way to youngest students.

During Covid-19 lockdown, schools moved to the distance learning environment. This has caused a rapid spread of a digital learning environment.

25.4   Requirements for Future Development of Science Education

Science Education is becoming a key element for the reduction of skills mismatch, given the increasing demand for scientific and technological occupations. Only with a strong and robust science education, students are free to choose scientific careers to fill the labor market's needs.

For this reason, special attention should be driven to early education (pre-school and primary education) especially when referring to the gender gap. This will require a qualitative and quantitative improvement in the teachers training. In particular, regarding the pre-school and primary school teachers, their university degree should include the specialization either in Science or Literature. This will allow them to preferably teach Mathematics and Science only with strong educational basis.

Equally, special training about Science Education and their teaching technique should be reinforced for upper secondary teachers during their university study. Currently, new teachers have to follow some mandatory courses about pedagogy, psychology, anthropology and teaching technique. These are general courses and do not refer specifically to Science Education.

Teaching methodologies in Italy are generally quite old and grounded on one-way transmission from teacher to students, as documented by OECD- Talis (OECD, 2020a). This is true also of Science Education, where teachers rely typically on front explanations and home assignment rather than on understanding of physical phenomena through direct experimentation at school. A few attempts are being made to try to redefine the teaching standards in Science: an example is the project on Physics by CERN and Fondazione Agnelli aimed at lower secondary schools. The project is aiming at fostering scientific culture and transmit science and technology concepts in an engaging way, exploiting and promoting Inquiry Based Learning as a possible different approach to study science with respect to traditional frontal teaching.

Demographic trends suggest that the overall number of students will fall by 1 million (about an eighth) by 2030, starting from early grades: this implies a reduction of 65,000 jobs among teachers. Hence, in the next decade new entrants in schools will be limited to a partial replacement of retiring teachers. This implies the strong need to update and retrain existing Science teachers, especially as far as their teaching methodologies are concerned, in order to improve students' achievements.

25.5   Discussion and Conclusion

In order to enhance the level of knowledge and competencies in Sciences by Italian students, the starting point has to be the improvement in the quality of Science educators. Currently, as we saw, Science teachers do not possess all the teaching skills necessary to engage and motivate students in the field at different grades. Policies to upgrade Science teachers' skills should include at least three aspects.

First, as far as existing teachers are concerned, they need a massive and compulsory training campaign to enlarge the set of teaching tools at their disposal to teach Sciences, especially focused on designing and running practical experiments. This implies a huge investment from the Government to finance such training, possibly thanks to Next Generation EU. Secondly, science teacher skills improvement has to be assessed before returning them to the classroom: in case they fail to achieve an adequate level, they will not be allowed to resume teaching and will have to undertake further training until they reach a satisfactory standard. The assessment outcome should also be a component of teachers' advancement along a career ladder, which does not yet exist in Italy but is currently under discussion. Thirdly, teachers' wages should be redefined in order to recognize effort and competencies on the part of teachers. Currently the only criterion for teachers' wage increase is seniority: as said, a career mechanism is under discussion. A greater progression in wages could help to motivate teachers who invest more heavily in their training and foster the importance and the key role of science educators.

As far as new teachers are concerned, an important implication is that, considering the high demand for science and technical figures in the private sector, not enough young graduates are motivated to go into teaching. As clearly shown in "Gli insegnanti nella scuola italiana" (Argentin, 2018, Fig. 5.3), the probability to become a teacher is strongly negatively related to the final graduation mark, used as proxy of their competencies: hence, best graduates shy away from the teaching profession. This is especially true for Science teachers. Incentives, such as wage premia, should be considered in order to attract the best graduates into the profession.

All in all, planning for the long-term professional enhancement of Science Educators should become a key priority to support the future development of the country.

References

ALMALAUREA. (2020). *Condizione occupazionale dei Laureati - Rapporto 2020 | AlmaLaurea.* https://www.almalaurea.it/universita/occupazione/occupazione18/volume

ANVUR. (2018). *Rapporto Biennale sullo Stato del Sistema Universitario e della Ricerca 2018.*

Argentin, G. (2018). *Gli insegnanti nella scuola italiana* (il Mulino-Volumi). https://www.mulino.it/isbn/9788815275493

EUROSTAT. (2020a). *Early leavers from education and training by sex.* https://ec.europa.eu/eurostat/web/products-datasets/-/sdg_04_10

EUROSTAT. (2020b). *Population.* https://ec.europa.eu/eurostat/cache/metadata/en/demo_pop_esms.htm

EUROSTAT. (2020c). *Unemployment statistics - Statistics Explained.* https://ec.europa.eu/eurostat/statistics-explained/index.php/Unemployment_statistics

ISTAT. (2018). *Statistics and Data.* http://dati.istat.it/?lang=en

ISTAT. (2020a). *Annual national accounts.* https://www.istat.it/en/archivio/247355

ISTAT. (2020b). *Nidi e servizi educativi per l'infanzia.* https://www.istat.it/it/archivio/244116

ISTAT. (2020c). *Occupati e disoccupati (dati provvisori).* https://www.istat.it/it/archivio/246805

ISTAT. (2020d). *Statistiche culturali.* https://www.istat.it/it/archivio/239547

Istituto Nazionale per la VALutazione del Sistema educativo di Istruzione e di formazione. (2020). *INVALSI.* https://www.invalsi.it/invalsi/index.php

MIUR. (2004). *Piano Lauree Scientifiche.* https://www.pianolaureescientifiche.it/

MIUR. (2018a). *Indicazioni Nazionali.* http://www.indicazioninazionali.it/

MIUR. (2018b). *Percorsi di studio e formazione.* Scuola Secondaria Di Secondo Grado. https://www.miur.gov.it/scuola-secondaria-di-secondo-grado

MIUR. (2020a). *Esplora Dati.* https://dati.istruzione.it/espscu/index.html?area=anagScu

MIUR. (2020b). *Ricerca avanzata per docenti.* https://cercauniversita.cineca.it/php5/docenti/cerca.php

MIUR. (2020c). *Scuola, quest'anno nelle classi oltre 8,3 milioni di studentesse e studenti. Stampa e Comunicazione Ministero Dell'Istruzione e Ministero Dell'Università e Della Ricerca.* https://www.miur.gov.it/web/guest/-/scuola-quest-anno-nelle-classi-oltre-8-3-milioni-di-studentesse-e-studenti

OECD. (2018a). *PISA Test.* https://www.oecd.org/pisa/test/

OECD. (2018b). *Science performance (PISA) (indicator).* https://doi.org/10.1787/91952204-en

OECD. (2019). Gross domestic spending on R&D (indicator). *OECD Data.* https://doi.org/10.1787/d8b068b4-en

OECD. (2020a). *TALIS - The OECD Teaching and Learning International Survey*. http://www.oecd.org/education/talis/

OECD. (2020). Education at a Glance 2020. *OECD*. https://doi.org/10.1787/69096873-en

TIMSS & PIRLS International Study Center. (2015). *TIMSS 2015*. https://timssandpirls.bc.edu/timss2015/index.html

Maria Giulia Ballatore is a Research Fellow at the Department of Mathematical Sciences of the Politecnico di Torino, Italy. In the meantime, she is a PhD student in Engineering Education at School of Electrical, Electronics and Communications Engineering. Her research interests lie in the fields of engineering education, development and standardisation of learning technology, spatial abilities, gamification and gender issue.

Anita Tabacco received the Ph.D. in Mathematics in 1986, at Washington University in St. Louis (MO) USA. Currently she is full professor of Mathematical Analysis (since October 2002). Her mathematical research activities are related to real and complex analysis, functional analysis, harmonic analysis with particular interest in interpolation theory, theory of wavelets and applications to PDE's and integral equations. Moreover, she is working in the field of Engineering Education. Her research activities include the organization of conferences, workshops and graduate schools, the participation to numerous prestigious research projects both national and international. She is authors of numerous papers and has given many invited seminars and lectures both in Italy and abroad. She is deeply involved in the management of the university. She is now Vice Rector for Simplification; she has been Vice Rector for teaching from 2012 to 2018, and Vice Rector for recruitment from 2008 to 2012. She was also responsible for quality assurance in internal teaching organization.

Chapter 26
Science Education in Moldova

Roza Dumbraveanu

Abstract The concise state of the art in science education in the Republic of Moldova is presented alongside the description of the related environment variables that have an influence on it. The focus is on pre-university science curriculum, which is the basis in preparing students for their next pathways in life, whether they will continue to study science at tertiary level, will become scientists, will use science in a related career, or will apply their science knowledge as non-scientist citizens of the twenty-first century. Over the last twenty years, the school education system in Moldova has undergone a number of changes. The intention of the government was to improve the quality of education and to follow the European recommendations, but in reality, the reforms faced implementation challenges. The paper describes some of these challenges, identifies the issues that need reconsideration, and presents the author' reflections on the development of the science curriculum.

Keywords Science curriculum reforms · Competence-based education · Student centred approach

26.1 Overview of the Country

26.1.1 Geographical Location, Population and Political System

The Republic of Moldova (Moldova) is a former Soviet Union (SU) Republic. It is located in the South-East of Europe, mostly between two rivers, the Dniester and the Prut. Moldova is bounded by Romania to the west and Ukraine to the north, east, and south. The area is equal to 33.8 thousand km². Moldova became an independent state in 1991, after a period of national movements for independence and the collapse

R. Dumbraveanu (✉)
Chisinau Ion Creanga State Pedagogical University, Republic of Moldova, Ion Creanga str. 1, MD 2069 Chisinau, Moldova
e-mail: dumbraveanu.roza@upsc.md

© The Author(s), under exclusive license to Springer Nature Singapore Pte Ltd. 2022   435
R. Huang et al. (eds.), *Science Education in Countries Along the Belt & Road*,
Lecture Notes in Educational Technology,
https://doi.org/10.1007/978-981-16-6955-2_26

of SU, retaining its Soviet-defined borders. In 1992, the territory on the east bank of the Dniester River, named Transnistria, declared its independence and became an internationally unrecognized state, named the Pridnestrovian Moldavian Republic.

The territory of current Moldova was part of a Romanian principality (Principality of Moldova) from 1359 until 1538, when it became a vassal state of the Ottoman Empire. In 1812, the Russian Empire annexed the part of medieval Moldova from the Ottoman Empire and renamed the annexed region "Bessarabia". After World War I, the area became part of the Romanian State until the German-Russian non-aggression pact allowed the Soviet Union to re-annex Bessarabia in June 1940. In August 1940, the Soviet authorities created the Moldavian Soviet Socialist Republic (MSSR), encompassing mostly the interwar Bessarabia annexed territory and a strip of land on the eastern bank of the Dniester from an earlier-established (1924) Moldavian Soviet Socialist Autonomic Republic (MSSAR). At the same time, the southern and northern areas of Bessarabia and a big part of the MSSAR territory were incorporated into Ukraine.

The total population of Moldova is a variable due not only to intrinsic demography, but to an intense migration flow. According to the National Bureau of Statistics of the Republic of Moldova, the total resident population as for January 1, 2020, was 2.64 million persons (Valcov, et al., 2020).

The Republic of Moldova has a mixed system of government. It has a dual executive consisting of a President, Prime Minister, Council of Ministers, and a unicameral legislature. Judicial review of the constitutionality of laws is performed by the Constitutional Court (MECC, 2012).

26.1.2 Current Situation of Economic, Technologies and Cultural Development

Moldova has a small lower-middle-income economy. Although it is among the poorest countries in Europe, it has made some progress in promoting inclusive growth since the early 2000s. The economy has developed by an average of 4.6% annually in the past 20 years, driven by consumption and fueled by remittances (The World Bank, 2020).

Integration with Europe was on successive governments' policy reform agendas, but reforms on paper faced implementation challenges. A vulnerable political system, a polarized society, low productivity, demographic problems, skills mismatches, and a high instability to both climate-related and external factors are Moldova's biggest economic challenges (The World Bank, 2020).

Economic development is based on accumulation of capital, labor force and its productivity level, which includes and other parameters, such as technologies, efficient governance, skills, etc. The Government has made an attempt to outline development of Republic of Moldova in the context of the National Development Strategy, "Moldova 2020" (Government of Moldova, 2012) that had identified four critical issues as development priorities: education, roads, access to finance, and business environment. The priority related to education dimension was formulated as follows: aligning the education system to labor market needs in order to enhance labor productivity and increase employment in the economy.

The main economic activities with share in the generation of the GDP (2019) are: agriculture, forestry and fishing (9.9% GDP), manufacturing industry (10.9% GDP), construction (8.6% GDP), wholesale and retail trade (15,8% GDP), accommodation and food service (1.1% GDP), transport (4.8% GDP), information and communication (4,7% GDP), public administration (3.5% GDP); education (4.5% GDP); health and social assistance (3.7% GDP), professional, scientific and technical activities (1.9% GDP). From the distribution of employed persons by economic activities, it results that every 5th person is active in the agricultural sector (21.0%) (Valcov, et al., 2020).

National Program in Research and Development (R&D) had established the following strategic areas: (a) Materials, technologies and innovative products; (b) Health and biomedicine; (c) Biotechnology; (d) Energetic efficiency and valorization of renewable energy sources; (e) National heritage and societal development. The state of the art in R&D field is described in a report of the Academy of Science (Academia de Științe a Moldovei, 2020). Among achievements: scientific and cultural events, the publication of papers in national scientific journals, recognized internationally: Surface Engineering and Applied Electrochemistry (SCOPUS), Computer Science Journal of Moldova (WoS, Emerging Sources Citation Index); Chemistry Journal of Moldova (WoS, Master Journal List; SCOPUS); Emerging Sources Citation Index); Stratum Plus (SCOPUS, WoS), Quasi-groups and Related Systems (SCOPUS).

Unfortunately, the rate of the implementation of the innovations in the economy is low; the number of researchers is diminishing; the percentage of young researchers is low. The total number of researchers per 1 million people is 4.5 times lower than EU average. This gap is likely to widen, given the trends of emigration of talented young researchers and low attractiveness of scientific careers in Moldova. The expenses for research and innovation per capita are circa 6.6 Euro, 80 times lower than EU average (Academia de Științe a Moldovei, 2020).

The share of students and PhD students in sciences and engineering is significantly lower than previously and below the EU average. During the 2019 period, 58% of PhD degrees were awarded in law, economics, pedagogy and medical sciences, while much less degrees were awarded in natural sciences (9%) or engineering (7%) (Academia de Științe a Moldovei, 2020); the science fields lack therefore gradually qualified researchers.

26.2   Overview of the Education Development

26.2.1   Education System and Policy

The structure of the Education System of Moldova is presented in the Table 26.1.

The Education System has undergone a continuous reform process, starting from the declaration of the independence of Moldova. The current policies and standards in education are outlined in a series of general documents related to education or specific ones that describe the standards and the guidelines for different levels of education:

- The Education Code of the Republic of Moldova approved by the Parliament in 2014 (MECC, 2014).
- Education 2020—a strategy approved by the Ministry of Education, Research and Culture (MECC) in 2012, for the period 2012–2020 (MECC, 2012).
- National Qualification Framework approved by the MECC (MECC, 2010).
- The framework plan for general education approved each year by the MECC (MECC, 2019).

These documents state that the concept and the objectives of the education policy are in line with the experiences of the European countries, correspond to the integration tendency into European Education space, and are based on the relevant interconnection between specific national context and the European and global tendencies of the education development (MECC, 2012).

Table 26.1 The structure of the education system

The name/grades (ISCED level)	Age
Early childhood education (level 0)	3–6
Primary education/grades I to IV (level 1)	7–10
Lower secondary education—gymnasium/(grades V to IX) (level 2)	11–15
Upper secondary education (level 3): lyceum education, grades X–XII secondary vocational education (for graduates of gymnasium)	16–18
Secondary vocational education (for graduates of general secondary schools and lyceums) (level 4)	>16
Postsecondary vocational education (exclusive the first two years for graduates of gymnasium) (level 5)	>16
Bachelor higher education (cycle I) (level 6)	>19
Master higher education (cycle II) (level 7) Long first degree program equivalent at level 7 (integrated higher education, inclusive medicine and pharmacy)	>23
Doctorate (cycle III) (level 8) Post-doctorate	

26.2.2   *Statistics on the National Education*

In the 2019/2020 academic year, the education process was organized in 1373 educational institutions, including 1255 primary and secondary general education institutions with 333.1 thousand pupils (84%) and 27.4 thousand (73%) teachers; 91 technical schools with 43.6 thousand students (82%) and 3.9 thousand (80%) teachers; 27 higher education institutions (HEIs) with 56.8 thousand students (53%) and 4.3 thousand teachers (66%). The numbers in the brackets indicate the percentage related to the 2010 year.

The distribution of pupils in the primary and secondary general education by levels in the 2019/2020 academic year was as follows: 41.7%—primary education; 47.7%—gymnasiums; 10.6%—lyceums.

Compared to 2010, there was a decrease in the number of students and teachers; the most affected being the higher education (MECC, 2020; Valcov et al., 2020).

The quantitative and qualitative analysis of the dropout rates in general education was performed in (Gremalschi, 2015). The considered reasons for dropout:

1.   pupils consider that the subjects they study are too complicated for them;
2.   pupils got annual marks below the threshold needed for the national examinations;
3.   pupils did not pass the national assessment.

The data showed that the dropout rate was 2.49% for the primary education level, the main reason being the one mentioned at point 3: pupils did not pass the final exam, because they skipped the classes or are from the vulnerable families.

The dropout rate was 5.9% for students at the end of the gymnasium level, the main reasons being 2 and 3.

The dropout rate for students for lyceums is much higher: 49.42% in 2013/2014 academic year, and 32.7% in 2018/2019. The author (Gremalschi, 2015) considers that the reason is the inaccesibility of curricula, in terms of complexity. In fact, there are some other reasons: the quality of textbooks; the inappropriate way how the subjects are presented; the assessment tasks that do not correspond to the real life problems; the students' lack of motivation.

26.2.3   *Educational Research and International Collaboration*

Educational research is performed mainly at universities and at the Institute of Science Education (ISE). The specific of the country that inherited the education and the research system from SU is the existence of the Academy of Science that concentrated the main research activities in Physics (Institute of Applied Physics), Chemistry (Institute of Chemistry), and Biology (Institute of Microbiology and Biotechnology). These Institutes have doctoral schools, but they do not investigate science education.

At the same time, there exist professional bodies that include experts and academics from each science domain, which may have science education interests, like the Association of graduates in Physics, Society of Chemistry. Some research in science education is performed at the university faculties, which is mostly related to the teaching training and not significant in determining the visions and the directions of the science education.

The ISE interests are concentrated on pedagogy and psychology, being grouped into the following units: Professional Development Management; Quality of Education; Psychology; Non-formal Education. Its collaborators consider as their prerogative the responsibility to guide and to elaborate the school curriculums for all levels and disciplines, thinking that the theories of learning are the main points from where to start the curriculum developments and promoting their beliefs and interpretations of the educational approaches to the curriculum and textbook authors, and to teachers. Under these circumstances, the science education curriculum is a derivate from the general framework, outlined by the education sciences' representatives, not by the professional bodies in science.

International scientific and education cooperation is a priority for Moldova in the endeavor to integrate the country in the European Research Area and in the European Education Area. The Republic of Moldova became associated with the Seventh EU Framework Program for Research &Innovation (FP7) in 2012 and to Horizon 2020, starting from 2014. During the period 2014–2019, Moldova participated in 53 projects with an EU contribution of € 6.18 million. Considering the limited number of researchers and the limited intensity of the research effort in the country, this performance situates Moldova in a leading position among the Eastern Partnership countries as concerns participation in these programs (Räim et al., 2016).

International collaboration in education is assured by Tempus and Erasmus + Programs that involve Higher Education Institutions. The schools started to participate in the Twinning projects and also in STEM projects. During 2014–2019, 17 Capacity Building in Higher Education Projects were implemented in Moldovan HEIs, 17 Jean Monnet projects were awarded to Moldova HEIs and NGOs. Nearly 2,500 students and academic staff have benefitted from Erasmus + academic mobility between Moldova and the EU.

Over 3,000 young people from Moldova took part in EU funded non-formal education projects which organize short exchanges, trainings, common events and provide volunteering opportunities (EU4Moldova, 2020).

26.3 Current Situation of Science Education

26.3.1 Policies and Standards

The policy for science education is the same as for the other domains of education. The standards are outlined in the documents:

- National curriculum (for primary education, gymnasiums, lyceums), elaborated by a development team and approved by the MECC (MECC, 2019).
- Curriculum for tertiary education, elaborated by each tertiary institution, and approved by the MECC.

The study programs for pre-university education underwent several major changes. After receiving the statute of an independent country, Moldova continued for almost 10 years to use as an education legacy the textbooks from the Soviet Union and from Romania. A new national curriculum based on objectives has been elaborated in 2000 and implemented until 2010, when the curriculum was radically changed again with the financial support of the World Bank and Soros foundation, Moldova. The reason was the shift to competence-based education, in order to be in line with European tendencies.

A history of the elaboration of the National Curriculum as well as an analysis of the implementation of the competence-based education was performed in the work "Developing key-competences in general education: challenges and limits" (Gremalschi, 2015), supported by the Soros Foundation Moldova. According to Gremalschi (2015), the main problem in the competence-based approach is the transposition of the competences into curricula. But the difficulty is not only the approach used for covering all key-competences by the curricula. More important is the appropriate understanding of the competence concept and how to implement it into teaching and learning and assessment. Unfortunately, a part of the academic community, including the curriculum developers and the textbooks' authors, misunderstood this relation, as well as the relation between key-competences and the specific competences, alongside the student-centered learning approach.

Some other constraints of the 2010 curricula mentioned in MECC (2014), Gremalschi (2015) were:

- a high level of curriculum theorization;
- a low level of relevance and practical applicability of the curricular content for the learners' personal, social and professional life;
- an excessive accent of the formative and summative assessment on knowledge and memorizing content instead of competence assessment.

An evaluation of the competence-based curriculum in general education was made in 2018 by a team from ISE and MECC, having as the objective the argumentation of the new changes in the curriculum (Pogolṣa et al., 2018). This study included results of questionnaires distributed to samples from different target-groups: pupils, teaching and management staff, parents; the templates for the final reports; the expert teams' reports on each discipline with conclusions and recommendations. The reading of the study is a difficult task, as the structure, analysis, terminology, and language differ from the similar documents written in English with clear logic and coherence. The authors of this study had concluded that the curriculum should be revised and rewritten and had proposed some recommendations.

New National Curriculum for general education was elaborated and approved by MECC in 2019 (MECC, 2019).

The curriculum for Higher Education has been radically transformed in 2005 when the Republic of Moldova signed the Bologna declaration.

26.3.2   Curriculums, Digital Resources and Teacher Training

The National Science curriculum, version 2019, has the following structure for all school levels: preliminary, conceptual benchmarks, administration of the discipline, specific competences, units of competences; units of content, learning activities and products, methodological benchmarks for teaching and learning and assessment, bibliography.

The science subjects are included in the study programs as follows: the discipline Science (three years, starting from the second grade, 100 academic hours in total) for the primary education; Physics (231 h, four years), Chemistry (165 h, three years starting from the 7th grade) and Biology (231 h, four years) for gymnasiums; Physics and Astronomy (297/231 h, three years), Chemistry (264/165, three years) and Biology (264/165 h, three years) for lyceums. The first number in the brackets refers to the real profile lyceums, the second—to the humanistic ones. The number of academic hours allocated for studying science disciplines is quite low in comparison to:

- time allocated by the European countries (Gremalschi, 2015);
- time really needed to develop the stated competences (Gremalschi, 2015);
- time allocated for humanistic disciplines (data from MECC (2019));
- time allocated in the previous study programs (before 1991 year).

A key feature of effective science education is the selection and implementation of relevant materials for students, including educational digital resources that comprise digital courses, tutorials, learning objects, learning and assessment tasks, and other educational aids, starting from digitalized versions of handbooks and ending with interactive simulations and collaboratively generated content on virtual platforms. The availability of digital content that covers all grade levels and all subject areas and that can be delivered to all learners is a challenge both for the HEIs and for pre-university institutions. Though the policy documents have set objectives for 2020 to train teachers in the effective use of ICT, to integrate ICT into education, to elaborate digital educational content, these targets were not completely achieved. The digital literacy picture reveals a distribution of teachers with different levels of skills: technological enthusiastic teachers at the competence edge and the majority at the very basic skills level. Therefore, only a small part of them are able to elaborate digital resources for teaching and learning and assessment.

Science teachers use more frequently World Wide Web public resources: videos from YouTube, images, presentations for explaining new concepts and lesson topics; web sites with lesson plans and learning tasks' worksheets for students; freely available software for creating quizzes and tests for formative and summative assessments.

The use of simulations, of educational interactive software or virtual labs for demonstration of natural phenomena or processes is quite limited, as not all teachers are accounted with these kinds of resources or because they are not freely available on the web.

To enrich the existing digital resources and to help teachers from schools, different state and private organizations during pandemic period came with several initiatives and projects in order to elaborate digital content. The projects involve Universities, the National Association of ICT Companies, the Academy of Innovation and Change through Education, some private companies, the Chisinau City Hall, MECC. The results are video and/or text lessons and other digital resources, made available online, that are supposed to be used by teachers and students as teaching and learning resources. This process is ongoing and the research is needed to conclude on the impact and relevance of these resources on teaching and learning science subjects.

Science teacher training is a big problem for Moldova. The teaching profession has to face up to demographic changes. About half of the teachers are aged over 50 years and 17% are retired people. Moldova experiences difficulties in attracting good school graduates, especially for science teachers. Migration flows add more problems. A considerable part of teachers left the country for a better life in Europe. This confronts the education with the reality of lacking qualified teaching staff for pre-university institutions.

Initial teacher training is based on two cycles study system (Bachelor and Master). The primary science teachers are trained within the faculties of primary education, 3 or 4 (for those who study a foreign language, additionally) years. Graduates of the first cycle are allowed to teach in the primary school (all subjects). The master's degree is not compulsory to become a primary teacher, though it is recommended. There are several HEIs, that offer study programs for primary teachers and as usual, there are no problems with the students' enrolments.

The worse is the situation with the science teachers for gymnasiums and lyceums. The Education Code stipulates that the lyceum teachers should hold a master's degree. There is a shortage of science teachers in general, and of graduates of the Bachelor level in the science domain, in particular. Only a few HEIs offer study programs in physics, chemistry, and biology with an integrated compulsory teacher training module (60 ECTS) at the first cycle. Graduates of these study programs are not eager to become teachers and as a result the Master programs in these domains are not demanded. Schools are happy to accept Bachelor degree graduates which level of qualification is low.

In-service teachers have different possibilities for continuous professional development: short term courses, seminars, and workshops offered by some HEIs, ISE, and other institutions.

26.3.3   Student Assessment and Achievement

Student assessment includes formative and summative methods. One may judge achievements having the results of the summative assessment at the end of the semester or of the academic year. Each HEI and school keeps records on students' achievements and makes reports to MECC, but these results are not made public. The assessment methods usually used in science education in schools are tests, solving of problems, the reports from the laboratory works, portfolios.

The national evaluations consist of tests, in written form, at the end of primary education, gymnasiums, and lyceums, but only the results for lyceums are published. The methodology of these evaluations is established by MECC, which is entitled to select the assessed disciplines, to approve the assessment subjects and criteria, as well as the assessment standards. The assessment subjects for the science disciplines do not evaluate the declared competences, but only the students' knowledge, in the best cases up to the application level.

The Republic of Moldova had participated several times in the international assessment program TISS in 1999 (8th grade), in 2003 (4th and 8th grades), and in 2007.

The pupils gained a score of 459 for Science in 1999, closed to the international average (488). Almost the same position was maintained in 2003 with the score 496 (4th grade) and 471 (8th grade), slightly above the average (Gremalschi, 2015). The results for 2007 are not available.

Also the Republic of Moldova has participated three times (2009, 2015, 2018) in the Program for International Student Assessment (PISA), which is a triennial survey of 15-year-old students that assesses the extent to which they have acquired the key knowledge and skills essential for full participation in society, organized by OECD.

Students from Moldova showed lower score than the OECD average in all domains: reading, mathematics and science. The score for science was 428, the same as in 2015, and 15 points higher than in 2009. Compared to the OECD average, in 2018, a smaller proportion of students in Moldova performed at the highest levels of proficiency (Level 5 or 6) in at least one subject; at the same time, a smaller proportion of students achieved a minimum level of proficiency (Level 2 or higher) in at least one subject. Some 57% of students in Moldova attained Level 2 or higher in science (OECD average: 78%). At a minimum, these students can provide possible explanations in familiar contexts or draw conclusions based on simple investigations. In Moldova, only 1% of students were top performers in science, meaning that they were proficient at Level 5 or 6 (OECD average: 7%). These students can creatively and autonomously apply their knowledge of and about science to a wide variety of situations, including unfamiliar ones (Avvisati et al., 2019).

26.3.4   Science and Technology Venues and Centers

The opportunities for students to explore the scientific phenomena and to do experiments within the formal establishments are limited. The universities offer possibilities for students to work in their laboratories according to the schedule of the study programs. These facilities are open to school students during open days just for visiting or during special organized national contests for lab activities or for the training of the students for international contests.

The interested pupils have the chance to develop their curiosity as extra-curricular activities within different programs and projects that were initiated within partnerships between education institutions from Moldova and International bodies. Some of the most interesting initiatives are listed below.

The "RoboClub" robotics study program[1] was launched in March 2014 with the financial support of the United States Development Agency (USAID) and the Government of Sweden, in partnership with the Ministry of Education and the National Association of Information and Communication Technology Companies (ATIC). This program has several components:

- Endowment of educational institutions with LEGO® MINDSTORMS® EV3 sets. The schools are selected on competition basis, and equipment comes largely from international donations.
- Organization of training classes in formal or non-formal format as extra-curricular activities.
- Annually national competitions "First Lego League". The winner teams have participated in international competitions and achieved good results. For example, the team from Moldova had won three gold medals at the international robotics competition in 2019.

GirlsGoIT[2] program, launched in Moldova in 2015, and led by the not-for-profit organization TEKEDU, encourages girls and young women to go in the field of technology from a young age, choosing STEM (science, technology, engineering and mathematics) education path and empowering girls and young women in and through technology to have better future education and employment opportunities.

Tekwill[3] is a hub that has co-working spaces, tech labs (IoT, 3D printing) and community events. It aims to improve the skills needed within the IT sector, thereby creating high quality and well paid jobs, and deterring the emigration of talented young people. Tekwill encourages local startups and existing companies to expand and attract international IT companies to invest in Moldova. Tekwill was created with the support of the United States Agency for International Development (USAID) and the Government of Sweden through SIDA/Swedish International Development Agency in the framework of the project "Development of Moldova ICT Excellence

---

[1] https://mecc.gov.md/ro/content/fost-lansat-proiectul-pilot-roboclub.

[2] https://girlsgoit.org/local-clubs.html.

[3] https://www.tekwill.md/.

Center" implemented by Moldova's Association of ICT Companies in partnership with Technical University of Moldova.

The "Future Classroom lab" is a program[4] that aims to develop the digital skills and competencies needed for the twenty-first century, as well as to increase students' interest in careers in science, technology and engineering by promoting creative, collaborative, self-motivating and project-based training methods that integrate knowledge and skills in various subjects, especially STEAM. The program brings together the joint efforts of the Orange Moldova Foundation, the Moldova Competitiveness Project, funded by the United States Agency for International Development (USAID) and the Government of Sweden, together with the MECC.

26.3.5 Utilizing Emerging Technologies

The use of technologies and digital resources in science education got an impetus during the pandemic as the teachers were forced to teach at distance. The available technologies in schools comprise desktop computers, laptops, TV sets, video-projectors, white-boards. MECC underwent efforts to complement the existing hardware with new equipment from national funds and from international donations. The software component is still a weak point both as applications and as digital resources.

We must distinguish between the use of technologies in the general education system and in the projects and programs mentioned in the Sect. 26.3.4. These projects attract the most talented and curious students, who are trained by experienced trainers, including school teachers, according to special programs, which differ from those in the school curriculum in terms of content, approaches, and advanced technical facilities. These activities are alternative approaches that stimulate interest and encourage engagement, and are valuable for improving science literacy and the development of science process skills for some young people. The competitions and contests organized within these programs at national and international levels and the obtained results denote the fact that students have the potential and the ability to study science and to get involved in solving real life problems.

It is required to expand these experiences to all schools. There is a need to develop teachers' skills, especially with regard to science content and pedagogy of science content, and in facilitating hands-on activities for science lessons, as well as on the implementation of modern technologies to enhance student learning of science. Many teachers are ill-informed about current developments in science, about inquiry- and problem-based learning, about emerging technologies potential, and, being themselves disappointed due to poor technological or social conditions, they can hardly be expected to promote effective science education.

---

[4] https://www.clasaviitorului.md/.

26.4   Requirements for Future Development of Science Education

It is known that there are challenges with science education in many countries of the world (Science Education in Europe: National Policies, Practices and Research, 2011). Moldova is not an exception in this regard, having a number of extra problems. Such problems lead to a progressive loss of students' interest in science. As a consequence, a small number of students are choosing to study the sciences at higher levels and those who have chosen science-based study programs encounter difficulties.

Continuous curricular reforms, starting from 1991, did not improve science education. The success of science education depends on the following major components: academic staff, the study programs or curricula, available resources. These components are interdependent to a high extent and the challenge for the current situation in science education is: from which to start the improvements nowadays? Though the two studies (Gremalschi, 2015; Pogolşa et al., 2018) had identified a part of the existing problems of the competence-based curriculum approved in 2010 and slightly had mentioned some problems in science education (Gremalschi, 2015), the new version did not solve them: at the contrary, other problems appeared.

Elaboration of the study programs requires a clear vision of the role of science in the current society affected by severe calamities, disasters, and survival issues, at a global level, as well as in the considered country, in particular. There are some peculiarities in science education that depends on the country-specific, but they are minor in comparison to the common science education challenges.

The formulated and the developed competences for the science domain in secondary education should establish the minimum level of attainment for the graduates, as the science-literate persons for the 21st Century. This is the "science as culture" argument; that science is as worth studying in itself. The school should educate humans that care about the living environment and the future of the earth and of civilization. At the same time, these competences should lay a foundation for the small proportion of individuals who will become future scientists; should provide a background for professionals, which careers need science education (engineers, medical doctors, agronomists, etc.), professionals required by the future structure of the economy and the labor market. At present, neither educational, nor sociological research has been carried out in Moldova, which would have had as an object of study the correctness and the relevance of the defined competencies in the school curriculum; the constructive alignment between those competences, the learning activities, the assessment tasks, and the designed resources; the achievements and the failures and their real reasons.

The main actors in the elaboration, implementation, monitoring, evaluation and revision of the curricula up to now are only the education researchers, teachers from the general and the university education, without significant representation of the direct and indirect beneficiaries of the educational system: students, science researchers, business environment, professional associations, and employers. The study (Gremalschi, 2015) revealed, that the institutional framework of the curricular development processes does not guarantee the participation of the representatives of several approaches in the field of education sciences and that sometimes the curricular policy documents are based only on the "achievements of the local science" and of one "national school", which volens-nolens overlooked the experience of the European countries.

Also, there is no clear separation of the competencies of the actors involved in the curricular development the elaboration, implementation, monitoring, and evaluation of the curriculum being carried out, de facto, by some and the same institutions and even by some and the same persons (Gremalschi, 2015), and sometimes their level of expertise is questionable.

The distinct roles as curriculum developers, elaborators of textbooks and didactic resources, curriculum, textbooks' & educational resources' evaluators should be assigned by real competition for the competent experts with international experience in science and education domains.

It is necessary that both policy and curricula documents be reviewed and changed. It is embarrassing to say that these documents lack the quality and the rigors of the International ones. Though the descriptive text mentions the European policies and practices, there are no references to documents, neither to concrete provisions, neither to the key role of science education as it is recommended by EU policies.

The MECC should extend the institutional framework for the elaboration of key documents, should ensure the involvement of all interested actors and, most importantly, not limit itself to national experience and expertise. Although in the case of secondary education, unlike higher education, European policies do not involve the creation of a common European space of education, the general school in the Republic of Moldova should be synchronized with those in European countries, curricular harmonization being one of the first steps in this direction (Gremalschi, 2015). Curricular harmonization should follow from the recognition at the education policy level of the priority of science education, as the country need for a skilled population, competent in science, technology, engineering and mathematics, not only for comprising the driving force for economic prosperity, but for having a science and technology literate society.

The decades of continuous curricula reforms did not prove the achievement of the declared intentions. Irrelevant design of the curricula, misunderstanding of concepts, lack of constructive alignment in teaching & learning & assessment, low-quality textbooks are a few of the reasons that require harmonization of curricular standards, contents, and assessment in science subject areas, as a way to solve the existing deficiencies and improving science education.

The limited appreciation of science disciplines and their importance to the vitality of the country's economy and future well-being has the consequence the lack of resources, both financial and educational, linked with a shortage of adequately trained teachers. Lack of finances is recognized in the policy documents, as the allocated resources are by far not sufficient and the decision-makers make reference to extra-budget funds, like international projects and sponsorships. This is one more important reason to use the developed science curricula and the textbooks from other countries.

It could be more efficient to establish national and international partnerships within some projects or programs, and to promote best practices of science education, to implement the proved teaching approaches and resources, including freely available, than to spend money for developing inappropriate content, trying to reinvent the wheel, in the ambition to develop curricula without sufficient competences, and without considering the best experiences.

26.5   Discussion and Conclusion

The human brain is a mysterious organ, which has an amazing characteristic to learn all the time. The current theories did not completely discover the mystery of brain learning. The existing learning theories used in formal education try to follow this complex process, and to help to develop a nation's capacity building, as education in any country can assure its citizens' well-being, can develop the economy, can build and maintain a healthy and a safe living environment, enhancing the ability and the performance of the citizens for the future.

The impact of science on people's lives is tremendously important nowadays. Consequently, scientific literacy for all has become the overarching objective of science education throughout the world; the science education mission is crucial in let students' brain to learn and to see science and technology in a wider context—as endeavors with important consequences for people and other living things—and that they learn to connect their knowledge of science and technology to the world beyond the school.

The brief description of science education in Moldova in this chapter revealed serious obstacles in the process of helping the brains of the young generation to acquire a body of knowledge on the natural and physical world and to be prepared for life and society challenges.

The science theories and concepts undergo changes but, for the most part, the basic ideas of science, like the cellular basis of life, the laws of energy, and the particle theory of matter, have proven to be stable in time and are the same for all countries. Therefore, the pedagogical approach and the curriculum for science education should be compatible and comparable for different countries, especially as regards the "big ideas" (Tracy, 2018). Big ideas go beyond discrete facts or skills to focus on larger concepts, principles, or processes. Big ideas are the broad, important understandings that students should retain long after they have forgotten many of the details of something that they have studied. Big ideas describe aspects of the fundamental concepts that are addressed at each grade level (UNESCO International Bureau of Education, 2016). Or, this "big idea" approach is blurred in the science curriculum in Moldova, both as concern the science content and as the learning approach as well. The good intention to implement the big idea of the competence-based education and student-centered learning turned into the fragmentation of content, in losing the logic of science concepts and relationships, of discovering the complexity and beauty of science progressively. The competence approach and the student-centered approach was misunderstood by the curriculum developers and the authors of the textbooks.

It can be stated that the competence-based curriculum was declared, but not properly implemented; the curriculum authors did not take into account the curriculums from European countries nor the current curriculums from the former Soviet Republics. As a result, the curricular policy documents are heterogeneous and the teaching and the management staff are disoriented to some extent. There are confusions on the relation between key-competences and the specific competences; on the definition of the specific competences for each scientific domain and course; on measurement criteria for competences at each education level. The curriculum authors introduced the term "subcompetence" in the 2010 version of the curriculum, which was criticized by a part of the academics. In the 2019 version, the term "subcompetence" was substituted by the term "unit of competence" with the meaning of the competence developed when studying a "unit of content" (several related topics). In fact, the units of competence are a kind of objectives related to the topics. The concept of learning outcomes is missing in general education and is used only in some HEIs.

The current textbooks in science education are of little help for learning science. The units of the content contain a small amount of text with definitions and some scratch data and fragments of information, with no comprehensive coherent explanations of concepts, phenomena, laws, or processes and a lot of tasks and questions, the answers for which students should find somewhere else. In this way the student-centered approach was very often extrapolated by the textbook authors to an excessive set of questions, for "discovering" nature's laws, instead of preparing students' background for scientific inquiry and stimulating them to ask questions about the investigated items, developing gradually students' research skills. An example is the theme Earth Planet, third grade, where the two blurred visions about Earth's origins are described in an axiomatic way, and pupils are asked to argue pro or contra these visions and to present different versions from other resources, including the Internet.

The assessment tasks are overloaded with tests that include multiple-choice questions, completing missing words in sentences, establishing connections between fragments of sentences, etc. This kind of assessment can measure knowledge of some facts, at the low level of cognition, in the best case, but cannot certify the competence development. Therefore, the constructive alignment between learning outcomes, learning activities, and assessment tasks is barely respected.

The challenges in science education presented shortly in this chapter must be discussed more broadly by the country's academic community, relevant staff from the universities, science researchers, professional associations, and subject school teachers. An analysis of science education and of the process of the curriculum elaboration as well as the experiences of curriculum implementation and the gained results of different countries is more than requested.

References

Academia de Ştiinţe a Moldovei. (2020). Raport asupra stării stiinţei în Republica Moldova în anul 2019. https://asm.md/raportul-asupra-starii-stiintei-din-republica-moldova-anul-2019.

Avvisati, A., Echazarra, A., Givord, P.and Schwabe, M. (2019). Programme for International student assessment (PISA). Results from PISA 2018. OECD. Directorate for Education and Skills. http://www.oecd.org/pisa.

EU4Moldova. (2020). Erasmus + Moldova. https://www.eu4moldova.md/en/content/erasmus-mol dova.

Government of Moldova. (2012). Moldova 2020. National development strategy: 7 solutions for economic growth and poverty reduction. https://cancelaria.gov.md/sites/default/files/document/attachments/1100271_en_moldova_2020_e.pdf

Gremalschi, A. (2015). Formarea competenţelor-cheie în învăţământul general: Provocări şi constrângeri: Studiu de politici educaţionale. Inst. de Politici Publice. https://www.soros.md/files/publications/documents/Studiu%20Formarea%20Competentelor-Cheie.pdf.

MECC. (2019). Curriculum naţional. Aria curriculară Matematică şi ştiinţe. https://mecc.gov.md.

MECC. (2020). Analiza statistică generală, anul de studii 2019–2020 (domeniul Educaţie). https://mecc.gov.md/sites/default/files/analiza_statistica_generala_anul_de_studii_2019-2020_dome niul_educatie_0.pdf.

MECC. (2014). Codul Educaţiei al Republicii Moldova. Monitorul Oficial, nr. 319–324 din 24.10.2014.

MECC. (2010). National Qualifications Framework. http://edu.gov.md/ro/content/cadrul-national-al-calificarilor-0.

MECC.(2019). Planul cadru pentru învăţământul primar, gimnazial şi liceal 2029–2020. https://mecc.gov.md.

MECC. (2012). Strategia de dezvoltare a educaţiei pentru anii 2014–2020 „Educaţia-2020”. https://mecc.gov.md/sites/default/files/1_strategia_educatia-2020_3.pdf.

Moldovan Political System. https://moldovanpolitics.com/political-system-2.

Pogolşa, L. et al. (2018). Institutul de Ştiinţe ale Educaţiei. Evaluarea curriculumului naţional în învăţământul general. Studiu.

Räim, T., et. al. (2016). European Commission. Peer Review of the Moldovan Research and Innovation system. https://rio.jrc.ec.europa.eu/sites/default/files/report/Moldova-PSF_PR-KIAX16004ENNOP.pdf.

Science Education in Europe: National Policies, Practices and Research. (2011) Education, Audiovisual and Culture Executive Agency. http://eacea.ec.europa.eu/education/eurydice.

Strategia Cercetării-Dezvoltării a Republicii Moldova pînă în 2020. Academia de Ştiinţe a Moldovei. https://asm.md/sites/default/files/2020.07/Raport_asupra_starii_stiintei_2019_ASM_Guvern_07_07_2020.pdf

The World Bank. (2020). Country context. Moldova. https://www.worldbank.org/en/country/mol dova/overview

Tracy, Ch. (2018) Guidelines for future physics curricula. School Science Review, 100 (370), 36 - 43. https://www.ase.org.uk/resources/future-of-science-curriculum.

Valcov, V., et al. (2020). Moldova in figures. Statistical pocket-book. National Bureau of Statistics of the Republic of Moldova. https://statistica.gov.md/public/files/publicatii_electronice/Moldova_in_cifre/2020/Breviar_2020_en.pdf

UNESCO International Bureau of Education. (2016). What makes a quality curriculum? https://unesdoc.unesco.org/.

Roza Dumbraveanu is doctor in physics and mathematics, associate professor at the Chisinau Ion Creanga State Pedagogical University, faculty Education Sciences and Informatics. She had graduated the Faculty of Physics at the State University of Moldova. Roza

Dumbraveanu has experience in elaboration of curriculum, teaching aids, digital resources, e-courses, in e-learning, in assessessment of study programs. She had actively participated in a series of national and international projects, related to curriculum development, using ICT in education, internationalization of Higher Education. Roza Dumbraveanu is member of the programme committees of several International conferences: IMCL: Mobile and Computer Aided Learning, ELETE: E-learning and mobile learning on telecommunications; CSEDU: Computer supported education.

Chapter 27
Education and Science in Montenegro as a Base for the General Development and Well-Being of the Society

Tamara Milić

Abstract The legal, strategic framework in the field of education and science is affirmative, open to change. The reform directions are modern, contemporary and dominantly follow the democratic trends and values of the EU. The values of tolerance, democracy and inclusion are strong. The Montenegrin education system consists of the following levels of education: preschool education, nine-year elementary education, general secondary education (gymnasiums) and secondary vocational education, higher education and adult education. Teaching is held according to subject programs. Inclusive education of children with special educational needs is an imperative and based on an Individual Development Education Program. Higher education in Montenegro can be acquired at the: university, faculty, art academy and college. The MEIS application is used to collect data and for the evidence based planning. All activities are implemented in international partnership, initiatives, programs and support. All levels and areas of education are strategic based. There is a great focus on the professional development of teachers in order to ensure the quality of education, primarily due to the results of the PISA test, which are not satisfactory. Significant research, innovation and activities in the field of technology have begun, but they are still in their initial phase. They should be supported, promoted, but also used for the common good and for the development and implementation of policies based on science, research and needs, primarily in the direction of general competencies.

Keywords Preschool education · Elementary education · Secondary education · Inclusive education · Research · Science

T. Milić (✉)
Ministry of Education Montenegro, Vaka Đurovića bb, 81000 Podgorica, Montenegro
e-mail: t.milic@mps.gov.me

453
R. Huang et al. (eds.), *Science Education in Countries Along the Belt & Road*,
Lecture Notes in Educational Technology,
https://doi.org/10.1007/978-981-16-6955-2_27

Table 27.1 General data

Citizens, census 2011	620 029
Montenegrin area, km^2	13 888
Capitol	Podgorica
Royal Capitol	Cetinje
Number of municipalities	24
Number of settlements, Census 2011	1 307

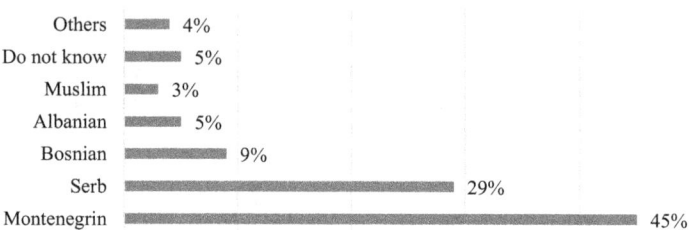

Fig. 27.1 National structure

27.1 Overview of the Country

27.1.1 Geographical Location, Population and Political System

Montenegro is located in Southeast Europe. Its area is 13,812 km^2; the mainland is 13,812 km^2 and 293 km along the coast.

The Constitution defines Montenegro as an independent and sovereign state, with a republican form of government. Montenegro is a civil, democratic, environmental and social justice state, based on the rule of law (Table 27.1).

The capital of the country is Podgorica, and the Royal capital is Cetinje. The official currency is the euro. Based on the last Census data from 2011 Montenegro has 620,029 inhabitants[1] (Fig. 27.1).

Legislative power is exercised by the Parliament, the executive by the Government, and the judicial—the Court. The official language is Montenegrin. Cyrillic and Latin alphabets are equal. Serbian, Bosnian, Albanian and Croatian are also in official use.

---

[1] The census is conducted every 10 years.

27.1.2   *Current Situation of Economy, Technologies and Cultural Development*

Montenegro began the transition process in the early 1990s, but could not achieve the level of national competitiveness due to international sanctions, wars in the region, and the severance of ties with the former Yugoslav market (Uvalic, 2003).

In the period from the beginning of the twentieth century, Montenegro took over part of the conduct of economic policy. The changeover from the dinar to the German mark and then the euro in 2002 formally confirmed the country's economic independence and the stability of macroeconomic indicators.

After the renewal of independence in 2006, a period of accelerated economic development began for Montenegro. Average GDP growth rate for the period 2006–2018 was about 5% (10% in 2018).

At the moment, Montenegro is in the process of European integration, and since 2017 it has been a member of the NATO Alliance.

In recent years, the Montenegrin economy has recorded strong economic growth. Gross domestic product (GDP) has grown at an average rate of 3.9% over the last five years, while in the period 2017–2019 the average growth rate was 4.4%. Montenegro's economy is mostly service based and overall tertiary sector contributes 54.5% to the GDP and employs three-quarters of the workforce. Tourism, which alone provides one-quarter of the GDP, is the third largest sector and consumes 34% of total investment. Industry represents around 16% of the country's GDP and employs 17% of the workforce. The manufacturing sector is still underdeveloped and accounts for only 4% of GDP. Agriculture accounts for 6.7% of the GDP and 7.8% of the workforce, but it is important also for tourism development.[2]

One of the characteristics of Montenegro's labour market is a lower activity rate compared to EU countries and also countries from the region, but labour market trends are improving during the few previous years. In addition, the biggest problem is unemployment of youth. The average wage in June 2020 was 778 EUR (gross), while the minimum wage in Montenegro is 222 EUR.

Even though a small economy, Montenegro has increased its R&D investments to 0.38% of GDP in 2015, with government being the major investor (0.22 of GDP), followed by business sector (0.11), universities (0.03) and then investments from abroad (0.02). Montenegro has 1766 registered researchers. The gender balance of researchers is very well balanced (50% male and female). According to the European Innovation Scoreboard Montenegro is recognized as high Innovation-friendly environment. Other strong components include broadband penetration, innovators, and attractive research system and employment share services (Ministry of Science, 2019a).

Montenegro has been working on development of Science Technology Parks and technology incubators, during the last several years. Following a feasibility study, Montenegro established 'impulse centre' incubators in Nikšić and Bar and in 2020

---

[2] MONSTAT, GDP data.

Table 27.2 ICT overview[4]

Internet access at home	With internet	Without internet	Do not know
	72,2%	27%	0,8%
Internet devices	Personal computer	Laptop	Mobile phone
	46,6%,	65%	79,2%
Internet connection type	ADSL	Mobile broadband	Narrowband
	75,5%	48%	8,8%
Average internet usage	Almost every day	Once a week	Less than a week
	87,6%	10,9%	1,4%
ICT usage in enterprises	Using the computer	Using the internet	With Web site, home page
	98,5%	99,2%	80%

national STP, headquartered in Podgorica, with the University of Montenegro and the Municipality. The main priority sectors in which science, innovation and research, would be focused are: sustainable agriculture, energy and health tourism, thereby strengthening the competitiveness and internationalization of the economy (Ministry of Science, 2019b).

In addition, regarding the information and communication technologies percent of ICT usage in 2018 is shown in Table 27.2.[3]

The goal of the cultural policy of Montenegro is to develop and promote contemporary cultural and artistic creativity, activities and life, valorise heritage, contribute to the preservation and affirmation of various identities. At the national level, 11 public institutions have been established to perform cultural activities, four in the field of cultural and artistic creation: the Montenegrin National Theatre and the Royal Theatre "Zetski dom", the Music and the Center for Contemporary Art. Seven of them are in the field of cultural heritage: the National Museum of Montenegro, the National Library "Đurđe Crnojević", the Library for the Blind, the Montenegrin Cinematheque, the Maritime Museum, the Natural History Museum and the Center for Conservation and Archaeology. In accordance with the principle of decentralization, the obligation of the state to provide conditions for the development of culture at the local level is prescribed (Ministry of Culture, 2016).

According to data from 2018 creative industries contribute 1.5% to gross value added (GVA). In addition 2.031 business entities or 3.5% of all registered business entities (legal entities and entrepreneurs) is in this sector while 7.252 or 3.6% of the total number of employees in business entities at the level of Montenegro. The sector contributes with 1.4% to the total revenues of business entities and with 2.2% in total exports of services.

In this area Montenegro continuously participates in the programs of international organizations: UNESCO, The Council of Europe, the European Commission and the Regional Cooperation Council.

---

[3] MONSTAT.
[4] MONSTAT.

Table 27.3 Educational levels and institutions

Preschool education	21 public institutions	23 private preschool institutions	
Elementary education	163 public schools	4 private schools	
Secondary education	47 public secondary schools	One private gymnasium	
Resource centres	RC for Hearing and Speech disabilities	RC for Physical and Sight disabilities	RC for Intellectual disabilities and Autism
Adult education providers	109		

27.2 Overview of the Education development

27.2.1 Education System and Policy

The Montenegrin education system consists of the following levels of education: preschool education (involves children up to the age of six), elementary education (nine-year compulsory and free of charge), general secondary education (gymnasiums) and secondary vocational education, higher education and adult education. The system comprises of: 21 public and 23 private preschool institutions; 163 publics, 4 private elementary schools; 47 public secondary schools (gymnasiums, secondary vocational and mixed schools) and one private gymnasium; three resource centers and 109 licensed adult education organizers (Table 27.3).

Preschool education involves children up to the age of 6 (until they start attending elementary school) in preschool institutions that can be state-owned (public) and private.

Elementary education in Montenegro is carried out according to the provisions of the General Law on education[5] and the Law on elementary education.[6] Nine-year elementary school education is compulsory and free of charge and carried out in three cycles (3 + 3 + 3) for children from 6 to 15 years. During the first cycle, all lessons in all subjects are taught by a teacher, and a preschool teacher engaged for the matter of adaptation. Grades are descriptive. During the second cycle, subject teachers are introduced to the teaching process. During the third cycle, all lessons are taught by subject teachers. Grades are numeral (scale from 1 to 5). At the end of each cycle, there is an external assessment.

---

[5] "Official Gazette of Montenegro", no. 64/02, 31/05, 49/07, "Official Gazette of Montenegro", no. 04/08, 21/09, 45/10, 45/11, 37/13, 47/17.

[6] "Official Gazette of Montenegro", no. 64/02, 49/07, "Official Gazette of Montenegro", no. 45/10, 37/13, 47/17.

General secondary education (gymnasiums) is carried out in line with the provisions of the General Law on education and of the Law on gymnasium.[7] General secondary school education lasts for four years. It is carried out in gymnasiums, mixed secondary schools, which, apart from their gymnasium curriculum, also implement vocational secondary school programs. Out of the total number of secondary school students in Montenegro, 33% are acquiring gymnasium education. Gymnasiums curriculum consist of three components: compulsory subjects, compulsory elective subjects and compulsory elective contents.

Secondary vocational education is carried out in line with the provisions of the General Law on education and of the Law on vocational education.[8] Vocational education can be carried out as:

– two years' elementary vocational education;
– three or four years' secondary vocational education;
– higher vocational education (lasting two years, as extension of the secondary vocational education).

The Montenegrin education system for children with special education needs (SEN) sets as it first choice and imperative the inclusive education in regular schools.[9] A base for the work with SEN children is the Individual Development Education Program. The term Special Educational Needs is used at a pre-university education level and it includes children with: disabilities[10] and difficulties.[11] Special institutions have been reformed into Resource Centres that provide inclusive services: early intervention, individual support and work with children, trainings, mobile outreach activities.

Higher education in Montenegro can be acquired at the: university, faculty, art academy and college. There are four universities and three faculties. The condition is a certificate of accreditation from the Agency for Control and Quality Assurance of Higher Education and a license to work issued by the Ministry of Education. Study programs that can be implemented at public and private institutions of higher education are: bachelor, 180 ECTS; master, 120 ECTS and doctoral, 180 ECTS. A new study model has been introduced, $3 + 2 + 3$ with free bachelor and master studies at public institutions of higher education, i.e. the University of Montenegro. From 2020/21 school year, second cycle studies, i.e. master studies, are free of charge. Compulsory practical classes in the amount of 25% are introduced.

---

[7] "Official Gazette of Montenegro", no. 64/02, 49/07, "Official Gazette of Montenegro", no. 45/10, 37/13, 47/17.

[8] "Official Gazette of Montenegro", no. 64/02, 49/07, "Official Gazette of Montenegro", no. 45/10, 37/13, 47/17.

[9] "Official Gazette of Montenegro", no. 45/10, 47/17.

[10] Physical, intellectual, sensory, autism and combined.

[11] Speech and language difficulties, behavioural, learning, serious chronic and long-lasting illnesses, emotional, social, language and cultural depravation.

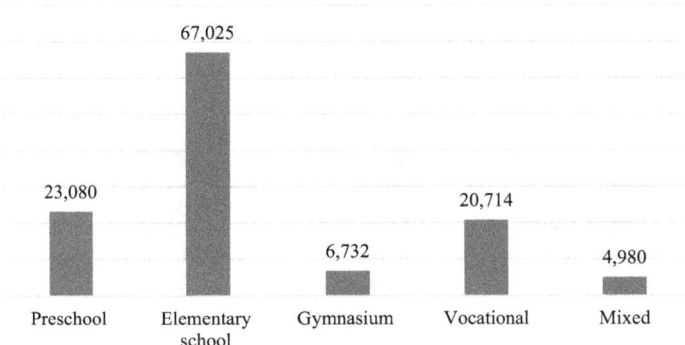

Fig. 27.2 Enrolment – level of education.[13]

Adult education can be organized by licensed providers. Publicly valid educational programs for elementary, general and vocational education are adapted to adults in terms of scope, organization and duration. Non-formal education programs are for: professional qualifications, retraining, additional training, and specialization. At present, there are 130 education programs for the acquisition of professional competencies and 117 for the acquisition of key skills and competencies.

27.2.2 Statistics on the National Education

The number of children enrolled in preschool education in the 2019/20 school year is 23.080. The percentage of children from age 3 up to age 6 (to enrolment in school) for the 2018/2019 school year is 72,62%.[12] In the 2019/20 school year, 67.025 students attended elementary education. There are a total of 27.446 students in secondary schools. Of this number, in gymnasium –6.732, vocational schools – 20.714, of which in art (fine art, music) is 236, and in mixed (gymnasiums and vocational) –4.980 students (Fig. 27.2).

A total of 2.616 employees are in preschool institutions, of which 1.304 are teaching staff. There is a total of 7.280 employees in elementary schools, of which 5.171 teachers, and in secondary schools a total of 3.406, and teachers –2.656 (Fig. 27.3).

The Ministry of Education in cooperation with UNICEF has improved the Montenegrin Education Information System—MEIS application, as well as the accompanying procedures at the school level in order to identify the risk of early school leaving. The focus in on the children who are at risk of dropping out of school, as well as on children who miss a lot of school (especially from Roma and Egyptian—RE population and on certain number of students excluded during the

---

[12] MONSTAT.

[13] Montenegrin Education Information System—MEIS.

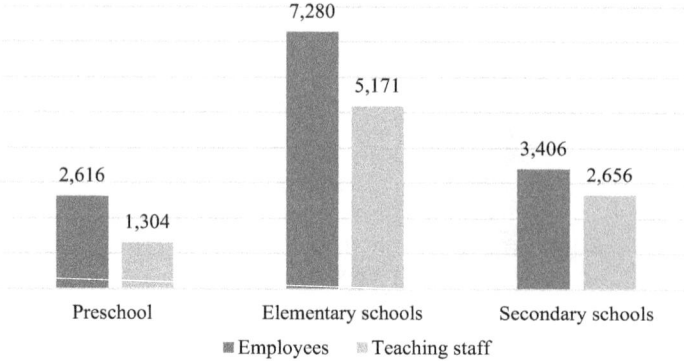

Fig. 27.3 Employees in education institutions.[14]

school year due to absences, behaviour, etc.). For example, in 2015/2016 school year, out of a total of 28,106 students in secondary schools, 153 students, or 0.54%, were no longer in the education system by the end of the school year. Furthermore, the number of excluded students in the same year was 1,125. Although the expressed percentage and numbers are not considered as high and worrying, it creates an unrealistic picture of the total percentage of „drop-out" because these students represent the so-called „hidden drop out ". For all these reasons, risk indicators, warning criteria, measures to be taken have been done.

27.2.3 Educational Research and International Collaboration

The Bureau for Educational Services determines and ensures the quality of education in schools and performs developmental, advisory, research and professional work by levels of education. One of the basic functions is research and advisory work. The Department for Research and Development of the Education System performs the following tasks: monitoring, analysis and development of the education system; research and monitoring of experiments in the field of education, all with the aim of proposing evidence-based measures for the development of different education levels, new teaching technologies and their application; evaluations of the effectiveness of new teaching methods and forms of work; planning the implementation of reform and program solutions; implementation of new educational programs; monitoring the inclusion of children from marginalized groups in the education system; comparative analysis of the application of new educational programs with the existing ones, etc.[15]

---

[14] MEIS.

[15] www.zzs.gov.me

Key international partners in the education system are UNICEF, UNDP, SAVE the children, UNODC, KulturKontakt, Roma Education Fund- REF, European Training Foundation—ETF, Council of Europe, European Commission and DEU, UNESCO. They are predominantly focused on initiatives and programs that lead to quality, equality, equity of education, early development, inclusion, participation of marginalized groups, innovative practices, strengthening teachers, competencies, democratic practices, atmosphere in schools.

In the field of science, through international cooperation as well as cooperation with donors, in Montenegro work has been done on providing access for domestic scientific community to large-scale international research infrastructures, which are equipped with the latest technology (CERN, EMBL, EMBO, ESA, etc.). We emphasize some of them. The WISE Center for Research and Innovation in the Western Balkans has a focus on research excellence; Networks of Excellence; Technology transfer; Development of start-up companies. The ERA represents a unique framework for cooperation in EU research and innovation, in which the mobility of knowledge, researchers and technologies is enabled. Montenegro has a status of an observer and actively participates in the work of the European Research and Innovation Advisory Board (ERAC).

27.3   Current Situation of Science Education

27.3.1   Policies and Standards

The crucial policies of the education system in general are oriented to the quality, development, achievement, equity, equality, access, continuity, participation, innovations (Table 27.4).

The program for the development and support of talented students 2020–2022 is essentially aimed at improving the support of talented students, monitoring the work, teacher competencies. The focus is on competitions, awards, scholarships, research camps, international Olympiads, study visits to science centres, fairs and events.

In the area of science, the key policies are:

The Strategy on scientific-research activity (2017–2021) defines: Development of human resources and research capacities; international cooperation and networking; the synergy between science and economy. They are focused on the: affirmation of the research profession and the creation of a scientific research critical mass; development of international cooperation in order to improve the quality of scientific research work; motivation and innovation towards commercialization.

Goals of Innovation Strategy (2016-2020) are: Increasing the capacity for innovation and technological development (infrastructure, human resources, monitoring); Strengthening financing and participation of innovation system; Strengthening innovations in the economic sector.

Table 27.4 Strategies and goals

Strategy	Key goals
The Strategy on early and preschool education 2021–2024	Coverage, quality, early development of children from vulnerable groups
The Strategy on General secondary education (2015–2020)	Classes of philological, mathematical and sports gymnasiums, in English
The Strategy of vocational education (2015–2020)	Professional training, independent living, equal opportunities, employment and social inclusion
The Strategy on inclusive education (2019–2025)	Accessibility, equity, equality, continuity and quality
The Strategy on higher education (2016–2020)	Quality, practice, mobility, entrepreneurial-innovative-research activities
The Strategy on adult education (2015–2025)	Social inclusion, lifelong learning, employability, mobility, competences
The Strategy on teacher education (2017–2024)	Initial and continuing teacher education, support for professional development

The Smart Specialization Strategy (2019–2024) sets out development priorities aimed at building a competitive advantage, linking research and innovation strengths with the needs of the economy, responding to opportunities and market development. Smart specialization leads to the increases the competitiveness of the economy by concentrating and linking research and innovation resources, encourage innovation, thus contributing to economic growth and the overall progress of society.

The Law on Academic Integrity regulates moral and professional principles that academic, research and other staff and students must comply to. Academic integrity is academic behaviour that ensures the preservation of academic integrity, dignity of the profession, standards and quality of work and products of work, the spirit of equal cooperation with all participants in the academic process, truth-orientation as a fundamental value and respect for legal regulations as the basis of responsibility of members of the academic community in accordance with the principles of academic integrity. The principles of academic integrity are: honesty, objectivity, openness, freedom in teaching and research, and accountability to the academic community and society.

In practice connection between education and science is functioning on the following ways (Table 27.5).

Ministry of Science also supports various programs of the NGO dealing with the promotion of science, such as summer schools, seminars and workshops, through annual call for proposals, worth around 100.000 € in total.

Table 27.5 Projects and activities in science education area

Project	Activities
Scientific Expedition & workshops[a]	Science promotion - The programme is conceived as a moving expedition across Montenegro, with young people in primary and secondary schools as the target group
Hands on Particle Physics International Masterclasses	Annually more than 13.000 high school students in 60 countries come to one of 225 nearby universities or research centres. Lectures conducted by scientists give insight in topics and methods of basic research at the fundaments of matter and forces, enabling the students to perform measurements on real data from particle physics experiments themselves. At the end of each day the participants join in a video conference for discussion of their results
LEGO League	Organized by NGO Young Inventors, in the last 10 years—students develop and apply STEM
Board of European Students of Technology, branch in Montenegro	Local and regional competitions in engineering[c]
First Tech Challenge - led by NGO Montenegro Robotics[b]	Pupils develop creativity, programming, constructing skills

[a]Started on 27 February 2020
[b]Montenegro joined World Robot Olympiad, led by the Machine Engineering faculty of the University of Montenegro - this year's competition was cancelled due to Covid-19
[c]http://best.ac.me/?page_id=398

27.3.2 Curriculums, Digital Resources and Teacher Training

In Montenegro there are developed plan and program for every subject. The subject program consists of: determination, nature and purpose of the subject; Number of classes by years of education and forms of teaching; Outcomes: activities, concepts, correlation; Didactic recommendations; Knowledge Standards and Exam Catalogue. The concept of outcomes implies a precise indication of the knowledge, skills and values that are developed through the program. The outcome should enable knowledge, development of cognitive and social skills, and formation of attitudes and values that would be applied in life situations. Concretization enables the planning of teaching that serves to verify outcomes of knowledge (not simple recognition and reproduction but everyday application).

The analysis of general education curricula: for elementary school, gymnasium, teaching faculties in relation to key competencies (within the program "My values and virtues", UNICEF Montenegro), indicates the current situation and offers recommendations for updating the curriculum. It was noticed that the teaching is focused on the contents of the program, less on mastering them. The subject programs do not show the connection of key competencies for the twenty-first century with the academic achievement of students and their educational role in schooling. Study programs on the Teachers Faculties do not have written outcomes related to the training of future teachers for the development of their and pupils' key competencies for the twenty-first century. The formulations of goals and outcomes are insufficiently precise and clear. There is no internal coherence in the curricula; there is no harmonization of goals and outcomes both within and between different educational levels, but also between the elements of the curriculum in individual subjects. In the curricula of teaching faculties, the following is neglected: the development of key competencies of students and preparation for the profession. Competences required for employment are insufficiently developed horizontally in the programs by levels, and viewed vertically, the higher the level of education, the fewer key competencies (mostly in elementary school, least at university) and there is no coherence and consistency in their development (Pešikan & Lalovi, 2016).

The System of Professional Development of Teachers is seen as: "the process of improving the skills and competencies of teachers with the aim of improving the quality of teaching/educational work, learning/development and achievement of students/children". The authors of the program are selected through a public competition that is announced every other year. The National Council for Education makes a decision on the selection of programs after the Commission, formed by the Bureau for Educational Services, i.e. the Centre for Vocational Education, professionally evaluates and proposes programs. (Milić, 2018).

The renewal of the work license can be achieved after completing 16 h of professional training in priority and 8 h in other areas of professional training, conducted over a period of five years.[16]

The most suggested forms and activities of professional development are training, counselling, observation in the school, ICT learning, coaching, collaborative learning, realization of experimental and model classes or activities, conducting action research. The participation in the work of professional networks, focus groups, round tables, panel discussions, presentations, debates, conferences, etc. is encouraged.

For teachers, a school portal with links related to preschool, inclusive education, and Internet safety is offered as an additional tool in work and communication. It contains numerous materials that serve to inform teachers, but also the preparation and implementation of the teaching process.

Key ongoing projects in this area are:

The "Schools for the 21st Century" project, funded by the British government, has been implemented with the British Council for three years. It addresses the skills necessary for the twenty-first century that will meet the challenge of the new digital economy. Teacher training is aimed at being able to transfer the acquired knowledge to elementary schools' pupils. The project has so far enabled students to establish micro bite clubs and use micro bite devices for the first time. (British Council, 2020).

The project "Digital Classroom" started with the company M-tel, which envisages the development of digital textbooks as a supplement to textbooks. (Dautović, 2020).

---

[16] Rulebook on detailed conditions, manner and procedure of issuing and license renewal.

The project "Integration of key competencies in the education system in Montenegro" is implemented through the IPA program. The goal is to improve the quality of education through the development of a framework of key competencies, the improvement of initial teacher training and continuing professional development of teachers. A special focus is on the disciplines in the field of science, technology, engineering and mathematics (STEM), which are key to successfully following the current trends and technologies of the modern age (Ministarstvo prosvjete, 2020).

27.3.3 Student Assessment and Achievement

Montenegro joined PISA research with the Agreement on Participation in PISA (2004). The PISA 2006 was a trial, and since 2009 Montenegro has been regularly participating in the PISA.

At the PISA 2012 the average (OECD) achievement in mathematical literacy was 494, and Montenegrin students had 410 points (level one). The most complex tasks of levels five and six[17] in Montenegro are solved by 1% of students. In the area of reading literacy, the percentage of functionally illiterate students (below level 2) is higher than the percentage of OECD countries. The achievement in the field of scientific literacy is 410 points. An extremely small percentage of students (0.4%) have developed their scientific literacy to the highest level of achievement (level five and six). As many as 80% of students failed to solve medium-difficulty tasks, and less than 1% solved complex tasks.

In the PISA 2015 participated 5665 fifteen-year-olds from 49 secondary schools and 15 elementary schools from Montenegro: 32.18% students from high schools, 61.47% from four-year vocational schools, 5.1% from three-year vocational schools and 1.24% from elementary schools. It was conducted in Montenegrin and Albanian. In the PISA 2015 main areas were natural sciences (physics, chemistry, biology). By ranking countries according to the average result in reading literacy, Montenegro is in 48th place with 427 points. The average achievement in mathematical literacy in Montenegro on the PISA 2015 test is 418, which is 72 points lower than the OECD average (Paljević and others 2015).

Regarding TIMMS Montenegro, currently waiting for the first time results—it would be promoted in December 2020 and official report would be available next year.

---

[17] Level 6: students generalize, make complex conclusions and apply knowledge. The lower level limit of achievement in the field of mathematics is 669.3 points; reading 698, and scientific literacy 707.9 points. Level 5: students know how to choose, compare, evaluate, discuss solutions, find information, research, connect and evaluate. The level lower in the field of mathematical is 607, reading 626, and scientific literacy is 633.3.

27.3.4 Science and Technology Venues and Centres

Montenegro currently has one science and technology park, established at the end of 2019 and expected to be operational to the end of 2021. Science and Technology Park Montenegro wants to become generator of innovation processes in Montenegro, by supporting creative, innovative and high technology-based companies. Primary mission of Science and Technology Park Montenegro will be support of entrepreneurial companies through processes of incubation (focus on developing and strengthening Start-up companies/teams), funding, training, mentoring and integration with the ecosystems of technological innovations, both regional and global, cooperation with business and academic communities (Drobnjak, 2020).

The Technopolis Innovation and Entrepreneurship Centre is in Nikšić and is a place of support provision for innovative ideas and processes. With its capacities, it represents a suitable place for the processes of incubation and pre-acceleration of business ventures.

Within the formal education system, there is significant support for knowledge transfer provided by the University of Montenegro and the University of Donja Gorica (UDG), which as part of their activities work on the promotion of knowledge, science, and scientific research.

In recent years, the active role of the business sector has been noticed, especially when it comes to innovation activities.

One of our particularly important undertakings that involves international partnerships is the "Open Science Days" Festival, which has taken on a traditional character over the years. The festival is organized with the aim of promoting science in society, increasing its visibility and especially bringing closer science and research to young generations. This is the biggest promotional event in the field of science on the territory of Montenegro featuring a series of exhibitions, lectures, panel discussions, interactive workshops, as well as a number of other promotional contents in all cities of Montenegro.

The best Montenegrin students have an opportunity to be a part of the CERN[18] Student Summer School—to be included in research teams, listen the lectures, visit labs and facilities, and participate in workshops facilitated by scientists that are leaders in their fields of research.

27.3.5 Utilizing Emerging Technologies

Government of Montenegro is in negotiations with a renowned technology provider to establish an education and training centre in AR/VR, with the aim to upskill its population in a large spectrum of fields, and promote these technologies for innovation development. Agreement is expected until the end of 2020.

---

[18] *CERN—The European Organization for Nuclear Research—Founded in 1954, the CERN laboratory sits astride the Franco-Swiss border near Geneva.*

When it comes to emerging technologies in Montenegro, their use has not yet taken root in the right way. It is evident that there is a great need for this type of technology, and it is to be expected that in the near future these technologies will get their place.

Science Technology Park Montenegro is currently actively working on the establishment of an AVR[19] centre, which will make a significant contribution to the popularization of this type of technology.

NGO Education Improvement Institute has been developing innovative products in VR technologies, aimed at STEM education.

Several start-ups in the field of Edu-Tech are under development in Montenegro (You-Learn platform, Clockwork Briefcase etc.).

27.4 Requirements for Future Development of Science Education

Modern means and equipment, new laboratories with advanced technologies are necessary to encourage scientific education, but an adequate exchange of knowledge and experiences certainly seems crucial.

Also there is public awareness, competent teachers, researchers, curious, talented and supported pupils and students.

There must be strengthen cooperation between two sectors: education and science, and more focus on competences in technology, as well as in the natural sciences. Pupils should have more opportunities to have scientific knowledge, practical experience, and applicable skills.

It is also important to support the implementation of the Gifted Students Program, which would promote science, recruit, empower and orient future dedicated researchers and innovators.

This requires financial resources, conditions, infrastructure, equipment, devices, labs, public awareness campaign, communication strategies, and orientation towards the quality of general education, pre-service and in-service teachers' training, preparation and motivation of the researchers, flexible curricula, general and specific working environment.

27.5 Discussion and Conclusion

Legally, the strategic framework is set as affirmative, modern, contemporary, within the democratic values of the EU and open to change.

It is important to continue the trend of development, harmonization according to quality, competencies. Also, to improve a culture of responsibility, participation.

---

[19] Augmented and Virtual Reality.

Educational and science environment should be supported by conditions, services, resources (human, infrastructural, financial), for learning and research equally accessible to all.

Competencies need to be improved (especially for the work in the twenty-first century), methodology and approaches, responsibility and motivation of teaching staff.

It is necessary to continuously monitor, improve the quality of education, introduce innovative models - practical guidance, application of assistive technology, improvement of digital competencies at all levels.

A general curriculum should be oriented towards areas, not subjects, which will be more flexible, enable outcomes that are easily individualized, and oriented to the key competences.

It is necessary to have more connections between education and science, joint activities, programs, initiatives.

Leadership, policy creation, implementation and practice has to be based on science, research. In addition, a culture of the process of evaluation should be encouraged.

Results, products, innovations should be promoted, made transparent, visible, as well as a significance, importance, effects and benefits of education, science, technology, and importance of scientific research.

Improve the cooperation of partners (educational institutions, local community, entrepreneurs, services, NGOs), especially for the development of missing services and initiatives.

References

British Council. (2020). *Škole za 21. vijek: unapređenje nastave u učionici*, Retrieved from https://www.britishcouncil.me/

Dautović, E. (2020). *Počinje digitalizacija u obrazovanju, Preduzetnica.me*, Retrieved from http://preduzetnica.me/

Drobnjak, R. (2020). *We will open the doors of the Science and Technology Park on time, if not sooner*, Portal CDM. Retrieved from https://www.ntpark.me/en/news/

Milić S. (2018). *Kontinuirani profesionalni razvoj vaspitača u Crnoj Gori*, UNICEF.

Milović N., Radović M. (2018). *Bazna studija: Privreda, za potrebe izrade Prostornog plana Crne Gore Ministarstvo održivog razvoja i turizma*, Podgorica.

Ministarstvo kulture. (2016). *Program razvoja kulture, 2016–2020*, Cetinje, Montenegro.

Ministarstvo prosvjete. (2017). *Strategija obrazovanja nastavnika u Crnoj Gori 2017–2024*, Podgorica, Montenegro.

Ministarstvo prosvjete. (2014). *Analiza obrazovnih postignuća crnogorskih učenika na PISA testiranju 2012*, Podgorica, Montenegro.

Ministry of Scince. (2019). *Strategy for scientifc research activity 2017–2021*, Podgorica, Montenegro.

Ministry of Scince. (2019). *Smart Specialisation Strategy 2019–2024*, Podgorica, Montenegro

Statistical Office of Montenegro—MONSTAT. (2019). *Statistical annual-book*, Podgorica, Montenegro.

Paljević D., Nenadović D., Čarapić T., Vujošević T. (2015). *PISA 2015. U Crnoj Gori – Rezultati,* Ispitni centar Crne Gore, Podgorica, Montenegro.

Pešikan A., Lalović Z. (2016). *Obrazovanje za život: ključne kompetencije za 21. vijek u kurikulumima u Crnoj Gori,* Unicef, Podgorica.

Uvalic, M. (2003). *Economic Transition in Southeast Europe.* University of Perugia.

World Travel & Tourism Council. (2018). *Travel & tourism economic impact 2018, Montenegro.*

Tamara Milić is Head of Unit for preschool and inclusive education in the Ministry of Education of Montenegro. Basic profession is psychology, Master degree on Education Policy. Trained for the forensic methods in psychology, specialized instruments for the personal assessment (NEOPI-R, Rorschach), in Psychoanalytic psychotherapy, Rational Emotional Therapy - RET method, ABA method, and for the SPSS / specialized statistical program. Main activities and responsibilities are in the educational policy of preschool, SEN and other at risk children - direct communication regarding implementation of proposed measures with preschool institutions, elementary and secondary schools (teachers, professionals, parents, NGOs, etc.). Main areas of work are in early child development, child and rights of persons with disabilities, inclusive education. There is a numerous of projects, as well as publications and researches in the field of education students with special education needs, individualization, transition, innovative services, assistive technology, SEN children Career Guidance and Counselling, etc. A significant number of activities coordinated and realized in cooperation with international organization and donors (UNICEF, UNESCO, Save the children, Council of Europe, European Training Foundation - ETF, UNODC, OECD, KulturKontakt, Roma Education Fund).

Chapter 28
Science Education in Slovenia

Eva Klemenčič, Andrej Flogie, and Robert Repnik

Abstract This chapter provides a short overview of the education system in Slovenia with a focus on science education. Institutions for basic and upper secondary education collaborate with universities and research centers on a national level in the development and research projects that aim to enhance students' scientific and mathematics literacy, entrepreneurship competences and encourage the use of e-learning environments. According to international student assessments, PISA and TIMSS, Slovenia ranks above the OECD average in mathematics and natural science. There are many competitions, science fairs, summer schools, and workshops organized on a national level. Based on the analysis of the study program Subject teacher at the University of Maribor, we can conclude trainee teachers obtain good theoretical knowledge in the field of natural science and adequate competences for the sensible use of ICT in teaching. In the future, greater focus should be on e-learning, e-teaching, and distance learning, as well as introducing AI and VR in education.

Keywords Slovenia · ICT · Natural science education · Subject teacher

E. Klemenčič (✉) · A. Flogie · R. Repnik
Faculty of Natural Sciences and Mathematics, University of Maribor, Koroška cesta 160, 2000
Maribor, Slovenia
e-mail: eva.klemencic@um.si

A. Flogie
e-mail: andrej.flogie@um.si

R. Repnik
e-mail: robert.repnik@um.si

© The Author(s), under exclusive license to Springer Nature Singapore Pte Ltd. 2022   471
R. Huang et al. (eds.), *Science Education in Countries Along the Belt & Road*,
Lecture Notes in Educational Technology,
https://doi.org/10.1007/978-981-16-6955-2_28

28.1 Overview of the Country

28.1.1 Geographical Location, Population and Political System

Slovenia, a country located in central Europe, covers 20,271 square kilometers but it is topographically diverse as it connects the Mediterranean Sea, Alps, Karst, and Pannonian lowlands (Lavrencic et al., 2020). Its neighboring countries are Austria to the north, Croatia to the south, Italy to the west, and Hungary to the east. The official language is Slovene, which is a South Slavic language and has preserved the dual grammatical number. Slovenia has 2,097,195 citizens, approximately 103.4 citizens per square kilometer, with the densest population in the capital city Ljubljana. The average age of Slovenian citizens is 43.5 years (Statistical Office of the Republic of Slovenia [SORS], 2020).

After World War II, Slovenia became a part of the Federal People's Republic of Yugoslavia. In December 1990, more than 88% of the electorate voted for a sovereign and independent country. Slovenia declared its independence in 1991, following by the Ten-Day War. In 1992, the United Nations accepted the Republic of Slovenia as a member. Slovenia constitution established a parliamentary form of government. Slovenia is a unitary multiparty republic with two legislative houses, National Council and National Assembly. In May 2004, Slovenia joined the European Union and entered NATO (Strlič & Osterman, 2011). Regardless of its small size, Slovenia has 212 local administrative units.

28.1.2 Current Situation of Economic, Technologies and Cultural Development

Most of the economy in Slovenia is privatized. The main sources of income represent manufactures of automotive parts, household appliances, and pharmaceuticals. The share of enterprises with innovation activity is high, around 49%. The government allocates 2% of GDP on research and development, 1.4% in the business sector, 0.3% in the government sector, and 0.2% in the higher education sector. Agriculture and forestry contribute only around 1.4% to total GDP. Slovenia is not fully self-sufficient in terms of food and energy and had met 47.9% of domestic energy demand with domestic energy resources, mainly petroleum products and nuclear.

Lately, the global pandemic COVID-19 influenced Slovenia's economy, which led to an increase of the unemployment rate to 5.2% and a decrease of the annual GDP volume growth, which is 4.1% (SORS, 2020).

28.2   OVERVIEW of the Education Development

28.2.1   Education System and Policy

The education system in Slovenia comprises optional preschool education, compulsory nine years long basic education, typically from the age of six to the age of fifteen, following by the optional upper secondary and tertiary education (European Commission, 2019).

In the school year 2019/2020, 87,708 children enrolled in preschool education. Slovenia has not yet achieved the European strategic goal for education and training, which aims to the inclusion of 95% of four- and five-year-olds in preschool education by 2020. On average, an educator to child ratio is 1:7.5 (Kozmelj, 2020a).

According to the legislation of the Republic of Slovenia, all children must receive basic general education. Compulsory basic education consists of three educational cycles and is conducted by public and private schools, schools for children with special needs, and adult education organizations. The average class size is 19.2 students, which is below the OECD average. In the school year 2019/2020, 190,156 children enrolled, 5.7% of them were in programs for children with special needs (Kozmelj, 2020a). All students attend science-related subjects distributed over a 9-year curriculum. Subjects are the following: Learning about the environment, Natural science & technology, Natural Science, Physics, Chemistry, and Biology. In addition, they can attend elective subjects such as Astronomy, Projects in physics and technology, Electrical engineering and robotics electronics, Projects in physics and ecology, and others (Ministry of education, science, and sports [MESS], 2020).

The upper secondary education divides into two categories. The general upper secondary education with curriculum over four years aims to prepare students to enroll in the higher education programs and concludes by general national exam. In addition to the general program, there is also a classical program with a focus on Latin, History, and Philosophy, and a sports program for young professional athletes (2TM, 2020). Vocational and technical upper secondary education has a range of programs with different lengths and conclude with the final exam. The goal is that students obtain specific qualifications for vocation. Although the upper secondary education is optional, 72,738 of students were enrolled in the school year 2019/2020, which is 90.6% of residents aged 15 to 18, among whom 65.3% attended vocational and technical programs. The number of all students is decreasing, but in the following years, the trend will probably change (Kozmelj, 2020a).

Tertiary education includes higher vocational colleges and higher education, which follows Bologna Declaration and consists of three cycles: first cycle (bachelor), second cycle (master) and third cycle (doctoral). Some study programs are integrated combining first and second cycle. Around 45.7% of residents aged 19 to 24 are enrolled in tertiary education. In the academic year 2019/2020, the number of enrolled students increased by 1% for the first time in 10 years to 76,728. The vast majority, 63.5%, present students enrolled in the first cycle and 14% enrolled in higher vocational education (Kozmelj, 2020b).

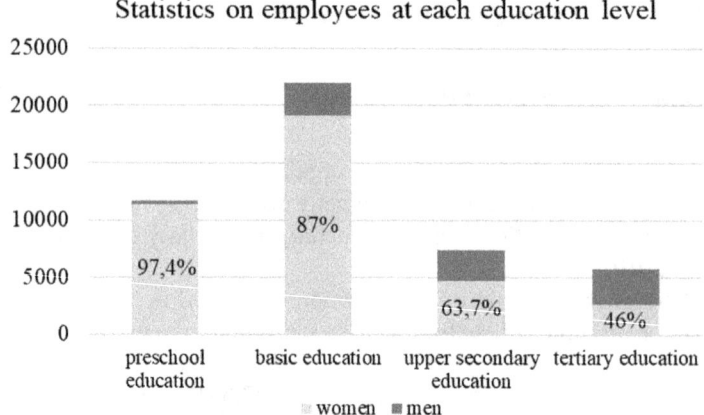

Fig. 28.1 The structure of employees at preschool, basic, upper secondary and tertiary education. Most employees in education are women, except at the tertiary education level. Data by the SORS (2020)

Figure 28.1 presents the structure of employees in each education level for the school year 2019/2020. The overall number of employees in basic education level increased by 5% according to the previous school year (Kastelic & Tuš, 2020), which corresponds to the continuous growth of the number of children enrolled in basic education since the school year 2010/2011. In 2016, the study by Ploj Virtič et al. (2016) warned about the low numbers of students in the science education program, which is leading to the shortage of science teachers that is already noticeable. For example, the enrollment of students in educational technics and technology decreased by 70% since 2007. If all students will successfully complete their studies, 40% of the necessary number of teachers of technics and technology is covered.

The last comprehensive evaluation of the whole education system in Slovenia, *The White Paper on Education*, was published in 2011. It includes guidelines, goals, and strategic challenges in preschool, basic and upper secondary education, guidelines for students with special needs, talented students, students at ethnically mixed areas, and adult education. Besides, it includes a section on private schools and kindergarten, and on music schools, as well as on the education of professional staff and their development (National expert team, 2011). In 2020, a new White Paper on Education is in preparation with a focus on solutions for the low motivation of students.

28.2.2 *Educational Research and International Collaboration*

Educational research is the main aspect of the Educational Research Institute established in 1965. Its activities are to develop scientific methodologies, research different educational sciences, and encourage interdisciplinary and multidisciplinary connections with other institutions. Their basic and applied studies cover the sociology of education and knowledge, pedagogical anthropology, comparative education, philosophy of education, and more (Educational Research Institute, 2019). They collaborate with experts at international research and education centers as well as with universities in Slovenia. Currently, they collaborate with the National education institute Slovenia and the University of Maribor within a research and development project *Innovative learning environments supported by the ICT*, which aims to develop sensible, comprehensive, and active use of interactive tools, services, and mobile devices in schools (Flogie et al., 2016). An open, creative, innovative, and sustainable learning environment supported by ICT is also envisioned in the "Strategic guidelines for further implementation of ICT in the Slovenian education" adopted in 2016. The systematic integration of ICT in education started in 1994 within the project *RO – Computer Literacy*. This project aimed to equip schools with hardware and software, educate teaching staff about the sensible pedagogical use of ICT, and update didactic materials and textbooks. Slovenia continued the ICT implementation in schools within the projects *E-education, E-competences for teachers in bilingual schools, E-books, E-schoolbag*, and others (Svetec et al., 2020). Nowadays, the use of ICT in education in Slovenia is following the average use of ICT in classrooms of EU countries (Pelgrum, 2001; Wastiau et al., 2013). The survey of ICT in education indicates that students in Slovenia are in digitally supportive schools in terms of equipment and teaching staff (European Commission, 2013).

In Slovenia, there are three public universities, the University of Ljubljana, the University of Maribor, and the University of Primorska, which all offer education research study programs.

The University of Ljubljana was founded in 1919 and is the oldest higher education institution in Slovenia. It connects 23 faculties and 3 art academies and covers all research areas by the FRASCATI and ISCED classification in education and art. Currently, around 38,000 students are enrolled (2020). In 2019, the Academic Ranking of World Universities ranked the University of Ljubljana among the top 600 and the University of Maribor among the top 800 universities (ARWU, 2019). The University of Maribor is the second-largest public tertiary education institution founded in 1975. It integrates 17 faculties and has approximately 13 500 students (University of Maribor, 2019). From 2007 to 2011, the University of Maribor carried out the project Development of Science Competences (Repnik et al., 2010). In the frame of this project, innovative teaching strategies, and new didactic materials and models were developed for implementation in natural science-related activities in preschool, basic and upper secondary education.

All public and private tertiary education institutions are actively engaged in students and staff mobility programs, as well as research projects and programs at the Slovenian research agency (ARRS) and European social fund. Research projects and programs are conducted in collaborations with other research institutes in Slovenia and across the world. Based on the Slovenian Current Research Information system (SICRIS, 2020), there are currently active 1026 research organizations, 1607 research groups, 15,964 researchers, 521 research projects, and 335 research and infrastructure programs. Among all active, ten projects and eight programs are in educational studies. Aside from ARRS projects and programs, tertiary education institutions participate in other educational development projects. For example, the project *Innovative Learning and Teaching in Higher Education* connects all three public universities and includes the formation of the so-called multipliers, who are trained at foreign higher education institutions and disseminate the obtained knowledge and experiences into Slovenian higher education (INOVUP, 2019). Besides, multipliers gain new acquaintances, which can lead to new international collaborations in educational studies and other research areas. In collaboration with the National education institution, the teaching staff is actively engaged in project *Scientific and mathematical literacy: the development of critical thinking and problem-solving*, which aims to implement active didactic approaches, critical thinking, and solving complex authentic problems by using ICT to enhance positive attitude, understanding and literacy (Bone, 2019). Another joint project, *Strengthening the entrepreneurship competences, and promoting a flexible transition between education and environment at basic education/upper secondary education,* focuses on training teachers to design activities that enhance the development of entrepreneurship competences. Aside from projects, the National education institution provides professional help in formative assessment and e-learning, publishes manuals, guidelines, organizes seminars and conferences, and collaborates with international organizations.

28.3 Situation of Science Education

28.3.1 Policies and Standards

At the national level, the Ministry of Education, Science, and Sport has the main role and responsibility for policies, legislation, regulation, strategies, and standards of the education system. It regulates finances and administration, supervises employment procedures, and assesses the outcomes of the education system through statistics, national assessments, and evaluation and research studies. Rules on norms and standards for the implementation of the primary school programme (2007) and educational programs in secondary education (2010) determine finances, systematizations of work positions and workload, lesson duration, class size, compulsory and optional activities, and more. Changes in programs and curricula are implemented, monitored, and evaluated according to the Regulations of Modernisation of Educational Activities (2003). The recent update of curricula includes the use of ICT. The incorporation of ICT in the education process encourages the teaching and learning process (Tearle, 2004) and enables positive learning outcomes, especially from the science teacher's perspectives (Jimoyiannis & Komis, 2007). Besides, digital and ICT literacies are important life skills that affect success in different areas of life. Nevertheless, it is important to point out ICT is only recommendations and is used in class according to the personal preferences of a teacher.

The work of the Ministry of Education, Science and Sport is professionally supported by three national agencies: the National Education Institute Slovenia, the Institute of the Republic of Slovenia for Vocational Education and Training and the Slovenian Institute for Adult Education.

28.3.2 *Curriculums, Digital Resources and Teacher Training*

There are different teacher training programs at universities in Slovenia. To become an educator in kindergarten, one must obtain a degree in preschool education at the tertiary level. Teaching staff for the first five years of basic education level must obtain a degree in elementary education. For the next four years of basic education and upper secondary education, teachers are trained for specific subjects. Other teacher training programs are in Fine Arts education, Music education, Sports coaching, and Inclusion in Education. To teach at schools, graduates must also pass the professional exam in the field of education by the Ministry of Education, Science, and Sport.

In the following, we analyze the curricula of the science teacher-training program at the Faculty of Natural Science and Mathematics (2019), which is a unified master's program *Subject teacher* and lasts 5 years. The program obtained the accreditation in 2015 by the National Agency for Quality Assurance in Higher Education. At the enrollment, students select two field-orientations (biology, technology, mathematics, computing, chemistry, and physics). For analysis, we divide subjects into three categories. The first category includes general pedagogical subjects, which are the same for all orientations. Compulsory general pedagogical subjects are the following: Information and communications technology, Communication and rhetoric for science and technology students, Pedagogy, Psychology of development and learning, Didactics, and Working with children with special needs. Besides, students must select two elective subjects, among which two of them include ICT in greater shares (E-learning by design and Computer-supported laboratory work). Others include ICT as a method of teaching and assessment. Basic theory subjects in the field of both selected orientations form the second category. Basic theory subjects in the field of both selected orientations form the second category. For example, if one field-orientation is physics, students have courses in Mechanics, Electromagnetism, Oscillation and Waves, Thermodynamics, Modern Physics, Physics experiments, Computer in physics, Complex systems, Measurements in Physics, Applied Physics, and Environmental physics. Courses are distributed during the entire study program. Students first obtain basic theoretical knowledge then extend their knowledge in laboratory work, where they acquire laboratory skills essential for autonomous execution of physics experiments, and at last, apply their knowledge in other fields of physics. The third category connects all orientation-specific didactic subjects. In physics, those divide further into four types: Astronomical didactic subjects, Didactics of physics with practicum for basic education and upper secondary education, peda-gogical practice for teaching physics, and elective courses. Within these subjects, students learn didactic methods, approaches, and strategies, as well as the reason-able use of ICT in class, acquire teaching skills, and deepen their knowledge of experimentation at school. They learn how to safely design and execute experi-mental work, including the use of a computer with interfaces and sensors or tablets with applications. In science education, teachers need to know how and when to use a specific didactic method or approach to enhance reasoning in laboratory and field-work. Besides theoretical field-related subjects and specialized-didactic subjects, it is important trainee teachers obtain in-depth knowledge of ICT, computer sciences and multimedia in order to achieve sufficient level of digital literacy. Therefore, it is important curricula in teacher training programs include the ICT (Gerlič, 2005). The analyzed study program follows these guidelines.

28.3.3 Student Assessment and Achievement

Slovenia regularly participates in international comparative studies in the field of education on students' knowledge and literacy, such as PISA and TIMSS. Interna-tional studies are an important source of comparative findings on the functioning and effectiveness of the educational system.

In the 2018 PISA study, 6401 Slovenian students participated, and all upper secondary education institutions were included. In science literacy, students achieved 507 points, which is 6 points less than in PISA 2015 but still above the OECD average of 489 points. Countries in the same level of science literacy are Poland, New Zealand, Great Britain, Netherland, Germany, Australia, and the USA. Students achieved the highest points, 509, in mathematics literacy, which is 20 points above the OECD average. Interestingly, the gender gap is not present in mathematics literacy but is statistically significant and one of the largest among the OECD members for science literacy (Schleicher, 2019).

The release of international results of the 2019 TIMSS assessment will be in December 2020. Therefore, we analyze results from 2015. Results show, students in 4th and 8th grade of basic education achieved scores above the TIMSS average in both areas, mathematics and natural science. On average, we notice that the scores improve through time. TIMSS Advanced evaluates the knowledge in mathematics and physics at the end of upper secondary education level. Students that attended advanced mathematics achieved the highest, 549 points, and those who attended basic level mathematics achieved 460 points, which is below the TIMSS average. In physics, students achieved 531 points, 17% achieved the highest level of knowledge, and 8% achieved the highest score.

In physics, students achieved 531 points, 17% achieved the highest level of knowl-edge and 8% achieved the highest score. In Slovenia, the share of students achieving a high level and a basic level of knowledge is greater than in other countries. Detailed results show students achieved good results in mechanics and thermodynamics but have relatively weak knowledge in wave and nuclear physics. In cognitive domains, students are weak in knowing and reasoning but better in applying knowledge. Both in mathematics and in physics, the gender gap is statistically significant, as girls achieved a lower score than boys (Japelj Pavešić & Svetlik, 2016).

We can speculate about the reasons behind these successes. In our opinion, the following contribute:

- Instead of one *Science* subject, content divides among Physics, Chemistry, Biology, and Technics and Technology.
- Students have the opportunity to attend a variety of science-related competitions and workshops at regional and national levels.
- Teachers are encouraged to use ICT, innovative approaches, and focus on procedural knowledge and the development of competences. External national assessments of students' knowledge are prepared, organized, and evaluated by the National Examinations Centre (2020). Each year, national assessments are carried out in the 6th and 9th grades of basic education, at the completion of secondary vocational and technical education (so-called *vocational matura*) and at the end of general upper secondary education (so-called *matura*). The exam in the 6th grade of a basic education level assesses the knowledge in Slovene, Mathematics, and foreign language. In 9th grade, students take an exam in Slovene, Mathematics, and the third subject, which changes through the years. At the end of upper secondary education, students are required to take the external national exam simultaneously, following the same procedures, regulations, and assessment criteria. If they pass, they obtain a certificate of achievement of necessary standards of knowledge and gain the opportunity to continue education at the tertiary level. External national assessment in Physics consists of two parts. The first part is a written exam with two sections: the multiple-choice section (35% of the total score) and structured tasks (45% of the total score). Students can work on six structured tasks (measurements, mechanics, thermodynamics, electromagnetism, oscillations, wave and optics, modern physics & astronomy) but choose three for evaluation. The second part is laboratory work with reports, which presents 20% of the total score.

In basic and upper secondary education, the assessment methods are in the teacher's domain. Often, knowledge is evaluated by written and oral exams. The oral exam usually focuses on the theory of specific contents, while the written exam consists of structured tasks, multiple-choice questions, and open-ended questions. In upper secondary education, laboratory work in Chemistry and Physics is also evaluated based on reports.

In tertiary education, students need to fulfill all the obligations of the study program. Methods of assessments of knowledge are explicitly determined in the curriculum. Different approaches vary among subjects, programs, and cycles of tertiary education level. Generally, students can complete their obligations by the final exam, written or oral, typically at the end of a course or semester. In some subjects, students can pass more tests or home assignments throughout the whole course, which counts as a final exam. In other subjects, tests or home assignments are required to take the final exam. Beside written and oral exams, typical assessments methods are defense of laboratory work and presentation of seminar or project work. For students that train to become teachers, the assessment method is also the implementation of a lesson in school. All rules on examination and assessment of knowledge are public and can differ amongst universities.

28.3.4 Outreach Science Education

Throughout the whole education, students can participate in knowledge and skills competitions, camps, workshops, and summer schools organized by public and private organizations. Association for technical culture of Slovenia (ATCS, 2020) organizes regional and national competitions in Chemistry, Biology, Logic, Natural Science, Computer programming, Construction & processing technology, Modeling (aviation, rocket, ship, car & radiogoniometry), and Innovative technologies (ICT, multimedia, video editing, internet of things, digital photography, robotics, 3D programming). Students compete individually. The typical assessment method is a theoretical written test combined with practical work. ATCS also organizes the competition Youth researchers, where students prepare and present their research project, which can be an individual or teamwork. In addition, ATCS offers creative summer schools, workshops, and research camps throughout the whole year. The society of mathematicians, physicists, and astronomers of Slovenia unite pedagogues, researchers, and students with a goal to popularize science among the public. It organizes competitions in knowledge in Mathematics, Physics, Astronomy, Business and Recreational mathematics, and Natural science, and publishes the newsletter Obzornik and the journal Presek (DMFA, 2020). In 2017, the Association of Natural Science and Mathematics of Slovenia (Association of Natural Science & Mathematics of Slovenia, 2017) was established. It offers physics camps, seminars, and organized practice for competitions.

Formal and informal science education intertwine. Usually, teachers assist students to prepare them for competitions, supervise their project and research work, and suggest them extracurricular activities. It is a common practice the teachers reward students for their success at competitions with bonus points at formal subjects. Besides, we notice that participation in informal activities in science education, such as competitions and summer schools, often results in students continuing their formal education in science.

28.3.5 Utilizing Emerging Technologies

The field of Artificial intelligence (AI) has been present in the Slovenian educational policy in the last decade The Slovenian Ministry of Education, Science, and Sport intensively supports research and development projects in the field of Innovative teaching and learning methods supported by modern technologies. Slovenian national project Innovative Pedagogy supported by ICT and AI (Flogie & Aberšek, 2019) is one of the good examples that show us, how technology and AI can support and speed up the process of the digital school transformation. Artificial intelligence in Education will be also one of the priorities of the upcoming Slovenian Presidency of the Council of the EU in 2021. Currently, the major role in AI in Slovenia has the Artificial Intelligence Laboratory at the Jozef Stefan Institute, which has developed a range of software tools. AI Laboratory participates in projects, for example, *The Event Registry* and *COPCAMS*, and in the development of the *OECD AI Observatory*. In March 2020, the Slovenian government signed the contract of the establishment of the *International Research Center for Artificial Intelligence (IRCAI) under the Auspices of UNESCO* (Minevich, 2020).

AI is not yet formally incorporated in science education and is not included within curricula. There are also raising ethical questions about AI and its use in education, which need to be answered. Slovenia has a clear intention in the field of AI in education, that the following recommendations:

- explicitly defining ethical behaviour,
- crowdsourcing human morality, and
- making AI systems more transparent,

should be seen as a starting point for developing ethically aligned AI systems in Education. Failing to imbue ethics into AI systems, we may be placing school system in the dangerous situation of allowing algorithms to decide what's best for the society.

Nevertheless, the unintentional use of the 1st wave of AI (internet AI), for example, speech to text and text to speech algorithms, predictive and search suggestions are present in schools and in our private life. Some ongoing pilot projects also aim to introduce Innovative pedagogy virtual reality (VR) in education. VR and AI is currently included in some higher education courses in computer, technology and mechanical engineering. During the digital transformation process of the Education (reform), which is already underway, all those contents and goals will be included into the K12 curriculum.

28.4 Requirements for Future Development of Science Education

In science education, it is important teachers have the flexibility to introduce novelties and findings. One of such novelty is the so-called m-learning, which refers to the use of mobile technology in didactics activities (Korucu & Alkan, 2011). In addition to ICT, which is well incorporated in the analyzed study program The Subject Teacher presented in Sect. 3.2, future science education will require in-depth knowledge of e-learning as a supportive system for distance learning. Teachers mainly use ICT in the classroom to improve motivation and assist in understanding. In distance learning, ICT serves as the main communication path between teachers and students. Synchronous use allows two-way communication and incorporates audio–video conference calls. In asynchronous forms, students can participate at their own time independently through didactic materials prepared in advance. In our opinion, distance learning approaches and methods, as well as training in the use of software are required in teacher training for future development of science education. Next, it is of our belief competences will be even of greater importance than today. Therefore, it is reasonable to acquaint students in teacher training programs with competences and with strategies to encourage the development of competences in class through different activities. In science education, a special focus must be on science, technology, and mathematical competences, digital competence, entrepreneurship competences, and communication competences. Trainee teachers should consciously include activities that enhance the development of competences. Based on the analysis of the study program The Subject Teacher, we can conclude it already has subjects with a focus on competences.

The third crucial requirement for the future development of science education is to provide enough science teachers and students in science education programs. Currently, we face a decrease in students in science education programs, which leads to a shortage of teachers. The latter would be even more prominent in the future with the retirements of in-service teachers. To solve this problem, it is important to improve interest, motivation, and a positive attitude towards science education. In addition, we suggest financial support for students of the deficit programs.

28.5   Discussion and Conclusion

The Slovenian education system provides a good baseline in science education. Science-related subjects are incorporated over the whole duration of education and are backed by ICT. Teachers and educators are included in the development and evaluation studies and projects on a national level to enhance scientific, mathematical, and digital literacy, and encourage the development of competences, which will have an even greater role in the future. Competences are essential for lifelong learning and the increase in future employability, as new discoveries require the development of new professions. In addition, they collaborate in projects that aim to implement flexible forms of learning and create innovative learning environments.

Throughout the first seven years of basic education students have one, integrated subject in natural science. Later, students have compulsory classes in physics, chemistry, biology, and technics and technology. Students can also take optional classes, such as astronomy, physics projects, and programming, and attend science workshops, fairs, summer schools and competitions in knowledge. At international assessments, PISA and TIMSS, students' performance is above the OECD average for natural science, mathematics, and physics. In addition, those two assessments point at a problem of a significant gender gap in achievements, which need to be addressed in future. Despite the above-average achievements of students in science and mathematics at international assessments, there is a low interest in studies of science education.

Based on the analysis of the study program Subject teacher, we show that trainee-teachers acquire all the necessary knowledge and experience from the selected specialized field, didactics, and pedagogy. During their studies, they have opportunities to collaborate in scientific research, projects, and organizations of competitions and workshops for schools. The study program incorporates ICT and innovative and flexible approaches and emphasizes the importance of competences. In future, it would be reasonable to include distance learning and innovative teaching and learning methods supported by ICT and AI.

References

ARWU. (2019). Academic ranking of world universities 2019. Retrieved from http://www.shangh airanking.com/ARWU2019.html.

Association of Natural Science and Mathematics of Slovenia. (2017). *Activites.* https://nmk.fnm. um.si/.

ATCS. (2020). *Association for technical culture of Slovenia.* Retrieved fromhttps://www.zotks.si/.

Bone, J. (2019). *Activities in scientific and mathematical literacy at faculties and the National Education Institute Slovenia. Analysis.* Ljubljana, Slovenia: ZRSŠ. https://www.zrss.si/pdf/dej avnosti_na-ma_poti_fakultete.pdf [in Slovene].

DMFA. (2020). *The society of mathematicians, physicists and astronomers of Slovenia.* DMFA. https://www.dmfa.si/En/Default.aspx.

Educational Research Institute. (2019, July). *Educational research institute: A few facts* [Brochure]. https://www.pei.si/ISBN/PI_facts_ang.pdf.

European Comission. (2013, February). *Survey of Schools: ICT in education.* Luxembourg. Retrieved from https://ec.europa.eu/digital-single-market/sites/digital-agenda/files/KK-31-13-401-EN-N.pdf.

European Commission. (2019, December). *EURYDICE.* Organisation of the education system and of its structure. Retrieved from https://eacea.ec.europa.eu/national-policies/eurydice/content/org anisation-education-system-and-its-structure-77_en.

Faculty of Natural Sciences and Mathematics. (2019, February). *Unified master's program The Subject Teacher.* https://www.fnm.um.si/index.php/2016/01/18/univerzitetni-tudijski-program-prve-stopnje-izobraevalna-biologija-dvopredmetni-tudijski-program-2/ [in Slovene].

Flogie, A., Šverc, A., & Vičič Krabonja, M. (2016). *Inovative pedagogy 1:1 in light of 21st century competences.* Maribor [in Slovene].

Flogie, A., & Aberšek B. (2019). *The impact of innovative ICT education and AI on the pedagogical paradigm: Newcastle upon Tyne.* Cambridge Scholars Publishing.

Gerlič, I. (2005). Organization: Use of information and communication technology in Slovenian schools. *Organizacija (Kranj), 38*(8), 383–385.Retrieved from https://www.dlib.si/stream/URN: NBN:SI:DOC-YL4SEA8U/fcff6844-10b2-4342-9527-75f078c95f05/pdf [in Slovene].

INOVUP (2019). *Innovative learning and teaching in higher education.* Retrieved from http://www. inovup.si/en/about#purpose-and-objectives.

Japelj Pavešić, B., & Svetlik, K. (2016). Knowledge of pre-university mathematics and physics in Slovenia and around the world: results of TIMSS advanced 2015. Ljubljana, Slovenia: Pedagoški inštitut. http://timsspei.splet.arnes.si/files/2017/06/13-TA15-preduniverzitetna.pdf [in Slovene].

Jimoyiannis, A., & Komis, V. (2007). Examining teachers' beliefs about ICT in education: Implications of a teacher preparation programme. *Teacher Development, 2*, 149–173. https://doi.org/10.1080/13664530701414779.

Kastelic, N., & Tuš, J. (2020, June). Statistical office Republic of Slovenia. Slightly more teachers in basic and in upper secondary education. https://www.stat.si/StatWeb/en/News/Index/8907.

Korucu, A., & Alkan, A. (2011). Differences between m-learning (mobile learning) and e-learning, basic terminology and usage of m-learning in education. *Procedia - Social and Behavioral Sciences, 15*, 1925–1930. https://doi.org/10.1016/j.sbspro.2011.04.029.

Kozmelj, A. (2020a, May). *Statistical office Republic of Slovenia.* In 2019/20, too, more basic school pupils and fewer upper secondary school students than in the previous years. Retrieved from https://www.stat.si/StatWeb/en/News/Index/8854.

Kozmelj, A. (2020b, July). *Statistical office Republic of Slovenia.* For the first time in ten years, the number of students up again. Retrieved from https://www.stat.si/StatWeb/en/News/Index/8802.

Lavrencic, K., Allcock, J., Gosar, A., & Barker, T. (2020, August). *Encyclopedia Britannica, Inc.* Slovenia. Retrieved from https://www.britannica.com/place/Slovenia.

Minevich, M. (2020, April). *Forbes.* Here's how Slovenia is shaping the new human centric society and pioneering the world in AI. Retrieved from https://www.forbes.com/sites/markminevich/2020/04/13/heres-how-slovenia-is-shaping-the-new-human-centric-society-and-pioneering-the-world-in-ai/.

Ministry of education, science and sports. (2020, May). *Gov.si.* Programs and curricula of the basic education level. Retrieved from https://www.gov.si/teme/programi-in-ucni-nacrti-v-osnovni-soli/ [in Slovene].

National Examinations Centre. (2020). *RIC.* General information. Retrieved from https://www.ric.si/ric_eng/general_information/.

National expert team. (2011, June). White paper on education in Slovenia [White paper]. Ministry of education and sport. Retrieved from http://pefprints.pef.uni-lj.si/1195/1/bela_knjiga_2011.pdf.

Pelgrum, W. (2001). Obstacles to the integration of ICT in education: Results from a worldwide educational assessment. *Computer & Education, 37*, 163–178. https://doi.org/10.1016/S0360-1315(01)00045-8.

Ploj Virtič, M., Dolenc, K., Aberšek, B., Šorgo, A., Kocijančič S. (2016). *The role and importance of technical education in primary school: who will teach technics and technology in 2020?* Maribor: Faculty of Natural Sciences and Mathematics. Retrieved from https://www.fnm.um.si/wp-content/uploads/2016/12/Vloga-in-pomen-tehniskega-izobrazevanja-Koncna.pdf [in Slovene].

Regulations of Modernisation of Educational Activities, Publ. L. No. 011–03–10/2003, 13 (2003). Retrieved from https://www.uradni-list.si/1/objava.jsp?sop=2003-01-0561.

Repnik, R., Gerlič, I., Grubelnik, V., & Ferk, E. (2010). *Definition of natural science competencies: a scientific monograph.* Maribor: Faculty of Natural Sciences and Mathematics.

Rules on norms and standards for the implementation of the primary school programme, Publ. L. No. 2007–3311–0026 (2007). http://www.pisrs.si/Pis.web/pregledPredpisa?id=PRAV7973.

Rules on norms and standards for the implementation of educational programmes in secondary education, Publ. L. No. 2010-3311-0027 (2010). Retrived from http://pisrs.si/Pis.web/pregledPredpisa?id=PRAV10249&d-49683-p=2&d-49681-o=2&d-49681-p=1&d-49681-s=2.

Schleicher, A. (2019). *OECD.* PISA 2018: Insights and Interpretations. Retrieved from https://www.oecd.org/pisa/PISA%202018%20Insights%20and%20Interpretations%20FINAL%20PDF.pdf.

SICRIS. (2020). *SICRIS.* Information system on research activity in Slovenia. Retrieved from https://www.sicris.si/public/jqm/memo.aspx?lang=slv&opdescr=presentation&opt=6&subopt=1#kratka.

Statistical Office of the Republic of Slovenia [SORS]. (2020). *Stat.si.* Retrieved from https://www.stat.si/StatWeb/en.

Strlič, N., & Osterman, J. (2011). *dvajset.si.* Republic of Slovenia: 20 years of independence [Brochure]. Retrieved from http://www.dvajset.si/fileadmin/dokumenti/PDF/20let.pdf [in Slovene].

Svetec, M., Repnik, R., Arcet, R., & Klemenčič, E. (2020). Educational technology at the study program of educational physics at the University of Maribor in Slovenia. *The role of technology in education*, 1-16. https://doi.org/10.5772/intechopen.85081.

Tearle, P. (2004). A theoretical and instrumental framework for implementing change in ICT in education. *Cambridge Journal of Education, 34*(3), 331–351. https://doi.org/10.1080/030576 4042000289956.

University of Ljubljana. (2020). *Publications*. University in numbers 2019 [Brochure]. Retrieved from https://www.uni-lj.si/media_center/publications/.

University of Maribor. (2019). *University of Maribor* [Brochure]. Retrieved from https://www.um.si/en/about/About/Pages/default.aspx.

Wastiau, P., Blamire, R., Kearney, C., Quittre, V., Van de Gaer, E., & Monseur, C. (2013). The use of ICT in education: A survey of schools in Europe. *The European Journal of Education, 48*(1), 11–27. https://doi.org/10.1111/ejed.12020.

2TM. (2020). *Education in Europe*. Educational system of Slovenia. Retrieved from https://2tm.si/slovenian-education-system/?lang=en.

Dr. Eva Klemenčič is an assistant professor at the Faculty of Natural Science and Mathematics, University of Maribor. Her research areas are liquid crystals and applicative physics. She studies the electrocaloric effect in liquid crystals, topological defects, and structural and phase transitions. She collaborates in didactics projects on natural science literacy and entrepreneurship competences. She is also a commission member at the competition of young researchers in Physics, organized by the Association for Technical Culture of Slovenia.

Andrej Flogie holds a PhD in Educational Science. He is Assistant Professor at the Faculty of Natural Sciences and Mathematics of the University of Maribor, Slovenia, and Director of Anton Martin Slomšek Institute, Slovenia. He is the author of numerous publications in various fields, with most of his research in recent years focused around teaching and learning processes, cognitive science, and creating and developing intelligent learning environments and AI. He is also an adviser for the Education digitalization of the minister in the cabinet of the Ministry of Education, Science and Sport.

Dr. Robert Repnik is an associate professor of physics at the Faculty of Natural Science and Mathematics, University of Maribor, and a member of the Computational Intensive Complex Systems research team. His research areas are liquid crystals and didactics of physics. In the field of liquid crystals, he studies surface phenomena and topological defects in a plan parallel liquid crystal cell, and structural and phase transitions. In didactics of physics, he studies the effectiveness of implementations of modern contents and technologies in physics class, the development of teaching astronomy, and didactic approaches in the use of ICT. He was the coordinator of the project Development of Natural Science Competences. His other engagements are the popularization of science, where he participates in the organization of different workshops, competitions, and science fares. He is also active in the Association for Technical Culture of Slovenia.

Chapter 29
Science Education in Turkey

Cengiz Hakan Aydin, Sibel Kaya, Eda Atasoy, and Merve Diyarbakirli

Abstract Turkey is located in the middle of two continents, Asia and Europe, and position itself more a European country then Asian. The education system in Turkey is quite centralized and standardized. The curriculum provided by the Ministry of National Education intends to provide equal education opportunity to all. However, large student body, increasing number of refugees especially from Syria and Afghanistan, highly competitive structure of the education system, shortage of awareness towards science education and infrastructure are among the challenges the country faces. The main goals of science education in primary and secondary schools include gaining scientific knowledge in astronomy, biology, physics, chemistry, earth science and engineering; using science process skills and twenty-first century skills in understanding nature and solving daily problems; understanding scientific method and ethics in science; analyzing the relationship between science, environment and economic development. The EBA platform provides some digital tools and content to help teachers in their science education classrooms. STEM classes and programs in especially private schools as well as science education centers established by MoNE, NGOs and the municipalities are some of the strategies in the country to overcome challenges and develop the science education. The science clubs can be an alternative mean to widen the science education to more students and raise an awareness.

Keywords Science education in Turkey · Science education centers · Science clubs · Awareness towards science education · Infrastructure for science education

C. H. Aydin (✉) · E. Atasoy · M. Diyarbakirli
Anadolu University, Eskisehir, Turkey
e-mail: chaydin@anadolu.edu.tr

E. Atasoy
e-mail: ekaypak@anadolu.edu.tr

M. Diyarbakirli
e-mail: mervebakkal@anadolu.edu.tr

S. Kaya
Kocaeli University, Kocaeli, Turkey

29.1   Overview of the Country

29.1.1   *Geographical Location, Population and Political System*

Located at the crossroads of Asia and Europe, Turkey; that is the Republic of Turkey is a transcontinental country. Asian part of Turkey is separated from European part by the Bosphorus; located in İstanbul, the Dardanelles and the Sea of Marmara. Turkey has borders with 8 countries: Iraq and Syria to the south, Iran, Azerbaijan and Armenia to the east, Georgia to the north-east, Bulgaria and Greece to the north-west, and surrounded by 4 seas: the Mediterranean, the Aegean, the Sea of Marmara, and the Black Sea. Turkey's land area is 783.562 square kilometers, which makes it the 37th largest country in the world. It roughly resembles a rectangle and is 1,660 km wide. Turkey's territory lies between latitudes 35° and 43° N, and longitudes 25° and 45° E. There are 7 geographical regions in Turkey: Black sea, Marmara, Mediterranean, Aegean, Central Anatolia, Eastern Anatolia, as well as Southeastern Anatolia. Turkey consists of 81 cities. While the capital city of Turkey, Ankara is located in Central Anatolia, the largest city, Istanbul is located in Marmara (Table 29.1).

Departing from statistics provided by TURKSTAT (TUIK), the country's population is 83 million 154 thousand 997 people according to the 2019 population censuses. The most densely populated cities in Turkey are İstanbul (with the population of 15 million 519 thousand 267 people), Ankara (with the population of 5 million 639 thousand 76 people), İzmir (the population of 4 million 367 thousand 251 people), and Bursa (the population of 3 million 56 thousand 120 people). People within the 15–64 age group constitute 67.8 percent, the 0–14 age group 23.1 percent, and senior citizens aged 65 years or older 9.1 percent of the total population. Also, there are 1 million 531 thousand 180 foreigners living in Turkey.

The Republic of Turkey was founded on 29th of October, 1923 by Mustafa Kemal Atatürk, who was later elected as the first president of Turkey. During his governance between 1923 and 1938, Atatürk initiated a large number of reforms, including the ones in education, science, and social life. These reforms, adopting Western thought, philosophy and customs, all together contributed to the development of Turkey. Between the years of 1923 and 2018, a parliamentary representative democracy was ruled in Turkey. However, as a result of the referendum in 2017, a presidential system was adopted in the country and in the presidential election in 2018, Recep Tayyip Erdoğan became the first president elected by direct voting. In the new system, the office of Prime Minister was abolished, and its powers and the Cabinet's powers were transferred to the President. Now, Turkey acts as a unitary centralized state.

Turkey has been an active member of the UN, NATO, the IMF, the World Bank, the OECD, OSCE, BSEC, OIC, G20 and the Council of Europe. In 2005, Turkey started accession negotiations with the European Union, but later Turkey's accession talks were suspended by the EU governments upon the European Parliament's call.

Table.29.1 Results of population censuses, 1935–2000 and results of address based population registration system, 2007–2019 (TURKSTAT, 2020)

Population by years, age group and sex, 2015–2019

Year	Age group	Total	Male	Female	Total (%)	Male (%)	Female (%)
2015	Total	78 741 053	39 511 191	39 229 662	100.0	100,0	100,0
	0–14	18 686 220	9 695 191	9 191 029	24.0	24.5	23.4
	15–64	53 359 594	26 972 556	26 387 036	67.6	68.3	67.3
	65 +	6 495 239	2 843 442	3 651 797	8.2	7.2	9.3
2016	Total	79 614 671	40 043 650	39 771 221	100.0	100.0	100.0
	0–14	13 925 782	9 715 020	9 210 762	23.7	24.3	23.2
	15–64	54 237 586	27 409 238	26 828 348	68.0	68.4	67.5
	65 +	6 651 503	2 919 392	3 732 111	8.3	7.3	9.4
2017	Total	80 610 525	40 535,135	40 275 390	100.0	100.0	100.0
	0–14	19 033 486	9 769 101	9 264 387	23.6	24.1	23.0
	15–64	54 831 652	27 732 601	27 149 051	67.9	68.4	67.4
	65 +	6 395 365	3 033 433	3 861 952	3.5	7.5	9.6
2018	Total	82 003 682	41 139 980	40 663 902	100.0	100.0	100.0
	0–14	19 134 329	9 846 565	9 337 764	23.4	23.9	22.9
	15–64	55 633 349	28 123 283	27 510 066	67.6	68.4	67.3
	65 +	7 166 204	3 170 132	4 016 072	8.8	7.7	9.8
2019	Total	83 154 997	41 721 136	41 433 661	100.0	100.0	100.0
	0–14	19 212 345	9 859 547	9 352 798	23.1	23.6	22.6
	15–64	56 391 925	28 524 329	27 867 596	67.6	68.4	67.3
	65 +	7 550 727	3 337 260	4 213 467	9.1	8.0	10.2

29.1.2   Current Situation of Economic, Technologies and Cultural Development

Recognized as a newly industrialized country, Turkey is considered as the world's 19th largest economy by GDP and 13th largest economy by PPP GDP. Turkish economy is classified as an emerging market economy by IMF. GDP Growth Rate in Turkey has been 0.95 percent on average between 1998 and 2020, reached its highest, 5.80 percent in the last quarter of the year 2016, and recorded low, -11 percent in the mid of the year 2020 (the sharpest contraction in Turkish economy on record, apparently caused by corona virus pandemic) (Trading Economics, 2020). According to the statistics released by TURKSTAT (January 2020) in Turkey labor participation rate is 51%, unemployment rate is 13.8%, non-agricultural unemployment rate is 15.7%, and employment rate is 44%. Turkey is among the world's leading manufacturers in agricultural products, textiles, motor vehicles, ships and other transportation equipment, construction materials, consumer electronics and white goods. Turkey's largely free-market economy is driven by its industry and recently service sectors, although its traditional agriculture sector still accounts for about quarter of employment. The country's major industries are steel/metallurgy, textile and clothing, petroleum products, food, and automotive. 16% of Turkish labor force works in agriculture, 20.7% in industry, 5.2% in construction, and 58.1% in services (TURKSTAT, 2020). Another sector which has been a big source of income for Turkey is tourism. The annual income from tourism is average 30 million USD per year. In 2019, the number of departing visitors is reported as 51 million (TURKSTAT, 2020).

In the Eight Five-Year Development Plan, the vision of Turkey's national science and technology policy is set as: (a) to create a country which has a mastery on the initiation of science and technology serving economic and social benefits, (b) to earn worldwide respect on the promotion of technology, and (c) to become a recognized information society. In this plan, biotechnology, genetic engineering, software, information and communication technologies, space science and technologies, nuclear technology, technologies for exploitation of seas and marine animals, clean energy technologies have been identified as priority areas (Sekizinci Bes Yıllık Kalkınma Planı, 2000).

In respect to the national development plan, Turkey has been attributing a great importance to R&D activities in recent years. In the country, TUBITAK acts as the leading agency for developing science and technology as well as spearheading innovations. During 2019, Turkey initiated significant high-tech developments by starting to work on a flying car, indigenous electric vehicle, and a laser gun. On December 2019 Turkey presented the first prototype of the indigenous electric vehicle to the public. Also, in 2019 numerous technological events hosting millions of people were held in the country, including Teknofest (Turkey's largest aviation and technology event), WIN EURASIA (Eurasia's biggest industrial fair), the 14th International Defense Industry Fair (IDEF'19), international entertainment and games expos such as Gaming Istanbul (GIST) and GameX, and International Automotive Engineering Conference (IAEC) (Hurriyet Daily News, 2019).

Further, according to the Information Society Statistics (2020), there is a high increase on the usage of ICTs in both enterprises and individuals between the years of 2004 and 2020. In 2020, the internet usage is 94.9 percent in enterprises and 79 percent in households and individuals. 90.7 percent of households has access to the internet (Table 29.2).

Turkey has a cosmopolitan culture, including a mixture of elements from different cultures such as Ottoman, Turkic, Anatolian, Eastern, and Western. The reason behind this mix is the westernization process of Turkey, started in the period of Ottoman Empire and today still goes on and Turkish people's being an immigrant society. Right now, Turkish culture can be described as a product of efforts to become a "modern" Western state while preserving traditional religious and historical values. From the past, Turkey has contributed to the world culture by visual arts (The Tortoise Trainer by Osman Hamdi Bey; Antibes and Disasters of War by Abidin Dino),literature (Awakening by Namık Kemal; The Poet's Marriage by Şinasi; My Name Is Red,

Table.29.2 Use of Information and Communication Technology (ICT) in Enterprises, Survey on Information and Communication Technology (ICT) Usage Survey in Households and by Individuals (TURKSTAT, 2020)

Information Society Statistics, 2004–2020

	2004	2005	2006	2007	2008	2009	2010	2011	2012	2013	2014	2015	2016	2017	2018	2019	2020
ICT usaga in Enterprises																	
Computer Usage	–	87,8	–	88,7	90,6	90,7	92,3	94,0	93,5	92,0	94,4	95,2	95,9	97,2	97,0	96,7	–
Internet Access	–	80,4	–	85,4	89,2	88,8	90,9	92,4	92,5	90,8	89,9	92,5	93,7	95,9	95,3	94,9	94,9
Having Website	–	48,2	–	63,1	62,4	58,7	52,5	55,4	58,0	53,8	58,6	65.5	66,0	72,9	66,1	51,5	53,7
ICT Usage in Households and Individuals																	
Computer Usage (Total)	23,6	22,9	–	33,4	38,0	40,1	43,2	46,4	48,7	49,9	53,5	54,8	54,9	56,6	59,6	–	–
Internet Usage (Total)	18,8	17,6	–	30,1	35,9	38,1	41,6	45,0	47,4	48,9	53,8	55,9	61,2	66,6	72,9	75,3	79,0
Households with access to the Internet	7,0	8,7	–	19,7	25,4	30,0	41,6	42,9	47,2	49,1	60,2	69,5	76,3	80,7	83,8	88,3	90,7

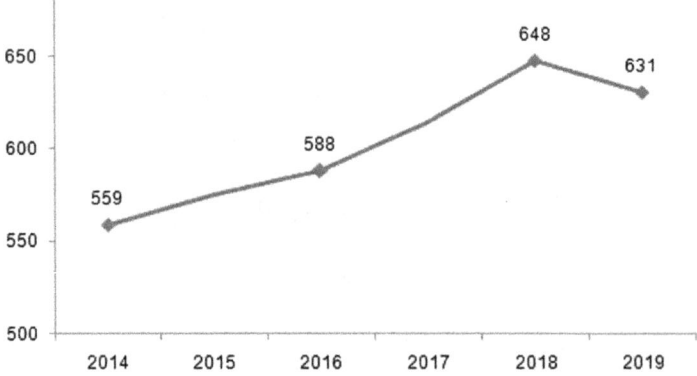

Fig. 29.1 Cultural employment rates, ages 15 + , 2014–201 (TURKSTAT, 2020)

The Museum of Innocence, and The Red-haired Woman by Orhan Pamuk, 2016 Nobel prize winner), music (Tarkan and Sertap Erener as pop stars; Kerem Görsev as jazz and blues musician; Fazıl Say and Idil Biret as western classical music pianists), architecture (Sultan Ahmed Mosque, Stari Most, Selimiye Mosque, and Blue Mosque by Mimar Sinan) and cinema (Three monkeys, Once upon a Time in Anatolia, Winter Sleep by Nuri Bilge Ceylan; The Road, Hope, The Wall by Yılmaz Güney; The Ignorant Fairies, Red Istanbul, Facing Windows by Ferzan Özpetek).

According to the statistics released by TURKSTAT (2020), the cultural employment rate in Turkey is 631 thousand people (Fig. 29.1).

From these men accounted for 53.2% of cultural employment, and women 46.6%. 61.0% of those in cultural employment were in the 30–54 age group, 29.2% in the 15–29 age group, and 9.7% in the age group 55 and above.

When compared to the previous years, in 2019 there has been a decrease in the number of movie theatres (1.1% to 2 826 thousand), the number of seats in movie theaters (1.4% to 337 thousand 914), cinema audience (12.8% to 56 million 479 thousand 209 people). On the other hand, there has been an increase in the theater seats (12.8% to 9 thousand 796 people), the number of shows held in theater halls (0.2% to 33 thousand 835 people), and theater audience number (0.7% to 7 million 899 thousand 547people).

29.2 Overview of the Education Development

29.2.1 Education System and Policy

All the educational activities in Turkey are conducted on a central level by the Ministry of National Education (MoNE). Education system basically consists of four levels: pre-school, primary, secondary and higher education. Besides MoNE, there are several bodies that help shape education policies and conduct educational activities. These institutions can be listed as following:

- The National Council of Education
- The Board of Education
- The Directorate for Strategy Development
- The Directorate for Guidance and Inspection
- The Vocational Education Council
- The Council of Higher Education (YÖK)
- The Assessment, Selection and Placement Centre (ÖSYM)
- The National Council for Teacher Training.

When the general structure of the education system in Turkey is taken into consideration, it can be easily seen that the system basically consists of four levels. The first level is pre-school education. Early childhood education in Turkey covers the nursery and day care centers for the children of 0–36 months. Preschool education is conducted for the children of 36–66 months in kindergartens and nursery schools. Primary school education consists of two stages. The first stage "primary school" which lasts for four years (1st, 2nd, 3rd and 4th grades), covers the children of 66 months of to 10-year-olds. The second stage which is "middle school" (5, 6, 7 and 8 grades) also lasts for 4 years and it covers children between the ages of 11 and 14. The second level is the secondary education, which also lasts for 4 years (9, 10, 11 and 12th grades) and covers the students between the ages of 14 and 18.. Within the secondary education, the education is conducted at various types of high schools such as Anatolian High School, Science High School, School of Fine Arts, Sports High School, School of Social Sciences, the Anatolian Religious High Schools and Technical/Vocational High School. Since 2012–2013 academic year, these two levels of education (three stages as primary, middle and high school) have been compulsory for all the children in Turkey.

At tertiary level, higher education covers the students over the age of 17. In Turkey, after it changed its policies towards higher education, the number of state universities which was 53 in 2003 has reached 129. Along with state universities, by 2020, there are 72 foundation universities and 5 vocational schools. Turkey offers associate, bachelors, masters and doctoral degrees provided by more than 200 universities in total under the supervision and regulations of the Council of Higher Education (YÖK).

The general structure of the education system is summarized in a chart given below by the Council of Higher Education in its report regarding the Higher Education System in Turkey in January 2019 (Fig. 29.2).

In terms of educational policies, Turkey is known to have three strategic guidelines which are the Strategic Plan for the Ministry of National Education (2010–14), the Tenth Development Plan (2014–18); and the Lifelong Learning Strategy Paper. These strategic plans put forward the medium- and long-term education goals; and an overall government strategy. The Turkish government aims to reach certain goals by 2023 in education. By 2023, the government and the Ministry of National Education aim to: (a) have a society of educated individuals; (b) complete FATIH project, which

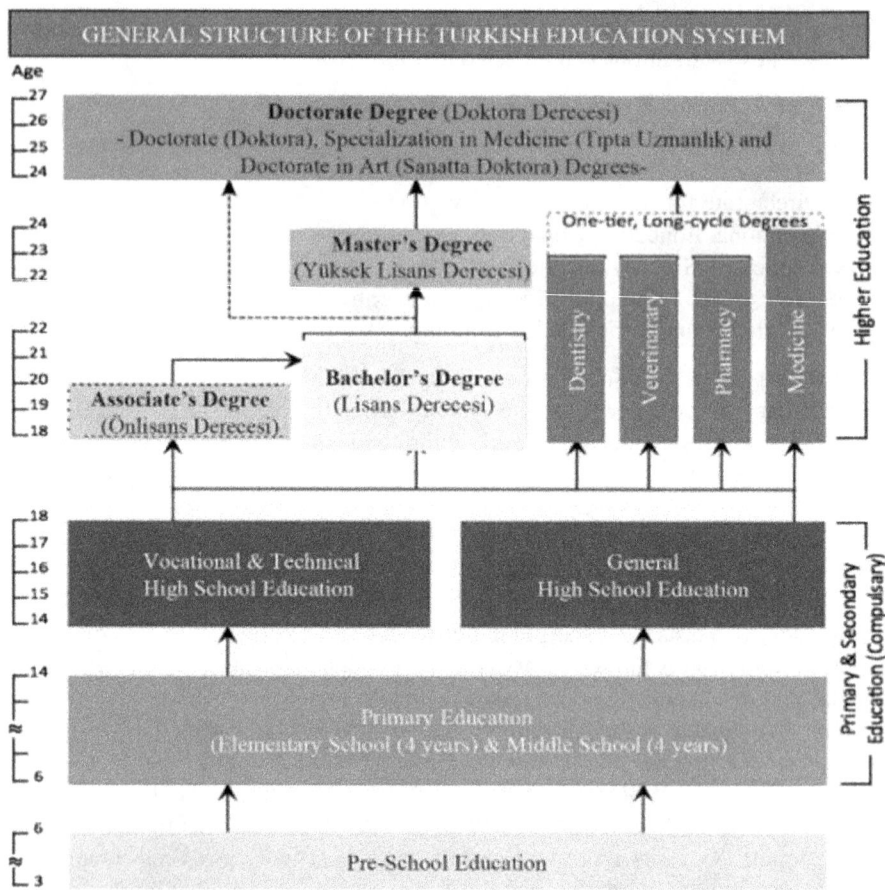

Fig. 29.2 The chart demonstrating the general structure of the education system in Turkey by CoHE (YOK, 2020)

aims to equip each classroom with an interactive white board and each student with a tablet computer; (c) increase participation rates in compulsory education to 100%; (d) promote vocational education; (e) implement reform of the YÖK; (f) increase the number of private universities; (g) improve the quality of universities; (h) increase the number of academics in universities; (i) implement a policy of language learning; (j) terminate gender and regional disparities; and (k) prepare students for upper education and the future in a more flexible structure.

29.2.2   *Statistics on the National Education*

According to the statistics provided by the Ministry of National Education in 2019–2020 academic year, there are 1,117,686 teachers working in different levels and institution in the first three levels (preschool, primary and secondary education) of formal education in Turkey. The details regarding the numbers of teachers by levels and institutions are demonstrated in the table below (Table 29.3).

In addition, MoNE presented the numbers of students in the first three levels of education in its report in 2019–2020 academic year. In Turkey, with its young population, there are 18,241,881students studying are preschools, primary, middle and high schools.

The statistics related to higher education is provided by the Council of Higher Education. In its report on Higher Education System in Turkey in January 2019, it was stated that there were 7.560.371 students who are studying in any of the present higher education institutions during the academic year of 2017–2018 (Table 29.4).

It was also stated in the report that during the same academic year, there were 158.098 faculty members working at higher education institutions in Turkey (Table 29.5).

Table.29.3 Number of teachers and students in K12

Grade levels	Number of teachers	Number of students
Pre-school	56,218	1,629,720
Primary school	309,247	5,279,945
Middle school	371,590	5,701,564
High school	380,631	5,630,652
Total	1,117,686	18,241,881

Table.29.4 Number of students in higher education

Grade levels	Number of teachers
Associate degree	2,768,757
Bachelor's degree	4,241,841
Master's degree	454,673
Doctoral degree	95,100
Total	7.560.371

Academic rank	Number
Table.29.5 Number of students in higher education	
Professor	24,640
Associate Professor	14,456
Faculty Member, PhD	37,520
Others	81,482
Total	158,098

29.2.3 Educational Research and International Collaboration

Educational research in Turkey is mainly conducted at university and financed by some governmental, private and international organizations. TÜBİTAK (the Scientific and Technological Research Council of Turkey) supports universities and individual researchers via various funds, scholarships or projects. EU agencies are also supportive for Turkish researchers by providing funds for educational research projects. As new trends and issues emerge in educational research, Ministry of Education has begun to fund and take part in several different research projects such as EBA which is an educational content network. TÜBA (Turkish Academy of Sciences) is another body that supports educational research in Turkey. Even though there is some participation of non-governmental organizations in educational research, the funds and projects can be considered to be very limited.

The Council of Higher Education is also responsible for international collaboration with educational bodies and institutions. The Council of Higher Education gives importance to collaboration and states this in its reports regularly:

> Owing to the fact that education is a crucial tool for building a global community, cooperation among countries can help foster integration, share knowledge and best practices and also solve problems. In order to achieve our internationalization strategy, we do not only focus on improving current relations with our partners but also establish a bond with new players.

Therefore, there are several protocols, projects and collaborative work carried out under the supervision of Ministry of National Education and the Council of Higher Education such as Erasmus and Mevlana exchange projects.

29.3 Current Situation of Science Education

29.3.1 Policies and Standards

Since Turkey has a centralised education system, Ministry of National Education (MoNE) determines the implementation of science education policy in primary and secondary schools. At the tertiary level, policies are implemented by another central body, the Council of Higher Education. Curriculum development, textbook approval and the framework for assessment practices are all determined centrally. Policies and standards are revised regularly in order to meet the demands of scientific and technological developments. In 2018, Turkey announced Education Vision 2023 reform. According to this reform, schools and teachers will have more autonomy in reconstructing and implementing curriculum. Schools are expected to reduce the hours of compulsory lessons and create a flexible and modular course schedule, to organize curriculum according to interests and abilities of children and to establish designs and skills workshops (MoNE, 2018a).

The recent revision of science curriculum, as in other subjects, have sought to shift the focus of learning from the rote learning of content knowledge towards a competency and skill-based approach that challenges learners to apply what they know and can do in real-world contexts (Kitchen et al., 2019). The main goals of science education in primary and secondary schools include gaining scientific knowledge in astronomy, biology, physics, chemistry, earth science and engineering; using science process skills and twenty-first century skills in understanding nature and solving daily problems; understanding scientific method and ethics in science; analyzing the relationship between science, environment and economic development (MoNE, 2018b).

The science curriculum places student in the center. Through active participation, students are expected to formulate questions, design inquiry, collect data, analyze and reach conclusions. Science teachers are responsible for providing resources and a rich environment for students to reach their goals. Teachers are expected to have an interdisciplinary vision in teaching science that includes integrating science with technology, engineering and mathematics. Science learning can occur outside of school as well as inside. Therefore, science centers, museums, zoos, botanical gardens or factories can be visited for enriched first-hand experience (MoNE, 2018b).

29.3.2   Curriculums, Digital Resources and Teacher Training

According to 2018 revised national curriculum, students start taking Science in third grade. In grades 1 and 2, science education is provided through a course called the Knowledge of Life, in which science is integrated with social studies. In grades 1 through 4, Science is taught by homeroom teachers. From grade 5, it is taught by subject area teachers. There are four main learning areas in grades 3 to 8: Earth and the Universe, Living Things and Life, Physical Processes, Matter and Its Properties. Each of the four learning areas includes several units, which are developed in a spiral format. In other words, students learn same topics in more in depth as they move across grades. In addition to these four learning areas, Science-Engineering-Entrepreneurship has been integrated into the curriculum in all grades. Accordingly, students are expected to design projects and share with the audience through science fairs (MoNE, 2018b).

Science textbooks are distributed by the MoNE to all teachers and students free of charge. In addition, they are digitally available on Education Information Network (EBA) which is the official digital education platform in Turkey. Other popular digital education platforms are *khanacademy.org.tr*, *morpakampus.com* and *okulistik.com*. There are resources for both students and teachers in these platforms. The resources include course videos, animations, documentaries, experiments, interactive applications, tests and more. During Covid-19 pandemic, online learning was supplemented by a program of educational broadcasting across six national public television channels with content for children from early childhood education to upper secondary level.

The educational pathway for preschool, elementary and middle school teachers include a degree from a teacher education program. For high school teachers, there are two sources: either a degree from a teacher education program or a degree from a four-year college and teaching certificate. Specific to science education, graduates with a bachelor's degree in Biology, Chemistry and Physics form the Faculty of Arts and Sciences are eligible to teach these subjects after receiving a teaching certificate. Teaching certificates are provided by the MoNE in association with the universities' teacher education programs. All teacher candidates for public schools must enter a state exam which is held every year. This multiple-choice exam covers subject knowledge as well as pedagogy and culture. Candidates are interviewed after the exam. Based on exam and interview results, teachers are appointed by the MoNE as trainees. After one year, trainees are evaluated and if they meet the requirements, become full-time teachers. Private school teachers go through separate selection procedures. Each private school make their own selection which usually involves interview and mock teaching.

29.3.3 Student Assessment and Achievement

Turkey participated in Trends in International Mathematics and Science Study (TIMSS) assessments in 1999, 2007, 2011, 2015 and 2019 at eighth grade and in 2011, 2015 and 2019 at fourth grade level. According to the most recent report (TIMSS 2015), Turkey ranked 35th among 47 countries at fourth grade level and 21st among 39 countries at eighth grade level in Science. Although there have been improvements in recent assessments in terms of average score, Turkey is still below the international average (Mullis et al., 2016).

Turkey has also been participating in Program for International Student Assessment (PISA) since 2003. In the most recent PISA assessment in 2018, Turkey ranked 39th among 79 countries with the average score of 468. Unfortunately, the OECD reported that students' academic proficiency differed widely across schools, regions and by socio-economic background in Turkey (OECD, 2020). Students who attend science high schools scored almost as high as the top scoring country, China (provinces) in Science. However, according to 2018 statistics, the share of science high schools is only 4% among all the high schools in Turkey (MoNE, 2019).

Science high schools accept students with top scores on the High School Placement Examination (LGS). These schools play an important role in encouraging students to pursue careers in science, technology, engineering, and mathematics (STEM). The goal of science high schools is to provide students with a strong foundation in mathematics and science therefore the curriculum involves considerably more science and mathematics lessons compared to other high schools (IEA, 2019).

At the end of eighth grade, students are assessed nationwide. This multiple-choice exam (LGS) not only determines student placement at selected high schools but also provides information about schools and regions in terms of their achievement at the end of primary education. In recent years, open-ended diagnosis assessment projects have been conducted in smaller scales. One of these projects is called Monitoring Academic Skills (ABIDE). The main purpose of this project is to measure students' upper-level cognitive skills such as critical thinking and problem solving in science, mathematics and Turkish through open-ended questions (MoNE, 2015).

The Student Learning Achievement Monitoring assessment (2019), which assesses curriculum outcomes, was introduced to provide schools with diagnostic information on students' strengths and weaknesses in Turkish, mathematics and science at grades 4, 7 and 10. The assessment includes background questionnaires for students, teachers and principals. As of early 2020, the computer-based assessment had been piloted for grade 7. Further feasibility studies are in progress for primary and upper secondary students (grades 4 and 11). An initial review of the pilot supported the validity and reliability of the study (OECD, 2020).

The current Science curricula in both primary and secondary schools contain a great number of student assessment instruments. Teachers are encouraged to use formative and multifaceted learning assessment methods, rather than typical paper and pencil tests. Assessment instruments include written and oral examinations, projects, performance assessment, portfolios, journal writing, checklists, observations, posters, peer evaluations, and self-evaluations. Primary and secondary school students are also monitored by e-school system. With this system, teachers keep track of students' academic and personal interests such as hobbies and books they read.

At the end of secondary school, a national university placement examination determines the institution and program for which students are accepted. With limited places available in the most prestigious upper secondary schools and bachelor's programs, the placement exams are very competitive, creating a high level of stress among students. At the tertiary level, since 2001, Turkey has been progressing towards the Bologna model, and the National Qualifications Framework is aligned with the European Higher Education Area (OECD, 2020). As part of this process, online learning, student mobility, student-centered instruction are emphasized. In terms of assessment, lowering the share of final exams and including diverse assessment techniques, such as projects, seminars and oral examinations are highly encouraged. However, recent reports show that the process of change within Turkey has been top-down, driven by the Council of Higher Education and student-centered instruction and assessment have not been fully internalized (Onursal-Beşgül, 2017).

29.3.4 Science and Technology Venues and Centers

In 1963, the Scientific and Technological Research Council of Turkey (TUBITAK) was established as an autonomous institution to advance science and technology and support research. The council is responsible for promoting, developing, organizing, conducting, and coordinating scientific research in accordance with national and global priorities and targets. It also acts as an advisory agency to the Turkish government on science and technology issues. The council organizes Science Olympics every year for upper secondary students. The main goal is to promote interest in science and mathematics related fields. Students take a two-tier test in Physics, Chemistry, Biology, Mathematics and Information Technology (IT). Students who have been successful in these tests are rewarded in varying degrees of funds. Furthermore, a team of top scoring students participate in international Science Olympics (TUBITAK, 2020).

TUBITAK has also been a pioneer in disseminating and supporting science centers across Turkey. These centers aim to boost interest in science among all ages, make science and technology easy to understand, connect science and technology with everyday life and conduct experiments in interactive, multisensory environments. To this date, there are 6 science centers directly funded by TUBITAK and 14 science centers funded by municipalities and private organizations in major cities of Turkey (TUBITAK, 2019). The numbers are likely to increase in the coming years.

In 2007, the Ministry of National Education launched science and arts centers in order to support gifted and talented students. Science and arts centers aim to provide high quality, advanced level education for these students, starting as early as 1st grade. There is at least one science and arts center in each province of Turkey. Science and arts centers are independent educational institutions designed to enable gifted students to develop awareness of their individual talents, discover their creativity and problem-solving skills without interrupting their formal education (IEA, 2019).

There are also private venues that promote science learning. In 2000, Space Camp Turkey was opened in Izmir as a privately funded organization. It's the first and only space camp in Turkey and attracts international visitors as well as locals. In this camp, through interactive, space-related simulations, both youth and adults learn about communication, teamwork, and leadership in a dynamic environment. All activities implemented in Space Camp Turkey are organized around STEAM (Science, Technology, Engineering, the Arts, and Mathematics) education standards. Programs at Space Camp Turkey focus on simulators to give participants the sensations of working and living in space.

29.3.5 Utilizing Emerging Technologies

With the digital transformation of industries, a new era called Industry 4.0 has started. In line with the emergence of Industry 4.0, a new period called Education 4.0 took effect concurrently. Education 4.0 refers to a period in which digital transformation and innovation started to dominate educational practices. Though slow to meet the requirements of Education 4.0, Turkey launched number of initiatives in education to catch up with digital transformation (Keser & Semerci, 2019).

In 2010, Turkey launched the Movement to Enhance Opportunities and Improve Technology (FATIH) project, aiming to extend and enhance the use of technology in teaching and learning. FATIH has evolved into a longer-term program that seeks to foster digital skills and improve access to Information and Communication Technologies (ICT) in schools. By 2019, Nearly 450 000 interactive white boards were installed in classrooms, more than 1.4 million tablet computers were distributed to upper secondary level students and teachers, and approximately 1 million teachers had enrolled in either online or onsite professional development (OECD, 2020).

From 2018, with the introduction of the Education Vision 2023, the development of the Education Information Network (EBA) gained importance. EBA is the official national digital education platform, providing interactive and subject-specific digital content for students and teachers across Turkey from pre-school to upper secondary education. Features include a smart content recommendation system, gamified features, and the EBA Portfolio where students can display their achievements and their work. For older students, an EBA Academic Support feature has been developed to provide adaptive individualized learning based on artificial intelligence assisted analytics. Furthermore, to support teachers, professional development and digital library resources are available, and the EBA Professional Development Platform provides online support for teachers' continuous professional development. EBA has been crucial especially during COVID-19 pandemic to provide distance learning (OECD, 2020).

In recent years coding education and robotics gained interest in Turkey as part of developments in STEM fields (Turan & Aydoğdu, 2020). In 2012, Ministry of Education incorporated a course named "Information Technologies and Software" in school curriculum. Within this course, students receive coding education starting from the fifth grade. However, coding applications start as early as pre-school. One of the most popular coding tools *Scratch* is widely used in Turkey. Turkey is second only to the USA in number of Scratch users (Demir & Seferoğlu, 2019).

Another emerging technology in Education 4.0 is augmented reality (AR) in which physical and virtual objects are combined in a mixed space. In recent years, AR has gained interest in science teaching. Applications such as *Space Craft 3D* and *Anatomy 4D* are the most downloaded mobile applications. However, in a survey conducted in 2019, only 25% of teachers in Turkey reported that they use AR in their classrooms. They stated that they lack knowledge and skills to use AR in their classrooms (Boz, 2019).

Finally, short-term summer and winter camps for students from all ages have become widely popular in Turkey in order to make STEM fields more interesting and accessible. These camps are funded by private organizations or TUBITAK and usually organized by universities. One of these camps, called Summer Science Academy, has been organized by prominent Turkish universities, such as Bogazici, Bilkent, Hacettepe and Middle East Technical University. In these camps, students and researchers meet with the specialists and experience the applications of robotics and artificial intelligence (AI) (Science Academy, 2020).

29.4   Requirements for Future Development of Science Education

In order to develop the science education in Turkey, the requirements can be classified under awareness and infrastructure. Although there is not any scientific indicator or evidence, it can, still, be claimed that the Corona Virus Pandemic has cultivated a significant *awareness* towards importance of science and science education. However, we must confess that this tendency is more toward health sciences and some engineering but not really towards basic science fields, such as physics, biology and chemistry. One evidence or indicator of this awareness can be the field preferences of the students who are seeking for higher education. Among those students who scored high in the university entrance exam, a great majority prefers medical education, nursing, psychology and engineering in computer sciences. On the other hand, basic science programs such as physics, biology and chemistry are not popular among young generations and in many universities these programs have been closed. That is why one of the first requirements for the future development of science education is to raise an awareness among youngsters toward all the fields of science. MoNE (2016) supports the idea of raising an awareness towards STEM fields by creating special programs for STEM education in schools and train teachers toward integration of STEM programs or activities into their regular programs.

Raising an awareness is not enough if there is no *infrastructure*, which does not only refer to hardware, software, equipment and labs but also human resources like teacher support. Due to the massive structure of the education system (nearly 18 million students and 2 million teachers), it is quite difficult to provide sufficient lab environment in every school and enough equipment in those labs. Although the current program recommends and requires practical science experiments that can be conducted without advance or even basic lab equipment almost all levels, from kindergarten to 12the grade, the shortage of equipment, lab environments, hardware and software as well as large class sizes and highly test-preparation structure of the educational system barrier the development of science education in Turkey. Shortage of support for science teachers is also another barrier. Many science teachers are alone in schools and do not get the support they need to guide and facilitate their students to experiment the science topics and develop an awareness toward science. So, the MoNE should collaborate with NGOs and other stakeholders to overcome the barriers and build a sufficient infrastructure for teachers and students. Also, there should be accessible, flexible and sustainable professional development as well as support systems for teachers to serve better to their students.

29.5 Discussion and Conclusion

Not only science education but in general whole education system in Turkey have been struggling with several important challenges. The large body of students, nearly 18 million, in K12 is by itself already a big challenge. Providing quality education to this massive group is hard to accomplish job for the MoNE. Special STEM classes in some schools and the science centers, where extracurricular STEM activities are offered to some talented students, competitions funded by TUBITAK or other organizations are among the strategies employed to overcome the barrier and promote the development of science education.

Another major challenge is the highly competitive structure of the education system. Students need to take centralized exams and get higher scores to be able to continue their education in better schools and universities. Although major centralized exams are at the end of 8th grade (High School Entrance Exam) and 12th grade (the University Entrance Exam), the competition starts in primary school grades. The courses including those related to science designed in a way that supports students' preparation to these exams. This structure is a barrier for spending time on science experiments, hypothesis generation and testing, and other science related activities. In general, teachers in 5th, 6th, 9th, and 10th grades only try to offer a science education as it should be as much as they can, but others mainly focus on exam preparation.

The shortage of awareness toward basic science fields and infrastructure is also another challenge for the MoNE. Especially science teachers should be considered as change agents to raise an awareness among students and parents toward basic science fields as well as other STEM fields. They should have re-training (professional development) opportunities on how to integrate and conduct easy to do, daily life experiments into their teaching. Also, a support system for science teachers should be established for those teachers who are struggling with different problems, including shortage of equipment and knowhow.

In every school, students are encouraged to take parts in various club activities besides their regular courses. Although the science club is one of the most widely seen clubs in schools, activities in the science clubs are limited. These clubs can also serve as a mean to reach more students to cultivate scientific thinking, questioning, and critical thinking. Series of activities that can be completed with easy to find equipment along with a structured implementation plans and models for different circumstances can be designed and shared with the teachers who are leading these clubs. Reginal and national collaboration activities can also help students interact and collaborate with their peers, showcase their artifacts and experiences in the club activities. Collaboration with the NGOs, such as the Education Voluntaries Foundation of Turkey (TEGV), can also help schools and teachers to implement the club activities in the schools. If participation to these kinds of club activities can be considered as one of the criteria for entering the further education institutions (science high schools or higher education programs), the students and parents' interests can be elevated and the science education in Turkey can be developed.

References

Bilim Ve Teknoloji Özel İhtisas Komisyonu Raporu. (2000). Sekizinci beş yillik kalkinma plani. Retrieved September 23, 2020, from http://www.sbb.gov.tr/wp-content/uploads/2018/11/08_Bil imVeTeknoloji.pdf.

Boz, M. S. (2019). *Eğitimde artırılmış gerçeklik uygulamalarının değerlendirilmesi*. Ankara: Milli Eğitim Bakanlığı, Yenilik ve Eğitim Teknolojileri Genel Müdürlüğü.

Demir, Ö., & Seferoğlu, S. S. (2019). Developing a Scratch-based coding achievement test. *Information and Learning Sciences*, 383–406.

Hurriyet Daily News. (2019). Turkey spearheads major high-tech developments in 2019. Retrieved Retrieved September 23, 2020, from https://www.hurriyetdailynews.com/turkey-spearheads-major-high-tech-developments-in-2019-150419.

IEA (2019). *TIMSS 2015 encyclopedia*. Retrieved September 22, 2020, from http://timss2015.org/encyclopedia/countries/turkey/.

Keser, H., & Semerci, A. (2019). Technology trends, education 4.0 and beyond. *Contemporary Educational Researches Journal*, *9*(3), 39–49.

Kitchen, H., Bethell, G., Fordham, E., Henderson, K., & Li, R. R. (2019). *OECD reviews of evaluation and assessment in education: Student assessment in Turkey*. OECD Publishing. 2, rue Andre Pascal, F-75775 Paris Cedex 16, France.

Mullis, I. V. S., Martin, M. O., Foy, P., & Hooper, M. (2016). *TIMSS 2015 international results in science*. Retrieved from http://timssandpirls.bc.edu/timss2015/international-results/.

MoNE (2015). Minitoring Academic Skills (ABIDE). Retrieved September 22, 2020, from http://abide.meb.gov.tr/.

MoNE. (2018a). *Education vision 2023*. MoNE Publications.

MoNE (2018b). *Science curriculum*. Retrieved September 22, 2020, from http://mufredat.meb.gov.tr/.

MoNE (2019). *PISA 2018 Turkey primary report*. Retrieved September 22, 2020 from, http://www.meb.gov.tr/meb_iys_dosyalar/2019_12/03105347_PISA_2018_Turkiye_On_Raporu.pdf.

OECD (2020). *Education policy outlook: Turkey*. Retrieved September 22, 2020, from http://www.oecd.org/education/policy-outlook/country-profile-Turkey-2020.pdf.

Onursal-Beşgül, Ö. (2017). Translating norms from Europe to Turkey: Turkey in the Bologna process. *Compare: A Journal of Comparative and International Education*, *47*(5), 742–755.

Science Academy (2020). *Summer schools*. Retrieved September 22, 2020, from https://yazokulu.bilimakademisi.org/.

Trading Economics. (2020). Turkey GDP growth rate. Retrieved September 23, 2020, from https://tradingeconomics.com/turkey/gdp-growth#:~:text=GDP%20Growth%20Rate%20in%20Turkey,the%20second%20quarter%20of%202020.

TUBITAK (2019). *Science centers*. Retrieved September 22, 2020, from https://bilimmerkezleri.tubitak.gov.tr/.

TUBITAK (2020). *National science olympics*. Retrieved September 22, 2020, from https://www.tubitak.gov.tr/tr/olimpiyatlar/icerik-ulusal-bilim-olimpiyatlari.

Turan, S., & Aydoğdu, F. (2020). Effect of coding and robotic education on pre-school children's skills of scientific process. *Education and Information Technologies*, 1–11.

Cengiz Hakan Aydin, PhD is a full professor in Anadolu University of Turkey where he has been offering courses in the field of open and distance learning since early 1990s such as Designing Open and Distance Learning, Instructional Design, and Research on ODL. He had also served as instructional designer in the Open Education System of the University until September 2017 and as the dean of one of the major ODL faculty of the University between 2013-2016. His current research interest focuses on design and development of ODL environments, integration of new technologies into ODL, Open Education Resources (OER) and Massive Open Online

Courses (MOOCs). Professor AYDIN served as the President and the Board Member of the International Division of AECT, as one of the board member of the International Council for Educational Media (ICEM). Professor AYDIN has been a member of editorial boards of the journals, the International Review of Research in Open and Distance Learning (IRRODL), and Educational Media International (EMI). He is currently serving as the editor of the book review section of IRRODL, a member of the steering committee of OpenupEd, and a member of the ICDE's Advocacy Committee for OER.

Sibel Kaya received her MEd from the University of Pittsburgh and PhD from the Florida State University in the USA with specialization in elementary education. She worked with Dr. Carol Connor in Individualizing Student Instruction Project funded by National Institute for Child Health and Human Development. Currently, she is an associate professor in Faculty of Education at Kocaeli University, Turkey. She teaches Science Methods, Science Laboratory, Research Methods and Teaching Practicum courses at the undergraduate level. At the graduate level, she teaches Research Methods and Statistics. Her research interests include science teaching and learning and classroom discourse. She authored several peer-reviewed journal articles and book chapters on science teaching and learning. She also participated in projects regarding science classroom discourse, science fairs and teachers' professional development.

Eda Atasoy is a lecturer of English at Anadolu University. She gained her Bachelor's degree in the field of English Language Teaching from Anadolu University in 2008 and her Master's Degree in the field of English Language Teaching from Bilkent University in 2012. She is still doing her PhD, in the field of Distance Education at Anadolu University. Her interest areas are open and distance learning, dark web, research methods, open and distance learning theories, and rhizomatic learning.

Merve Diyarbakirli is an academician with nine years of experience working as an English Instructor at Anadolu University, School of Foreign Languages, Eskisehir, Turkey. She holds a BA degree in English Language Teaching from Middle East Technical University, Ankara, Turkey. She completed several international certificate programs such as CELTA from Cambridge University, Blended Learning in Language Education Certificate from NILE Norwich, the UK, and Teaching English to Teens and Pre-teens Program by University of Oregon, the USA. She is currently doing an MA degree in Distance Education at Anadolu University, Turkey. She specializes in educational technologies and have been responsible for training her colleagues about recent educational technologies and application for the last six years. Her main areas of academic interest include teacher education, distance education practices, open educational resources and educational technologies. She is a life-long learner and is always looking for new opportunities to improve herself professionally. She is a positive team member in her workplace and uses her practical skills to solve problems and encourage colleagues. She is an enthusiastic maker, a cyclist and a runner. She likes to spend her free time with upcycling or do-it-yourself projects and by going on long bike rides. She is also an animal lover and inspired daily by her little dog "Luca".

Part V
Conclusion

Chapter 30
A Summary of Science Education in Countries Along the Belt and Road: Insights and Recommendations

Xiangling Zhang, Ahmed Tlili, Lixin Zhu, Yao Song, and Tianyue Sun

Abstract This chapter is a summary of the previous chapters. First, it introduces the territories, populations, and economic levels of the Belt and Road countries selected in this book. Secondly, the educational development of various countries is introduced, including educational policies and statistics related to educational development. Next, it comes to the development of science education in these countries, analyzing the achievements of science education development, the status and innovation of science education policies, resources, as well as the application of new technology. Then, it discusses the needs of the development of science education, and finally puts forward suggestions for the Belt and Road countries to promote the development of science education.

Keywords Science education · Belt and road · Educational policy · Educational resources · Educational technology

30.1 Country Profiles

30.1.1 Territory and Population

There are a large number of the Belt and Road countries, with huge differences in terms of territory, population and economic development. Pakistan, Philippines and Egypt have more than twice the population of other Belt and Road countries (without counting China), among which Pakistan has the largest population. Algeria, Saudi Arabia, Sudan, and South Africa have larger territories, while others have smaller.

X. Zhang (✉)
ShiFang Street No. 2, Huangsi Street, Xicheng District Beijing 100120, China
e-mail: zhangxiangling@bjie.ac.cn

A. Tlili · L. Zhu · Y. Song
Beijing Normal University, HaiDian District, No.19 Xinjiekouwai Street, Beijing 100875, China

T. Sun
525 West 120th Street, New York, NY 10027, USA

© The Author(s), under exclusive license to Springer Nature Singapore Pte Ltd. 2022   509
R. Huang et al. (eds.), *Science Education in Countries Along the Belt & Road*,
Lecture Notes in Educational Technology,
https://doi.org/10.1007/978-981-16-6955-2_30

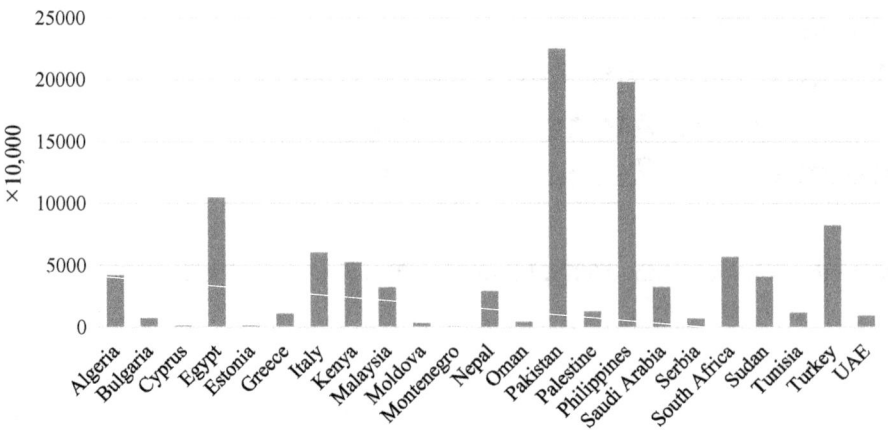

Fig. 30.1 Population of some belt and road countries in 2019

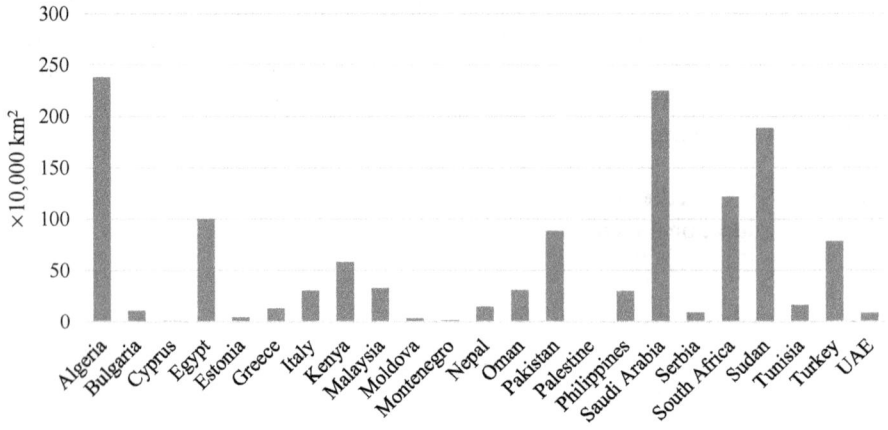

Fig. 30.2 Territory area of some belt and road countries

The population and territory area of some Belt and Road countries mentioned in this book are respectively shown in Figs. 30.1 and 30.2.

30.1.2 Gross Domestic Product

The GDP can reflect the economic changes and development status of a country. The GDP of some countries along the Belt and Road from 2012 to 2019 is shown in Table 30.1.

Among the above-mentioned countries, Turkey and Saudi Arabia have higher GDP than other countries. Turkey's GDP from 2012 to 2019 has decreased; while

Table 30.1 GDP of some belt and road countries from 2012 to 2019 (Unit: billion USD) Data

Country (Sorted by the initial)	2012	2013	2014	2015	2016	2017	2018	2019
Algeria	209.1	209.8	213.8	166.0	160.0	170.0	175.4	171.2
Bulgaria	54.0	55.6	56.9	50.6	53.8	59.0	66.2	68.6
China	8,532	9,570	10,476	11,062	11,233	12,310	13,895	14,280
Cyprus	25.0	23.9	23.2	19.8	21.0	22.7	25.3	24.9
Egypt	279.1	288.4	305.6	239.4	332.4	235.7	249.7	303.1
Estonia	23.2	25.3	26.8	23.0	24.3	26.9	30.6	31.5
Greece	242.1	238.5	235.3	195.3	192.7	199.6	212.1	205.3
India	1,828	1,857	2,039	2,104	2,295	2,651	2,701	2,871
Italy	2,087	2,141	2,159	1,836	1,876	1,957	2,091	2,005
Kenya	50.4	55.1	61.4	64.0	69.2	79.0	87.8	95.5
Libya	81.9	65.5	41.1	27.8	26.2	37.9	54.1	52.1
Malaysia	314.4	323.3	338.1	301.4	301.3	319.1	358.7	364.7
Moldova	8.7	9.5	9.5	7.7	8.1	9.7	11.5	12.0
Montenegro	4.1	4.5	4.6	4.1	4.4	4.9	5.5	5.5
Nepal	21.7	22.2	22.7	24.4	24.5	29.0	33.1	34.2
Oman	76.5	78.6	80.7	68.4	65.4	70.6	79.8	76.3
Pakistan	224.4	231.2	244.4	270.6	278.7	204.6	314.6	278.2
Philippines	261.9	283.9	297.5	306.4	318.6	328.5	346.8	376.8
Saudi Arabia	736.0	746.6	756.6	654.3	645.0	688.6	786.5	793.0
Serbia	43.3	48.4	47.1	39.7	40.7	44.2	50.6	51.5
Slovenia	46.6	48.4	50.0	43.1	44.7	48.5	54.1	54.2
South Africa	396.3	366.8	350.9	317.6	296.4	349.6	368.3	351.4
Sudan	48.2	51.2	61.7	64.5	52.8	45.0	34.5	32.3
Tunisia	45.0	46.3	47.6	43.2	41.8	39.8	39.8	39.2
Turkey	880.6	957.8	939.0	864.3	869.7	859.0	778.4	761.4
UAE	374.6	390.1	403.1	358.1	357.0	385.6	422.2	421.1

Source World Bank https://data.worldbank.org.cn/

Saudi Arabia's GDP has basically maintained growth. There is not much difference among the GDP of the Philippines, Malaysia, Egypt, and the UAE from 2012 to 2019, and they are basically in a growth trend. The GDP trends of some countries are shown in Fig. 30.3.

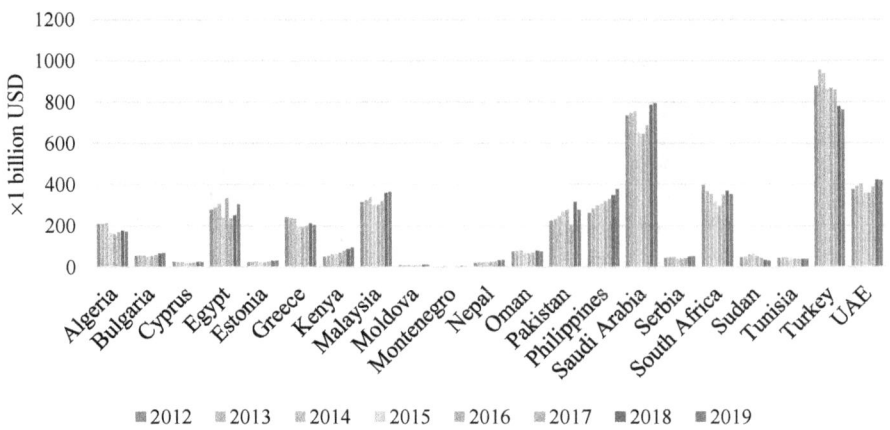

Fig. 30.3 GDP histogram of some belt and road countries from 2012 to 2019

30.2   Overview of Educational Development

30.2.1   Education Systems and Policies

In some of the Belt and Road countries selected in this book, the education system generally includes preschool education, elementary education, junior high school education, high school education, high school non-higher education, higher education, special education, and adult education. Compulsory education generally includes elementary education and secondary education, with a specific length of 8 to 13 years, and the average length of compulsory education is about 10 years. Among these Belt and Road countries, Turkey, Saudi Arabia, Kenya, and the Philippines have the longest compulsory education period of 12 years. However, Croatia and Serbia have the shortest compulsory education period, only 8 years.

Most of the Belt and Road countries provide a longer period of free education than compulsory education, among which Turkey, the UAE, Estonia, Bulgaria, Tunisia, and Poland provide longest period of free education, 16 years, while Croatia is the lowest with only 8 years. The average period of free education in these countries is about 12 years. Table 30.2 illustrates the length of free education and compulsory education in some Belt and Road countries.

30.2.2   Statistics on the Development of Education

The government expenditure on education as a percentage of GDP reflects, to a certain extent, the government's emphasis on education and the level of effort the whole society has made to develop education of the country.

Table 30.2 Educational systems and policies of some belt and road countries (Total Number of Years Guaranteed by Legal Framework). Data *Source* UNESCO Institute of Statistics http://data.uis.unesco.org/

Country	Free education	Compulsory education
Turkey	16	12
Bulgaria	16	11
UAE	16	10
Estonia	16	9
Tunisia	16	9
Slovenia	13	9
Cyprus	12	12
Kenya	12	12
Philippines	12	12
Saudi Arabia	12	12
Greece	12	9
Italy	12	9
Serbia	12	8
Malaysia	11	11
Nepal	10	10
Oman	10	10
Pakistan	10	10
China	9	9
Libya	9	9
Montenegro	9	9
South Africa	9	9

According to the collected data, the average public education expenditure of the selected Belt and Road countries accounts for 4.59% of GDP. According to the minimum standard of 4% of GDP advocated by UNESCO for government education expenditure (Education 2030, UNESCO), about half of the countries are higher than this level. The proportion of government expenditure on education in GDP in some countries is shown in Table 30.3.

Academic qualification education refers to general education and adult academic education, excluding various non-academic training. General education includes: ordinary elementary schools, ordinary secondary schools (junior high schools, high schools), secondary vocational schools (vocational junior high schools, vocational high schools), secondary professional schools, technical schools, and ordinary higher schools. It is used to reflect the educational level of a new labor force in a country or region.

In some Belt and Road countries, Estonia has the highest per capita education years at 14.05 years, and Pakistan has the lowest, at 5.02 years. The average per capita education years in the enumerated countries is 10.75 years. The average years of education in more countries are given in Table 30.4.

Table 30.3 Government expenditure on education as a percentage of GDP in some belt and road countries (2018).

Country	Percentage (%)
Tunisia	6.20 (2016)
South Africa	6.16
Cyprus	5.78 (2017)
Kenya	5.31
Nepal	5.1
Estonia	4.97
Montenegro	4.5 (2016)
Malaysia	4.48
Bulgaria	4.09
China	4.04 (2019)
Italy	4.04 (2017)
Slovenia	3.94 (2017)
Serbia	3.59
Pakistan	2.9 (2017)

Data Source UNESCO Institute for Statistics http://data.uis.unesco.org/

Table 30.4 Average years of education in some belt and road Countries (2017).

Country	Average years of education
Estonia	14.05 (2018)
China	13.6 (2019)
Slovenia	12.77
UAE	12.55 (2018)
Bulgaria	11.36
Serbia	11.16
Cyprus	10.38
Malaysia	10.37
Greece	10.26 (2016)
Saudi Arabia	10.23
Italy	10.19 (2015)
South Africa	10.15
Oman	9.56 (2015)
Philippines	8.45
Turkey	8.28
Tunisia	7.22 (2016)
Pakistan	5.02

Data Source UNESCO Institute of Statistics http://data.uis.unesco.org/

Table 30.5 Net enrollment rate for primary school by belt and road countries (2017).

Country	Rate of female (%)	Rate of male (%)	Total rate (%)
Montenegro	98.54	100	99.99
China	99.91	99.96	99.94
Slovenia	95.72	99.55	99.77
Malaysia	100	99.31	99.65
Cyprus	99.18	99.66	99.42
Tunisia	100	98.05	98.99
UAE	97.46	100	98.74
Greece	98.77	98.38	98.57
Serbia	98.07	98.24	98.16
Saudi Arabia	97.37	98.91	98.15
Estonia	97.97	97.72	97.84
Oman	100	94.60	97.19
Italy	96.68	97.01	96.85
Nepal	92.62	100	96.30
Philippines	95.90	96.48	96.20
Turkey	94.74	95.33	95.04
South Africa	89.93	88.17	88.99
Bulgaria	86.78	86.61	86.70
Sudan	61.02	62.36	61.70

Data Source UNESCO Institute of Statistics http://data.uis.unesco.org/

The net enrollment rate refers to the ratio of the number of school-age students to the total population of school-age age. It is necessary to consider the age of the students in school, and only count the number of students in the same age group as the denominator. Among the selected Belt and Road countries, Montenegro has the highest net enrollment rate for primary school of 99.99%, and Sudan has the lowest, which is 61.70%. The average of these countries is 95.83%, and about 80% of countries are above average (Table 30.5).

30.3   Current Situation of Science Education

30.3.1   Assessment and Achievement of Science Education

Scientific literacy generally refers to the knowledge of key scientific concepts and the understanding of scientific processes, which includes the application of science to cultural, political, social and economic issues (Miller, 1983). At present, the issue of scientific literacy is becoming more and more important in education. In current school education, students have gradually tended to learn through exploratory learning rather than memorizing facts. This also means that understanding the process of science and the application of scientific concepts is one of the central goals of current education. As the concept of core literacy has been put forward and deepened, countries around the world are paying more and more attention to scientific literacy. How to scientifically and rationally evaluate the scientific literacy of students is one of the important challenges facing science education, and the current internationally authoritative scientific literacy evaluation. The project accordingly provides us with a reliable basis for evaluating the current state of science education in Belt and Road countries especially the scientific literacy of students.

Currently, there are mainly two major internationally influential and well-designed student scientific literacy assessment projects: one is the Program for International Student Assessment (PISA) hosted by the Organization for Economic Co-operation and Development (OECD), and the other is the Trends in International Mathematics and Science Study (TIMSS) organized by The International Association for the Evaluation of Educational Achievement (IEA). Among them, PISA mainly evaluates the reading, science and mathematics of 15-year-old students, which is held every three years; TIMSS evaluates the scientific literacy and mathematics literacy of students in fourth and eighth grades, which is held every four years.

(1)   PISA.

The most recent PISA test was conducted in 2018, and the evaluation results were released in December 2019. Among the 79 countries or regions participating in the PISA 2018 evaluation, 50 of the Belt and Road countries are included: China, Morocco, South Korea, Singapore, Malaysia, Brunei, UAE, Turkey, Qatar, Saudi Arabia, Azerbaijan, Georgia, Kazakhstan, Thailand, Indonesia, Philippines, Cyprus, Russia, Austria, Greece, Poland, Serbia, Czech Republic, Bulgaria, Slovakia, Albania, Croatia, Bosnia and Herzegovina, Montenegro, Estonia, Lithuania, Slovenia, Hungary, North Macedonia (formerly Macedonia), Romania, Latvia, Ukraine, Belarus, Moldova, Malta, Portugal, Italy, Luxembourg, New Zealand, Chile, Uruguay, Peru, Costa Rica, Panama, Dominica.

After sorting out, the scores of the science project evaluation in PISA 2018 of the countries in this book are shown in Fig. 30.4. Among them, the horizontal axis represents the average score of the country, and the vertical axis represents the standard deviation of its score, which measures the dispersion of the students' score of the country relative to its mean.

As shown in Fig. 30.4, in terms of the average score of scientific literacy, Estonia is in the leading position, while the Philippines is relatively behind other countries; in terms of the degree of dispersion of scores, the level of scientific literacy of Malaysia students is generally closer to the average. The distribution of literacy levels is relatively uneven among countries.

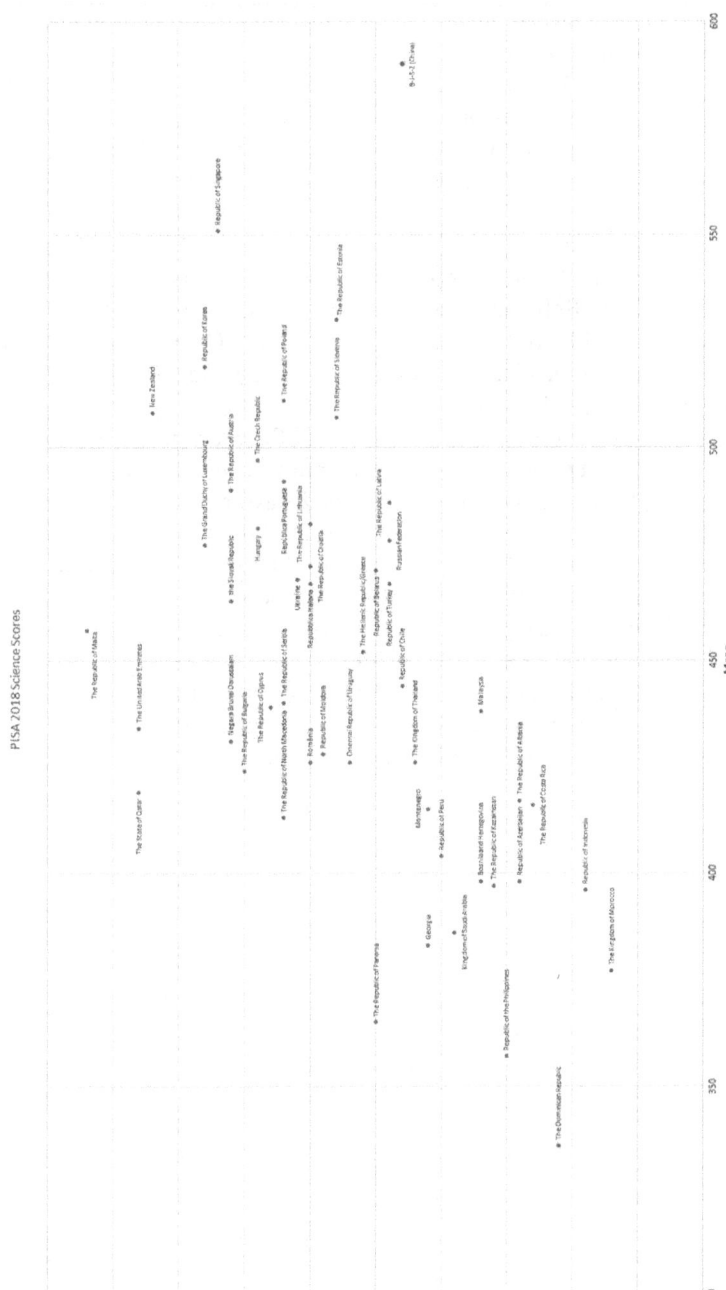

Fig. 30.4 PISA 2018 science scores

(2)   TIMSS.

The most recent TIMSS test was conducted in 2019, and the evaluation results are expected to be released in December 2020. Therefore, we will refer to the results of the last evaluation, TIMSS 2015. The TIMSS assessment is for fourth-grade and eighth-grade students. Among the 47 countries participating in the fourth-grade student assessment, 29 belong to the Belt and Road countries, namely: Morocco, South Korea, Singapore, UAE, Kuwait, Turkey, Qatar, Oman, Saudi Arabia, Bahrain, Iran, Georgia, Kazakhstan, Indonesia, Cyprus, Russia, Poland, Serbia, Czech Republic, Bulgaria, Slovakia, Croatia, Lithuania, Slovenia, Hungary, Portugal, Italy, New Zealand, Chile; Among the 39 countries participating in the evaluation of eighth grade students, 26 belong to the Belt and Road countries: South Africa, Morocco, Egypt, South Korea, Singapore, Malaysia, UAE, Kuwait, Turkey, Qatar, Oman, Lebanon, Saudi Arabia, Bahrain, Iran, Georgia, Kazakhstan, Thailand, Russia, Lithuania, Slovenia, Hungary, Malta, Italy, New Zealand, Chile.

After sorting out, the scores of the fourth and eighth grades in the TIMSS 2015 science literacy assessment project of the above-mentioned countries are shown in the figure. Among them, the horizontal axis represents the average score of the country, and the vertical axis represents the standard deviation of the sample average score.

As shown in Fig. 30.5, in the TIMSS 2015 science literacy assessment for fourth-grade students, out of the countries mentioned in this book, Bulgaria, Slovenia and Serbia have very close average scores, and are all in a relatively leading position. By contrast, Saudi Arabia is ranked in a low posititon.

As Fig. 30.6 shows, the TIMSS 2015 science literacy assessment for eighth grade students, Slovenia ranks high in average scores, while the two countries with the lowest scores are in Africa, namely South Africa and Egypt.

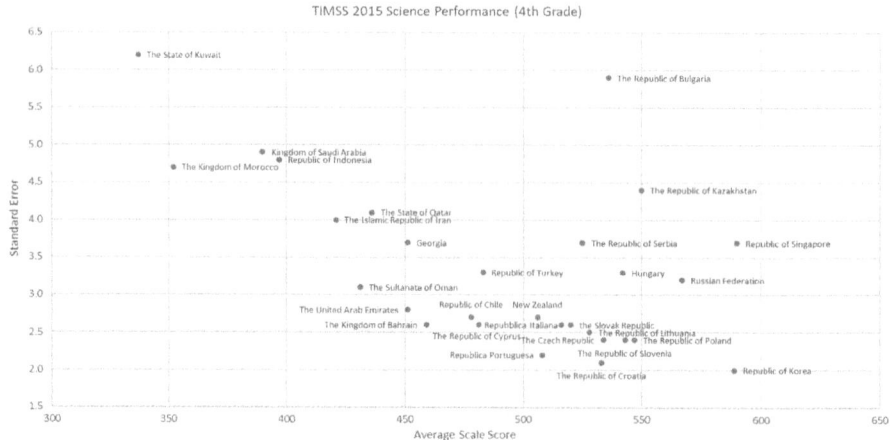

Fig. 30.5 TIMSS 2015 science performance (4th Grade)

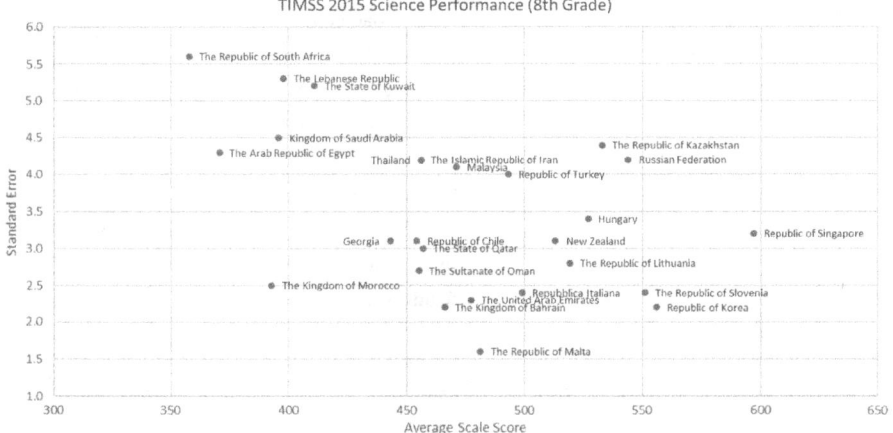

Fig. 30.6   TIMSS 2015 science performance (8th Grade)

(3)   Comparison of scientific evaluation.

Among the Belt and Road countries participating in the PISA 2018 assessment, the five countries with the highest average scientific scores are Singapore, Estonia, South Korea, Poland, and New Zealand, and the five countries with the lowest average scientific scores are Dominica, the Philippines, Panama, Morocco, and Georgia. Among the Belt and Road countries that participated in the TIMSS 2015 evaluation, the countries with an average scientific score of more than 550 for fourth-year students are Singapore, South Korea, Russia, and Kazakhstan. The countries with an average scientific score of less than 400 are Kuwait, Morocco, Saudi Arabia, Indonesia. The countries with an average science score of more than 550 for eighth graders are Singapore, South Korea, Slovenia, while the countries with an average science score of less than 400 are South Africa, Egypt, Morocco, Saudi Arabia, and Lebanon.

Table 30.6 shows the top 10 and bottom 10 Belt and Road countries with PISA 2018 scientific scores. Among them, Estonia and Slovenia ranked in the top 10, while Saudi Arabia and the Philippines performed poorly. Among the top ten PISA 2018 scientific score average scores, the Belt and Road countries occupy three seats, namely Singapore ranked second, Estonia ranked fourth, and Korea ranked seventh. In the bottom ten, the Belt and Road countries occupy eight seats, namely Kazakhstan, Indonesia, Saudi Arabia, Georgia, Morocco, Panama, the Philippines and Dominica. The above is also a microcosm of the PISA scientific achievements of the Belt and Road countries. Judging from the ranking of all countries/regions participating in the PISA 2018 scientific test, the Belt and Road countries are polarized, and most of them are located at the backward pole. Of the 39 countries that ranked at the bottom 50%, 32 belongs to the Belt and Road countries, accounting for about 82%.

Similar to PISA, TIMSS 2015 fourth-grade students' average scores of science scores among the top ten Belt and Road countries, accounting for five places: Singapore ranked first, South Korea ranked second, Russia ranked fourth, Kazakhstan

Table 30.6 PISA 2018 scientific scores of the belt and road countries

Top 10 scientific scores			Bottom 10 scientific scores		
Country	Average score	PISA ranking	Country	Average score	PISA ranking
Singapore	551	2/78	Azerbaijan	398	67/78
Estonia	530	4/78	Bosnia and Herzegovina	398	67/78
Korea	519	7/78	Kazakhstan	397	69/78
Poland	511	11/78	Indonesia	396	70/78
New Zealand	508	12/78	Saudi Arabia	386	71/78
Slovenia	507	13/78	Georgia	383	73/78
Czech Republic	497	21/78	Morocco	377	74/78
Portugal	492	26/78	Panama	365	76/78
Austria	490	28/78	Philippines	357	77/78
Latvia	487	29/78	Dominica	336	78/78

ranked eighth, and Poland ranked ninth. Among the top ten average scores of eighth grade students in science, the Belt and Road countries also occupy five seats: Singapore ranked. South Korea ranked fourth, Slovenia ranked fifth, Russia ranked seventh, and Kazakhstan ranked ninth. In the TIMSS science test for fourth-grade students, the bottom ten countries were all from the Belt and Road countries; and among the bottom ten in science scores of eighth-grade students, the Belt and Road countries occupy eight seats. Judging from the rankings of all countries/regions participating in the TIMSS 2015 scientific test, most of the Belt and Road countries are still in a relatively lagging position. Among the 24 countries with the bottom 50% of the fourth-grade students' scores, 18 belong to the Belt and Road countries. "One Road" countries account for 75%; among the 20 countries with the bottom 50% of the eighth-grade students, 18 belong to the Belt and Road countries, accounting for 90%. Table 30.7 respectively gives the top 10 and bottom 10 Belt and Road countries of the 4th grade and the 8th grade in TIMSS 2015.

From a distribution point of view, both the PISA 2018 and TIMSS 2015 evaluation results indicate that the economic, political, and culturally stable Belt and Road countries have better scientific performance of students, which is clearly reflected in the developed countries, such as Slovenia, Italy and Turkey. On the contrary, whether there are major contradictions or obstacles in any aspect of economy, politics, or culture, the scientific performance of elementary and middle school students in the corresponding country would also be affected. The ranks of some Arab or African countries, including Oman, Saudi Arabia, Egypt and South Africa are low. This reflects the inadequacy of their science education level, but the economy is not restricting a country's science education, the only key factor for development.

Table 30.7   TIMSS 2015 scientific scores of the Belt and Road countries

4th grade

Top 10 scientific scores			Bottom 10 scientific scores		
Country	Average score	TIMSS ranking	Country	Average score	TIMSS ranking
Singapore	590	1/47	Bahrain	459	38/47
Korea	589	2/47	UAE	451	39/47
Russia	567	4/47	Georgia	451	39/47
Kazakhstan	550	8/47	Qatar	436	41/47
Poland	547	9/47	Oman	431	42/47
Slovenia	543	11/47	Iran	421	43/47
Hungary	542	12/47	Indonesia	397	44/47
Bulgaria	536	16/47	Saudi Arabia	390	45/47
Czech Republic	534	17/47	Morocco	352	46/47
Croatia	533	18/47	Kuwait	337	47/47

8th Grade

Top 10 Scientific Scores			*Bottom 10 Scientific Scores*		
Country	*Average score*	*TIMSS ranking*	*Country*	*Average score*	*TIMSS ranking*
Singapore	597	1/39	Iran & Thailand	456	27/39
Korea	556	4/39	Oman	455	29/39
Slovenia	551	5/39	Chile	454	30/39
Russia	544	7/39	Georgia	443	31/39
Kazakhstan	533	9/39	Kuwait	411	33/39
Hungary	527	12/39	Lebanon	398	34/39
Lithuania	519	15/39	Saudi Arabia	396	35/39
New Zealand	513	16/39	Morocco	393	36/39
Italy	499	20/39	Egypt	371	38/39
Turkey	493	21/39	South Africa	358	39/39

30.3.2   Science Education Policies and Standards

The policy of science education plays a vital role in the development of science education and the improvement of science literacy of students. Although different countries and regions have different levels of education, most of them have recognized such importance and have formulated corresponding policies and standards.

In most countries, citizens have the opportunity to receive education of several different academic stages, and the policy of science education is student-centered, focusing on scientific literacy and students' abilities and scientific computer technology. For example, Turkey's science education policies and standards are moving closer to quality education. As part of the development of STEM, Turkey has incorporated courses of coding education and robotics into the school curriculum. Moreover, students' multiple abilities and qualities are focused more in scientific education.

A country's science education policies and standards have a profound impact on the educational development, and can have a deeper influence on the stability and well-being of the citizens. There are several innovative policies and standards of science education in the Belt and Road countries. First, a number of national departments work together to improve the level of national science education, and enhance teachers' scientific literacy. For instance, Greece has always taken it as an important topic to improve science education, and has formulated many plans and projects to identify the best practices to promote science education. Second, the countries attach great importance to STEM education for students, and pay attention to the use of emerging technology in science education.

30.3.3 Science Education Curriculums and Digital Resources

The Belt and Road countries attach great importance to the development of science education courses. The science education courses are rich in content, have sound curriculum standards, and have clear training goals. And there are digital resources supporting the course, so that students can browse the digital resources easily.

The Estonian science education curriculum consists of a general part and appendices. The general part outlines the basic values of education, learning and education goals, learning concepts and learning environment, an overview of the abilities required by each school level, learning organization, assessment and completion of classes and schools, etc. The learning and education objectives include 8 general abilities required in all subject areas and are developed in all subjects through extracurricular activities. These general abilities are (Basic school national curriculum, 2011): (1) cultural and value skills, (2) social and civic skills, (3) autonomy, (4) learning skills, (5) communication skills, (6) mathematics, science and technology skills, (7) entrepreneurial skills, (8) digital capabilities.

The science courses in Malaysian schools are compulsory courses for grades 1–3 (level 1), grades 4–6 (level 2) and grades 1–3 (junior high school). There are three main types of primary school education in Malaysia. They are national schools, national schools (Chinese) and national schools (Tamil), and their teaching media are Malay, Chinese and Tamil, respectively. However, the science courses of these three schools are different. At the same time, when students converge in secondary education, Malay is used as a medium for curriculum teaching. The science standard curriculum for grades 1–6 used in Malaysia is compiled by the Curriculum Development Department of the Ministry of Education of Malaysia. The mission of the Ministry of Education is to compile science textbooks for use in Malaysian schools. For science textbooks used in elementary schools, the Textbook Office will contact the

Institute of Language and Literature to publish relevant textbooks. As for science and pure science textbooks, the Department of Textbooks requires Malaysian publishers to submit sample chapters of each science syllabus, and a committee composed of experts in the corresponding professional fields provide their feedback accordingly.

Educational courses and digital resources in some countries also have extremely innovative practices, such as: e-schoolbag, the largest digital resource library in Estonia, Khan Academy, the digital education platform in Turkey, and DOST, the first digital science library in the Philippines.

30.3.4   Outreach Science Education

Outreach science education can make students and the public become interested in science in many ways. Providing off-campus science education activities can also deepen the understanding of science and its applications and cultivates valuable communication skills. Effective scientific promotion is interesting and beneficial because it is very important to the development of future citizens and technological progress.

Out-of-school science education is inseparable from the Informal Learning Environment, which can narrow the knowledge gap between school science and the life world, and build a bridge between the knowledge framework and the knowledge context. In an informal learning environment, collaborative partnerships between school teachers and educators are essential, such as sharing curriculum goals and activity frameworks as well as jointly promoting project development.

Due to the complexity of political, economic, cultural and other practical factors, the development of science education in the Belt and Road countries is uneven. Specifically, countries such as Libya and Sudan are suffering from unstable political situations and even wars. There are serious deficiencies in science education from policies to systems, from facilities to teachers, and it is difficult to implement science courses in a practical way. Some countries, such as Pakistan, whose overall economic level is backward and trapped in regional political and religious conflicts, also face obstacles in terms of policies and teachers. On the contrary, similar situations are relatively rare in countries with relatively stable political, economic, and cultural aspects. Although these countries have many problems and challenges in science education, they have basically formulated relatively complete policies and curriculum standards, and have a relatively abundant team of teachers and resources to ensure the smooth development of science education activities inside and outside the school.

Some off-campus practice projects for science education are quite innovative and have effectively promoted the cultivation of responsible citizens' scientific literacy. Some projects have also established an effective connection between school education and off-campus education, such as the Space Camp Turkey project in Turkey, the IKAMVAYOUTH project in South Africa, and the SETAC project in Italy. The Space Camp project in Turkey gives participants the feeling of living in space by simulators.

Table 30.8 Stages of the application of emerging technologies in ALECSO countries

Stage of the application of emerging technologies	Countries
Be aware of the role of AI, Internet of Things and blockchain	UAE
Already put AR/VR into use	UAE, Oman, Egypt
Plan to use AR/VR	Libya, Tunisia, Sudan
Hold a positive attitude towards new technologies	Palestine, Saudi Arabia

30.3.5 Application of Emerging Technologies in Science Education

In recent years, it is generally believed that the world is currently undergoing a new technological revolution, and the material basis of the new technological revolution is the emerging technology group. Emerging technologies are not one or two, but an organic group connected, promoted, and restricted each other.

The application of emerging technologies in science education constructs virtual or advanced learning situations for learners, helping learners understand the learning content in a multi-faceted manner and stimulate motivation for learning, as well as helping learners have a deeper understanding of the learning content.

More than half of the countries have applied VR, AR, and artificial intelligence to science education, and many countries attach great importance to the application of emerging technologies in science education. Take the ALECSO countries introduced in this book as examples. Table 30.8 summarizes the stages of the application of emerging technologies in education that the countries are at.

Some countries have devoted a lot to the development of science and technology education, and remarkable results have been achieved in the application of various emerging technologies. For example, the UAE has invested huge sums of money to cooperate with UB-TECH to build artificial intelligence teaching laboratories, and has built many robotic laboratories and intelligent platforms. Estonian schools are equipped with smart devices. However, some schools cannot afford the price of emerging technologies. Good news is that almost all countries have a positive attitude towards the application of new technologies in education.

30.4 The Development Needs of Science Education in the Belt and Road Countries

According to the feedback from experts from various countries, the development needs of science education in some of the Belt and Road countries can be summarized.

At the government and policy level, some Belt and Road countries need to formulate a unified and complete science education policy from top to bottom. This demand is not only particularly important for the relatively backward Belt and Road countries, but also applicable to all countries in the world. For example, Pakistan's need for a unified policy is very urgent. They regard science education as the first step to improve education policies, especially in primary and secondary levels. Furthermore, the provincial and federal governments also need to implement a unified policy in science education. Similarly, Nepal also needs multi-party collaboration (government, schools, universities, private sector and international) to plan and implement STEM education in its country. In addition, Montenegro emphasized the need to strengthen cooperation between the two sectors of education and science, which requires financial resources and national strategic support.

At the level of infrastructure and resources, as the economic development level of the Belt and Road countries is relatively lagging, they generally need to further improve the infrastructure and the construction of science education online platforms and resources. Laggard countries such as Pakistan emphasize that material and technical resources will help the implementation of educational policies and the implementation of inquiry-based teaching methods, thereby providing a good learning environment for science education. Therefore, these countries need to improve the foundations such as laboratories and science centers. Nepal is also facing similar problems as Pakistan. Its university science laboratory equipment is backward, and the scientific equipment and laboratories provided to some schools are not well used because these schools lack well-trained teachers, support, resources and maintenance budget. In addition, the huge structure of the Turkish education system (nearly 18 million students and 2 million teachers) makes it difficult to provide enough laboratory environments and enough equipment in every school. Although the current plan recommends and requires that students at almost all levels from kindergarten to 12th grade can carry out scientific experiments without advanced or even with basic laboratory equipment. Other countries, such as Oman, emphasize the need to provide students with scientific demonstrations, simulation and modeling programs, field visits to the natural environment, videos, and opportunities to visit science museums. They also need to adopt new methodological applications and innovations, such as teaching games and electronic laboratories, assistive technology for disabled students, intelligent robots, online forums, and social networks, are applied to teaching.

At the level of science teachers, many Belt and Road countries have emphasized that a strong reserve of teachers is critical to the development of their country's science education. For example, Estonia is facing the problem of aging science teachers. This is found in physics. It is also seen that older science teachers face difficulties at incorporating the emerging technologies or new teaching methods into their classroom despite their rich teaching experience. Turkey's science teachers face the obstacle of technical and they do not get enough guidance to help their students conduct experiments on science subjects and develop their understanding of science. The Ministry of Education should therefore cooperate with different stakeholders to establish sufficient infrastructures for teachers and students, as well as barrier-free, flexible and sustainable professional development and support systems. This can help teachers better provide learning experiences. The shortage of science teachers is more serious in countries with relatively backward economic development.

At the level of student awareness, some Belt and Road countries have expressed the need to strengthen students' understanding of the importance of science and science education. For example, in Turkey, basic science courses such as physics, biology, and chemistry are not popular among the younger generation, and some universities have even closed these courses. Therefore, they regard raising young people's awareness of various fields of science as one of the primary requirements for the development of future science education. The Ministry of Education also supports the creation of special programs for STEM education in schools to raise awareness of STEM fields and train teachers to STEM projects or activities. Similarly, Slovenia also believes that they need to increase students' interests, motivations and positive attitudes towards science. In Estonia, in addition to mathematics, science and technology ability, and digital ability, science education also emphasizes cultural value ability, social citizenship ability, autonomous ability, learning ability, communication ability and entrepreneurial ability. Therefore, science education also focuses on cultivating citizens who can use scientific thinking and creativity in daily life. Italy emphasizes the need to consider the impact of gender differences on science education, so it needs to focus on the two early education stages of preschool and primary education.

In summary, in future cooperation, regular meetings, exchanges in learning communities, training systems, standards, and resource sharing can all become the foundation for the sustainable development of science education in the Belt and Road countries. From the perspective of cooperation and sharing, many countries have relatively complete national standards and unified policies for science courses, which can provide references for the Belt and Road countries, such as Pakistan and Nepal. In addition, provide cross-country trainings and courses could help to promote collaboration related to science education along the Belt and road.

30.5 Conclusion

Based on the research of the current educational situations and the needs of the development of science education in countries along the Belt and Road, in order to better promote the coordinated development of science education and build a science education cooperation development platform serving the Belt and Road countries, the following development suggestions are put forward:

Firstly, establish a coordination mechanism to optimize resource allocation. According to the existing Belt and Road Science Education Coordination Committee of the Association for Science and Technology, it is clear that the liaison leaders of various countries hold regular high-level meetings to clarify the direction of cooperation and development of science education in the Belt and Road countries.

Secondly, more focus should be paid to the cultivation of science education talents. Strengthen the training and exchange of science education talents, and develop diversified Belt and Road training mechanisms, such as the development of science education exchange programs for foreign students, the Belt and Road national science education research special fund, and jointly promote the Belt and Road national science education Quality improvement.

Thirdly, vigorously promote the construction of educational informatization and open educational resources. Well-developed countries and regions can make full use of educational information technology, open up channels for cooperation and sharing among countries in science education, and help underdeveloped countries to train teachers and improve their informatization capabilities. In addition, countries actively participate in the creation and practice of open educational resources for science education, including online auxiliary platforms and scientific experiment resources, virtual experiment platforms, and scientific teaching aids.

References

Basic school national curriculum (2011). Retrieved from https://www.riigiteataja.ee/akt/114022 018008.
Miller, J. D. (1983). Scientific literacy: A conceptual and empirical review. *Daedalus, 112*(2), 29–48.

Correction to: Science Education in Countries Along the Belt & Road

Ronghuai Huang, Bing Xin, Ahmed Tlili, Feng Yang, Xiangling Zhang, Lixin Zhu, and Mohamed Jemni

Correction to:
R. Huang et al. (eds.), *Science Education in Countries Along the Belt & Road*, Lecture Notes in Educational Technology, https://doi.org/10.1007/978-981-16-6955-2

In the original version of the book, the following corrections are have been incorporated:

In chapter "Science Education in Pakistan: Existing Situation and Perspectives for Planner", Order of the authors "Muhammad Yasir Mustafa, Afaq Ahmed, Ali Gohar Qazi" have been changed to "Muhammad Yasir Mustafa, Ali Gohar Qazi, Afaq Ahmed" in the Frontmatter, Table of Contents and Chapter.

In chapter "Science education in ITALY", the authors' "Martino Bernardi and Andrea Gavosto" have been removed in the Frontmatter, Table of Contents and Chapter.

The correction chapters and the book has been updated with the changes.

The updated versions of these chapters can be found at
https://doi.org/10.1007/978-981-16-6955-2_19
https://doi.org/10.1007/978-981-16-6955-2_25

CPSIA information can be obtained
at www.ICGtesting.com
Printed in the USA
BVHW051955240123
657000BV00008B/77